Photosynthesis Bibliography

volume 10/1 1979

References no. 36588-40373 / ABA-ZWA

Editors Z. Šesták & J. Čatský

Dr W. Junk Publishers –The Hague 1982

Contributors:
Z. Šesták
J. Čatský
I. Tichá
J. Pospišilová
J. Solárová
D. Hodáňová
J. Zime

ISBN–13:978–90–6193–049–5   e–ISBN–13:978–94–009–7975–8
DOI: 10.1007/978–94–009–7975–8

PREFACE

The bibliography includes papers in all fields of photosynthesis research - from studies of model biochemical and biophysical systems of the photosynthesis mechanism to primary production studied by the so-called growth analysis. In addition to papers devoted entirely to photosynthesis, papers on other topics are included if they contain data on photosynthetic activity, photorespiration, chloroplast structure, chlorophyll and carotenoid synthesis and destruction, *etc.*, or if they contain valuable methodological information (measurement of selected environmental factors, leaf area, *etc.*). In many branches it has been difficult to define the limits of interest for photosynthesis researchers. This problem has arisen *e.g.* in topics dealing with the transfer of gases, where - in addition to the papers on carbon dioxide transfer - some papers on water vapour transfer are included, these being of general application or bringing new approaches. On the other hand, many papers dealing with the anatomy and physiology of stomata have been omitted, if the aspect of carbon dioxide or water vapour exchange has not been discussed.

This volume contains references to papers published in the year 1979, and, similarly to Vol. 9, also addenda including references published in the preceding period (*i.e.* 1966 to 1978). The numbers of these additional references are labelled with an asterisk (*) in the list of references.

To maximize the value of the bibliography the references are arranged alphabetically by authors' names, and each volume is provided with three indexes. The Authors' Index contains all names of authors, co-authors and editors. The Subject Index covers primary items chosen according to their interest for photosynthesis researchers. Starting with Vol. 6, the Subject Index has been newly arranged and enlarged. It contains more details on the electron transport chain, carbon fixation pathways, gas exchange on leaf and canopy level, *etc.*, and also on internal and environmental factors affecting photosynthesis and related processes. In the Plant Index, the most important crop plants and selected plant types and groups are indexed.

Cumulative indexes accompany Volumes 1, 5, and then every fifth volume, *i.e.* Volumes 10, 15, *etc.*

We have tried to cover fully the relevant papers which have appeared in the most important scientific periodicals and books. Articles published in local journals, mimeographed booklets, *etc.*, were chosen mostly from reprints and lists of publications received directly from the authors. Only abstracts published in regular journals were included.

Since some 4000 relevant papers are currently published every year and included in this bibliography, and since almost all citations have been checked with the originals, collecting and preparing for publication of such a large amount of material would have been impossible without the collaboration of the authors of the relevant publications. The courtesy of those authors who have already supplied us with reprints is highly appreciated.

We acknowledge with thanks the cooperation of our colleagues from the Institute of Experimental Botany of the Czechoslovak Academy of Sciences in Prague, especially Mrs. **DRAHOMÍRA TĚŽKÁ**, Mrs. **LUDMILA HÁVOVÁ**, Mrs. **LENKA KOLČABOVÁ** and Mrs. **MARIE MANDLOVÁ** who helped in preparing the card material. The librarian of our Institute, Mrs. **ZORA ZAWOYSKA** helped us with checking the references.

Dr. Z. ŠESTÁK and Dr. J. ČATSKÝ

Institute of Experimental Botany
Czechoslovak Academy of Sciences

Flemingovo n. 2
CS-160 00 PRAHA 6
Czechoslovakia

INSTRUCTIONS FOR USE

All references are arranged alphabetically according to the authors' names and the
year of publication. They are numbered and these numbers are used in the indexes.
In case of a book title, the number is preceded by B. An asterisk preceding the number denotes the reference published in the preceding period (1966 - 1978).

The references contain the original unshortened title of the paper (book).
English, French and German titles are cited in the original language. Titles in other
languages are supplemented with a translation in English (sometimes using the title
of the respective English abstract or a shortened title with omitted deadweight
words). Titles of Japanese, Chinese etc. papers are given in English translation only.
The journals' names are abbreviated mainly according to the "Style Manual for Biological Journals" (Second Edition, Amer. Institute of Biological Sciences, Washington,
D.C. 1964), e.g. :

Abhandlungen	chinese	Industry	Publishers
Abstract	Chromatography	inorganic	quantitative
Abteilung	Commission	Institute	Quarterly
Academy	Communication	international	Radiation
Acta	comparative	Investigation	Radiobiology
Africa	Comptes rendus	italian	Rastenii
agricultural	Conference	Izvestiya	Recherche
Agriculture	Congress	Jahrbuch	Report
Agronomy	Contribution	japanese	Research
Akademie (-emiya)	Cytochemistry	Japan	Review
Algology	Cytology	Journal	royal
allgemeine	czechoslovak	Klasse	russian
american	Dendrology	Laboratory	russkii
America	Department	Landwirtschaft	scandinavicus
analytical	Deutschland	Letters	Science
Anatomy	Disease	Limnology	Section
angewandte	Dissertation	Magazin	Series (-iya)
Annals	Doklady	marine	Society
annual	Dopovidi	Mathematics	sovetskii
anorganisch (-nic)	Ecology	Microbiology	soviet
applied	Education	miscellaneous	special
Arbeit	Embryology	molecular	SSSR
Archiv	Encyclopedia	Monograph	Station
Atmosphere	Engineer	moskovskii	Supplement
atomic	Enzymology	Mycology	Survey
Australia	european	national	Symposium
Beiheft	experimental	natural	technical
Belgique	Experiment	Naturforschung	Technology
Bericht	Faculty	neerlandicus	Tijdschrift
biochemical	Federation	Netherland	Transaction
Biochemistry	Fizika	New Zealand	Travail (-aux)
biokhimicheskii	Fiziologiya	nuclear	tropical
Biokhimiya	Forestry	Oceanography	Trudy
biological(-ogicheskii) Forschung		Optics	ukrainian
Biology (-ogiya)	Foundation	organic	UK
biophysical	France	original	US, USA
Biophysics	Gazette	Otdelenie	USSR
Bodenkunde	general	Pathology	University
bolgarskii	genetical	Pflanzen-	végétal
botanical (-anicheskii) Genetics		Philosophy	Virology
Botany	Gesellschaft	physical	Virusforschung
british	Giornale	Physics	Volume
Bulletin	helveticus	physiological	Weekblad
Canada	Histochemistry	Physiology	Wetenschappen
cellular (-ulaire)	Histology	Phytopathology	Wissenschaft
central	Horticulture	Plant (-arum)	Zeitschrift
chemical	hungaricus	polish	Zeitung
Chemistry	Husbandry	Proceedings	Zentralblatt
chimicus	imperial	Publication	Zhurnal

The numbers at the end of each reference of a journal article denote : volume (issue) : first page - last page, year of publication. The number of issue is given only in the journal where each issue is paginated separately.

Book titles are cited according to the title page, not to the book jacket or cover (if the names of the editors are not given on the title page, they are not cited in the reference). The publishing house, place and year of publication are included.

Brackets at the end of the reference give bibliographic details and explanations to the contents, not given in the original. The following abbreviations are used most often :

ab	abstract	Ital.	Italian
Arm.	Armenian	Jap.	Japanese
Belorus.	Belorussian	Latv.	Latvian
Bil	biliproteins	Lithu.	Lithuanian
Bulg.	Bulgarian	Norweg.	Norwegian
Car	carotenoids	PC	paper chromatography
CC	column chromatography	PhAR	photosynthetically active radiation
Chin.	Chinese		
Chl	chlorophyll	Pol.	Polish
Croat.	Croatian	Ps	photosynthesis
Cyt	cytochromes	R	Russian
Dan.	Danish	Roum.	Roumanian
E	English	Span.	Spanish
F	French	Swed.	Swedish
G	German	TLC	thin-layer chromatography
GC	gas chromatography	Tr	transpiration
Georg.	Georgian	Ukr.	Ukrainian
Hung.	Hungarian	Uz.	Uzbeg
IRGA	infra-red gas analyser		

The transliteration of Cyrillic characters is in accordance with the BSI-ASA/ /SC-Z39 draft table, *i.e.* :

Translit.	Cyrill.	Translit.	Cyrill.
a	а	p	п
b	б	r	р
ch	ч	s	с
d	д	sh	ш
e	е	shch	щ
ė	э	t	т
f	ф	ts	ц
g	г	u	у
i	и	v	в
ĭ	й	y	ы
k	к	ya	я
kh	х	yu	ю
l	л	z	з
m	м	zh	ж
n	н	''	ъ
o	о	'	ь

Several exceptions apply for Ukrainian, Belorussian and Serbian:

Translit.	Cyrill.		Translit.	Cyrill.	Translit.	Cyrill.
Ukrainian:		Serbian:	c̨	ц	j	j
y	и		č	ч	lj	љ
i	і		ć	ħ	nj	њ
ï	ї		dj	ђ	š	ш
Belorussian:			dž	џ	ž	ж
ŭ	ў		h	х		

Authors' names are presented in spelling used in the original paper. If this spelling does not correspond to the original spelling used by the author (*e.g.* Russian papers of English authors), one spelling is referred to the other in the Authors' index.

Printers' errors in the original papers are marked by underlining the respective words (letters).

ERRATA

Reference No./page	For	Read

Volume 1
6591/p.357   1966.   1968.

Volume 5/1
20337/p.109   Scient. agr.   Sci. Agr.

Volume 6
Authors' index/p.233   KSĚNZEK, O.S.   KSĚNZHEK, O.S.

Volume 7
Authors' index/p.231   WILLIAMS, P.L.Le B.   WILLIAMS, P.J. leB.

Volume 8
Authors' index/p.260   KSENZEK, O.S.   KSENZHEK, O.S.
             /p.271   PINTHUIS, M.J.   PINTHUS, M.J.

Volume 9
35202/p.157   1 – 4,   94 – 98,
35203/p.157   1 – 6,   70 – 75,
Plant index/p.376 *Synechococcus*   *Anacystic*   *Anacystis*

36588 - ABAYCHI, J.K., RILEY, J.P. : The quantitative determination of phytoplankton pigments by high performance liquid chromatography. - Anal. chim. Acta *107* : 1 - 12, 1979.

36589 - ABDOURAKHMANOV, I.A., GANAGO, A.O., EROKHIN, Yu.E., SOLOV'EV, A.A., CHUGUNOV, V.A. : Orientation and linear dichroism of the reaction centers from *Rhodopseudomonas sphaeroides* R-26. - Biochim. biophys. Acta *546* : 183 - 186, 1979.

36590 - ABDULLAEV, Kh.A., USMANOV, P.D., TAGEEVA, S.V. : Sistema fotosinteticheskikh membran i évolyutsiya khloroplastov. [System of photosynthetic membranes and evolution of chloroplasts.] - Zh. obshch. Biol. *40* : 43 - 59, 1979. [In R, ab : E.]

*36591 - ABDULLAEV, M.A. : Vliyanie predposevnogo oblucheniya semyan tomatov na rost, razvitie i kachestvo rassady v zavisimosti ot intervala vremeni mezhdu oblucheniem i posevom. [Effect of pre-sowing irradiation of tomato seeds on growth, development and quality of seedlings in dependence on the time interval between irradiation and sowing.] - Tr. azerb. nauch.-issled. Inst. Zemled. *14* : 286 - 289, 1968. [Car; in R, ab : Azerb.]

*36592 - ABELIOVICH, A., WEISMAN, D. : Role of heterotrophic nutrition in growth of the alga *Scenedesmus obliquus* in high-rate oxidation ponds. - Appl. environ. Microbiol. *35* : 32 - 37, 1978. [Chl.]

36593 - ABER, J.D. : Foliage-height profiles and succession in northern hardwood forests. - Ecology *60* : 18 - 23, 1979. [Leaf area index.]

36594 - ABER, J.D. : A method for estimating foliage-height profiles in broad-leaved forests. - J. Ecol. *67* : 35 - 40, 1979.

*36595 - ABILOV, Z.K., GASANOV, R.A. : The source of the long-wavelength fluorescence of chlorophyll *in vivo*. - Stud. biophys. *68* : 155 - 158, 1978.

36596 - ABRANYI, A. : Fénymérés a Martonvásári fitotronban. [Light measurement in the Martonvásár phytotron.] - Bot. Közlem. *66* : 243 - 246, 1979. [In Hung., ab : E, R.]

36597 - ABRAVANEL, G., BORDERIES, G., CAILLIAU, L., CAVALIÉ, G. : Isolement des cellules foliaires du soja: Activité photosynthétique des suspensions obtenues. - Plant Sci. Lett. *16* : 171 - 180, 1979.

36598 - ABROS'KINA, L.S., VOROB'EVA, L.M., KVITKO, K.V. : Lyuminestsentsiya khlorofilla v mutantakh khlorelly i khlamidomonady. [Fluorescence of chlorophyll in *Chlorella* and *Chlamydomonas* mutants.] - Fiziol. Rast. *26* : 383 - 393, 1979. [In R, ab : E.]

36599 - ABUL-FATIH, H.A., BAZZAZ, F.A., HUNT, R. : The biology of *Ambrosia trifida* L. III. Growth and biomass allocation. - New Phytol. *83* : 829 - 838, 1979. [Growth analysis.]

*36600 - ABU-SHAKRA, S.S., PHILLIPS, D.A., HUFFAKER, R.C. : Nitrogen fixation and delayed leaf senescence in soybeans. - Science *199* : 973 - 975, 1978. [Chl.]

36601 - ACEVEDO, E., FERERES, E., HSIAO, T.C., HENDERSON, D.W. : Diurnal growth trends, water potential, and osmotic adjustment of maize and sorghum leaves in the field. - Plant Physiol. *64* : 476 - 480, 1979. [Dry-matter accumulation.]

36602 - ACOCK, B., CHARLES-EDWARDS, D.A., SAWYER, S. : Growth response of a *Chrysanthemum* crop to the environment. III. Effects of radiation and temperature on dry matter partitioning and photosynthesis. - Ann. Bot. *44* : 289 - 300, 1979.

36603 - ADAMS, C.A., RINNE, R.W. : Carbon dioxide fixation in developing soybean seeds. - Plant Physiol. *63* (Suppl.) : 17, 1979.

36604 - ADAMS, M.S., FAYYAZ, M.M. : Temperature acclimation of net photosynthesis in relation to growth of a cold hardy *Chrysanthemum*. - Oecologia *39* : 239 - 247, 1979.

*36605 - ADAMS, M.W.W., HALL, D.O. : Physical and catalytic properties of the hydro-
genase of *Rhodospirillum rubrum*. - In : SCHLEGEL, H.G., SCHNEIDER, K. (ed.) :
Hydrogenases: Their Catalytic Activity, Structure and Function. Pp. 159 -
- 169. Verlag Erich Goltze KG, Göttingen 1978.

36606 - ADAMS, M.W.W., HALL, D.O. : Properties of the solubilized membrane-bound
hydrogenase from the photosynthetic bacterium *Rhodospirillum rubrum*. - Arch.
Biochem. Biophys. *195* : 288 - 299, 1979.

36607 - ADAMS, M.W.W., RAO, K.K., HALL, D.O. : The photoactivated production of molecular hydrogen over prolonged periods catalysed by platinum or hydrogenase. -
Photobiochem. Photobiophys. *1* : 33 - 41, 1979.

36608 - ADAMS, S.M., KAO, O.H.W., BERNS, D.S. : Psychrophile C-phycocyanin. - Plant
Physiol. *64* : 525 - 527, 1979.

36609 - ADEPIPE, N.O., TINGEY, D.T. : Ozone phytotoxicity in relation to stress ethylene evolution and stomatal resistance in cowpea (*Vigna unguiculata*) cultivars. - Z. Pflanzenphysiol. *93* : 259 - 264, 1979.

36610 - ADHIKARY, S.P., PATTNAIK, H. : Effect of hormones on the growth and chlorophyll content of *Westiellopsis prolifica* JANET. - Curr. Sci. *48* : 23 - 24,
1979.

36611 - ADLER, K., BRECHT, E., MEISTER, A., SCHMIDT, O., SÜSS, K.-H. : Die Chloro-
plasten-Thylakoid-Membran : Biogenese, Pigmentorganisation, Protein-Funkti-
onsbeziehungen und Degeneration während der Seneszenz. Eine Übersicht. -
Kulturpflanze *27* : 13 - 48, 1979.

36612 - ADMAN, E.T. : A comparison of the structures of electron transfer proteins. -
Biochim. biophys. Acta *549* : 107 - 144, 1979.

*36613 - AGAPOVA, M.V. : Osobennosti rosta i razvitiya plodovo-yagodnykh kul'tur v
Predural'e. [Peculiarities of growth and development of fruit-berry cultures
in Predural'e.] - In : Vliyanie Fiziko-Khimicheskikh Faktorov Sredy na Rasteniya. Pp. 3 - 19. Permsk. gos. Univ. Im. A.M. Gor'kogo, Perm' 1978. [Ps;
In R.]

36614 - AHARONI, N., ANDERSON, J.D., LIEBERMAN, M. : Production and action of ethylene in senescing leaf discs. Effect of indoleacetic acid, kinetin, silver
ion, and carbon dioxide. - Plant Physiol. *64* : 805 - 809, 1979. [Chl.]

36615 - AHARONI, N., LIEBERMAN, M. : Ethylene as a regulator of senescence in tobacco leaf discs. - Plant Physiol. *64* : 801 - 804, 1979. [Chl.]

36616 - AHARONI, N., LIEBERMAN, M., SISLER, H.D. : Patterns of ethylene production
in senescing leaves. - Plant Physiol. *64* : 796 - 800, 1979. [Chl.]

36617 - AHMED, A.M., HEIKAL, M.D., SHADDAD, M.A. : Growth, photosynthesis and fat
content of some oil producing plants as influenced by some salinization
treatments. - Phyton *19* : 259 - 267, 1979.

36618 - AHO, N., DAUDET, F.-A., VARTANIAN, N. : Évolution de la photosynthèse nette
et de l'efficience de la transpiration au cours d'un cycle de dessèchement
du sol. - Compt. rend. Acad. Sci. Paris, Sér. D *288* : 501 - 504, 1979.

36619 - AIKAZYAN, V.Ts., NALBANDYAN, R.M. : Copper-containing proteins from *Cucumis
sativus*. - FEBS Lett. *104* : 127 - 130, 1979. [Plastocyanin.]

36620 - AIKING, H., SOJKA, G. : Response of *Rhodopseudomonas capsulata* to illumination and growth rate in a light-limited continuous culture. - J. Bacteriol.
139 : 530 - 536, 1979. [Chl.]

36621 - AITKEN, A. : Purification and primary structure of cytochrome *c*-552 from the
cyanobacterium, *Synechococcus* PCC 6312. - Europe. J. Biochem. *101* : 297 -
308, 1979.

36622 - AIZAWA, M., HIRANO, M., SUZUKI, S. : Photoelectrochemical oxygen evolution
from water by a manganese chlorophyll-liquid crystal electrode. - Electro-
chim. Acta *24* : 89 - 94, 1979.

*36623 - AKAO, S., CHATTERTON, N.J., CARLSON, G.E., HUNGERFORD, W.E. : [Studies on the
translocation of photosynthates in alfalfa. Part 1. The comparison of nitrogen, phosphorus and carbohydrates in high- and low-yield clones of alfalfa
differing in tolerance to frequency of cutting.] - Shikoku Nogyo Shikenjo Ho-
koku [Bull. Shikoku agr. Exp. Sta.] 30 : 129 - 134, 1977. [In Jap., ab : E.]

36624 - AKAZAWA, T. : Ribulose-1,5-bisphosphate carboxylase. - In : GIBBS, M., LATZKO,
E. (ed.) : Photosynthesis II. (Encycl. Plant Physiol. N.S. Vol.6.) Pp. 208 -
- 229. Springer-Verlag, Berlin - Heidelberg - New York 1979.

36625 - ÅKERLUND, H.-E., ANDERSSON, B., PERSSON, A., ALBERTSSON, P.-Å. : Isoelectric
points of spinach thylakoid membrane surfaces as determined by cross partition. - Biochim. biophys. Acta 552: 238 - 246, 1979.

36626 - AKHMANOV, S.A., BORISOV, A.Yu., DANIELIUS, R., GADONAS, R., KOZLOVSKII, V.S.,
PISKARSKAS, A., RAZZHIVIN, A.P. : Spectroscopy of photoreaction centers by
tunable picosecond parametric oscillators. - In : Laser Spectroscopy. Vol.4.
Springer Ser. Opt. Sci. 21 (Proceedings of the IV International Conference
on Laser Spectroscopy). Pp. 387 - 397. Springer-Verlag, Berlin - Heidelberg -
- New York 1979.

36627 - AKHMANOV, S.A., BORISOV, A.Yu., DANIELIUS, R.V., GADONAS, R.A., KOZLOWSKI,
V.S., PISKARSKAS, A.S., RAZJIVIN, A.P. : The number of excitation light pho-
tons per reaction centre separates the "photosynthetic" region from the "nonlinear" one in picosecond spectroscopy. - Stud. biophys. 77 : 1 - 3, 1979.

*36628 - AKIMOVA, T.V., DMITRIEV, V.P., NYUPPIEVA, K.A. : Vliyanie temperatury i osve-
shchennosti na soderzhanie pigmentov v list'yakh ogurtsa. [Effects of temperature and illuminance on the pigment content in cucumber leaves.] - In :
Ekologo-Fiziologicheskie Mekhanizmy Ustoĭchivosti Rasteniĭ k Deĭstviyu Ekstremal'nykh Temperatur. Pp. 74 - 80, 165. Karel'skiĭ Filial Akad. Nauk SSSR,
Petrozavodsk 1978. [In R.]

*36629 - AKIMOVA, T.V., POPOV, E.G. : Vliyanie temperatury na fotosintez i dykhanie
rasteniĭ ogurtsa. [Effect of temperature on photosynthesis and respiration
of cucumber plants.] - In : Ekologo-Fiziologicheskie Mekhanizmy Ustoĭchivosti
Rasteniĭ k Deĭstviyu Ekstremal'nykh Temperatur. Pp. 68 - 74, 165. Karel'skiĭ
Filial Akad. Nauk SSSR, Petrozavodsk 1978. [In R.]

36630 - AKITA, S., TANAKA, I. : Studies on the mechanism of differences in photosynthesis among species V. Stomatal response in high oxygen concentration and
its effect on the rate of apparent photosynthesis. - Jap. J. Crop Sci. 48 :
470 - 474, 1979.

36631 - AKULOVICH, N.K., LYAKHNOVICH, Ya.P., PARSHYKAVA, T.A. : Zmyanenne ul'trastruktury ètyyaplastaŭ yachmenyu pry dèzaktyvatsyi protakhlarafilidu nagrevannem
i dzeyannem lipazy. [Changes in structure of barley etioplasts during protochlorophyllide desactivation and lipase action.] - Vestsi Akad. Navuk belarus. SSR, Ser. biyal. Navuk 1979 (1) : 39 - 46, 138, 1979. [In Belorus.,
ab : E, R.]

36632 - AKULOVICH, N.K., NIKOLAEVA, L.F., PARSHIKOVA, T.A., PORSHNEVA, E.B. : Vliyanie nagrevaniya na sootnoshenie form protokhlorofillovogo pigmenta ètiolirovannykh list'ev i soderzhanie v nikh vosstanovlennykh piridinnukleotidov.
[Influence of heat-treatment on the ratio of protochlorophyll forms and on
reduced pyridine nucleotides content in etiolated leaves.] - Vest. mosk. gos.
Univ. Ser.16 - Biol. 1979 (2) : 56 - 62, 1979. [In R, ab : E.]

36633 - ALASAARELA, E. : Ecology of phytoplankton in the north of the Bothnian Bay.
- Acta bot. fenn. 110 : 63 - 70, 1979. [Primary production.]

36634 - ALASAARELA, E. : Spatial, seasonal and long-term variations in the phytoplanktonic biomass and species composition in the coastal waters of the Bothnian
Bay off Oulu. - Ann. bot. fenn. 16 : 108 - 122, 1979. [Chl.]

36635 - ALASAARELA, E. : Phytoplankton and environmental conditions in central and
coastal areas of the Bothnian Bay. - Ann. bot. fenn. 16 : 241 - 274, 1979.
[Chl.]

36636 - ALBERGONI, F.G., BASSO, B., TOSO, S. : Considerations on CO_2 effluxes in Zea
mays L. and Trifolium repens L. - Maydica 24 : 113 - 124, 1979. [Canopy CO_2
profiles.]

36637 - **ALBERS, D.J., CARPENTER, S.B.** : Influence of site, environmental conditions, mulching, and herbaceous ground cover on survival, growth, and water relations of european alder seedlings planted on surface mine spoil. - In : CARPENTER, S.B. (ed.) : Symposium on Surface Mining Hydrology, Sedimentology and Reclamation. Pp. 23 - 32. Ores Publications, Lexington 1979. [Stomatal resistance.]

36638 - **ALEKSIDZE, G.N., RACHVELISHVILI, E.V., POTSKHVERIYA, A.M.** : K izucheniyu vredonosnosti plodovykh listovykh tlei v Gruzii. [Damage caused by fruit aphids in Georgia.] - Soobshch. Akad. Nauk gruz. SSR *93*(1) : 193 - 195, 1979. [Chl; in R, ab : E, Georg.]

36639 - **ALHADEFF, M., CORONADO, R., FIGUEROA, N., SCHIFF, J.A.** : Loss and re-formation of protochlorophyll(ide) [P(ide)] in dark grown non-dividing *Euglena gracilis* var. *bacillaris*. - Plant Physiol. *63* (Suppl.) : 98, 1979.

*36640 - **ALI, H.C., WILLIAMS, R.L., JOHNSON, M.W.,Jr.** : The relationships of leaf area to grain yield and other factors in corn (*Zea mays* L.). - Z. Pflanzenzücht. *80* : 320 - 325, 1978.

*36641 - **ALIEV, D.A.** : O fiziologicheskikh osnovakh primeneniya mikroélementov pod khlopchatnik. [Physiological principles in the use of trace elements for cotton crops.] - Tr. azerb. nauch.-issled. Inst. Zemled. *14* : 251 - 256, 1968. [Photosynthates; in R, ab : Azerb.]

*36642 - **ALIEV, D.A.** : Dinamika soderzhaniya khlorofilla u ovoshchnykh kul'tur pri razlichnykh usloviyakh mineral'nogo pitaniya. [Dynamics of chlorophyll content in vegetable cultures under different mineral nutrition.] - Temat. Sb. Trudov azerb. nauch.-issled. Inst. Zemled. *16* : 152 - 162, 1976. [In R, ab : Azerb.]

36643 - **ALIEV, D.A., KAZIBEKOVA, E.G.** : Struktura fotosinteziruyushchei sistemy posevov pshenitsy kak uslovie ispol'zovaniya énergii solnechnoi radiatsii. [The structure of photosynthesizing system of wheat stands as a condition of solar energy utilization.] - Vest. sel'skokhoz. Nauki *1979*(5) : 43 - 48, 1979. [In R, ab : E.]

36644 - **ALJUBURI, H., HUFF, A., HSHIEH, M.** : Enzymes of chlorophyll catabolism in orange flavedo. - Plant Physiol. *63* (Suppl.) : 73, 1979.

36645 - **ALSCHER-HERMAN, R., JAGENDORF, A.T., GRUMET, R.** : Ribosome-thylakoid association in peas. Influence of anoxia. - Plant Physiol. *84* : 232 - 235, 1979.

36646 - **ALTMAN, J.A., BEDDARD, G.S., PORTER, G.** : Energy transfer in a model of the photosynthetic unit. - In : Chlorophyll Organization and Energy Transfer in Photosynthesis. Pp. 191 - 200. Excerpta Medica, Amsterdam - Oxford - New York 1979.

36647 - **AMAGASA, T., YOSHIDA, S.** : Transfer of label from aspartate to malate by the cell-free extract of *Sedum mexicanum* leaves. - Plant Cell Physiol. *20* : 1191 - - 1197, 1979.

B36648 - **AMBERGER, A.** : Pflanzenernährung. Ökologische und Physiologische Grundlagen. (Uni-Taschenbücher 846.) - Verlag Eugen Ulmer, Stuttgart 1979. [Ps.]

36649 - **AMBLER, R.P., DANIEL, M., HERMOSO, J., MEYER, T.E., BARTSCH, R.G., KAMEN, M.D.**: Cytochrome c_2 sequence variation among the recognised species of purple nonsulphur photosynthetic bacteria. - Nature *278* : 659 - 660, 1979.

36650 - **AMBLER, R.P., MEYER, T.E., KAMEN, M.D.** : Anomalies in amino acid sequences of small cytochromes *c* and cytochromes *c'* from two species of purple photosynthetic bacteria. - Nature *278* : 661 - 662, 1979.

*36651 - **AMESZ, J.** : Absorption difference spectroscopy of photosynthetic material. - Zagad. Biofiz. wspól. *3* : 157 - 163, 1978.

36652 - **AMESZ, J.** : Structure and function of the photosynthetic membrane. - Progr. Bot. *41* : 55 - 70, 1979.

*36653 - **AMLA, D.V.** : Stability, adsorption and growth characteristics of cyanophage AS-1. - Adv. Cyanophyte Res. *1978* : 131 - 139, 1978. [Ps inhibitors.]

36654 - **AMLA, D.V.** : Characteristics of pigment mutants of *Anacystis nidulans* : Ultra-violet sensitivity and multiplication of cyanophage AS-1. - Biochem. Physiol. Pflanzen *174* : 678 - 684, 1979. [Chl, Car, biliproteins.]

36655 - **ANDERSEN, A.** : The influence of temperature on photosynthetic rate in *Dieffen-bachia maculata* (LODD.) 'Exotica perfection'. - Årsskr. kgl. Vet. Landbohøjsk. *1979* : 15 - 24, 1979.

⋇ 36656 - **ANDERSEN, D.C., ARMITAGE, K.B.** : Caloric content of rocky mountain subalpine and alpine plants. - J. Range Manage. *29* : 344 - 345, 1976.

36657 - **ANDERSEN, J.M., JACOBSEN, O.S., GREVY, P.D., MARKMANN, P.N.** : Production and decomposition of organic matter in eutrophic Frederiksborg Slotssø, Denmark. - Arch. Hydrobiol. *85* : 511 - 542, 1979. [Chl.]

⋇36658 - **ANDERSEN, W.R., TINGEY, S.V., RINEHART, C.A.** : Differences in catalytic acti-vities of ribulose 1,5-bisphosphate carboxylase in various genetic lines of barley. - In : SIEGELMAN, H.W., HIND, G. (ed.) : Photosynthetic Carbon Assi-milation. P. 415. Plenum Press, New York - London 1978.

⋇36659 - **ANDERSON, J.E.** : Transpiration and photosynthesis in saltcedar. - Hydrol. Water Resour. Arizona Southwest *7* : 125 - 131, 1977.

36660 - **ANDERSON, J.M., BARRETT, J.** : Chlorophyll-protein complexes of brown algae : P700 reaction centre and light-harvesting complexes. - In : Chlorophyll Orga-nization and Energy Transfer in Photosynthesis. Pp. 81 - 104. Excerpta Medi-ca, Amsterdam - Oxford - New York 1979.

36661 - **ANDERSON, J.W., HOUSE, C.M.** : Polarographic study of oxaloacetate reduction by isolated pea chloroplasts. - Plant Physiol. *64* : 1058 - 1063, 1979.

36662 - **ANDERSON, J.W., HOUSE, C.M.** : Polarographic study of dicarboxylic-acid-depen-dent export of reducing equivalents from illuminated chloroplasts. - Plant Physiol. *64* : 1064 - 1069, 1979.

36663 - **ANDERSON, L.E.** : Interaction between photochemistry and activity of enzymes. - In : GIBBS, M., LATZKO, E. (ed.) : Photosynthesis II.(Encycl. Plant Physiol. N.S. Vol.6.) Pp. 271 - 281. Springer-Verlag, Berlin - Heidelberg - New York 1979.

36664 - **ANDERSON, L.E., CHIN, H.-M., GUPTA, V.K.** : Modulation of chloroplast fructose--1,6-bisphosphatase activity by light. - Plant Physiol. *64* : 491 - 494, 1979.

36665 - **ANDERSON, L.E., HANSEN, M.J., ANDERSON, J.B.** : Light-dark modulation of phos-phoglucomutase activity in pea leaf chloroplasts. - Plant Physiol. *63* (Suppl.) : 2, 1979.

36666 - **ANDERSON, L.E., HEINRIKSON, R.L.** : Chloroplast and cytoplasmic enzymes. VIII. Amino acid composition of the pea leaf aldolases. - Plant Physiol. *64* : 404 - - 405, 1979.

36667 - **ANDERSON, L.E., MANABE, K.** : Disulfide-linked peptides in the chloroplast thylakoid membrane. - Biochim. biophys. Acta *579* : 1 - 9, 1979.

36668 - **ANDERSSON, L., EGNÉUS, H.** : Chlorophyll composition in an isolated chlorophyll *a*-protein complex of photosystem I. - Physiol. Plant. *47* : 11 - 14, 1979.

36669 - **ANDONOVA, P., MEKHANDZHIEVA, A.** : Nyakoi promeni v plastidnite pigmenti na lyutsernata pod vliyanie na razlichnata vodoobezpechenost i nachini na napo-yavane. [Some changes in alfalfa plastid pigments resulting from varying wa-ter supply and irrigation method.] - Fiziol. Rast. (Sofia) *5*(1) : 33 - 42, 1979. [In Bulg., ab : E, R.]

36670 - **ANDRÉ,M.,DAGUENET,A.,MASSIMINO,J.,MASSIMINO,D.,RICHAUD,C.:** Le laboratoire C₂₃A. Un outil au service de la physiologie de la plante entière II.- Possibilités de la mini-informatique et premiers résultats. - Ann. agron. *30* : 153 - 166, 1979. [Ps.]

36671 - **ANDRÉ, M., GERBAUD, A.** : Consommation d'oxygène pendant la photosynthèse chez *Zea mays*. - Compt. rend. Acad. Sci. Paris, Sér. D *289* : 793 - 796, 1979.

36672 - **ANDRÉ, M., THOMAS, D.A., WILLERT, D.J von, GERBAUD, A.** : Oxygen and carbon dioxide exchanges in Crassulacean-acid-metabolism-plants. - Planta *147* : 141 - - 144, 1979.

36673 - **ANDREEVA, A.S., TIKHONOV, A.N., RUUGE, É.K.** : O vliyanii predystorii osveshcheniya na kinetiku okislitel'no-vosstanovitel'nykh prevrashcheniĭ P700 v
list'yakh bobov. [Influence of preillumination history on P700 redox transitions in broad bean leaves.] - Biofizika *24* : 548 - 549, 1979. [In R, ab : E.]

36674 - **ANDREEVA, A.S., VESELINOVA, Yu.M.** : Vliyanie usloviĭ adaptatsii na medlennuyu
induktsiyu fluorestsentsii v list'yakh vysshikh rasteniĭ. [Effect of adaptation conditions on slow fluorescence induction in leaves of higher plants.] -
Biofizika *24* : 175 - 177, 1979. [In R, ab : E.]

36675 - **ANDREEVA, N.E., CHIBISOV, A.K.** : Énergetika reaktsii fotookisleniya khlorofilla. [Energetics of the reaction of chlorophyll photooxidation.] - Dokl.
Akad. Nauk SSSR *248* : 1253 - 1256, 1979. [In R.]

36676 - **ANDREEVA, T.F., STROGONOVA, L.E., STEPANENKO, S.Yu., MAEVSKAYA, S.N., PROTA-
SOVA, N.N., MURASHOV, I.N.** : Zavisimost' aktivnosti fotosinteticheskogo apparata i rostovykh protsessov ot intensivnosti sveta i kontsentratsii CO_2 pri
dlitel'nom vozdeĭstvii étikh faktorov. [Dependence of the activity of photosynthetic apparatus and growth processes on illuminance and CO_2 level in
long-term action.] - Fiziol. Rast. *26* : 1156 - 1162, 1979. [In R, ab : E.]

36677 - **ANDREO, C.S., RAVIZZINI, R.A., VALLEJOS, R.H.** : Sulphydryl groups in photosynthetic energy conservation. V. Localization of the new disulfide bridges formed by *o*-iodosobenzoate in coupling factors of spinach chloroplasts. - Biochim.
biophys. Acta *547* : 370 - 379, 1979.

36678 - **ANDREW, M.H., NOBLE, I.R., LANGE, R.T.** : A non-destructive method for estimating the weight of forage on shrubs. - Aust. Rangel. J. *1* : 225 - 231,
1979.

36679 - **ANDREWS, T.J., ABEL, K.M.** : Photosynthetic carbon metabolism in seagrasses.
^{14}C-labeling evidence for the C_3 pathway. - Plant Physiol. *63* : 650 - 656,
1979.

36680 - **ANITOFF, O.E.** : Laser photoinduced changes in the high frequency dielectric
constant of chloroplasts and dyes. - In : **JENNINGS, B.R.** (ed.) : Electro-
-Optics and Dielectrics of Macromolecules and Colloids. Pp. 393 - 398. Plenum Press, New York 1979.

36681 - **ANPILOGOVA, N.N., LIMAR', R.S.** : Znachenie chasteĭ kolosa v formirovanii zer-
na pshenitsy. [Role of parts of the ear in wheat grain formation.] - Byull.
vses. nauch.-issled. Inst. Rastenievod. im. N.I. Vavilova *87* : 3 - 7, 1979.
[In R.]

*36682 - **ANTIPOV, N.I.** : O proniknovenii uglekislogo gaza cherez kutikulu rasteniĭ.
[CO_2 transfer through plant cuticle.] - Dokl. mosk. Obshch. Ispyt. Prirody,
Zool. Bot. *1978* (2) : 61 - 62, 1978. [In R.]

36683 - **ANTLFINGER, A.E., DUNN, E.L.** : Seasonal patterns of CO_2 and water vapor exchange of three salt marsh succulents. - Oecologia *43* : 249 - 260, 1979.

36684 - **ANTONIELLI, M., VENANZI, G.** : Structural properties of the rachis and hypso-
phyll of the maize ear. - Plant Sci. Lett. *15* : 301 - 304, 1979. [Chloro-
plast.]

36685 - **ANTONYUK, V.O., LUTSIK, M.D., BALUSHCHAK, I.M.** : Zminy aminokyslotnogo skladu
soku kalankhoe pirchastogo i kalankhoe degremona v protsesi fotosyntezu i
konservuvannya roslyn. [Changes in amino acid composition of juice from *Kalanchoë pinnata* and *Kalanchoë daigremontiana* during photosynthesis and plant
preservation.] - Farmats. Zh. (Kiev) *1979* (1) : 51 - 55, 1979. [In Ukr., ab :
E.]

36686 - **AOKI, S.** : [$^{14}CO_2$ fixation in leaf discs of *Camellia sinensis* (L.) O. KUNTZE.]
- Chagyo Gijutsu Kenkyu [Study of Tea] *56* : 1 - 5, 1979. [In Jap., ab : E.]

36687 - **AOKI, S.** : [Effects of temperature, CO_2 concentration and light intensity on
the oxygen inhibition of photosynthesis in tea leaf discs and cells.] - Chagyo Gijutsu Kenkyu [Study of Tea] *56* : 6 - 9, 1979. [In Jap., ab : E.]

36688 - **AOKI, S.** : [Single cell isolation from tea leaves and their photosynthetic
properties.] - Jap. J. Crop Sci. *48* : 343 - 349, 1979. [In Jap., ab : E.]

36689 - **APASHEVA, L.M., BUDZHIASHVILI, D.M., NAĬDICH, V.I., SHEVCHENKO, V.A.** : Prichiny ustoĭchivosti shtammov khlorelly k fizicheskim i khimicheskim faktoram sredy. [Principles of the resistance of *Chlorella* strains to physical and chemical factors of environment.] - Izv. Akad. Nauk SSSR, Ser. biol. *1979* (4) : 621 - 624, 1979. [Ps; in R, ab :·E.]

36690 - **APEL, K.** : Phytochrome-induced appearance of mRNA activity for the apoprotein of the light-harvesting chlorophyll *a/b* protein of barley (*Hordeum vulgare*). - Europe. J. Biochem. *97* : 183 - 188, 1979.

36691 - **APEL, P.** : Leitbündeldichte und Nettoassimilationsrate bei Sommergerste. - Arch. Züchtungsforsch. *9* : 179 - 184, 1979.

36692 - **APEL, P.** : Leitbündeldichte und Stomatafrequenz von Gramineen-Arten mit C_3-beziehungsweise C_4-pathway der Photosynthese. - Kulturpflanze *27* : 91 - 95, 1979.

36693 - **APEL, P., OHLE, H.** : CO_2-Kompensationspunkt und Blattanatomie bei Arten der Gattung *Moricandia* DC.(*Cruciferae*). - Biochem. Physiol. Pflanzen *174* : 68 - - 75, 1979.

36694 - **APEL, P., PEISKER, M.** : Pflanzenarten mit intermediärer Merkmalsausprägung in bezug auf den C_3- und C_4-pathway der Photosynthese. - Kulturpflanze *27* : 49 - 66, 1979.

36695 - **APLIN, P.S., HILL, D.J.** : Growth analysis of circular lichen thalli. - J. theor. Biol. *78* : 347 - 363, 1979.

36696 - **APPIANO, A., D'AGOSTINO, G., PENNAZIO, S.** : Development of dimorphic chloroplasts in a C_4 dicotyledon, *Gomphrena globosa* L., in relation to plastochron age. - J. submicroscop. Cytol. *11* : 479 - 488, 1979.

36697 - **ARADHYA, R.S., MADHAVAMENON, P.** : Induced mutagenesis in finger-millet (*Eleusine coracana* GAERTN.) with gamma-rays and ethyl methane sulphonate - II. Chlorophyll mutation frequency and spectrum. - Environ. exp. Bot. *19* : 123 - - 126, 1979.

36698 - **ARAI, K., KONO, Y.** : [Development of the rice panicle. II. Influences of nitrogen supply at heading on the pattern ȯf accumulation of dry matter and nitrogen in the caryopses at different positions on panicle.] - Jap. J. Crop Sci. *48* : 335 - 342, 1979. [In Jap., ab : E.]

36699 - **ARATA, H., NISHIMURA, M.** : Energetic coupling in the primary processes of photosynthesis in *Chromatium*. pH dependence of delayed fluorescence, electron transfer and degree of coupling. - J. Biochem. (Tokyo) *85* : 485 - 494, 1979.

36700 - **ARBI, N., SMITH, D., BINGHAM, E.T.** : Dry matter and morphological responses to temperatures of alfalfa strains with differing ploidy levels. - Agron. J. *71* : 573 - 577, 1979. [Growth analysis.]

36701 - **ARDITTI, J.** : Aspects of the physiology of orchids. - Adv. bot. Res. *7* : 421 - - 655, 1979. [Ps, Chl.]

36702 - **ARGYROUDI-AKOYUNOGLOU, J.H., AKOYUNOGLOU, G.** : The chlorophyll-protein complexes of the thylakoid in greening plastids of *Phaseolus vulgaris*. - FEBS Lett. *104* : 78 - 84, 1979.

36703 - **ARKIN, G.F., MONK, R.L.** : Seedling photosynthetic efficiency of a grain sorghum hybrid and its parents. - Crop Sci. *19* : 128 - 130, 1979.

36704 - **ARMITAGE, J.P., EVANS, M.C.W.** : Membrane potential changes during chemotaxis of *Rhodopseudomonas sphaeroides*. - FEBS Lett. *102* : 143 - 146, 1979. [Car.]

36705 - **ARMOND, P.A., BJÖRKMAN, O., STAEHELIN, L.A.** : Dissociation of supramolecular complexes in chloroplast membranes : a manifestation of heat damage to the photosynthetic apparatus. - Carnegie Inst. Year Book *78* : 153 - 157, 1979.

36706 - **ARMOND, P.A., HESS, J.L.** : Enhancement of high temperature stability of protein-protein interactions by deuterium oxide. - Carnegie Inst. Year Book *78* : 168 - 171, 1979.

36707 - **ARMSTRONG, F.A., HENDERSON, R.A., SYKES, A.G.** : Kinetic studies on reactions of iron-sulfur proteins. 2. An extension of the range of oxidants in the re-

action of reduced parsley 2-Fe ferredoxin and identification of specific binding sites using redox inactive $Cr(NH_3)_6^{3+}$ (and $Cr(en)_3^{3+}$). - J. amer. chem.
Soc. *101* : 6912 - 6917, 1979.

36708 - ARMSTRONG, W. : Aeration in higher plants. - Adv. bot. Res. *7* : 225 - 332,
1979. [Ps, resistances, photosynthates, O_2 and CO_2 transport.]

36709 - ARNESEN, U., HALLENSTVET, M., LIAAEN-JENSEN, S. : More about the carotenoids
of red algae. - Biochem. Syst. Ecol. *7* : 87 - 89, 1979.

36710 - ARNON, D.I., CHAIN, R.K. : Regulatory electron transport pathways in cyclic
photophosphorylation. Reduction of C-550 and cytochrome b_6 by ferredoxin in
the dark. - FEBS Lett. *102* : 133 - 138, 1979.

36711 - ARNTZEN, C.J., DITTO, C.L., BREWER, P.E. : Chloroplast membrane alterations
in triazine-resistant *Amaranthus retroflexus* biotypes. - Proc. nat. Acad. Sci
USA *76* : 278 - 282, 1979.

36712 - ARO, E.-M., VALANNE, N. : Effect of continuous light on CO_2 fixation and
chloroplast structure of the mosses *Pleurozium schreberi* and *Ceratodon pur-
pureus*. - Physiol. Plant. *45* : 460 - 466, 1979.

36713 - ARRON, G.P., SPALDING, M.H., EDWARDS, G.E. : Isolation and oxidative properties of intact mitochondria from the leaves of *Sedum praealtum*. A Crassulacean acid metabolism plant. - Plant Physiol. *64* : 182 - 186, 1979. [Chl.]

36714 - ARRON, G.P., SPALDING, M.H., EDWARDS, G.E. : Stoichiometry of carbon dioxide
release and oxygen uptake during glycine oxidation in mitochondria isolated
from spinach (*Spinacia oleracea*) leaves. - Biochem. J. *184* : 457 - 460, 1979.

36715 - ARRUDA, J.A. : A consideration of trophic dynamics in some tallgrass prairie
farm ponds. - Amer. Midland Natur. *102* : 254 - 262, 1979. [Ps.]

36716 - ARTECA, R.N., POOVAIAH, B.W. : Carbon dioxide fixation by potato roots (*Solanum tuberosum* L.) and its subsequent translocation. - Plant Physiol. *63*
(Suppl.) : 38, 1979.

36717 - ARTECA, R.N., POOVAIAH, B.W., SMITH, O.E. : Changes in carbon fixation, tu-
berization, and growth induced by CO_2 applications to the root zone of potato
plants. - Science *205* : 1279 - 1280, 1979.

36718 - ARYA, S.S., NATESAN, V., PARIHAR, D.B., VIJAYARAGHAVAN, P.K. : Stability of
β-carotene in isolated systems. - J. Food Technol. *14* : 571 - 578, 1979.

36719 - ASAMI, S., INOUE, K., AKAZAWA, T. : NADP-malic enzyme from maize leaf : regulatory properties. - Arch. Biochem. Biophys. *196* : 581 - 587, 1979.

36720 - ASAMI, S., INOUE, K., MATSUMOTO, K., MURACHI, A., AKAZAWA, T. : NADP-malic
enzyme from maize leaf : Purification and properties. - Arch. Biochem. Biophys. *194* : 503 - 510, 1979.

*36721 - ASANA, R.D., PARVATIKAR, S.R., SAXENA, N.P. : Studies in physiological ana-
lysis of yield. IX. Effect of light intensity on the development of the wheat
grain. - Physiol. Plant. *22* : 915 - 924, 1969. [Ps.]

*36722 - ASAY, K.H., MATCHES, A.G., NELSON, C.J. : Effect of leaf width on responses
of tall fescue genotypes to defoliation treatment and temperature regimes. -
Crop Sci. *17* : 816 - 818, 1977. [Dry-matter production.]

36723 - ASCENCIO, J., BOWES, G. : PEP carboxylase kinetics in *Hydrilla verticillata*
plants with high and low CO_2 compensation points (Γ). - Plant Physiol. *63*
(Suppl.) : 2, 1979.

36724 - ASHCROFT, W.J., MURRAY, D.R. : The dual functions of the cotyledons of *Acacia
iteaphylla* F. MUELL. (Mimosoideae). - Aust. J. Bot. *27* : 343 - 352, 1979.
[Ps.]

36725 - ASHTON, A.R., ANDERSON, L.E. : Reconstitution of the light activation system
for pea leaf chloroplast malate dehydrogenase. - Plant Physiol. *63* (Suppl.) :
24, 1979.

36726 - ASHTON, D.H., TURNER, J.S. : Studies on the light compensation point of *Eucalyptus regnans* F. MUELL. - Aust. J. Bot. *27* : 589 - 607, 1979.

36727 - ASHTON, F.M., GLENN, R.K. : Influence of chloro-, methoxy-, and methylthio-
-substitutions of bis(isopropylamino)-s-triazine on selected metabolic processes. - Pesticide Biochem. Physiol. *11* : 201 - 207, 1979. [Ps.]

36728 - ASLAM, M., HUFFAKER, R.C., RAINS, D.W., RAO, K.P. : Influence of light and
ambient carbon dioxide concentration on nitrate assimilation by intact barley seedlings. - Plant Physiol. *63* : 1205 - 1209, 1979. [Photosynthates.]

*36729 - ASLYNG, H.C. : Energiforsyning, vandforsyning og planteproduktion. [Energy,
water and plant production.] - Ugeskr. Agron. Hort. Forst. Lic. *1977* (41) :
883 - 889, 1977. [In Dan.]

36730 - ASTAUROVA, O.B., AFANASOVA, L.A., SALAMAKHA, O.V., YAZYKOV, A.A. : Isozyme
composition of malate dehydrogenase in two species of *Acetabularia*, in nor-
mal conditions and after nuclear implantation. - In : BONOTTO, S., KEFELI, V.,
PUISEUX-DAO, S. (ed.) : Developmental Biology of *Acetabularia*. Pp. 259 - 268.
Elsevier/North-Holland Biomedical Press, Amsterdam - New York - Oxford 1979.
[Chloroplast.]

36731 - ASTIER, C., VERNOTTE, C., DER-VARTANIAN, M., JOSET-ESPARDELLIER, F. :
Isolation and characterization of two DCMU-resistant mutants of the blue-gre-
en alga *Aphanocapsa* 6714. - Plant Cell Physiol. *20* : 1501 - 1510, 1979. [Ps.]

36732 - ASTON, M.J., LAWLOR, D.W. : The relationship between transpiration, root water uptake, and leaf water potential. - J. exp. Bot. *30* : 169 - 181, 1979.

36733 - ATRASHENOK, N.V., KHOTYLEVA, L.V., IL'CHENKO, V.P., RUBAN, V.V. : Issledovanie struktury khloroplastov kletok mezofilla lista tritikale. [Chloroplast
structure of triticale leaf mesophyll cells.] - Dokl. Akad. Nauk belorus.
SSR *23* : 941 - 943, 959, 1979. [In R, ab : E.]

36734 - ATTIWILL, P.M. : Nutrient cycling in a *Eucalyptus obliqua* (L'HÉRIT.) forest.
III Growth, biomass, and net primary production. - Aust. J. Bot. *27* : 439 -
458, 1979.

36735 - AUCLAIR, D. : A field technique for measuring $^{14}CO_2$ absorption by excised
or intact conifer needles. - Forest Sci. *25* : 72 - 80, 1979.

36736 - AUGUSTINE, J.J., STEVENS, M.A., BREIDENBACH, R.W. : Physiological, morphological, and anatomical studies of tomato genotypes varying in carboxylation
efficiency. - J. amer. Soc. hort. Sci. *104* : 338 - 341, 1979.

36737 - AUGUSTO, O., CILENTO, G. : Dark excitation of chlorophyll. - Photochem. Photobiol. *30* : 191 - 193, 1979.

36738 - AUSTENFELD, F.-A. : Nettophotosynthese der Primär- und Folgeblätter von
Phaseolus vulgaris L. unter dem Einfluß von Nickel, Kobalt und Chrom. -
Photosynthetica *13* : 434 - 438, 1979.

36739 - AVAKYAN, A.B., VENEDIKTOV, P.S., DOBRETSOV, G.E., RUBIN, A.B. : Issledovanie
stareniya khloroplastov metodom flyuorestsentnogo zonda. [Study of the chloroplast ageing by the method of fluorescent probe.] - Nauch. Dokl. vyssh. Shkoly, biol. Nauki *1979* (9) : 25 - 28, 1979. [In R.]

36740 - AVARMAA, R., KOCHUBEĬ, S., TAMKIVI, R. : Vremena zatukhaniya fluorestsentsii
khlorofilla v fotosistemakh 1 i 2 khloroplastov pri temperature 4,2 K. [Flu-
orescence quenching times of chlorophyll in photosystems 1 and 2 of chloroplasts at temperature 4.2 K.] - Eesti NSV Tead. Toim., Füüs. Mat. *28* (1) :
86 - 89, 1979. [In R.]

36741 - AVARMAA, R.A., KOCHUBEY, S.M., TAMKIVI, R.P. : Low-temperature fluorescence
decay and energy transfer in photosynthetic units. - FEBS Lett. *102* : 139 -
- 142, 1979.

36742 - AVERY, D.J., PRIESTLEY, C.A., TREHARNE, K.J. : Integration of assimilation
and carbohydrate utilization in apple. - In : MARCELLE, R., CLIJSTERS, H.,
VAN POUCKE, M. (ed.) : Photosynthesis and Plant Development. Pp. 221 - 231.
Dr. W. Junk bv. Publ., The ·Hague - Boston - London 1979.

36743 - AVI-DOR, Y., ROTT, R., SCHNAIDERMAN, R. : The effect of antibiotics on the
photocycle and protoncycle of purple membrane suspensions. - Biochim. biophys.
Acta *545* : 15 - 23, 1979.

*B36744 - **AVRON, M.** (ed.) : Proceedings of the Third International Congress on Photosynthesis. Volume I, II, III. - Elsevier Scientific Publishing Company, Amsterdam - Oxford - New York 1975.

36745 - **AVRON, M., SCHREIBER, U.** : Properties of ATP-induced chlorophyll luminescence
in chloroplasts. - Biochim. biophys. Acta *546* : 448 - 454, 1979.

36746 - **AXELSSON, L., KLOCKARE, B., SUNDQVIST, C.** : Oak seedlings grown in different
light qualities. 1. Morphological development. - Physiol. Plant. *45* : 387 -
- 392, 1979.

36747 - **AXELSSON, L., SELSTAM, E.** : Changes in the photodynamic properties of the
chlorophyll(ide) during the early stages of greening. - In : **APPELQVIST,
L.-Å., LILJENBERG, C.** (ed.) : Advances in the Biochemistry and Physiology
of Plant Lipids. Pp. 363 - 368. Elsevier/North-Holland Biomedical Press,
Amsterdam - New York - Oxford 1979.

36748 - **AYRES, P.G.** : CO_2 exchanges in plants infected by obligately biotrophic
pathogens. - In : **MARCELLE, R., CLIJSTERS, H., VAN POUCKE, M.** (ed.) : Photosynthesis and Plant Development. Pp. 343 - 354. Dr.W.Junk bv. Publ., The
Hague - Boston - London 1979.

36749 - **BACCARINI-MELANDRI, A., MELANDRI, B.A., HAUSKA, G.** : The stimulation of photophosphorylation and ATPase by artificial redox mediators in chromatophores
of *Rhodopseudomonas capsulata* at different redox potentials. - J. Bioenerg.
Biomembranes *11* : 1 - 16, 1979.

36750 - **BACH, S.D., JOSSELYN, M.N.** : Production and biomass of *Cladophora prolifera
(Chlorophyta, Cladophorales)* in Bermuda. - Bot. mar. *22* : 163 - 168, 1979.

36751 - **BACHOFEN, R.** : Labeling of membranes and reaction centers from the photosynthetic bacterium *Rhodospirillum rubrum* with fluorescamine. - FEBS Lett.
107 : 409 - 412, 1979.

36752 - **BADWAL, S.S., ASHWANI, SRIVASTAVA, K., CHAURASIA, B.D.** : Transgressive variation and interrelationship of seed yield with chlorophyll constituents in
Linum usitatissimum L. - Genet. agrar. *33* : 323 - 330, 1979.

36753 - **BAGNALL, D.** : Low temperature responses of three sorghum species. - In :
LYONS, J.M., GRAHAM, D., RAISON, J.K. (ed.) : Low Temperature Stress in Crop
Plants. The Role of the Membrane. Pp. 67 - 80. Academic Press, New York
1979.

*36754 - **BAHR, J.T.** : Activation of RuBP carboxylase by cyanate, $MnCl_2$, or $CaCl_2$. -
In : **SIEGELMAN, H.W., HIND, G.** (ed.) : Photosynthetic Carbon Assimilation.
Pp. 415 - 416. Plenum Press, New York - London 1978.

36755 - **BAIRD, B.A., HAMMES, G.G.** : Structure of oxidative- and photo-phosphorylation
coupling factor complexes. - Biochim. biophys. Acta *549* : 31 - 53, 1979.

36756 - **BAIRD, B.A., PICK, U., HAMMES, G.G.** : Structural investigation of reconstituted chloroplast ATPase with fluorescence measurements. - J. biol. Chem.
254 : 3818 - 3825, 1979.

36757 - **BAJRACHARYA, D., SCHROPFER, P.** : Effect of light on the development of glyoxysomal functions in the cotyledons of mustard (*Sinapis alba* L.) seedlings.
- Planta *145*: 181 - 186, 1979.

36758 - **BAKER, A.L., BAKER, K.K.** : Effects of temperature and current discharge on
the concentration and photosynthetic activity of the phytoplankton in the
upper Mississippi River. - Freshwater Biol. *9* : 191 - 198, 1979.

36759 - **BALANGÉ, A.P., ROLLIN, P.** : Régulation de la δ-aminolévulinate déshydratase
(E.C. 4.2.1.24) par le phytochrome dans les cotylédons de Radis. - Physiol.
vég. *17* : 153 - 166, 1979.

*36760 - **BALASUBRAMANIAN, V., SHANTHAKUMARI, P., SINHA, S.K.** : $^{14}CO_2$-fixation and
nitrate reductase activity *in vivo* in relation to hybrid vigour in maize. -
Indian J. exp. Biol. *15* : 780 - 782, 1977.

36761 - **BALLYEVA, O.B., LOMAZIN, A.G.** : Reparatsiya teplovogo povrezhdeniya v kletkakh lista tradeskantsii.I. Reparatsiya ul'trastruktury yadryshek i fototak-
sisa khloroplastov. [Repair of heat injury by leaf cells of *Tradescantia*.
I. Repair of nucleolar ultrastructure and phototaxis of chloroplasts.] - Tsitologiya *21* : 1170 - 1174, 1979. [In R, ab : E.]

36762 - **BALNOKIN, Yu.V., STROGONOV, B.P., KUKAEVA, E.A., MEDVEDEV, A.V.** : Zashchit-
naya funktsiya membran kletok *Dunaliella* pri vysokikh kontsentratsiyakh NaCl
v srede. [Protective function of *Dunaliella* cell membranes under high NaCl
concentrations in the medium.] - Fiziol. Rast. *26* : 552 - 559, 1979. [In R,
ab : E.]

36763 - **BALOUN, J.** : Chlortetracyklin a barva plastidů. [Chlortetracycline and the
colour of plastids.] - In : Rozvoj Farmacie v Rámci Vědecko-Technické Revolu-
ce. Pp. 27 - 30. Univ. Karlova, Praha 1979. [Chl, Car; in Czech.]

36764 - **BALTSCHEFFSKY, M., LUNDIN, A.** : Flash-induced increase of ATPase activity
in *Rhodospirillum rubrum* chromatophores. - In : MUKOHATA, Y., PACKER, L.
(ed.) : Cation Flux across Biomembranes. Pp. 209 - 218. Academic Press,
New York - San Francisco - London 1979.

36765 - **BAMBERG, E., APELL, H.-J., DENCHER, N.A., SPERLING, W., STIEVE, H.,
LÄUGER, P.** : Photocurrents generated by bacteriorhodopsin in planar bilayer
membranes. - Biophys. Struct. Mech. *5* : 277 - 292, 1979.

36766 - **BANERJI, D., SHARMA, V.** : Parallelism in Hill activity and anthocyanidin
content in *Euphorbia pulcherrima*. - Phytochemistry *18* : 1767 - 1768, 1979.

36767 - **BARANKIEWICZ, T.J., POPOVIC, R.B., ZALIK, S.** : Ribulose-1,5-bisphosphate
carboxylase and phosphoenolpyruvate carboxylase activity in barley and its
virescens mutant. - Biochem. biophys. Res. Commun. *87* : 884 - 889, 1979.

*B36768 - **BARBER, J.** (ed.) : The Intact Chloroplast. - Elsevier Scientific Publ. Company, Amsterdam - New York - Oxford 1976.

B36769 - **BARBER, J.** (ed.) : Photosynthesis in Relation to Model Systems. - Elsevier,
Amsterdam - New York - Oxford 1979.

36770 - **BARBER, J.** : Energy transfer and its dependence on membrane properties. -
In : Chlorophyll Organization and Energy Transfer in Photosynthesis. Pp.
283 - 304. Excerpta Medica, Amsterdam - Oxford - New York 1979.

36771 - **BARBER, J.** : Primary processes of photosynthesis : Structural and functional aspects. - Photochem. Photobiol. *29* : 203 - 207, 1979.

36772 - **BARBER, J.** : Studying photosynthesis with picosecond spectroscopy.-Photobiol.
Bull. *1* (5) : 84 - 100, 1979.

36773 - **BARBER, J., CHOW, W.S.** : A mechanism for controlling the stacking and un-
stacking of chloroplast thylakoid membranes. - FEBS Lett. *105* : 5 - 10,
1979.

36774 - **BARBER, J., SEARLE, G.F.W.** : Double layer theory and the effect of pH on
cation-induced chlorophyll fluorescence. - FEBS Lett. *103* : 241 - 245, 1979.

36775 - **BARBOUR, M.G., RADOSEVICH, S.R.** : C^{14} uptake by the marine angiosperm *Phyl-
lospadix scouleri*. - Amer. J. Bot. *66* : 301 - 306, 1979.

36776 - **BARDEN, J.A., FERREE, D.C.** : Rootstock does not affect net photosynthesis,
dark respiration, specific leaf weight, and transpiration of apple leaves. -
J. amer. Soc. hort. Sci. *104* : 526 - 528, 1979.

36777 - **BAR'ETAS, P.K., FAĬZIEV, Sh., IMAMALIEV, A.I.** : Vliyanie defoliantov na peredvizhenie assimilyatov iz list'ev khlopchatnika. [Effect of defoliants on
assimilate transport from leaves of growing cotton plants.] - Fiziol. Rast.
26 : 161 - 166, 1979. [In R, ab : E.]

36778 - **BARINOV, G.V.** : Kineticheskiĭ podkhod k issledovaniyu pervichnogo produktsi-
onnogo protsessa v okeane. [Kinetic approach to the investigation of primary
production process in the ocean.] - Izv. Akad. Nauk SSSR, Ser. biol. *1979* :
113 - 118, 1979. [In R, ab : E.]

36779 - BARKER, P., EDMISTON, J. : Detection of semiquinone intermediates in pigment
        leached from *Sinapis alba* L. - Biochem. Physiol. Pflanz. *174* : 425 - 430,
        1979.

36780 - BARLOW, H.W.B. : Sectorial patterns in leaves on fruit tree shoots produced
        by radioactive assimilates and solutions. - Ann. Bot. *43* : 593 - 602, 1979.

B36781 - BARON, W.M.M. : Organization in Plants. - Edward Arnold, London 1979. [Ps.]

36782 - BARR, R., CRANE, F.L. : Catechols stimulate ferricyanide reduction in chloro-
        plast Photosystem II. - Biochim. biophys. Acta *546* : 77 - 83, 1979.

36783 - BARR, R., MELHEM, R., LEZOTTE, A.L., CRANE, F.L. : Control of Photosystem II
        electron transport in spinach chloroplasts. - Plant Physiol. *63* (Suppl.) :
        54, 1979.

36784 - BARRACLOUGH, R., ELLIS, R.J. : The biosynthesis of ribulose bisphosphate
        carboxylase. Uncoupling of the synthesis of the large and small subunits in
        isolated soybean leaf cells. - Europe. J. Biochem. *13* : 165 - 177, 1979.

*36785 - BARRETT, J.E. III, AMLING, H.J. : Effects of developing fruits on production
        and translocation of ^{14}C-labeled assimilates in cucumber. - HortScience *13* :
        545 - 547, 1978.

36786 - BARSKIĬ, E.L., SAMUILOV, V.D. : Sdvigi polosy pogloshcheniya bakteriokhlo-
        rofilla pri 880 nm v khromatoforakh i subkhromatofornykh pigment-belkovykh
        kompleksakh *Rhodospirillum rubrum.* [Shifts of the bacteriochlorophyll absorp-
        tion band at 880 nm in chromatophores and subchromatophore pigment-protein
        complexes from *Rhodospirillum rubrum.*] - Biokhimiya *44* : 1805 - 1813, 1979.
        [In R, ab : E.]

36787 - BARSKY, E.L., SAMUILOV, V.D. : Blue and red shifts of bacteriochlorophyll
        absorption band around 880 nm in *Rhodospirillum rubrum.* - Biochim. biophys.
        Acta *548* : 448 - 457, 1979.

36788 - BARTA, A.L. : Effect of nitrogen supply on photosynthate partitioning, root
        carbohydrate accumulation, and acetylene reduction in birdsfoot trefoil. -
        Crop Sci. *19* : 715 - 718, 1979.

*36789 - BARTKOV, B.I. : Raspredelenie fotoassimilyatov po organam rasteniĭ. [Photo-
        synthates distribution in plant organs.] - In : Pogloshchenie i Peredvizhe-
        nie Veshchestv u Rasteniĭ. Pp. 39 - 55. Akad. Nauk SSSR, Vladivostok 1978.
        [In R.]

*36790 - BARTKOV, B.I. : Postuplenie postfotosinteticheskikh produktov v plody lipy.
        [Transfer of post-photosynthetic products into fruit of *Tilia.*] - In : Po-
        gloshchenie i Peredvizhenie Veshchestv u Rasteniĭ. Pp. 68 - 72, 81. Akad.
        Nauk SSSR, Vladivostok 1978. [In R.]

*36791 - BARTKOV, B.I., BARTKOVA, A.D., VOROB'EVA, S.M. : Ob ortostikhnosti transporta
        assimilyatov u soi. [Orthostichal photosynthate transport in soybean.] -
        In : Fiziologicheskie i Biokhimicheskie Issledovaniya Rasteniĭ na Dal'nem
        Vostoke. Pp. 14 - 17. Biol.-pochv. Inst. dal'nevostoch. Fil. sib. Otd. Akad.
        Nauk SSSR, Vladivostok 1970. [In R.]

*36792 - BARTKOV, B.I., SEMKIN, B.I., NARBUT, N.A., BARTKOVA, A.D., VORONKOVA, N.M. :
        Raspredelenie assimilyatov u soi v nachal'noĭ stadii naliva semyan. [Photo-
        synthate distribution in soybean in the primary phase of seed formation.] -
        In : Fiziologicheskie i Biokhimicheskie Issledovaniya Rasteniĭ na Dal'nem
        Vostoke. Pp. 8 - 14. Biol.-pochv. Inst. dal'nevostoch. Fil. sib. Otd. Akad.
        Nauk SSSR, Vladivostok 1970. [In R.]

36793 - BARTLETT, S.G., HARRIS, E.H., GRABOWY, C.T., GILLHAM, N.W., BOYNTON, J.E. :
        Ribosomal subunits affected by antibiotic resistance mutations at seven
        chloroplast loci in *Chlamydomonas reinhardtii.* - Mol. gen. Genet. *176* :
        199 - 208, 1979.

36794 - BARTLEY, M., HALLAM, N.D. : Changes in the fine structure of the desicca-
        tion-tolerant sedge *Coleochloa setifera* (RIDLEY) GILLY under water stress. -
        Aust. J. Bot. *27* : 531 - 545, 1979. [Chl.]

*36795 - BARYSHMAN, F.S., CHEPURNOĬ, V.S. : Vliyanie skumpii na fotosintez kashtana s"edobnogo. [The effect of sumac on sweet chestnut photosynthesis.] - Tr. kuban. sel'sko-khoz. Inst. *65* (93) : 89 - 93, 1973. [In R.]

36796 - BASHFORD, C.L., BALTSCHEFFSKY, M., PRINCE, R.C. : The phosphate potential and H$^+$/ATP ratio in *Rhodospirillum rubrum*. - FEBS Lett. *97* : 55 - 60, 1979.

36797 - BASHFORD, C.L., CHANCE, B., PRINCE, R.C. : Oxonol dyes as monitors of membrane potential. Their behavior in photosynthetic bacteria. - Biochim. biophys. Acta *545* : 46 - 57, 1979.

36798 - BASHFORD, C.L., PRINCE, R.C., TAKAMIYA, K.-I., DUTTON, P.L. : Electrogenic events in the ubiquinone-cytochrome b/c_2 oxidoreductase of *Rhodopseudomonas sphaeroides*. - Biochim. biophys. Acta *545* : 223 - 235, 1979.

*36799 - BASKIN, J.M., QUARTERMAN, E. : Light relations of *Psoralea subacaulis* T. & G. (*Leguminosae*). - Ecology *49* : 571 - 573, 1968. [Ps.]

36800 - BASSHAM, J.A. : The reductive pentose phosphate cycle and its regulation. - In : GIBBS, M., LATZKO, E. (ed.) : Photosynthesis II. (Encycl. Plant Physiol. N.S. Vol. 6.) Pp. 9 - 30. Springer-Verlag, Berlin - Heidelberg - New York 1979.

36801 - BASSI, P.K., EASTWELL, K.C., SPENCER, M.S. : Techiques for measurement of ethylene production and carbon dioxide fixation by intact shoots under controlled environmental conditions. - Plant Physiol. *63* (Suppl.) : 67, 1979.

36802 - BASSI, P.K., SPENCER, M.S. : A cuvette design for measurement of ethylene production and carbon dioxide exchange by intact shoots under controlled environmental conditions. - Plant Physiol. *64* : 488 - 490, 1979.

36803 - BASZYŃSKI, T., WAJDA, L., KRÓL, M., WOLIŃSKA, D., KRUPA, Z., TUKENDORF, A. : Chloroplast electron transport reactions of tomato leaves as affected by cadmium treatment. - In : Mineral Nutrition of Plants. Vol. II. Pp. 328 - 331. Publ. House Central Cooperative Union, Sofia 1979.

36804 - BATES, J.W. : The relationship between physiological vitality and age in shoot segments of *Pleurozium schreberi* (BRID.) MITT. - J. Bryol. *10* : 339 - 351, 1979.

*36805 - BATUEVA, R.A., AGAVERDIEV, A.Sh. : Okislitel'no-vosstanovitel'nye protsessy u ogurtsov posle deĭstviya na nikh kratkovremennykh zamorozkov. [Redox processes in cucumber after short-time frost periods.] - In : Vliyanie Fiziko--Khimicheskikh Faktorov Sredy na Rasteniya. Pp. 33 - 37. Permsk. gos. Univ. Im. A.M. Gor'kogo, Perm' 1978. [Ps; in R.]

36806 - BAUER, H. : Photosynthesis of ivy leaves (*Hedera helix* L.) after heat stress. III. Stomatal behaviour. - Z. Pflanzenphysiol. *92* : 277 - 283, 1979. [Ps.]

36807 - BAUER, H., SENSER, M. : Photosynthesis of ivy leaves (*Hedera helix* L.) after heat stress II. Activity of ribulose bisphosphate carboxylase, Hill reaction, and chloroplast ultrastructure. - Z. Pflanzenphysiol. *91* : 359 - 369, 1979.

36808 - BAUER, K., KÖCHER, H. : Die Wirkung 2.3-substituierter Naphthochinone auf einzellige Algen und isolierte Spinatchloroplasten. - Z. Naturforsch. *34 C* : 961 - 963, 1979.

36809 - BAUER, K., SEILER, W., GIEHL, H. : CO production by higher plants. - Z. Pflanzenphysiol. *94* : 219 - 230, 1979. [Ps.]

36810 - BAUMANN, G., GÜNTHER, G. : Effects of glyoxylate, glycine, and serine on the assimilation of $^{14}CO_2$ in mesophyll cells of *Chenopodium album*. - Biochem. Physiol. Pflanz. *174* : 160 - 168, 1979.

36811 - BAUMANN, G., GÜNTHER, G. : Effects of herbicides on $^{14}CO_2$ fixation in isolated mesophyll cells from *Beta vulgaris* (sugar beet) and *Chenopodium album*. - Biochem. Physiol. Pflanz. *174* : 723 - 725, 1979.

36812 - BAUSHER, M.G. : Changes in ATP levels and carbonic anhydrase in the presence of citrus blight. - Plant Physiol. *63* (Suppl.) : 115, 1979.

36813 - **BAUWE, H.** : Eine empfindliche Methode zur Ermittlung der Konzentration an Ribulose-1,5-bisphosphat-Carboxylase/Oxygenase in Blattextrakten. - Biochem. Physiol. Pflanz. *174* : 246 - 250, 1979.

36814 - **BAUWE, H., APEL, P.** : Biochemical characterization of *Moricandia arvensis* (L.) DC., a species with features intermediate between C_3 and C_4 photosynthesis, in comparison with the C_3 species *Moricandia foetida* BOURG. - Biochem. Physiol. Pflanz. *174* : 251 - 254, 1979.

36815 - **BAYLEY, H.** : Inhibitors of photosynthetic electron transport. The properties of diazidodialkylbenzoquinones. - Z. Naturforsch. *34 C* : 490 - 492, 1979.

36816 - **BAZIER, R., BURGHOFFER, C., COSTES, C.** : Répartition des lipides et des protéines lamellaires dans les chloroplastes d'espèces appartenant au genre *Triticum*. - Ann. Amélior. Plant. *29* : 665 - 682, 1979.

36817 - **BAZZAZ, F.A., CARLSON, R.W.** : Photosynthetic contribution of flowers and seeds to reproductive effort of an annual colonizer. - New Phytol. *82* : 223 - 232, 1979.

36818 - **BAZZAZ, F.A., CARLSON, R.W., HARPER, J.L.** : Contribution to reproductive effort by photosynthesis of flowers and fruits. - Nature *279* : 554 - 555, 1979.

36819 - **BAZZAZ, M.B., REBEIZ, C.A.** : Chloroplast culture - V. Spectrofluorometric determination of chlorophyll(ide) a and b and pheophytin (or pheophorbide) a and b in unsegregated pigment mixtures. - Photochem. Photobiol. *30* : 709 - 721, 1979.

36820 - **BAZZAZ, M.B., REBEIZ, C.A.**: Spectrofluorometric determination of chlorophyll(ide) a and b and pheo(phorbide) a and b in unsegregated pigment mixtures. - Plant Physiol. *63* (Suppl.) : 161, 1979.

36821 - **BEADLE, C.L., JARVIS, P.G., NEILSON, R.E.** : Leaf conductance as related to xylem water potential and carbon dioxide concentration in Sitka spruce. - Physiol. Plant. *45* : 158 - 166, 1979.

36822 - **BEAMS, H.W., KESSEL, R.G., SHIH, C.Y.** : Effects of ultracentrifugation on the mesophyll cells and chloroplasts of the spinach leaf, and on the cells and chloroplasts of entire duckweed plants. - Biol. cell. *35* : 87 - 96, 1979.

✶36823 - **BEARDEN, A.J., MALKIN, R.** : Primary processes in chloroplast photosynthesis : EPR studies of bound ferredoxin and P700. - Ann. N.Y. Acad. Sci. *222* : 858 - 870, 1973.

36824 - **BECK, E.** : Glycolate synthesis. - In : GIBBS, M., LATZKO, E. (ed.) : Photosynthesis II. (Encycl. Plant Physiol. N.S. Vol. 6.) Pp. 327 - 337. Springer-Verlag, Berlin - Heidelberg - New York 1979.

36825 - **BECKERSON, D.W., HOFSTRA, G.** : Effect of sulphur dioxide and ozone singly or in combination on leaf chlorophyll, RNA, and protein in white bean. - Can. J. Bot. *57* : 1940 - 1945, 1979.

36826 - **BECKERSON, D.W., HOFSTRA, G.** : Stomatal responses of white bean to O_3 and SO_2 singly or in combination. - Atmos. Environ. *13* : 533 - 535, 1979. [Stomatal resistance.]

36827 - **BECKERSON, D.W., HOFSTRA, G.** : Response of leaf diffusive resistance of radish, cucumber and soybean to O_3 and SO_2 singly or in combination. - Atmos. Environ. *13* : 1263 - 1268, 1979. [Stomatal resistance.]

36828 - **BEDDARD, G.S., FLEMING, G.R., PORTER, G., SEARLE, G.F.W., SYNOWIEC, J.A.** : The fluorescence decay kinetics of *in vivo* chlorophyll measured using low intensity excitation. - Biochim. biophys. Acta *545* : 165 - 174, 1979.

36829 - **BEDU, S.** : Étude comparative du système phosphohydrolasique de thylakoïdes de chloroplastes d'une plante calcifuge, le Lupin (*Lupinus Luteus* L.) et d'une plante calcicole, le Féverole (*Vicia Faba* L.). - Oecol. Plant. *14* : 61 - 73, 1979.

36830 - **BEDUNAH, D., TRLICA, M.J.** : Sodium chloride effects on carbon dioxide exchange rates and other plant and soil variables of ponderosa pine. - Can. J. Forest Res. *9* : 349 - 353, 1979.

36831 - **BEER, S., WAISEL, Y.** : Some photosynthetic carbon fixation properties of seagrasses. - Aquat. Bot. *7* : 129 - 138, 1979.

36832 - **BEEVERS, H.** : Microbodies in higher plants. - Annu. Rev. Plant Physiol. *30* :
159 - 193, 1979.

36833 - **BEHRENS, P.W., MARSHO, T.V.** : O_2 reduction in higher plants. - Plant Physiol.
63 (Suppl.) : 55, 1979.

36834 - **BEKÁREK, V., KAPLANOVÁ, M., SOCHA, J.** : Study of non-specific interactions
of carbonyl groups of chlorophyll *a* in solutions by IR spectroscopy. -
Studia biophys. *77* : 21 - 24, 1979.

36835 - **BEKASOVA, O.D., KOPELEVICH, O.V., SUD'BIN,A.I.** : Opredelenie opticheskikh
svoĭstv morskoĭ vody, soderzhaniya khlorofilla i vzvesi v poverkhnostnom sloe
okeana po spektral'nym znacheniyam yarkosti voskhodyashchego izlucheniya.
[Determination of optical properties of sea water, chlorophyll and suspended
matter concentrations in the upper layer of ocean based on spectral values
of upwelling radiance.] - Okeanologiya *19* : 233 - 238, 1979. [In R, ab : E.]

36836 - **BEKASOVA, O.D., SHUBIN, L.M., EVSTIGNEEV, V.B.** : Fikobilisomy iz sinezelenykh vodoroslеĭ *Aphanizomenon flos-aquae* i *Anabaena variabilis*. [Phycobili-
somes of blue-green algae *Aphanizomenon flos-aquae* and *Anabaena variabilis*.]
- Izv. Akad. Nauk SSSR, Ser. biol. *1979* : 198 - 207, 1979. [In R, ab : E.]

36837 - **BELANGER, F.C., REBEIZ, C.A.** : Chloroplast biogenesis XXVII. Detection of novel chlorophyll and chlorophyll precursors in higher plants. - Biochem. biophys. Res. Commun. *88* : 365 - 372, 1979.

36838 - **BELANGER, F.C., REBEIZ, C.A.** : Detection of novel chlorophylls and of their
precursors in higher plants. - Plant Physiol. *63* (Suppl.) : 96, 1979.

36839 - **BELAYA, G.A., MOROZOV, V.L.** : Ėkologo-fiziologicheskaya kharakteristika sub-
al'piĭskogo krupnotrav'ya na Kamchatke. [Ecophysiological characteristics
of subalpine broadleaf ecosystems in Kamchatka.] - Ėkol. Biol. vysokogor-
nykh Rast. *14* (2) : 17 - 23, 1979. [Ps; in R.]

*36840 - **BELIKOV, I.F., VORONKOVA, N.M.** : Postuplenie assimilyatov v semena iz list'-
ev i stvorok u soi. [Transport of photosynthates from leaves and pods into
seeds of soybean.] - In : Pogloshchenie i Peredvizhenie Veshchestv u Rasteniĭ.
Pp. 22 - 26, 80. Akad. Nauk SSSR, Vladivostok 1978. [In R.]

36841 - **BÉLIVEAU, R., BELLEMARE, G.** : Light-dependent phosphorylation of thylakoid
membrane polypeptides. - Biochem. biophys. Res. Commun. *88* : 797 - 803, 1979.

36842 - **BÉLIVEAU, R., BELLEMARE, G.** : Thylakoid membrane protein phosphorylation in
correlation with photosynthetic membrane activation. - Biochem. biophys.
Res. Commun. *91* : 1377 - 1382, 1979.

*36843 - **BELKIN, S., PADAN, E.** : Hydrogen metabolism in the facultative anoxygenic
cyanobacteria (blue-green algae) *Oscillatoria limnetica* and *Aphanothece halophytica*. - Arch. Microbiol. *116* : 109 - 111, 1978.

36844 - **BELL, A.D., ROBERTS, D., SMITH, A.** : Branching patterns : the simulation of
plant architecture. - J. theor. Biol. *81* : 351 - 375, 1979.

36845 - **BELL, J.N.B., RUTTER, A.J., RELTON, J.** : Studies on the effects of low levels of sulphur dioxide on the growth of *Lolium perenne* L. - New Phytol.
83 : 627 - 643, 1979. [Chl, growth analysis.]

36846 - **BELL, K., HIATT, H.D., NILES, W.E.** : Seasonal changes in biomass allocation
in eight winter annuals of the Mojave desert. - J. Ecol. *67* : 781 - 787,
1979.

*36847 - **BELOBRODSKAYA, L.K., IVANOV, B.N., MUZAFAROV, E.N.** : Zavisimost' skorosti fotosinteticheskogo ėlektronnogo transporta ot protonnogo obmena mezhdu khloroplastami i sredoĭ. [Dependence of the rate of photosynthetic electron transport on the proton exchange between chloroplast and medium.] - In : Biologiya i Nauchno-Tekhnicheskiĭ Progress. Pp. 81 - 83. Pushchino 1974. [In R.]

36848 - **BEN-AMOTZ, A.** : Hydrogen metabolism. - In : GIBBS, M., LATZKO, E. (ed.) :
Photosynthesis II. (Encycl. Plant Physiol. N.S. Vol. 6.) Pp. 497 - 506.
Springer-Verlag, Berlin - Heidelberg - New York 1979.

36849 - **BEN-BASSAT, D., ANDERSON, L.E.** : Bound and free glucose-6-phosphate dehydrogenase in the chloroplast. - Plant Physiol. *63* (Suppl.) : 8, 1979.

36850 - **BENECKE, U.** : Surface area of needles in *Pinus radiata* - variation with respect to age and crown position. - New Zeal. J. Forest. Sci. *9* : 267 - 271,
1979.

36851 - **BEN-GAD, D.Y., ALTMAN, A., MONSELISE, S.P.** : Interrelationships of vegetative
growth and assimilate distribution of *Citrus limettioides* seedlings in response to root-applied GA_3 and SADH. - Can. J. Bot. *57* : 484 - 490, 1979.

36852 - **BENGIS-GARBER, C., GROMET-ELHANAN, Z.** : Purification of the energy-transducing adenosine triphosphatase complex from *Rhodospirillum rubrum*. - Biochemistry *18* : 3577 - 3581, 1979.

36853 - **BENNERT, W.H., MOONEY, H.A.** : The water relations of some desert plants in
Death Valley, California. - Flora *168* : 405 - 427, 1979. [Ps.]

36854 - **BENNETT, J.** : Chloroplast phosphoproteins. Phosphorylation of polypeptides
of the light-harvesting chlorophyll protein complex. - Europe. J. Biochem.
99 : 133 - 137, 1979.

36855 - **BENNETT, J.** : Chloroplast phosphoproteins. The protein kinase of thylakoid
membranes is light-dependent. - FEBS Lett. *103* : 342 - 344, 1979.

36856 - **BENNETT, J.** : The protein that harvests sunlight. - Trends biochem. Sci. *4* :
268 - 271, 1979. [Chl.]

36857 - **BERÁNEK, V.** : Postavení listů v porostu a pokryvnost listoví (LAI) ve vztahu
k výnosu jarní pšenice. [Position of leaves in the stand and the leaf area
index in relation to the yield of spring wheat.] - Rost. Výroba (Praha) *25* :
255 - 264, 1979. [In Czech, ab : E, G, R.]

36858 - **BERCHTOLD, M., BACHOFEN, R.** : Hydrogen formation by cyanobacteria cultures
selected for nitrogen fixation. - Arch. Microbiol. *123* : 227 - 232, 1979.

B36859 - **BERDYKULOV, Kh.A.** : Fotosintez Mikrovodoroslĕĭ, Kul'tiviruemykh pod Otkrytym
Nebom. [Photosynthesis in Microalgae Cultivated under the Open Sky.] - FAN,
Tashkent 1979. [In R.]

36860 - **BERG, A.I., NOKS, P.P., KONONENKO, A.A., FROLOV, E.N., KHRYMOVA, I.N.,
RUBIN, A.B., LIKHTENSHTEĬN, G.I., GOL'DANSKIĬ, V.I., PARAK, F., BUKL, M.,
MËSSBAUÉR, R.** : Konformatsionnoe regulirovanie funktsional'noĭ aktivnosti
v fotosinteticheskikh membranakh purpurnykh bakteriĭ. [Conformational regulation of functional activity of photosynthetic membranes of purple bacteria.] - Mol. Biol. (Moskva) *13* : 81 - 89, 1979. [In R, ab : E.]

36861 - **BERG, A.I., NOKS, P.P., KONONENKO, A.A., FROLOV, E.N., USPENSKAYA, N.Ya.,
KHRYMOVA, I.N., RUBIN, A.B., LIKHTENSHTEĬN, G.I., KHIDEG, K.** : Konformat-
sionnaya podvizhnost' i funktsional'naya aktivnost' fotosinteticheskikh
reaktsionnykh tsentrov iz *Rhodopseudomonas sphaeroides*. [Conformational mobility and functional activity of photosynthetic reaction centres from *Rhodopseudomonas sphaeroides*.] - Mol. Biol. (Moskva) *13* : 469 - 477, 1979.
[In R, ab : E.]

36862 - **BERG, S.P., LUSCZAKOSKI, D.M., MORSE, P.D. II** : Spin label motion in the
internal aqueous compartment of spinach thylakoids. - Arch. Biochem. Biophys. *194* : 138 - 148, 1979.

36863 - **BERG, S.P., NESBITT, D.M.** : Chromium oxalate : A new spin label broadening
agent for use with thylakoids. - Biochim. biophys. Acta *548* : 608 - 615,
1979.

36864 - **BERG, S.P., NESBITT, D.M.** : Probing the aqueous interior of spinach thylakoids with the spin label TEMPAMINE. - Plant Physiol. *63* (Suppl.) : 53,
1979.

*36865 - **BERGMANN, H., LERCH, G., MÜNTZ, K., RAMON, J., TRAVIESO, A.** : Efecto fisio-
logico del cultivo de posturas de cafe en Cuba, al sol y bajo sombra.
[Physiological effects of cultivation of coffee stands in Cuba on sunlight
and in deep shade.] - Acad. Cienc. Cuba, Ser. biol. *25* : 3 - 27, 1970. [Ps,
Chl; in Span.]

36866 - BERGSTEIN, T., HENIS, Y., CAVARI, B.Z. : Investigations on the photosynthetic sulfur bacterium *Chlorobium phaeobacteroides* causing seasonal blooms in Lake Kinneret. - Can. J. Microbiol. *25* : 999 - 1007, 1979. [Ps.]

36867 - BERGUM, P.W., NADLER, K.D. : The inhibition of tetrapyrrole biosynthesis in greening barley by isonicotinic acid hydrazide (INH). - Plant Physiol. *63* (Suppl.) : 97, 1979. [Chl.]

*36868 - BERLAND, B.R., BONIN, D.J., MAESTRINI, S.Y. : Facteurs limitant la production primaire des eaux oligotrophes d'une aire côtière méditerranéenne (Calanque d'En-Vau, Marseille). - Int. Rev. ges. Hydrobiol. *63* : 501 - 531, 1978. [Chl.]

36869 - BERNARD, J.M., FITZ, M.L. : Seasonal changes in aboveground primary production and nutrient contents in a central New York *Typha glauca* ecosystem. - Bull. Torrey bot. Club *106* : 37 - 40, 1979.

36870 - BERNIER, G., SACHS, R.M. : Photosynthesis and flowering. - In : MARCELLE, R., CLIJSTERS, H., VAN POUCKE, M. (ed.) : Photosynthesis and Plant Development. Pp. 137 - 148. Dr.W.Junk bv. Publ., The Hague - Boston - London 1979.

36871 - BERRY, J., BADGER, M. : Measurement of [^{18}O] oxygen uptake during photosynthesis by intact leaves of C_3 and C_4 plants at controlled CO_2 concentration. - Plant Physiol. *63* (Suppl.) : 153, 1979.

36872 - BERRY, J.A., BADGER, M.R. : Direct measurement of photorespiration as a function of CO_2 concentration. - Carnegie Inst. Year Book *78* : 175 - 178, 1979.

*36873 - BERSENEVA, G.P., FINENKO, Z.Z., SERGEEVA, L.M. : Adaptatsiya morskikh planktonnykh vodoroslei k svetu. [Marine plankton algae adaptation to light.] - Okeanologiya *18* : 298 - 306, 1978. [Ps; in R, ab : E.]

36874 - BERSHTEĬN,B.I., OKANENKO, A.S. : Kaliĭ, fotosintez i metabolizm rasteniĭ. [Potassium, photosynthesis and metabolism of plants.] - Fiziol. Biokhim. kul't. Rast. *11* : 515 - 526, 1979. [In R, ab : E.]

*36875 - BERTHIER, J., LARPENT, J.P., LARPENT-GOURGAUD, M. : Light action on vegetative propagation in bryophytes. - J. Hattori bot. Lab. *41* : 193 - 203, 1976. [Ps.]

36876 - BEST, E.P.H., MEULEMANS, J.T. : Photosynthesis in relation to growth and dormancy in *Ceratophyllum demersum*. - Aquat. Bot. *6* : 53 - 65, 1979.

36877 - BETHLENFALVAY, G.J., NORRIS, R.F., PHILLIPS, D.A. : Effect of bentazon, a Hill reaction inhibitor, on symbiotic nitrogen-fixing capability and apparent photosynthesis. - Plant Physiol. *63* : 213 - 215, 1979.

36878 - BETHLENFALVAY, G.J., PHILLIPS, D.A. : Variation in nitrogenase and hydrogenase activity of Alaska pea root nodules. - Plant Physiol. *63* : 816 - 820, 1979. [H_2 production.]

36879 - BEVERSDORF, W.D. : Influence of ploidy level on several plant characteristics in soybeans. - Can. J. Plant Sci. *59* : 945 - 948, 1979. [Chloroplast.]

36880 - BEWLEY, J.D. : Physiological aspects of desiccation tolerance. - Annu. Rev. Plant Physiol. *30* : 195 - 238, 1979. [Ps.]

36881 - BHAGCHANDANI, P.M., THAKUR, P.C., SINGH, N. : Diallel-cross analysis of leaf size and yield in spinach. - Indian J. agr. Sci. *49* : 364 - 367, 1979.

36882 - BHARDWAJ, N. : Growth of *Tephrosia apollinea* and *T. hamiltonii* under reduced illuminance. - Photosynthetica *13* : 302 - 306, 1979. [Growth analysis.]

*36883 - BHARDWAJ, S.N., KARIVARATHA RAJU, T.V. : Translocation of photosynthates from the fruit walls and leaves during seed development of field pea (*Pisum arvense* L.). - Indian J. Plant Physiol. *15* : 38 - 55, 1972.

36884 - BHATIA, I.S., AHUJA, K.L., SUKHIJA, P.S. : Fatty acid synthesis in *Hydrilla* chloroplasts. - Physiol. Plant. *47* : 81 - 86, 1979. [Ps.]

*36885 - BHATT, M.V., PRASAD, H.N.V. : Synthetic studies in carotenoids. - In : CAMA, H.R., SASTRY, P.S. (ed.) : Vitamin and Carrier Functions of Polyprenoids. World Review of Nutrition and Dietetics. Vol.31. Pp. 141 - 148. S.Karger AG, Basel 1978.

36886 - BHATTATHIRI, P.M.A., DEVASSY, V.P. : Biological characteristics of the Laccadive Sea (Lakshadweep). - Indian J. mar. Sci. *8* : 222 - 226, 1979. [Ps, Chl.]

36887 - BIANCHI, A., STEGWEE, D. : Porphobilinogenase in greening leaves of *Phaseolus vulgaris*. - Z. Pflanzenphysiol. *91* : 377 - 383, 1979.

36888 - BICKEL, H., SCHULTZ, G. : Shikimate pathway regulation in suspensions of intact spinach chloroplasts. - Phytochemistry *18* : 498 - 499, 1979. [Plastoquinone.]

*36889 - BIDWELL, R.G.S. : The carbon dioxide machine, or plants on the make. - Proc. N.S. Inst. Sci. *28* : 26 - 34, 1977. [Ps.]

*36890 - BIEDERMANN, M. : Einwirkung von Detergenzien auf die Thylakoide von *Rhodospirillum rubrum*. - Arch. Mikrobiol. *75* : 171 - 178, 1971.

36891 - BINDER, A., BACHOFEN, R. : Isolation and characterization of a coupling factor I ATPase of the thermophilic blue-green alga (cyanobacterium) *Mastigocladus laminosus*. - FEBS Lett. *104* : 66 - 70, 1979.

36892 - BINDER, A., BACHOFEN, R. : Oxygen evolution and uptake as a measure of the light-induced electron transport in spinach chloroplasts. - In : CARAFOLI, E., SEMENZA, G. (ed.) : Membrane Biochemistry. A Laboratory Manual on Transport and Bioenergetics. Pp. 144 - 153. Springer-Verlag, Berlin - Heidelberg - New York 1979.

36893 - BINGHAM, G.E., COYNE, P.I. : Spectral distribution of dimmed HID lamps in a plant growth facility. - Agron. J. *71* : 513 - 515, 1979. [PhAR measurement.]

36894 - BINGHAM, S., SCHIFF, J.A. : Events surrounding the early development of *Euglena* chloroplasts. 15. Origin of plastid thylakoid polypeptides in wild-type and mutant cells. - Biochim. biophys. Acta *547* : 512 - 530, 1979.

36895 - BINGHAM, S., SCHIFF, J.A. : Events surrounding the early development of *Euglena* chloroplasts. 16. Plastid thylakoid polypeptides during greening. - Biochim. biophys. Acta *547* : 531 - 543, 1979.

36896 - BIRECKA, H., CHASKES, M.J., GOLDSTEIN, J. : Peroxidase and senescence. - J. exp. Bot. *30* : 565 - 573, 1979. [Chl.]

36897 - BIRMINGHAM, B.C., COLMAN, B. : Measurement of carbon dioxide compensation points of freshwater algae. - Plant Physiol. *64* : 892 - 895, 1979.

36898 - BISHOP, D.G., KENRICK, J.R., BAYSTON, J.H., MACPHERSON, A.S., JOHNS, S.R., WILLING, R.I. : The influence of fatty acid unsaturation on fluidity and molecular packing of chloroplast membrane lipids. - In : LYONS, J.M., GRAHAM, D., RAISON, J.K. (ed.) : Low Temperature Stress in Crop Plants: The Role of the Membrane. Pp. 375 - 389. Academic Press, New York - San Francisco - London 1979.

36899 - BISWAL, U.C., SINGHAL, G.S., MOHANTY, P. : Dark stress induced senescence of barley leaves: changes in chlorophyll *a* fluorescence of isolated chloroplasts. - Indian J. exp. Biol. *17* : 262 - 264, 1979.

36900 - BITTMAN, S., STEPPLER, H.A. : A gasometric apparatus for monitoring evaporation rate from plant tissues during transpiration and drying. - Can. J. Plant Sci. *59* : 545 - 548, 1979.

36901 - BJÖRKMAN, O., BADGER, M. : Time course of thermal acclimation of the photosynthetic apparatus in *Nerium oleander*. - Carnegie Inst. Year Book *78* : 145 - - 148, 1979.

36902 - BJÖRN, G.S. : Action spectra for *in vivo* and *in vitro* conversions of phycochrome *b*, a reversibly photometric pigment in a blue-green alga, and its separation from other pigments. - Physiol. Plant. *46* : 281 - 286, 1979.

*B36903 - BJÖRN, L.O. : Photobiologie. Licht und Organismen. - Gustav Fischer Verlag, Stuttgart 1975. [Ps.]

*B36904 - BJÖRN, L.O. : Light and Life. - Hodder and Stoughton, London - Sydney - Auckland - Toronto 1976. [Ps.]

36905 - BJÖRN, L.O., FORSBERG, A.S. : Imaging by delayed light emission (phytolumino-
graphy) as a method for detecting damage to the photosynthetic system. - Physiol. Plant. *47* : 215 - 222, 1979.

36906 - BJØRNLAND, T., TANGEN, K. : Pigmentation and morphology of a marine *Gyrodi-
nium (Dinophyceae)* with a major carotenoid different from peridinin and fuco-
xanthin. - J. Phycol. *15* : 457 - 463, 1979.

36907 - BLACK, V.J., UNSWORTH, M.H. : A system of measuring effects of sulphur dioxide on gas exchange of plants. - J. exp. Bot. *30* : 81 - 88, 1979. [Ps.]

36908 - BLACK, V.J., UNSWORTH, M.H. : Effects of low concentrations of sulphur dioxide on net photosynthesis and dark respiration of *Vicia faba*. - J. exp. Bot.
30 : 473 - 483, 1979.

36909 - BLACK, V.J., UNSWORTH, M.H. : Resistance analysis of sulphur dioxide fluxes
to *Vicia faba*. - Nature *282* : 68 - 69, 1979. [Ps.]

36910 - BLANCHET, R., GELFI, N. : Influence de réductions de la surface foliaire sur
la croissance, le développement et la production d'un Soja de type indéterminé
(*Glycine max* L. MERRIL, cv. Amsoy 71). - Compt. rend. Acad. Sci. Paris, Sér.D
289 : 299 - 302, 1979. [Ps.]

36911 - BLANKENSHIP, R.E., PARSON, W.W. : Kinetics and thermodynamics of electron
transfer in bacterial reaction centers. - In : BARBER, J. (ed.) : Photosynthesis in Relation to Model Systems. Pp. 71 - 114. Elsevier, Amsterdam - New
York - Oxford 1979.

36912 - BLANKENSHIP, R.E., PARSON, W.W. : The involvement of iron and ubiquinone in
electron transfer reactions mediated by reaction centers from photosynthetic
bacteria. - Biochim. biophys. Acta *545* : 429 - 444, 1979.

36913 - BLECKMAN, C.A., HULL, H.M., MORTON, H.L. : Ultrastructural effects of formulated picloram on leaflets of velvet mesquite and catclaw acacia. - Weed Res.
19 : 225 - 230, 1979. [Chloroplast.]

36914 - BLEIN, J.P., DUCRUET, J.M., GAUVRIT, C. : Identification et activité biologique d'une impureté du diuron technique. - Weed Res. *19* : 117 - 121, 1979.
[Ps.]

36915 - BLOK, M.C., VAN DAM, K. : Association of bacteriorhodopsin with lipid-impreg-
nated filters. Evidence for fusion of bacteriorhodopsin-containing vesicles
with the lipid phase of the filter. - Biochim. biophys. Acta *550* : 527 - 542,
1979.

36916 - BLOOM, A.J. : Salt requirement for Crassulacean acid metabolism in the annual succulent, *Mesembryanthemum crystallinum*. - Plant Physiol. *63* : 749 - 753,
1979.

36917 - BLOOM, A.J. : Diurnal ion fluctuations in the mesophyll tissue of the Crassulacean acid metabolism plant *Mesembryanthemum crystallinum*. - Plant Physiol.
64 : 919 - 923, 1979.

36918 - BLOOM, A.J., TROUGHTON, J.H. : High productivity and photosynthetic flexibi-
lity in a CAM plant. - Oecologia *38* : 35 - 43, 1979.

36919 - BLUM, H., SALERNO, J.C., RICH, P.R., OHNISHI, T. : Exchange integral for a
variety of tetranuclear ferredoxins. - Biochim. biophys. Acta *548* : 139 -
- 146, 1979.

36920 - BLUNDEN, G., JONES, E.M., PASSAM, H.C., METCALF, E. : Increases in chlorophyll retention times of limes after post-harvest immersion in N_6-benzylade-
nine and gibberellic acid. - Trop. Agr. *56* : 311 - 319, 1979.

36921 - BLYTHE, T.O., GROOMS, S.M., FRANS, R.E. : Determination and characterization
of the effects of fluometuron and MSMA on *Chlorella*. - Weed Sci. *27* : 294 -
- 299, 1979. [Ps.]

36922 - BOAG, T.S., BROWNELL, P.F. : C_4 photosynthesis in sodium-deficient plants. -
Aust. J. Plant Physiol. *6* : 431 - 434, 1979.

36923 - BOCHAROV, E.A., KLIMOV, S.V., DZHANUMOV, D.A. : Izmenenie soderzhaniya lipidov v khloroplastakh pervichnogo lista ozimoĭ pshenitsy v techenie sutok.

[Diurnal changes in lipid content in chloroplasts of primary leaf of winter wheat.] - Fiziol. Rast. *26* : 266 - 269, 1979. [In R, ab : E.]

36924 - BÖDDI, B., LÃNG, F., SOÓS, J. : A study of 650 nm protochlorophyll form in pumpkin seed coat. - Plant Sci. Lett. *16* : 75 - 79, 1979.

36925 - BODSON, M. : Photosynthèse et mise à fleurs chez *Sinapis alba*. - Physiol. vég. *17* : 424, 1979.

36926 - BODSON, M., BERNIER, G., KINET, J.M., JACQMARD, A., HAVELANGE, A. : Flowering of *Sinapis* as influenced by different treatments acting on photosynthetic activity. - In : MARCELLE, R., CLIJSTERS, H., VAN POUCKE, M. (ed.) : Photosynthesis and Plant Development. Pp. 73 - 82. Dr.W.Junk bv. Publ., The Hague - Boston - London 1979.

36927 - BOFFEY, S.A., ELLIS, J.R., SELLDÉN, G., LEECH, R.M. : Chloroplast division and DNA synthesis in light-grown wheat leaves. - Plant Physiol. *64* : 502 - - 505, 1979.

36928 - BÖGER, P. : Energieumwandlung durch photobiologische Wasserspaltung. - Umschau Wiss. Tech. *79* : 639 - 642, 1979.

36929 - BÖGER, P. : Transhydrogenase. - In : GIBBS, M., LATZKO, E. (ed.) : Photosynthesis II. (Encycl. Plant Physiol. N.S. Vol.6.) Pp. 399 - 409. Springer-Verlag, Berlin - Heidelberg - New York 1979.

36930 - BÖGER, P. : Utilization of solar energy by biological production of hydrogen - an approach from fundamental research. - Atomkernenergie Kerntechnik *33* (1) : 13 - 18, 1979.

36931 - BÖGER, P., KUNERT, K.-J. : Differential effects of herbicides upon trypsin- -treated chloroplasts. - Z. Naturforsch. *34C* : 1015 - 1025, 1979.

36932 - BOGOMOLNI, R.A., KLEIN, M.P. : Faraday rotation and photoconductivity of photosynthetic structures in microwave frequencies. - In : CHANCE, B., DEVAULT, D.C., FRAUENFELDER, H., MARCUS, R.A., SCHRIEFFER, J.R., SUTIN, N. (ed.) : Tunneling in Biological Systems. Pp. 405 - 416. Academic Press, New York - - San Francisco - London 1979.

36933 - BOGORAD, L. : The chloroplast, its genome and possibilities for genetically manipulating plants. - In : SETLOW, J.K., HOLLAENDER, A. (ed.) : Genetic Engineering. Vol.1. Pp. 181 - 203. Plenum Press, New York 1979.

36934 - BOHLING, H., HANSEN, H., SCHUNCK, G. : Ein verbessertes Verfahren zur Messung des Gasstoffwechsels von Pflanzenteilen während der Lagerung bei erhöhtem CO_2-Gehalt der Atmosphäre. - Angew. Bot. *53* : 53 - 58, 1979.

36935 - BÖHME, H. : Photoreactions of cytochrome b_{563} and f_{554} in intact spinach chloroplasts: Regulation of cyclic electron flow. - Europe. J. Biochem. *93* : 287 - 293, 1979.

36936 - BOHRA, D.R., SONI, S.R., SHARMA, B.D. : Ferns of Rajasthan - behaviour of chlorophyll and carotenoids in drought resistance. - Experientia *35* : 332 - - 333, 1979.

36937 - BOÏCHENKO, V.A., EFIMTSEV, E.I. : Ingibirovanie aktivnosti fotosistemy II u khlorelly pri vysokikh kontsentratsiyakh kisloroda. [Inhibition of photosystem II activity in *Chlorella* under high oxygen concentrations.] - Fiziol. Rast. *26* : 815 - 823, 1979. [In R, ab : E.]

36938 - BOLHÀR-NORDENKAMPF, H.R. : The possible mode of herbicidal action of atrazine basing on the gas exchange and the mode of plant damage after treatment. - Z. Naturforsch. *34C* : 923 - 925, 1979.

36939 - BOLLIG, I.C., WILKINS, M.B. : Inhibition of the circadian rhythm of CO_2 metabolism in *Bryophyllum* leaves by cycloheximide and dinitrophenol. - Planta *145* : 105 - 112, 1979.

36940 - BOLOGA, A.S. : Experiments on the photosynthesis rate in the alga *Cladophora vagabunda* L. under modified salinity and ionic ratios. - Rev. roum. Biol., Sér. Biol. vég. *24* : 127 - 131, 1979.

*36941 - BOLSENGA, S.J. : Photosynthetically active radiation transmission through
         ice. - NOAA tech. Memorandum *ERL GLERL-18* : 1 - 48, 1978.

 36942 - BOLSENGA, S.J. : Spectral distribution of radiation in the northern Great
         Lakes during winter. - J. Great Lakes Res. *4* : 226 - 229, 1978. [Method.]

 36943 - BOLT, J., SAUER, K. : Linear dichroism in light harvesting bacteriochloro-
         phyll proteins from *Rhodopseudomonas sphaeroides* in stretched polyvinyl alco-
         hol films. - Biochim. biophys. Acta *546* : 54 - 63, 1979.

 36944 - BOLTON, J.R., HALL, D.O. : Photochemical conversion and storage of solar ener-
         gy. - Annu. Rev. Energy *4* : 353 - 401, 1979.

*36945 - BONDYREV, M.I., OVSYANNIKOV, A.S. : Vliyanie plenkoobrazuyushchego komponenta
         insektitsidnykh sostavov na produktivnost' fotosinteza i rost plodov yabloni.
         [Effect of a film-forming component of insecticides on photosynthetic pro-
         ductivity and growth of apple fruits.] - Sb. nauch. Rabot vsesoyuz. nauch.-
         -issled. Inst. Sadovodstva (Michurinsk) *25* : 6 - 12, 1977. [In R.]

*36946 - BONZI, L.M., FABBRI, F. : Paracrystalline inclusions in the chloroplasts of
         sieve parenchyma cells of *Arum italicum* MILL. Preliminary report. - Caryolo-
         gia *31* : 129 - 136, 1978.

 36947 - BOOKJANS, G., BÖGER, P. : Algal ferredoxin-NADP$^+$ reductase with different
         molecular-weight forms. - Z. Naturforsch. *34C* : 637 - 640, 1979.

 36948 - BOOKJANS, G., BÖGER, P. : Complex-forming properties of butanedione-modified
         ferredoxin-NADP$^+$ reductase with NADP$^+$ and ferredoxin. - Arch. Biochem. Bio-
         phys. *194* : 387 - 393, 1979.

*36949 - BOOTH, C.R. : The design and evaluation of a measurement system for photo-
         synthetically active quantum scalar irradiance. - Limnol. Oceanogr. *21* :
         326 - 336, 1976.

 36950 - BORCHERS, C.A., GLAMM, A.B., SWANSON, C.A. : Source-sink patterns in relation
         to source strength. - Plant Physiol. *63* (Suppl.) : 34, 1979.

 36951 - BORCHERT, R. : Complete loss of stomatal functioning in aging leaves of tro-
         pical broadleafed trees. - Plant Physiol. *63* (Suppl.) : 60, 1979.

 36952 - BORISEVICH, G.P., NOKS, P.P., KONONENKO, A.A., RUBIN, A.B., VOZARI, É.: Po-
         lyarizatsiya fotosinteticheskikh membran i reaktsionnykh tsentrov iz *Rhodo-*
         pseudomonas sphaeroides, 1760-1 vo vneshnem èlektricheskom pole. [Electric
         field induced polarization of photosynthetic membranes and reaction centres
         of *Rhodopseudomonas sphaeroides,* 1760-1.] - Biofizika *24* : 843 - 848, 1979.
         [In R, ab : E.]

 36953 - BORISEVITCH, G.P., KONONENKO, A.A., RUBIN, A.B., KOCHUBEJ, S.M., SHADCHINA,
         T.M. : Effects of external electric fields on pea subchloroplast particles. -
         Plant Sci. Lett. *14* : 275 - 280, 1979.

 36954 - BORISEVITCH, G.P., LUKASHEV, E.P., KONONENKO, A.A., RUBIN, A.B. : Bacterio-
         rhodopsin (BR$_{570}$) bathochromic band shift in an external electric field. -
         Biochim. biophys. Acta *546* : 171 - 174, 1979.

 36955 - BORISOV, A.Yu. : Photosynthesizing organisms: converters of solar energy. -
         In : BARBER, J. (ed.) : Photosynthesis in Relation to Model Systems. Pp. 1 -
         - 26. Elsevier, Amsterdam - New York - Oxford 1979.

 36956 - BÖRNER, H. : Ursachen der selektiven Wirkung der Photosynthesehemmer Phenme-
         dipham und Bentazon. - Z. Naturforsch. *34C* : 926 - 930, 1979.

 36957 - BÖRNER, T., MANTEUFFEL, R., WELLBURN, A.R. : Enzymes of plastid ribosome-de-
         ficient mutants. Chloroplast ATPase (CF$_1$). - Protoplasma *98* : 153 - 161,
         1979.

 36958 - BORNMAN, C.H., HUBER, W. : *Nicotiana tabacum* callus studies. IX. Development
         in stressed explants. - Biochem. Physiol. Pflanzen *174* : 345 - 356, 1979.
         [Chl.]

* 36959 - BOROJEVIC, S. : Determination of optimal LAI and effective LAD for different
         wheat genotypes. - In : RAMANUJAM, S. (ed.) : Proceedings of the 5th Inter-
         national Wheat Genetics Symposium. Vol.2. Pp. 899 - 906. Indian Soc. Genetics
         Plant Breeding, New Delhi 1978.

36960 - BOTHA, F.C., BOTHA, P.J. : The effect of water stress on the metabolism of
         two maize lines. II. Effects on the rate of protein synthesis and chlorophyll
         content. - Z. Pflanzenphysiol. *94* : 179 - 183, 1979.

36961 - BOTTOMLEY, W., WHITFELD, P.R. : Cell-free transcription and translation of
         total spinach chloroplast DNA. - Europe. J. Biochem. *93* : 31 - 39, 1979. [Ps.]

36962 - BOUCHON, J. : Structure des peuplements forestiers. - Ann. Sci. forest. *36* :
         175 - 209, 1979. [Canopy structure.]

36963 - BOULTER, D., PEACOCK, D., GUISE, A., GLEAVES, J.T., ESTABROOK, G. : Relation-
         ships between the partial amino acid sequences of plastocyanin from members
         of ten families of flowering plants. - Phytochemistry *18* : 603 - 608, 1979.

36964 - BOURDU, R. : Structure and CO_2 assimilation of photosynthetic apparatus. -
         In : VAKLINOVA, S.G., VANKOVA-RADEVA, R., VASILEVA, V.S. (ed.) : Fotosinte-
         ticheskaya Assimilatsiya CO_2 i Fotodykhanie. Pp. 31 - 43. Izdatel'stvo bolgar-
         skoĭ Akademii Nauk, Sofiya 1979.

36965 - BOUSSIBA, S., RICHMOND, A.E. : Isolation and characterization of phycocyanins
         from the blue-green alga *Spirulina platensis*. - Arch. Microbiol. *120* : 155 -
         - 159, 1979.

36966 - BOUTON, J.H., BOLTON, J., BROWN, R.H. : Chromosome numbers of *Panicum* species
         differing in photosynthetic pathways. - Plant Physiol. *63* (Suppl.) : 38, 1979.

36967 - BOWES, G., HOLADAY, A.S., HALLER, W.T. : Seasonal variation in the biomass
         tuber density and photosynthetic metabolism of *Hydrilla* in three Florida
         Lakes. - J. aquat. Plant Manage. *17* : 61 - 65, 1979.

36968 - BOWES, J.M., CROFTS, A.R., ITOH, S. : A high potential acceptor for Photosys-
         tem II. - Biochim. biophys. Acta *547* : 320 - 335, 1979.

36969 - BOWES, J.M., CROFTS, A.R., ITOH, S. : Effects of pH on reactions on the donor
         side of Photosystem II. - Biochim. biophys. Acta *547* : 336 - 346, 1979.

36970 - BOXER, S.G., BUCKS, R.R. : A structural model for the photosynthetic reaction
         center. - J. amer. chem. Soc. *101* : 1883 - 1885, 1979.

36971 - BOXER, S.G., ROELOFS, M.G. : Chromophore organization in photosynthetic re-
         action centers: High-resolution magnetophotoselection. - Proc. nat. Acad.
         Sci. USA *76* : 5636 - 5640, 1979.

36972 - BOXER, S.G., WRIGHT, K.A. : Preparation and properties of a chlorophyllide-
         -apomyoglobin complex. - J. amer. chem. Soc. *101* : 6791 - 6794, 1979.

36973 - B.PAPP, L., PAPP, M., JAKUCS, P. : Mezőgazdasági és tölgyerdei növények kaló-
         riaértéke és energiatartalma. [Caloric values and energy content of crop and
         oak forest plants.] - Növénytermelés *28* : 155 - 161, 1979. [In Hung., ab : E.]

36974 - BRADFORD, K.J., HSIAO, T.C. : Alterations in leaf angle and stomatal conduc-
         tance during waterlogging are independent of leaf water potential. - Plant
         Physiol. *63* (Suppl.) : 88, 1979.

36975 - BRAND, J.J. : Role of Ca^{++} for photosynthesis in *Anacystis nidulans* membrane
         preparations. - Plant Physiol. *63* (Suppl.) : 40, 1979.

36976 - BRAND, J.J. : Spectral changes in membrane fragments and artificial liposomes
         of *Anacystis* induced by chilling. - Arch. Biochem. Biophys. *193*: 385 - 391,
         1979. [Car.]

36977 - BRAND, J.J. : The effect of Ca^{2+} on oxygen evolution in membrane preparations
         from *Anacystis nidulans*. - FEBS Lett. *103* : 114 - 117, 1979.

*36978 - BRÄNDÉN, C.I., LINDQVIST, Y. : Crystallographic and x-ray diffraction studies
         on glycolate oxidase from spinach. - In : SIEGELMAN, H.W., HIND, G. (ed.) :
         Photosynthetic Carbon Assimilation. P. 417. Plenum Press, New York - London
         1978.

36979 - BRANDT, P. : Evidence for regulative transactions between cytoplasmic and
         plastidial translation by *Euglena gracilis*. I. Quantitative changes of the
         spectrum of plastidial proteins by chloramphenicol or cycloheximide treat-
         ment. - Z. Pflanzenphysiol. *94* : 299 - 306, 1979. [Chl.]

36980 - BRANGEON, J., MUSTARDY, L. : The ontogenetic assembly of intra-chloroplastic lamellae viewed in 3-dimension. - Biol. cell. *36* : 71 - 80, 1979.

36981 - BRAUMANN, T., GRIMME, L.H. : Single-step separation and identification of photosynthetic pigments by high-performance liquid chromatography. - J. Chromatogr. *170* : 264 - 268, 1979.

36982 - BRAUNE, W. : *C*-phycocyanin - the main photoreceptor in the light dependent germination process of *Anabaena* akinetes. - Arch. Mikrobiol. *122* : 289 - 295, 1979. [Ps, Bil.]

36983 - BRAVDO, B., CANVIN, D. : Effect of carbon dioxide on photorespiration. - Plant Physiol. *63* : 399 - 401, 1979.

36984 - BRAY, E., BRENNER, M.L., PARSONS, L.R. : Abscisic acid levels and stomatal resistance of red-osier dogwood during cold acclimation. - Plant Physiol. *63* (Suppl.) : 104, 1979.

36985 - BRAYMAN, A., SCHAEDLE, M. : Stem photosynthesis in current-growth stems of *Populus tremuloides* MICHX. - Plant Physiol. *63* (Suppl.) : 74, 1979.

36986 - BREIDERT, D., SCHÖN, W.J. : Die Bildung und Speicherung von Kohlenhydraten und Proteinen in reifenden Gerstenkaryopsen. I. Methodenübersicht, Kohlenhydrate und Stickstoff-Fraktionen. - Angew. Bot. *53* : 65 - 81, 1979. [Photosynthate redistribution.]

36987 - BREITENBERGER, C.A., GRAVES, M.C., SPREMULLI, L.L. : Evidence for the nuclear location of the gene for chloroplast elongation factor G. - Arch. Biochem. Biophys. *194* : 265 - 270, 1979.

36988 - BREITENBERGER, C.A., MOORE, M.N., RUSSELL, D.W., SPREMULLI, L.L. : Purification of eukaryotic cytoplasmic elongation factor 2 and organellar elongation factor G by an affinity binding procedure. - Anal. Biochem. *99* : 434 - 440, 1979. [Chloroplast.]

36989 - BRETHERTON, G., HALLAM, N.D. : The movement of 2,4,5-trichlorophenoxyacetic acid into the leaves of *Rubus procerus* P.J.MUELL and its effect on chloroplast ultrastructure. - Weed Res. *19* : 307 - 313, 1979.

36990 - BRETON, J., GEACINTOV, N.E. : Chlorophyll orientation and exciton migration in the photosynthetic membrane. - In : Chlorophyll Organization and Energy Transfer in Photosynthesis. Pp. 217 - 236. Excerpta Medica, Amsterdam - Oxford - New York 1979.

36991 - BRETON, J., GEACINTOV, N.E., SWENBERG, C.E. : Quenching of fluorescence by triplet excited states in chloroplasts. - Biochim. biophys. Acta *548* : 616 - - 635, 1979.

36992 - BRETON-PROVENCHER, M., GAGNÉ, J.A., CARDINAL, A. : Estimation de la production des algues benthiques médiolittorales dans l'estuaire maritime du Saint--Laurent (Québec). - Natur. can. *106* : 199 - 209, 1979.

36993 - BREWER, P.E., ARNTZEN, C.J., SLIFE, F.W. : Effects of atrazine, cyanazine, and procyazine on the photochemical reactions of isolated chloroplasts. - Weed Sci. *27* : 300 - 308, 1979.

36994 - BREWSTER, J.L. : The response of growth rate to temperature in seedlings of several *Allium* crop species. - Ann. appl. Biol. *93* : 351 - 357, 1979.

36995 - BRIANTAIS, J.-M., VERNOTTE, C., PICAUD, M., KRAUSE, G.H. : A quantitative study of the slow decline of chlorophyll *a* fluorescence in isolated chloroplasts. - Biochim. biophys. Acta *548* : 128 - 138, 1979.

36996 - BRICKER, T.M., NEWMAN, D.W. : Changes in the chloroplast thylakoid polypeptide composition of senescing and re-greening soybean cotyledons. - Plant Physiol. *63* (Suppl.) : 73, 1979.

36997 - BRINCKMANN, E., WILLERT, D.J. von : Response of acid metabolism in CAM plants to environmental variables. - Naturwissenschaften *66* : 526 - 527, 1979.

36998 - BRINKMANN, G., SENGER, H. : Lichtabhängige Bildung der Thylakoidmembran während der Entwicklung des Photosyntheseapparates in der Pigmentmutante C-2A' von *Scenedesmus obliquus*. - Ber. deut. bot. Ges. *92* : 629 - 636, 1979.

36999 - BRITH-LINDNER, M., AVI-DOR, Y. : Interaction of ionophores with bacteriorho-
         dopsin. A flash photometric study. - FEBS Lett. *101* : 113 - 115, 1979.

37000 - BRITTON, G. : Carotenoid biosynthesis - a target for herbicide activity. - Z.
         Naturforsch. *34C* : 979 - 985, 1979.

37001 - BRITZ, S.J. : Chloroplast and nuclear migration. - In : HAUPT, W., FEINLEIB,
         M.E. (ed.) : Physiology of Movements. (Encycl. Plant Physiol. N.S. Vol.7.)
         Pp. 170 - 205. Springer-Verlag, Berlin - Heidelberg - New York 1979.

37002 - BRIX, H. : Effects of plant water stress on photosynthesis and survival of
         four conifers. - Can. J. Forest Res. *9* : 160 - 165, 1979.

37003 - BROCKMANN, H.Jr., BELTER, C. : Regioselektive Ringspaltung von Bacteriochlo-
         rophyll-Derivaten durch Photooxidation. - Z. Naturforsch. *34B* : 127 - 128,
         1979.

37004 - BROCKMANN, H.Jr., JÜRGENS, U., THOMAS, M. : Partialsynthese eines Bacterio-
         phäophorbid-*c*-methylesters. - Tetrahedron Lett. *1979* : 2133 - 2136, 1979.

37005 - BRODA, E., PESCHEK, G.A. : Did respiration or photosynthesis come first? -
         J. theor. Biol. *81* : 201 - 212, 1979.

37006 - BRODA, H., SCHWEIGER, G., KOOP, H.-U., SCHMID, R., SCHWEIGER, H.-G. : Chloro-
         plast migration : A method for continuously monitoring a circadian rhythm in
         a single cell of *Acetabularia*. - In : BONOTTO, S., KEFELI, V., PUISEUX-DAO, S.
         (ed.) : Developmental Biology of *Acetabularia*. Pp. 163 - 167. Elsevier/North-
         -Holland Biomedical Press, Amsterdam - New York - Oxford 1979.

37007 - BRODY, S.S., SINGHAL, G.S. : Spectral properties of chloroplast membranes
         as a function of physiological temperatures. - Biochem. biophys. Res. Commun.
         89 : 542 - 546, 1979.

37008 - BROOKS, A.S., LIPTAK, N.E. : The effect of intermittent chlorination of fresh-
         water phytoplankton. - Water Res. *13* : 49 - 52, 1979. [Ps, Chl.]

37009 - BROUERS, M. : Optical properties of *in vitro* aggregates of protochlorophylli-
         de in non-polar solvents. III. Infra-red spectra; visible absorption and flu-
         orescence of fractions obtained by differential centrifugation. - Photosynthe-
         tica *14* : 9 - 14, 1979.

37010 - BROUGHTON, W.J. : Effect of light intensity on net assimilation rates of ni-
         trate-supplied or nitrogen-fixing legumes. - In : MARCELLE, R., CLIJSTERS, H.,
         VAN POUCKE, M. (ed.) : Photosynthesis and Plant Development. Pp. 285 - 299.
         Dr. W. Junk bv. Publ., The Hague - Boston - London 1979.

37011 - BROUGHTON, W.J. : Relationship between carbon and nitrogen metabolism in
         Glycine max L.MERR. - In : Mineral Nutrition of Plants. Vol.I. Pp. 287 - 295.
         Publ. House Central Cooperative Union, Sofia 1979. [Growth analysis.]

37012 - BROVCHENKO, M.I., CHMORA, S.N., SLOBODSKAYA, G.A. : Vliyanie kisloroda na
         uglekislotnyĭ gazoobmen i metabolizm mezofilla list'ev sakharnoĭ svekly na
         svetu. [Effect of oxygen on carbon dioxide exchange and the metabolism of
         sugarbeet-leaf mesophyll in the light.] - Fiziol. Rast. *26* : 244 - 249, 1979.
         [In R, ab : E.]

37013 - BROWN, A.D. : Physiological problems of water stress. - In : SHILO, M. (ed.) :
         Strategies of Microbial Life in Extreme Environments. Pp. 65 - 81. Verlag
         Chemie, Weinheim - New York 1979. [Ps.]

37014 - BROWN, A.S., OFFNER, G.D., EHRHARDT, M.M., TROXLER, R.F. : Phycobilin-apopro-
         tein linkages in the α and β subunits of phycocyanin from the unicellular
         rhodophyte, *Cyanidium caldarium*. Amino acid sequences of ^{35}S-labeled chromo-
         peptides. - J. biol. Chem. *254* : 7803 - 7811, 1979.

37015 - BROWN, H.M, CHOLLET, R. : Quantitative dissociation of crystalline tobacco
         RuBP carboxylase into subunits under non-denaturating conditions. - Plant
         Physiol. *63* (Suppl.) : 65, 1979.

37016 - BROWN, J., BROWN, H.M., HESS, B., MUELLER, P., OESTERHELT, D., PARSON, W.,
         RÜPPEL, H., SPERLING, W., STIEVE, H. : Rhodopsin mediated processes. Group
         report. - In : GERISCHER, H., KATZ, J.J. (ed.) : Light-induced Charge Sepa-
         ration in Biology and Chemistry. Pp. 525 - 551. Verlag Chemie, Weinheim -
         - New York 1979. [Bacteriorhodopsin.]

37017 - BROWN, J.C., CATHEY, H.M., BENNETT, J.H., THIMIJAN, R.W. : Effect of light quality and temperature on Fe^{3+} reduction, and chlorophyll concentration in plants. - Agron. J. *71* : 1015 - 1021, 1979.

37018 - BROWN, J.C., FOY, C.D., BENNETT, J.H., CHRISTIANSEN, M.N. : Two light sources differentially affected ferric iron reduction and growth of cotton. - Plant Physiol. *63* : 692 - 695, 1979. [Chl.]

37019 - BROWN, J.S. : Spectral studies of triton-solubilized chlorophyll-proteins from *Euglena* and antenna chlorophyll from spinach. - Carnegie Inst. Year Book *78* : 189 - 194, 1979.

37020 - BROWN, R.H. : Gas analysis of the reduced O_2 response of photosynthesis in *Panicum milioides* and *Panicum schenckii*. - Plant Physiol. *63* (Suppl.) : 153, 1979.

37021 - BROWN, R.H., SIMMONS, R.E. : Photosynthesis of grass species differing in CO_2 fixation pathways. I.Water-use efficiency. - Crop Sci. *19* : 375 - 379, 1979.

37022 - BROWNELL, P.F. : Sodium as an essential micronutrient element for plants and its possible role in metabolism. - Adv. bot. Res. *7* : 117 - 224, 1979. [Ps, Chl, Bil.]

37023 - BROWSE, J.A. : An open-circuit infrared gas analysis system for measuring aquatic plant photosynthesis at physiological pH. - Aust. J. Plant Physiol. *6* : 493 - 498, 1979.

37024 - BROWSE, J.A., BROWN, J.M.A., DROMGOOLE, F.I. : Photosynthesis in the aquatic macrophyte *Egeria densa*. II.Effects of inorganic carbon conditions on ^{14}C fixation. - Aust. J. Plant Physiol. *6* : 1 - 9, 1979.

37025 - BROWSE, J.A., DROMGOOLE, F.I., BROWN, J.M.A. : Photosynthesis in the aquatic macrophyte *Egeria densa*. III Gas exchange studies. - Aust. J. Plant Physiol. *6* : 499 - 512, 1979.

37026 - BRUGNONI, G.P., MOSER, P., TREBST, A. : Site of action and quantitative structure-activity relationship of a series of herbicidal N-aryl-substituted 3,4-dimethyl-2-hydroxy-5-oxo-2,5-dihydro-pyrrolones. - Z. Naturforsch. *34C* : 1028 - 1031, 1979. [Ps.]

37027 - BRUINSMA, J. : Root hormones and overground development. - In : SCOTT, T.K. (ed.) : Plant Regulation and World Agriculture. Pp. 35 - 47. Plenum Press, New York - London 1979. [Dry-matter production.]

37028 - BRULFERT, J., ARRABAÇA, M.C., GUERRIER, D., QUEIROZ, O. : Changes in the isozymic pattern of phosphoenolpyruvate. An early step in photoperiodic control of Crassulacean acid metabolism level. - Planta *146* : 129 - 133, 1979.

37029 - BRUNSKILL, G.J., SCHINDLER, D.W., ELLIOTT, S.E.M., CAMPBELL, P. : The attenuation of light in Lake Winnipeg waters. - Fish. mar. Serv. Manuscr. Rep. *1522* : i - v, 1 - 79, 1979. [Chl.]

37030 - BRYANT, D.A., GUGLIELMI, G., TANDEAU DE MARSAC, N., CASTETS, A.-M., COHEN-BAZIRE, G. : The structure of cyanobacterial phycobilisomes : a model. - Arch. Microbiol. *123* : 113 - 127, 1979.

37031 - BRZOSKA, W. : Zeitbedingte Variabilität organspezifischer Energiegehalte in *Alnus viridis* (CHAIX) DC. - Verhandl. Ges. Ökol. (Münster 1978) *7* : 421 - - 427, 1979.

37032 - BUBLICHENKO, N.V., UMRIKHINA, A.V., KRASNOVSKIĬ, A.A. : Obrazovanie svobodnykh radikalov pri fotokhimicheskikh reaktsiyakh protokhlorofilla. [Formatio of free radicals in photochemical reactions of protochlorophyll.] - Biofizika *24* : 588 - 593, 1979. [In R, ab : E.]

37033 - BUCHANAN, B.B. : Ferredoxin-linked carbon dioxide fixation in photosynthetic bacteria. - In : GIBBS, M., LATZKO, E. (ed.) : Photosynthesis II. (Encycl. Plant Physiol. N.S. Vol.6.) Pp. 416 - 424. Springer-Verlag, Berlin - - Heidelberg - New York 1979.

37034 - **BUCHANAN, B.B., CRAWFORD, N.A., YEE, B.C., NISHIZAWA, A.N.** : New f- and m-type cytoplasmic thioredoxins from leaves. - Plant Physiol. *63* (Suppl.) : 3, 1979.

37035 - **BUCHANAN, B.B., WOLOSIUK, R.A., SCHÜRMANN, P.** : Thioredoxin and enzyme regulation. - Trends biochem. Sci. *4* : 93 - 96, 1979. [Ps.]

*37036 - **BUCHER, J.B., KELLER, T.** : Einwirkungen niedriger SO_2-Konzentrationen im mehrwöchigen Begasungsversuch auf Waldbäume. - VDI Ber. *314* : 237 - 242, 1978. [Ps.]

*B37037 - **BÜCHER, T., NEUPERT, W., SEBALD, W., WERNER, S.** (ed.) : Genetics and Biogenesis of Chloroplasts and Mitochondria. - North-Holland Publ.Co., Amsterdam - - New York - Oxford 1976.

37038 - **BUCHHOLZ, B., REUPKE, B., BICKEL, H., SCHULTZ, G.** : Reconstitution of amino acid synthesis by combining spinach chloroplasts with other leaf organelles. - Phytochemistry *18* : 1109 - 1111, 1979.

37039 - **BUCKLE, K.A., RAHMAN, F.M.M.** : Separation of chlorophyll and carotenoid pigments of *Capsicum* cultivars. - J. Chromatogr. *171* : 385 - 391, 1979.

*37040 - **BUCKLEY, C.E., HOUGHTON, J.A.** : A study of the effects of near UV radiation on the pigmentation of the blue-green alga *Gloeocapsa alpicola*. - Arch. Microbiol. *107* : 93 - 97, 1976.

*37041 - **BUDYKINA, N.P., BALAGUROVA, N.I.** : Kharakter izmeneniĭ fotosinteza i uglevodnogo obmena u kartofelya pod vliyaniem retardantov i kholodovoĭ zakalki. [Changes in photosynthesis and carbohydrate exchange of potato plants under the influence of retardants and chilling.] - Ėkologo-Fiziologicheskie Mekhanizmy Ustoĭchivosti Rasteniĭ k Deĭstviyu Ekstremal'nykh Temperatur. Pp. 112 - - 119. Petrozavodsk 1978. [In R.]

37042 - **BUGG, W., RIECK, C.E., COHEN, W.S., WHITMARSH, J.** : Inhibition of chloroplast electron transport by diphenyl ether herbicides. - Plant Physiol. *63* (Suppl.) : 41, 1979.

37043 - **BUGGELN, R.G.** : Photosynthesis and translocation in relation to growth in *Laminariales*. - In : MARCELLE, R., CLIJSTERS, H., VAN POUCKE, M. (ed.) : Photosynthesis and Plant Development. Pp. 251 - 261. Dr.W.Junk bv. Publ., The Hague - Boston - London 1979.

37044 - **BUGGELN, R.G., LUCKEN, S.** : Kinetic characteristics of photoassimilate translocation in *Alaria esculenta* (Laminariales, Phaeophyceae). - Planta *147* : 241 - 245, 1979.

37045 - **BUKHOV, N.G., KARAPETYAN, N.V.** : Fotoindutsirovannye izmeneniya pogloshcheniya pri 800 nm, svyazannye s funktsionirovaniem fotosistemy I. [Light-induced absorbance changes at 800 nm related with the functioning of photosystem I.] - Biokhimiya *44* : 705 - 710, 1979. [In R, ab : E.]

37046 - **BUKHOV, N.G., KARAPETYAN, N.V.** : Nezavisimost' ot temperatury skorosti bystroĭ fazy temnovogo vosstanovleniya $P700^+$. [Temperature independence of the rate of rapid phase of $P700^+$ dark reduction.] - Biofizika *24* : 806 - 810, 1979. [In R, ab : E.]

37047 - **BULYCHEV, A.A., KURELLA, G.A., PUCHKOVA, T.V., URAZMANOV, R.I.** : Fotoindutsirovannoe pogloshchenie fenazinmetosul'fata tilakoidami izolirovannykh khloroplastov. [Light-induced uptake of phenazine methosulfate by thylakoids of isolated chloroplasts.] - Fiziol. Rast. *26* : 20 - 27, 1979. [In R, ab : E.]

37048 - **BULYCHEV, A.A., KURELLA, G.A., URAZMANOV, R.I.** : Ėnergozavisimoe pogloshchenie fenazinmetosul'fata v izolirovannykh khloroplastakh. [Energy-dependent uptake on phenazine methosulphate in isolated chloroplasts.] - Biokhimiya *44* : 1460 - 1467, 1979. [In R, ab : E.]

37049 - **BUNCE, J.A., CHABOT, B.F., MILLER, L.N.** : Role of annual leaf carbon balance in the distribution of plant species along an elevational gradient. - Bot. Gaz. *140* : 288 - 294, 1979.

37050 - BURBAEV, D.Sh.,LEBANIDZE, A.V., MUKHIN, E.N., NEZNAĬKO, N.F. : Issledovanie
        metodom ÉPR dvukh form rastvorimogo ferredoksina iz list'ev gorokha. [ESR
        study of two forms of soluble ferredoxin from pea leaves.] - Biofizika 24 :
        15 - 20, 1979. [In R, ab : E.]

37051 - BURCZYK, J. : Carotenoids localised in the cell wall of Chlorella and Scene-
        desmus (Chlorophyceae). - Bull. Acad. pol. Sci., Sér. Sci. biol. 27 : 13 - 19,
        1979.

37052 - BURDETT, A.N. : A nondestructive method for measuring the volume of intact
        plant parts. - Can. J. Forest Res. 9 : 120 - 122, 1979.

37053 - BURDGE, E.L., WILSON, K.G. : The effect of amino acid analogues on pea chlo-
        roplast protein synthesis. - Plant Physiol. 63 (Suppl.) : 47, 1979.

37054 - BURIAN, K., SIEGHARDT, H. : The primary producers of the Phragmites belt,
        their energy utilization and water balance. - In : LÖFFLER, H. (ed.) : Neu-
        siedlersee: The Limnology of a Shallow Lake in Central Europe. (Monogr.biol.
        Vol. 37.) Pp. 251 - 272. Dr. W.Junk bv. Publ., The Hague - Boston - London
        1979.

37055 - BURKE, J.J., ARNTZEN, C.J. : Identification of the in vivo precursor to the
        chlorophyll a/b light-harvesting complex. - Plant Physiol. 63 (Suppl.) :
        27, 1979.

37056 - BURKE, J.J., STEINBACK, K.E., ARNTZEN, C.J. : Analysis of the light-harvest-
        ing pigment-protein complex of wild type and a chlorophyll b-less mutant of
        barley. - Plant Physiol. 63 : 237 - 243, 1979.

37057 - BURKE, S., ARONOFF, S. : Separation of radiochemically-pure chlorophylls a
        and b. - Chromatographia 12 : 808 - 809, 1979.

37058 - BURRIS, J.E., WEDGE, R., LANE, A. : Diurnal fluctuations in aquatic primary
        productivity. - Plant Physiol. 63 (Suppl.) : 2, 1979. [Ps.]

37059 - BURTON, J.W., WILSON, R.F., BRIM, C.A. : Dry matter and nitrogen accumulation
        in male-sterile and male-fertile soybeans. - Agron. J. 71 : 548 - 552, 1979.

*37060 - BURWELL, C.C. : Solar biomass energy : an overview of U.S. potential. - Sci-
        ence 199 : 1041 - 1048, 1978. [Ps.]

37061 - BUSCHMANN, C. : The influence of kinetin on the biosynthesis of chlorophyll.
        - In : MARCELLE, R., CLIJSTERS, H., VAN POUCKE, M. (ed.) : Photosynthesis
        and Plant Development. Pp. 193 - 203. Dr.W.Junk bv. Publ., The Hague - Bos-
        ton - London 1979.

37062 - BUSCHMANN, C., LICHTENTHALER, H.K. : The influence of phytohormones on pre-
        nyllipid composition and photosynthetic activity of thylakoids. - In :
        APPELQVIST, L.-Å., LILJENBERG, C. (ed.) : Advances in the Biochemistry and
        Physiology of Plant Lipids. Pp. 145 - 150. Elsevier/North-Holland Biomedical
        Press, Amsterdam 1979.

37063 - BUSEY, P., MYERS, B.J. : Growth rates of turfgrasses propagated vegetatively.
        - Agron. J. 71 : 817 - 821, 1979. [Dry-matter accumulation.]

37064 - BUTLER, W.L. : Tripartite and bipartite models of the photochemical appara-
        tus of photosynthesis. - In : Chlorophyll Organization and Energy Transfer
        in Photosynthesis. Pp. 237 - 256. Excerpta Medica, Amsterdam - Oxford - New
        York 1979.

37065 - BUTLER, W.L., TREDWELL, C.J., MALKIN, R., BARBER, J. : The relationship be-
        tween the lifetime and yield of the 735 nm fluorescence of chloroplasts at
        low temperatures. - Biochim. biophys. Acta 545 : 309 - 315, 1979.

37066 - BYKOV, O.D. : Dykhanie list'ev pshenitsy na svetu : Svyaz' s fotosintezom.
        [Respiration of wheat leaves in light : connection with photosynthesis.] -
        Byull. vsesoyuz. nauch.-issled. Inst. Rastenievod. N.I.Vavilova 87 : 7 - 12,
        1979. [In R.]

37067 - BYKOV, O.D. : Dykhanie list'ev pshenitsy na svetu : Sravnenie s vydeleniem
        CO_2 v poslesvetovoĭ period. [Respiration of wheat leaves in light : Compa-
        rison with post-illumination CO_2 burst.] - Byull. vsesoyuz. nauch.-issled.
        Inst. Rastenievod. N.I.Vavilova 87 : 12 - 18, 1979. [In R.]

*37068 - BYKOV, O.D., KOSHKIN, V.A., CHERNIKOV, V.A., IVANOV, A.S. : Avtomaticheskiĭ
pribor dlya izmereniya aktivnosti beta-radioaktivnykh preparatov. [An automatic apparatus for measuring the activity of beta-radioactive samples.] -
Tr. priklad. Bot. Genet. Selektsii *61* (3) : 13 - 16, 1978. [Ps measurement;
in R, ab : E.]

37069 - BYSTROVA, M.I., MAL'GOSHEVA, I.N., KRASNOVSKIĬ, A.A. : Issledovanie moleku-
lyarnogo mekhanizma samosborki agregirovannykh form bakteriokhlorofilla *c*.
[Molecular mechanism of self-assemblage of aggregated bacteriochlorophyll *c*.]
- Mol. Biol. (Moskva) *13*: 582 - 594, 1979. [In R, ab : E.]

37070 - CADÉE, G.C., HEGEMAN, J. : Phytoplankton primary production, chlorophyll and
composition in an inlet of the western Wadden sea (Marsdiep). - Neth. J. Sea
Res. *13* : 224 - 241, 1979.

37071 - CAI, Ke, ZHANG, Ying-huang : [Studies on plant growth regulators and transport of C^{14}-photosynthetic assimilates.] - Acta phytophysiol. sin. *5* (4) :
327 - 333, 1979. [In Chin., ab : E.]

37072 - CALDWELL, C.D., LE FEVRE, P.E., AIKMAN, D.P. : An open-circuit apparatus for
continuous determination of net ion uptake by seedlings grown hydroponically.
- Can. J. Bot. *56* : 2767 - 2772, 1978. [Ps.]

*37073 - CALE, W.G. Jr. : Modeling grassland primary productivity using piecewise sta-
tionary, piecewise linear mathematics. - Ecol. Model. *7* : 107 - 123, 1979.

37074 - CALLOW, J.A., CALLOW, M.E., EVANS, L.V. : Nutritional studies on the parasitic red alga, *Choreocolax polysiphoniae*. - New Phytol. *83* : 451 - 462, 1979.
[Ps.]

37075 - CALVAYRAC, R., BOMSEL, J.-L., LAVAL-MARTIN, D. : Analysis and characterization of 3-(3,4-dichlorophenyl)-1,1-dimethylurea (DCMU)-resistant *Euglena*. I.
Growth, metabolic and ultrastructural modifications during adaptation to
different doses of DCMU. - Plant Physiol. *63* : 857 - 865, 1979.

37076 - CALVAYRAC, R., LAVAL-MARTIN, D., DUBERTRET, G., BOMSEL, J.-L. : Analysis and
characterization of 3-(3,4-dichlorophenyl)-1,1-dimethylurea (DCMU)-resistant
Euglena. II. Modifications affecting photosynthesis during adaptation to
different doses of DCMU. - Plant Physiol. *63* : 866 - 872, 1979.

37077 - CALVAYRAC, R., LEDOIGT, G., LAVAL-MARTIN, D. : Analysis and characterization
of DCMU-resistant *Euglena gracilis*. III. Thylakoid modifications and dark
"recovery" of photosynthesis. - Planta *145* : 259 - 267, 1979.

37078 - CAMM, E., GREEN, B. : Extraction of Photosystem II chlorophyll-protein complexes by the non-ionic detergent octyl β-D-glucopyranoside. - Plant Physiol.
63 (Suppl.) : 28, 1979.

*37079 - CAMMACK, R., HALL, D.O. : Mechanisms of electron transfer as mediated by
iron-sulphur proteins. - Hoppe-Seyler's Z. physiol. Chem. *359* : 1068 - 1069,
1978.

37080 - CAMMACK, R., LUIJK, L.J., MAGUIRE, J.J., FRY, I.V., PACKER, L. : EPR spectra
of Photosystem I and other iron protein components in intact cells of cyanobacteria. - Biochim. biophys. Acta *548* : 267 - 275, 1979.

37081 - CAMMACK, R., RYAN, M.D., STEWART, A.C. : The EPR spectrum of iron-sulphur
centre B in photosystem I of *Phormidium laminosum*. - FEBS Lett. *107* : 422 -
426, 1979.

37082 - CAMPBELL, C.A., DAVIDSON, H.R. : Effect of temperature, nitrogen fertilization and moisture stress on growth, assimilate distribution and moisture use
by Manitou spring wheat. - Can. J. Plant Sci. *59* : 603 - 626, 1979.

37083 - CANNELL, R.Q., GALES, K., SNAYDON, R.W., SUHAIL, B.A. : Effects of short-term
waterlogging on the growth and yield of peas (*Pisum sativum*). - Ann. appl.
Biol. *93* : 327 - 335, 1979.

37084 - CANVIN, D.T. : Photorespiration : Comparison between C_3 and C_4 plants. - In : GIBBS, M., LATZKO, E. (ed.) : Photosynthesis II. (Encycl. Plant Physiol. N.S. Vol.6.) Pp. 368 - 396. Springer-Verlag, Berlin - Heidelberg - New York 1979.

*37085 - CAPLE, M.B., CHOW, H., STROUSE, C.E. : Photosynthetic pigments of green sulfur bacteria. The esterifying alcohols of bacteriochlorophylls c from *Chlorobium limicola*. - J. biol. Chem. 253 : 6730 - 6737, 1978.

37086 - CAPONE, D.G., PENHALE, P.A., OREMLAND, R.S., TAYLOR, B.F. : Relationship between productivity and $N_2(C_2H_2)$ fixation in a *Thalassia testudinum* community. - Limnol. Oceanogr. 24 : 117 - 125, 1979. [Ps.]

37087 - CARBON, B.A., BARTLE, G.A., MURRAY, A.M. : A method for visual estimation of leaf area. - Forest Sci. 25 : 53 - 58, 1979.

37088 - CARBONNEAU, A., CASTERAN, P. : Irrigation-depressing effect on floral initiation of Cabernet Sauvignon grapevines in Bordeaux area. - Amer. J. Enol. Viticult. 30 : 3 - 7, 1979. [Ps.]

37089 - CARLIER, M.-F., HAMMES, G.G. : Interaction of nucleotides with chloroplast coupling factor I. - Biochemistry 18 : 3446 - 3451, 1979.

37090 - CARLIER, M.-F., HOLOWKA, D.A., HAMMES, G.G. : Interaction of photoreactive and fluorescent nucleotides with chloroplast coupling factor I. - Biochemistry 18 : 3452 - 3457, 1979.

37091 - CARLSSON, R., SUNDQVIST, C. : Oxygen consumption stimulated by δ-aminolevulinic acid in darkness and during irradiation of dark grown wheat leaves. - Physiol. Plant. 47 : 105 - 111, 1979. [Chl.]

37092 - CARMELI, C., LIFSHITZ, Y., GUTMAN, M. : Divalent metal ions as modifiers of the nonlinear initial rates of ATPase activity in photosynthetic coupling factors. - In : MUKOHATA, Y., PACKER, L. (ed.) : Cation Flux across Membranes. Pp. 249 - 259. Academic Press, New York - San Francisco - London 1979.

37093 - CARMI, A., KOLLER, D. : Regulation of photosynthetic activity in the primary leaves of bean (*Phaseolus vulgaris* L.) by materials moving in the water-conducting system. - Plant Physiol. 64 : 285 - 288, 1979.

37094 - CARMI, A., SHOMER, I. : Starch accumulation and photosynthetic activity in primary leaves of bean (*Phaseolus vulgaris* L.). - Ann. Bot. 44 : 479 - 484, 1979.

37095 - CARNEVALE, J., COLE, E.R., CRANK, G. : Fluorescent light catalyzed autoxidation of β-carotene. - J. agr. Food Chem. 27 : 462 - 463, 1979.

37096 - CARPENTER, D.J., CARPENTER, S.M. : A comparison of optical and biochemical classifications of ocean waters. - Deep-Sea Res. 26 A : 763 - 774, 1979. [Chl.]

37097 - CARPENTER, E.J., WALSBY, A.E. : Gas vacuole collapse in marine *Oscillatoria* (*Trichodesmium*) *thiebautii* (*Cyanophyta*) and the effect of nitrogenase activity and photosynthesis. - J. Phycol. 15 : 221 - 223, 1979.

37098 - CARRIER, J.M., NEVE, N. : Oxidation-reduction states of pyridine nucleotides measured by an adapted enzymatic cycling method in maize leaves submitted to anoxia. - Photosynthetica 13 : 323 - 331, 1979.

37099 - CARRIER, J.-M., THIBAULT, P. : Cinétique de photoréduction du NADP chez la feuille de Maïs en anoxie. - Compt. rend. Acad. Sci. Paris, Sér. D 288 : 607 - 610, 1979.

37100 - CASADORO, G., RASCIO, N. : Patterns of thylakoid system formation. - J. Ultrastruct. Res. 69 : 307 - 315, 1979.

37101 - CASADORO, G., RASCIO, N. : Plastid ultrastructural features in the various tissues of sunflower leaves. - Cytobios 24 : 157 - 166, 1979. [Chl.]

37102 - CASTELFRANCO, P.A., PARDO, A.D., CHERESKIN, B.M., WEINSTEIN, J.D. : Mg-chelatase in developing chloroplasts. - Plant Physiol. 63 (Suppl.) : 98, 1979.

37103 - CASTELFRANCO, P.A., WEINSTEIN, J.D., SCHWARCZ, S., PARDO, A.D., WEZELMAN, B.E. : The Mg insertion step in chlorophyll biosynthesis. - Arch. Biochem. Biophys. 192 : 592 - 598, 1979.

37104 - CASTELLI, F., CHEDDAR, G., RIZZUTO, F., TOLLIN, G. : Laser photolysis studies of quinone reduction by pheophytin a in alcohol solution. - Photochem. photobiol. 29 : 153 - 163, 1979.

37105 - ČATSKÝ, J., TICHÁ, I. : CO_2 compensation concentration in bean leaves : effect of photon flux density and leaf age. - Biol. Plant. 21 : 361 - 364, 1979.

37106 - CATTANEO, A., KALFF, J. : Primary production of algae growing on natural and artificial aquatic plants : A study of interactions between epiphytes and their substrate. - Limnol. Oceanogr. 24 : 1031 - 1037, 1979.

37107 - CEDEÑO-MALDONADO, A., LIU, L.C., DELGADO, L. : Effect of photosynthetic inhibitor herbicides on nitrate reductase activity of non-target species. - J. Agr. Univ. Puerto Rico 63 : 412 - 414, 1979.

37108 - CERDA, A., BINGHAM, F.T., HOFFMAN, G.J., HUSZAR, C.K. : Leaf water potential and gaseous exchange of wheat and tomato as affected by NaCl and P levels in the root medium. - Agron. J. 71 : 27 - 31, 1979. [Stomatal resistance.]

37109 - CERDÁ-OLMEDO, E., TORRES-TARTÍNEZ, S. : Genetics and regulation of carotene biosynthesis. - Pure appl. Chem. 51 : 631 - 637, 1979.

*37110 - CESCON, B., SCARAZZATO, P. : Oceanographic features of the Gulf of Trieste. II. Chemical observations. - Rev. int. Océanogr. méd. $35/36$: 91 - 109, 1974. [Primary production.]

37111 - CESCON, B., SCARAZZATO, P. : Hydrological features of the Adriatic Sea during winter and spring. Chemical observations. - Boll. Geofis. teor. appl. 21 (81) : 13 - 37, 1979. [Primary productivity.]

37112 - CEULEMANS, R., IMPENS, I. : Study of CO_2 exchange processes, resistances to carbon dioxide and chlorophyll content during leaf ontogenesis in poplar. - Biol. Plant. 21 : 302 - 306, 1979.

37113 - CHABOT, B.F. : Metabolic and enzymatic adaptations to low temperature. - In : UNDERWOOD, L.S., TIESZEN, L.L., CALLAHAN, A.B., FOLK, G.E. (ed.) : Comparative Mechanisms of Cold Adaptation. Pp. 283 - 301. Academic Press, New York - San Francisco - London 1979. [Ps.]

37114 - CHABOT, B.F., JURIK, T.W., CHABOT, J.F. : Influence of instantaneous and integrated light-flux density on leaf anatomy and photosynthesis. - Amer. J. Bot. 66 : 940 - 945, 1979.

37115 - CHAIN, R.K., MALKIN, R. : On the interaction of 2,5-dibromo-3-methyl-6-isopropylbenzoquinone (DBMIB) with bound electron carriers in spinach chloroplasts. - Arch. Biochem. Biophys. 197 : 52 - 56, 1979.

37116 - CHAMPIGNY, M.-L., MOYSE, A. : Photosynthetic carbon metabolism in wild, primitive and cultivated forms of wheat at three levels of ploidy : Role of the glycolate pathway. - Plant Cell Physiol. 20 : 1167 - 1178, 1979.

37117 - CHANCE, B., DEVAULT, D., TASAKI, A., THORNBER, J.P. : The effects of high hydrostatic pressure of light-induced electron transfer and proton binding in $Chromatium$. - In : CHANCE, B., DEVAULT, D.C., FRAUENFELDER, H., MARCUS, R.A., SCHRIEFFER, J.R., SUTIN, N. (ed.) : Tunneling in Biological Systems. Pp. 387 - 403. Academic Press, New York - San Francisco - London 1979.

37118 - CHAO, Fu-hung, CHU, Chung-hsi, MAO, Ta-chang, HSU, Chun-hui, TAI, Yun-ling : [The effects of magnesium ion on the DCIP photoreduction activity and apparent absorption spectrum in chloroplasts of spinach and of the shade plants ($Chlorophytum$ $comosum$, $Malaxis$ $monophyllos$).] - Acta phytophysiol. sin. 5 : 19 - 26, 1979. [In Chin., ab : E.]

37119 - CHAPMAN, D.J., LEECH, R.M. : Changes in pool sizes of free amino acids and amides in leaves and plastids of Zea $mays$ during leaf development. - Plant Physiol. 63 : 567 - 572, 1979.

37120 - CHAPMAN, K.S.R., HATCH, M.D. : Aspartate stimulation of malate decarboxylation in Zea $mays$ bundle sheath cells : Possible role in regulation of C_4 photosynthesis. - Biochem. biophys. Res. Commun. 86 : 1274 - 1280, 1979.

37121 - CHARI, M., ROCHER, J.P. : Saccharide metabolism and photosynthetic adaptation in *Lolium multiflorum* and *L. perenne*. - Photosynthetica *13* : 307 - 314, 1979.

37122 - CHARLES-EDWARDS, D.A. : A model for leaf growth. - Ann. Bot. *44* : 523 - 535, 1979. [Ps.]

37123 - CHARLES-EDWARDS, D.A. : Photosynthesis and crop growth. - In : MARCELLE, R., CLIJSTERS, H., VAN POUCKE, M. (ed.) : Photosynthesis and Plant Development. Pp. 111 - 124. Dr.W.Junk bv. Publ., The Hague - Boston - London 1979.

37124 - CHATTERTON, N.J., SILVIUS, J.E. : Photosynthate partitioning into starch in soybean leaves as effected by length of the photosynthetic period. - Plant Physiol. *63* (Suppl.) : 65, 1979.

37125 - CHATTERTON, N.J., SILVIUS, J.E. : Photosynthate partitioning into starch in soybean leaves I. Effects of photoperiod versus photosynthetic period duration. - Plant Physiol. *64* : 749 - 753, 1979.

37126 - CHAVAN, R.R., KOTHARI, I.L., PATEL, J.D. : A simple area calculating device (ACD) for biological systems. - Curr. Sci. *48* : 792 - 793, 1979.

37127 - CHELM, B.K., HALLICK, R.B., GRAY, P.W. : Transcription program of the chloroplast genome of *Euglena gracilis* during chloroplast development. - Proc. nat. Acad. Sci. USA *76* : 2258 - 2262, 1979.

*37128 - CHEMARINA, O.V. : Intensivnost' fotosinteza u diploidnykh i triploidnykh klonov osiny. [Photosynthetic rate in diploid and triploid clones of *Populus tremula*.] - In : Genetika, Selektsiya, Semenovodstvo i Introduktsiya Lesnykh Porod. Pp. 69 - 75. Moskva 1974. [In R.]

37129 - CHEMERIS, Yu.K., GRISHINA, N.A., KALACHEV, V.A., VENEDIKTOV, P.S. : Issledovanie fotoindutsirovannogo okisleniya tsitokhroma *f* i izmeneniya absorbtsii v oblasti 520 nm v kul'ture sinkhronno delyashchikhsya kletok khlorelly. [Photoinduced oxidation of cytochrome *f* and the changes in absorption in the region of 520 nm in the culture of synchronously dividing cells of *Chlorella*.] - Nauch. Dokl. vyssh. Shkoly, biol. Nauki *22* (5) : 39 - 44, 1979. [In R.]

37130 - CHEMERIS, Yu.K., GRISHINA, N.A., VENEDIKTOV, P.S. : Razlichnyĭ kharakter sinteza khlorofillov *a* i *b* v techenie tsikla razvitiya khlorelly. [Differences in the synthesis of chlorophyll *a* and *b* in the course of *Chlorella* cell cycle.] - Fiziol. Rast. *26* : 378 - 382, 1979. [In R, ab : E.]

37131 - CHEN, J.-S., BLANCHARD, D.K. : A simple hydrogenase-linked assay for ferredoxin and flavodoxin. - Anal. Biochem. *93* : 216 - 222, 1979.

*37132 - CHEN, K. : Identification of two different chromosomes which code for specific small subunit polypepties of fraction I protein. - In : SIEGELMAN, H.W., HIND, G. (ed.) : Photosynthetic Carbon Assimilation. P. 417. Plenum Press, New York - London 1978.

37133 - CHEN, K., MEYER, V.G. : Mutation in chloroplast DNA coding for the large subunit of fraction 1 protein correlated with male sterility in cotton. - J. Hered. *70* : 431 - 433, 1979.

37134 - CHEN, S.S., BERNS, D.S. : Effect of plastocyanin and phycocyanin on the photosensitivity of chlorophyll-containing bilayer membranes. - J. Membrane Biol. *47* : 113 - 127, 1979.

37135 - CHEPKO, G., WEISTROP, J.S., MARGULIES, M.M. : The absence of cytoplasm ribosomes on the envelopes of chloroplasts and mitochondria in plants : implications for the mechanism of transport of proteins into these organelles. - Protoplasma *100* : 385 - 392, 1979.

37136 - CHERNIKOV, V.S. : Issledovanie promezhutochnykh produktov v reaktsii fotovosstanovleniya khlorofilla *a* i feofitina *a*. [Intermediate products of photoreduction reactions of chlorophyll *a* and pheophytin *a*.] - Biofizika *24* : 362, 1979. [In R.]

37137 - CHERNOKOLEV, A.T., KUKUSHKIN, A.K., SOLNTSEV, M.K. : Termolyuminestsentsiya subkhloroplastnykh chastits i deĭstvie ADF, ATF i razobshchiteleĭ na termolyuminestsentsiyu khloroplastov vysshikh rasteniĭ. [Thermoluminescence of

subchloroplast particles and effects of ATP, ADP and uncouplers on thermoluminescence of higher plant chloroplasts.] - Biofizika *24* : 342 - 343, 1979.
[In R, ab : E.]

*37138 - **CHESHEL', E.Ya., GIRS, G.I.** : Sostoyanie pigmentov i khlorofill-belkovogo
kompleksa khvoi sosny obyknovennoĭ v posadkakh Shirinskoĭ stepi. [State of
pigments and chlorophyll-protein complex of needles of *Pinus nigra* in forests of Shirinskaya steppe.] - In : Fiziologo-Biokhimicheskie Protsessy u
Khvoĭnykh Rasteniĭ. Pp. 48 - 55, 143. Akad. Nauk SSSR, Krasnoyarsk 1978.
[In R.]

*37139 - **CHETVERIKOVA, N.I., ZHEMCHUGOVA, V.P., ZMEEVA, V.N., CHERNODED, G.K.** :
Rol' ėkzogennykh i ėndogennykh faktorov v formirovanii plodov (na primere
rasteniĭ gorokha). [Role of exogenous and endogenous factors in fruit formation (on an example of pea).] - In : Pogloshchenie i Peredvizhenie Veshchestv
u Rasteniĭ. Pp. 27 - 34. Akad. Nauk SSSR, Vladivostok 1978. [Photosynthates
transport; in R.]

*37140 - **CHEVALLIER, D., NURIT, F., DOUCE, R.** : Interactions between mitochondria
and chloroplasts in moss spore cells. - In : DUCET, G., LANCE, C. (ed.) :
Plant Mitochondria. Pp. 349 - 356. Elsevier/North-Holland Biomedical Press,
Amsterdam 1978.

37141 - **CHICHEV, P.N.** : Vliyanie na ekstremni polozhitelni temperaturi v"rkhu fotosintezata. III. Skorost na CO_2 fiksatsiyata i razpredelenieto na ^{14}C v zavisimost ot prod"lzhitelnostta na posledeĭstvieto. [Effect of extreme positive
temperatures on photosynthesis. III. CO_2 fixation rate and ^{14}C distribution
in relation with after-effect duration.] - Fiziol. Rast. (Sofia) *5* (2) : 19 -
- 27, 1979. [In Bulg., ab : E, R.]

37142 - **CHOCK, J.S., MATHIESON, A.C.** : Physiological ecology of *Ascophyllum nodosum*
(L.) LE JOLIS and its detached ecad *scorpioides* (HORNEMANN) HAUCK (*Fucales,
Phaeophyta*). - Bot. mar. *22* : 21 - 26, 1979. [Ps.]

*37143 - **CHOLLET, R.** : Chemical modification of tobacco RuBP carboxylase by 2,3-buta-
nedione. - In : SIEGELMAN, H.W., HIND, G. (ed.) : Photosynthetic Carbon Assimilation. Pp. 417 - 418. Plenum Press, New York - London 1978.

*37144 - **CHOLLET, R.** : Chemical modification of tobacco RuBP carboxylase by cyanate. -
In : SIEGELMAN, H.W., HIND, G. (ed.) : Photosynthetic Carbon Assimilation.
Pp. 418 - 419. Plenum Press, New York - London 1978.

37145 - **CHOLLET, R.** : Inactivation of tobacco RuBP carboxylase by modification of
arginyl residues with 2,3-butanedione(BD) and phenylglyoxal(PGO). - Plant
Physiol. *63* (Suppl.) : 153, 1979.

37146 - **CHOU, Pei-chen, TAN, Keh-hui, LI, Liang-pi, CHANG, Cheng-tung, CHANG, Kue-
-cheng, LI, Shou-chuan** : [Temperature induced changes in photosynthetic ele-
tron transfer in chloroplasts of detached lettuce leaves and spinach in the
dark.] - Acta phytophysiol. sin. *5* : 41 - 48, 1979. [In Chin., ab : E.]

37147 - **CHOUDHURY, N.K., BISWAL, U.C.** : Changes in photoelectron transport of chloroplasts isolated from dark stressed leaves of maize seedlings. - Experientia *35* : 1036 - 1037, 1979.

37148 - **CHOUDHURY, N.K., BISWAL, U.C.** : Changes in the content of chlorophyll, protein and nucleic acids and in the efficiency of photoelectron transport of
chloroplasts during growth of maize seedlings. - Plant Sci. Lett. *16* : 95 -
- 99, 1979.

*37149 - **CHRISTELLER, J.T., LAING, W.A.** : A kinetic study of ribulose bisphosphate
carboxylase from the photosynthetic bacterium *Rhodospirillum rubrum*. - Biochem. J. *173* : 467 - 473, 1978.

37150 - **CHRISTIANSEN, M.N.** : Physiological bases for resistance to chilling. - Hort-
Science *14* : 583 - 586, 1979. [Ps.]

37151 - **CHRISTIN, M.S., MUZAFAROV, E.N., AKULOVA, E.A.** : Magnetische Mikro-Mischvor-
richtung zum Spektralphotometer SPECORD UV VIS zur Untersuchung des immobi-
lisierten Zytochroms C. - Jenaer Rundschau *24* : 176, 1979.

*37152 - CHRÔST, R.J., WAŻYK, M. : Primary production and extracellular release by
phytoplankton in some lakes of the Masurian Lake District, Poland. - Acta
microbiol. pol. *27* : 63 - 71, 1978. [Chl.]

37153 - CHUA, N.-H., BLOMBERG, F. : Immunochemical studies of thylakoid membrane polypeptides from spinach and *Chlamydomonas reinhardtii*. A modified procedure
from crossed immunoelectrophoresis of dodecyl sulfate-protein complexes. -
J. biol. Chem. *254* : 215 - 223, 1979.

37154 - CHUA, N.-H. , SCHMIDT, G.W. : Transport of proteins into mitochondria and
chloroplasts. - J. Cell Biol. *81* : 461 - 483, 1979.

*37155 - CHUNAEV, A.S., LIPKIND, B.I., KVITKO, K.V., GILLER, Yu.E. : Izmenchivost' so-
stoyaniya khlorofilla i karotinoidov v kletkakh *Chlamydomonas reinhardii*
137 C. I. Mutanty s narushennym sintezom khlorofilla. [Variability in chlorophyll and carotenoid states in cells of *Chlamydomonas reinhardii* 137 C. I.
Mutants with defect chlorophyll synthesis.] - Nauch. Dokl. vyssh. Shkoly,
biol. Nauki *1978* (6) : 38 - 43, 1978. [In R.]

37156 - CHUNAEV, A.S., LIPKIND, B.I., KVITKO, K.V., GILLER, Yu.E. : Izmenchivost'
sostoyaniya khlorofilla i karotinoidov v kletkakh *Chlamydomonas reinhardii*
137 C. II. Mutanty s oslablennym nakopleniem karotinoidov. [Variability of
chlorophyll and carotenoids in cells of *Chlamydomonas reinhardii* 137 C. II.
Mutants with reduced accumulation of carotenoids.] - Biol. Nauki *1979* (11) :
45 - 51, 1979. [In R.]

37157 - CIANZIO, S.R.de, FEHR, W.R., ANDERSON, I.C. : Genotypic evaluation for iron
deficiency chlorosis in soybeans by visual scores and chlorophyll concentration. - Crop Sci. *19* : 644 - 646, 1979.

37158 - CIFERRI, O., PASCQUALE, G.di, TIBONI, O. : Chloroplast elongation factors
are synthesized in the chloroplast. - Europe. J. Biochem. *102* : 331 - 335,
1979.

37159 - CLARK, J.R., HACKETT, W.P. : Distribution of soluble ^{14}C-labelled assimilates as related to distribution of ^{14}C-benzyladenine in shoot tips of *Hedera
helix*. - Physiol. Plant. *47* : 87 - 90, 1979.

37160 - CLARKE, R.H., HOBART, D.R., LEENSTRA, W.R. : The triplet state of the chlorophyll dimer. - J. amer. chem. Soc. *101* : 2416 - 2423, 1979.

37161 - CLARKE, R.H., LEENSTRA, W.R., HAGAR, W.G. : Observation of a triplet state
in chlorophyll protein 668 via optically detected magnetic resonance. - FEBS
Lett. *99* : 207 - 209, 1979.

37162 - CLAUSS, H. : Auslösung der circadianen Photosynthese-Rhythmik bei *Acetabularia* durch Blaulicht. - Protoplasma *99* : 341 - 346, 1979.

37163 - CLAUSSEN,W., LENZ, F. : Der Einfluß unterschiedlicher Lichtintensität auf
die Saccharose- und Stärkegehalte der Blätter und deren Bedeutung für die Regulierung der Netto-Photosyntheseraten bei Auberginen (*Solanum melongena* L.).
- Gartenbauwissenschaft *44* : 10 - 14, 1979.

37164 - CLAUSSEN, W., LENZ, F. : Die Bedeutung des Assimilatsstaus in den Blättern
für die Regulierung der Netto-Photosyntheseraten bei Auberginen (*Solanum melongena* L.). - Angew. Bot. *53* : 41 - 52, 1979.

37165 - CLAYTON, R.K. : Artificial reaction centres to mimic photosynthesis. - Nature
278 : 11, 1979.

37166 - CLAYTON, R.K., RAFFERTY, C.N., VERMEGLIO, A. : The orientations of transition moments in reaction centers of *Rhodopseudomonas sphaeroides*, computed
from data of linear dichroism and photoselection measurements. - Biochim.
biophys. Acta *545* : 58 - 68, 1979.

37167 - CLAYTON, R.K., VERMEGLIO, A. : Photochemical polarization of the bacterial
photosynthetic membrane. - In : CONE, R.A., DOWLING, J.E. (ed.) : Membrane
Transduction Mechanisms. Pp. 49 - 59. Raven Press, New York 1979.

37168 - CLÉMENT, B., TOUFFET, J. : Influence du substrat sur la productivité primaire d'une lande armoricaine. - Compt. rend. Acad. Sci. Paris, Sér. D *288* :
1219 - 1221, 1979.

37169 - CLIFFORD, P.E. : Source limitation of sink yield in mung beans. - Ann. Bot. *43* : 397 - 399, 1979.

37170 - CLIJSTERS, H., VAN ASSCHE, F., MARCELLE, R. : Fotosynthese en supra-optimale plantenvoeding met sporenelementen. [Photosynthesis and superoptimal plant nutrition with trace elements.] - Extern *8* : 293 - 305, 1979. [In Hol.]

37171 - CLOUGH, J.M., ALBERTE, R.S., TEERI, J.A. : Photosynthetic adaptation of *Solanum dulcamara* L. to sun and shade environments. II. Physiological characterization of phenotypic response to environment. - Plant Physiol. *64* : 25 - 30, 1979.

37172 - CLOUGH, J.M., PATTENDEN, G. : Naturally occurring poly-*cis* carotenoids. Stereochemistry of poly-*cis* lycopene and its congeners in "Tangerine" tomato fruits. - J. chem. Soc. *1979* : 616 - 619, 1979.

37173 - CLOUGH, J.M., TEERI, J.A., ALBERTE, R.S. : Photosynthetic adaptation of *Solanum dulcamara* L. to sun and shade environments. I. A comparison of sun and shade populations. - Oecologia *38* : 13 - 21, 1979.

*37174 - CLYMO, R.S. : A model of peat bog growth. - In : HEAL, O.W., PERKINS, D.F. (ed.) : Production Ecology of British Moors and Montane Grasslands. Pp. 187 - 223. Springer-Verlag, Berlin - Heidelberg - New York 1978. [Productivity.]

37175 - COCITO, C., TIBONI, O., VANLINDEN, F., CIFERRI, O. : Inhibition of protein synthesis in chloroplasts from plant cells by virginiamycin. - Z. Naturforsch. *34 C* : 1195 - 1198, 1979.

37176 - COCK, J.H., FRANKLIN, D., SANDOVAL, G., JURI, P. : The ideal cassava plant for maximum yield. - Crop Sci. *19* : 271 - 279, 1979. [Growth analysis.]

37177 - COCKBURN, W., TING, I.P., STERNBERG, L.O. : Relationships between stomatal behavior and internal carbon dioxide concentration in Crassulacean acid metabolism plants. - Plant Physiol. *63* : 1029 - 1032, 1979. [Ps.]

37178 - COCQUEMPOT, M.F., THOMAS, D., CHAMPIGNY, M.L., MOYSE, A. : Immobilization of thylakoids in porous particles and stabilization of the photochemical processes by glutaraldehyde action at subzero temperature. - Europe. J. appl. Microbiol. Biotech. *8* : 37 - 41, 1979.

37179 - CODD, G.A., COOK, C.M., STEWART, W.D.P. : Purification and subunit structure of D-ribulose 1,5-bisphosphate carboxylase from the cyanobacterium *Aphanothece halophytica.* - FEMS Microbiol. Lett. *6* : 81 - 86, 1979.

37180 - COGDELL, R.J. : Photochemical reactions centre of photosynthetic bacteria. - Biochem. Soc. Trans. *7* : 1228 - 1231, 1979.

37181 - COGDELL, R.J., THORNBER, J.P. : The preparation and characterization of different types of light-harvesting pigment-protein complexes from some purple bacteria. - In : Chlorophyll Organization and Energy Transfer in Photosynthesis. Pp. 61 - 79. Excerpta Medica, Amsterdam - Oxford - New York 1979.

37182 - COHEN, A.S., POPOVIC, R.B., ZALIK, S. : Effects of polyamines on chlorophyll and protein content, photochemical activity, and chloroplast ultrastructure of barley leaf discs during senescence. - Plant Physiol. *64* : 717 - 720, 1979.

37183 - COHEN, C.E., REBEIZ, C.A. : Spectrofluorometric characterization of the protochlorophyll species in etiolated tissues. - Plant Physiol. *63* (Suppl.) : 97, 1979.

37184 - COKE, L., SIONIT, N. : Stomatal behavior and gas exchange in leaves of cassava (*Manihot esculenta*) treated with abscisic acid and vomifoliol. - Plant Physiol. *63* (Suppl.) : 121, 1979.

37185 - COKER, G., SCHUBERT, K.R. : The role of dark CO_2 fixation in amino acid biosynthesis in soybean root nodules. - Plant Physiol. *63* (Suppl.) : 112, 1979.

37186 - COLE, R.M., COHEN, W.S. : Studies on the coupling of electron flow to phosphorylation in maize mesophyll chloroplasts. - Plant Physiol. *63* (Suppl.) : 30, 1979.

37187 - COLLATZ, G.J., BADGER, M., SMITH, C., BERRY, J.A. : A radioimmune assay for RuP$_2$ carboxylase protein. - Carnegie Inst. Year Book *78* : 171 - 175, 1979.

37188 - COLLOS, Y., SLAWYK, G. : ^{13}C and ^{15}N uptake by marine phytoplankton. I. Influence of nitrogen source and concentration in laboratory cultures of diatoms. - J. Phycol. *15* : 186 - 190, 1979.

37189 - COLMAN, B., MAWSON, B.T., ESPIE, G.S. : The rapid isolation of photosynthetically active mesophyll cells from *Asparagus* cladophylls. - Can. J. Bot. *57* : 1505 - 1510, 1979.

*37190 - COLMAN, R.L., O'NEILL, G.H. : Seasonal variation in the potential herbage production and response to nitrogen by kikuyu grass (*Pennisetum cladestinum*). - J. agr. Sci. *91* : 81 - 90, 1978.

37191 - COMBE, L. : Effet du gaz carbonique et de l'éclairement sur la croissance et la répartition des assimilats chez le Radis (*Raphanus sativus* L.). - Ann. agron. *30* : 217 - 231, 1979.

37192 - ÇONESA, A.P., METTAUER, H., HAEFLINGER, R., TRENDEL, R., TUAL, Y., GROSS, P. : Étude de la productivité de l'agrosystème betteravier en Alsace. Essai d'établissement d'un modèle empirique prédictif. - Ann. agron. *30* : 281 - 303, 1979.

37193 - CONJEAUD, H., MATHIS, P., PAILLOTIN, G. : Primary and secondary electron donors in Photosystem II of chloroplasts. Rates of electron transfer and location in the membrane. - Biochim. biophys. Acta *546* : 280 - 291, 1979.

37194 - CONNER, A.J., MAHANTY, H.K. : Growth responses of unicellular algae to polychlorinated biphenyls : New evidence for photosynthetic inhibition. - Mauri Ora *7* : 3 - 17, 1979. [Ps.]

37195 - CONTRERAS, S., SOTO, M.A., TOHA C., J.: Applied microalgae photosynthesis : discharge mechanisms in highly illuminated cells. - Biotechnol. Bioeng. *21* : 159 - 165, 1979.

37196 - CONWAY, H.L., WILLIAMS, S.C. : Sorption of cadmium and its effect on growth and the utilization of inorganic carbon and phosphorus of two freshwater diatoms. - J. Fish. Res. Board Can. *36* : 579 - 586, 1979.

37197 - COOMBS, J. : Enzymes of C$_4$ metabolism. - In : GIBBS, M., LATZKO, E. (ed.) : Photosynthesis II. (Encycl. Plant Physiol. N.S. Vol. 6.) Pp. 251 - 262. Springer-Verlag, Berlin - Heidelberg - New York 1979.

37198 - COOPER, J.L. : Growth and yield of a semi-dwarf and a standard height wheat cultivar in the Murrumbidgee irrigation area. - Aust. J. exp. Agr. anim. Husb. *19* : 554 - 558, 1979. [Leaf area index.]

37199 - COOPER, P.J.M. : The association between altitude, environmental variables, maize growth and yields in Kenya. - J. agr. Sci. *93* : 635 - 649, 1979. [Growth analysis.]

37200 - CORKER, G.A., HENKIN, B.M., SHARPE, S.A. : The contribution of *Rhodospirillum rubrum* ferredoxin II to *in vivo* EPR signals and its role in electron transport. - Photochem. Photobiol. *29* : 141 - 146, 1979.

37201 - COSTES, C. : Mutations induites et photosynthèse. - Bull. Soc. bot. France *126* : 53, 1979.

37202 - COSTES, C., BAZIER, R. : Heterogeneity of lipids in wheat freeze-dried chloroplast lamellae. - In : APPELQVIST, L.-Å., LILJENBERG, C. (ed.) : Advances in the Biochemistry and Physiology of Plant Lipids. Pp. 151 - 157. Elsevier/North-Holland Biomedical Press, Amsterdam 1979.

37203 - COSTES, C., BURGHOFFER, C., JOYARD, J., BLOCK, M., DOUCE, R. : Occurrence and biosynthesis of violaxanthin in isolated spinach chloroplast envelope. - FEBS Lett. *103* : 17 - 21, 1979.

37204 - CÔTÉ, R., LACROIX, G. : Influence de débits élevés et variables d'eau douce sur le régime saisonnier de production primaire d'un fjord subarctique. - Oceanol. Acta *2* : 299 - 306, 1979. [Chl.]

37205 - CÔTÉ, R., LACROIX, G. : Variabilité journalière de la chlorophylle a et des taux de production primaire dans le fjord du Saguenay. - Natur. can. *106* : 189 - 198, 1979.

37206 - COTTON, T.M., VAN DUYNE, R.P. : An electrochemical investigation of the redox properties of bacteriochlorophyll and bacteriopheophytin in aprotic solvents. - J. amer. chem. Soc. *101* : 7605 - 7612, 1979.

37207 - COUNCE, P.A., HOUSLEY, T.L. : The effect of cooling the transport path on the movement of ^{14}C-assimilates in corn. - Plant Physiol. *63* (Suppl.) : 45, 1979.

37208 - COWAN, D.A., GREEN, T.G.A., WILSON, A.T. : Lichen metabolism. 1. The use of tritium labelled water in studies of anhydrobiotic metabolism in *Ramalina celastri* and *Peltigera polydactyla*. - New Phytol. *82* : 489 - 503, 1979.

37209 - COWAN, D.A., GREEN, T.G.A., WILSON, A.T. : Lichen metabolism. 2. Aspects of light and dark physiology. - New Phytol. *83* : 761 - 769, 1979.

37210 - CRAIG, S.R., BUDD, K. : Chloride uptake by *Anacystis nidulans* (*Cyanophyceae*). - J. Phycol. *15* : 300 - 304, 1979. [Ps.]

37211 - CRAMER, W.A., WHITMARSH, J. : An application of electron transfer theory to a problem in chloroplast membrane topography. - In : CHANCE, B., DEVAULT, D.C., FRAUENFELDER, H., MARCUS, R.A., SCHRIEFFER, J.R., SUTIN, N. : Tunneling in Biological Systems. Pp. 363 - 370. Acad. Press, New York - San Francisco - London 1979.

37212 - CREACH, E. : Dark carbon dioxide fixation under aerobic and anaerobic conditions in maize leaves after preillumination in the absence of oxygen. Ribulose 1,5-bisphosphate can serve as a primary acceptor of carbon dioxide. - Plant Physiol. *63* : 788 - 791, 1979.

37213 - CREACH, E. : Enhanced dark carbon dioxide fixation in maize. Effect of the oxygen concentration during preillumination on $^{14}CO_2$ uptake and the intramolecular labeling pattern of malate and aspartate. - Plant Physiol. *64* : 435 - 438, 1979.

37214 - CRESPI, H.L., FERRARO, J.R. : Active site structure of bacteriorhodopsin and mechanism of action. - Biochem. biophys. Res. Commun. *91* : 575 - 582, 1979.

37215 - CRESPO, H.M., FREAN, M., CRESSWELL, C.F., TEW, J. : The occurrence of both C_3 and C_4 photosynthetic characteristics in a single *Zea mays* plant. - Planta *147* : 257 - 263, 1979.

37216 - CRESSWELL, C.F., TEW, A.J., LEWIS, O.A.M. : The regulation of carbon metabolism in C_4 photosynthetic plants by inorganic nitrogen. - In : HEWITT, E.J., CUTTING, C.V. (ed.) : Nitrogen Assimilation of Plants. Pp. 451 - 473. Academic Press, London - New York - San Francisco 1979.

*37217 - CRILEY, R.A., ANDERSEN, A. : Short term effect of four growth retardants on photosynthesis. - In : Proceedings of the Plant Growth Regulator Working Group. (Fourth Annu. Meet. 1977). Pp. 287 - 294. Great Western Sugar Co., Agricultural Research Center, Longmont 1977.

37218 - CROFTS, A.R. : Role of quinones in photosynthetic electron transport. - In : GERISCHER, H., KATZ, J.J. (ed.) : Light-Induced Charge Separation in Biology and Chemistry. Pp. 389 - 410. Verlag Chemie, Weinheim - New York 1979.

37219 - CROSTHWAITE, L., SHEEN, S.J., BURTON, H.R. : Alkylating activity in the extract and pyrolyzate of tobacco leaves varying in genotype and chemical treatment. - Tobacco Sci. *23* : 35 - 37, 1979. Tobacco int. *181* (7) : 110 - 112, 1979. [Chl.]

*37220 - CROW, T.R. : Biomass and production regressions for trees and woody shrubs common to Enterprise forest. - In : ZAVITKOVSKI, J. (ed.) : The Enterprise, Wisconsin, Radiation Forest. (Radioecological Studies, Pt. 2). Pp. 63 - 67. U.S. Energy Res. and Develop. Administ., Tech. Inform. Center, Oak Ridge 1977.

*37221 - **CROW, T.R.** : Effects of gamma radiation on the biomass structure of the arbo-
          real stratum in a Northern forest. - In : ZAVITKOVSKI, J. (ed.) : The Enter-
          prise, Wisconsin, Radiation Forest. (Radioecological Studies, Pt. 2.) Pp.
          69 - 78. U.S. Energy Res. and Develop. Administ., Tech. Inform. Center, Oak
          Ridge 1977.

37222 - **CROWTHER, D., MILLS, J.D., HIND, G.** : Protonmotive cyclic electron flow
         around photosystem I in intact chloroplasts. - FEBS Lett. *98* : 386 - 390,
         1979.

37223 - **CROZE, E., KELLY, M., HORTON, P.** : Loss of sensitivity to diuron after tryp-
         sin digestion of chloroplast photosystem II particles. - FEBS Lett. *103* :
         22 - 26, 1979.

37224 - **CRUIZIAT, P., THOMAS, D.A., BODET, C.** : Comparaison entre mesures locales
         et mesure globale de la résistance stomatique de feuilles de Tournesol (*He-
         lianthus annuus*). - Oecol. Plant. *14* : 447 - 459, 1979.

37225 - **CSORBA, I., BUZÁS, Z., POLYÁK, B., BOROSS, L.** : High-speed video-densitomet-
         ric determination of chlorophylls and carotenes separated by thin-layer chro-
         matography. - J. Chromatogr. *172* : 287 - 293, 1979.

37226 - **CULLEN, J.J., RENGER, E.H.** : Continuous measurement of the DCMU-induced fluo-
         rescence response of natural phytoplankton populations. - Mar. Biol. *53* :
         13 - 20, 1979.

37227 - **CUNNINGHAME, M.E., BOWES, B.G., HILLMAN, J.R.** : An ultrastructural study of
         foliar senescence in *Taxus baccata* L. - Ann. Bot. *43* : 527 - 528, 1979. [Chlo-
         roplast.]

37228 - **CURTIS, S.E., RAWSON, J.R.Y.** : Measurement of the transcription of nuclear
         single-copy deoxyribonucleic acid during chloroplast development in *Euglena
         gracilis*. - Biochemistry *18* : 5299 - 5304, 1979.

37229 - **CYBULSKY, D.L., NAGY, A., KANDEL, S.I., KANDEL, M., GORNALL, A.G.** : Carbonic
         anhydrase from spinach leaves. Chemical modification and affinity labeling. -
         J. biol. Chem. *254* : 2032 - 2039, 1979.

37230 - **CZARNOWSKI, M.** : Potential photosynthesis of tree leaves polluted by indus-
         trial emissions. - Bull. Acad. pol. Sci., Sér. Sci. biol. *27* : 605 - 612,
         1979.

37231 - **CZERPAK, R., CZECZUGA, B.** : Występowanie i biosynteza karotenoidów u bakte-
         rii. [Occurrence and biosynthesis of carotenoids in bacteria.] - Wiadom. bot.
         23 : 73 - 88, 1979. [In Pol.]

37232 - **CZUCHAJOWSKA, Z., NIEMTUR, S.** : Seasonal changes of chlorophyll and carote-
         noids content in needles of *Pinus nigra, Pinus strobus* and *Pinus silvestris*
         as influenced by industrial emissions. - Acta biol. (Katowice) *8* (Prace nauk.
         Uniw. śląsk. 278) : 182 - 190, 1979.

37233 - **CZUCHAJOWSKA, Z., STRĄCZEK, T.** : Seasonal changes of pigments concentration
         in the leaves of *Vaccinium myrtillus* and *Vaccinium vitis-idaea* and the influ-
         ence of industrial emissions. - Acta Soc. Bot. Pol. *48* : 551 - 558, 1979.

37234 - **DALEY, L.S., VINES, H.M., BIDWELL, R.G.S.** : Oxidation of phosphohydroxypyru-
         vate: Physiological implications in plants. - Can. J. Bot. *57* : 1 - 3, 1979.
         [Photorespiration.]

37235 - **DALGARN, D., MILLER, P., BRICKER, T., SPEER, N., JAWORSKI, J.G., NEWMAN, D.W.:**
         Galactosyl transferase activity of chloroplast envelopes from senescent soy-
         bean cotyledons. - Plant Sci. Lett. *14* : 1 - 6, 1979.

37236 - **DALGARN, D.S., NEWMAN, D.W.** : Lipids of dark-treated oat (*Avena sativa*) leaf
         chloroplast thylakoids. - J. exp. Bot. *30* : 551 - 556, 1979.

37237 - **DAMSZ, B.** : Biometric analysis of the thylakoid system development during chloroplast ontogenesis in leaves of two orchids, *Coelogyne cristata* and *Cymbidium insigne*. - Biochem. Physiol. Pflanzen *174* : 802 - 810, 1979.

37238 - **DANDONNEAU, Y.** : Concentrations en chlorophylle dans le Pacifique tropical sud-ouest: comparaison avec d'autres aires océaniques tropicales. - Oceanol. Acta *2* : 133 - 142, 1979.

*37239 - **DANNIGKEIT, W.** : CO_2-Assimilationshemmung und Metabolisierung von Bentazon bei unterschiedlich empfindlichen Pflanzen. - Z. Pflanzenkr. Pflanzensch. *84* : 540 - 546, 1977.

37240 - **DANTUMA, G., KLEIN HULZE, J.A.** : Production and distribution of dry matter, and uptake, distribution and redistribution of nitrogen in *Vicia faba major* and *minor*. - In : BOND, D.A. *et al.* : Some Current Research on *Vicia faba* in Western Europe. Pp. 396 - 406. Bari 1979.

37241 - **DARBYSHIRE, B., HENRY, R.J., MELHUISH, F.M., HEWETT, R.K.** : Diurnal variations in non-structural carbohydrates, leaf extension, and leaf cavity carbon dioxide concentrations in *Allium cepa* L. - J. exp. Bot. *30* : 109 - 118, 1979.

*37242 - **DAS, V.S.R., RAGHAVENDRA, A.S.** : C_4 photosynthesis and a unique type of Kranz anatomy in *Glossocordia boswallaea (Asteraceae)*. - Proc. indian Acad. Sci. Sect. B *84* : 12 - 19, 1976.

37243 - **DAS, V.S.R., RAGHAVENDRA, A.S.** : Antitranspirants for improvement of water use efficiency of crops. - Outlook Agr. *10* (2) : 92 - 98, 1979. [Ps.]

37244 - **DAS, V.S.R., VEERANJANEYULU, K.** : Leaf gas exchange characteristics of twentysix tropical weed and crop plants. - Plant Physiol. *63* (Suppl.) : 141, 1979.

37245 - **DAVENPORT, J.W., McCARTY, R.E.** : Transmembrane proton gradients drive ATP synthesis in thylakoids. - Plant Physiol. *63* (Suppl.) : 28, 1979.

37246 - **DAVIES, A.G., SLEEP, J.A.** : Inhibition of carbon fixation as a function of zinc uptake in natural phytoplankton assemblages. - J. mar. Biol. Assoc. U.K. *59* : 937 - 949, 1979.

37247 - **DAVIES, A.G., SLEEP, J.A.** : Photosynthesis in some British coastal waters may be inhibited by zinc pollution. - Nature *277* : 292 - 293, 1979.

37248 - **DAVIES, B.H.** : Solved and unsolved problems of carotenoid formation. - Pure appl. Chem. *51* : 623 - 630, 1979.

37249 - **DAVIES, D.D.** : The central role of phosphoenolpyruvate in plant metabolism. - Annu. Rev. Plant Physiol. *30* : 131 - 158, 1979.

37250 - **DAVIES, F.S., TERAMURA, A.H., BUCHANAN, D.W.** : Yield, stomatal resistance, xylem pressure potential, and feeder root density in three rabbiteye blueberry cultivars. - HortScience *14* : 725 - 726, 1979.

37251 - **DAVIS, D.J., KROGMANN, D.W., SAN PIETRO, A.** : Electron donation to Photosystem I : Comparison of plant and algal donors. - Plant Physiol. *63* (Suppl.) : 54, 1979.

37252 - **DAVIS, D.J., KROGMANN, D.W., SAN PIETRO, A.** : Localization of polylysine inhibition in a photosystem I subchloroplast particle. - Biochem. biophys. Res. Commun. *90* : 110 - 116, 1979.

37253 - **DAVIS, D.J., SAN PIETRO, A.** : Preparation and characterization of a chemically modified plastocyanin. - Anal. Biochem. *95* : 254 - 259, 1979.

37254 - **DAVIS, M.S., FORMAN, A., FAJER, J.** : Ligated chlorophyll cation radicals : Their function in photosystem II of plant photosynthesis. - Proc. nat. Acad. Sci. USA *76* : 4170 - 4174, 1979.

37255 - **DAVIS, M.S., FORMAN, A., HANSON, L.K., THORNBER, J.P., FAJER, J.** : Anion and cation radicals of bacteriochlorophyll and bacteriopheophytin *b*. Their role in the primary charge separation of *Rhodopseudomonas viridis*. - J. phys. Chem. *83* : 3325 - 3332, 1979.

37256 - **DAVIS, R.C., PEARLSTEIN, R.M.** : Chlorophyllin-apomyoglobin complexes. - Nature *280* : 413 - 415, 1979.

37257 - DAVTYAN, G.S., BABAKHANYAN, M.A. : Nutrition and productivity of tobacco
plants under open-air hydroponic conditions. - In : Mineral Nutrition of
Plants. Vol. I. Pp. 349 - 354. Publ. House Central Cooperative Union, Sofia
1979. [Ps, Chl, Car.]

37258 - DAVYDOVA, N.N., TRIFONOVA, I.S. : Diatomei planktona i donnykh otlozheniĭ i
soderzhanie khlorofilla v osadkakh dvukh raznotipnykh ozer Karel'skogo pere-
sheĭka kak pokazateli protsessa évtrofirovaniya. [Diatoms of plankton and
bed sediments and chlorophyll concentration in sediments of two different
types of lakes of Karelian isthmus as indicators of eutrophication process.]
- Bot. Zh. 64 : 1174 - 1183, 1979. [In R.]

37259 - DAWSON, J.O., GORDON, J.C. : Nitrogen fixation in relation to photosynthesis
in Alnus glutinosa. - Bot. Gaz. 140 (Suppl.) : S70 - S75, 1979.

37260 - DAY, W., PARKINSON, K.J. : Importance to gas exchange of mass flow of air
through leaves. - Plant Physiol. 64 : 345 - 346, 1979. [Stomatal resistance.]

37261 - DeBENEDETTI, E., JAGENDORF, A. : Inhibition of chloroplast coupling factor
ATPase by 5'-p-fluorosulfonylbenzoyl adenosine. - Biochem. biophys. Res.
Commun. 86 : 440 - 446, 1979.

37262 - DeBONTE, L.R., VAUGHN, K.C., BRICKER, T.M., WILSON, K.G. : Chlorophyll complex II deficiency in the citrine mutant of tomato. - Photosynthetica 13 :
332 - 336, 1979.

37263 - DECHEVA, R., KOSEVA, D. : S"d"rzhanie i dinamika na absolyutno sukho veshchestvo, khlorofil (a+b) i karotinoidi v listata na kazanl"shkata roza.
[Content and dynamics of the absolute dry matter, chlorophyll (a+b) and carotenoids in Kazanlyk rose leaves.] - Rasteniev"dni Nauki 16 (1) : 26 - 33,
1979. [In Bulg., ab : E, R.]

37264 - DE FILIPPIS, L.F. : The effect of heavy metals on the absorption spectra of
Chlorella cells and chlorophyll solutions. - Z. Pflanzenphysiol. 93 : 129 -
137, 1979.

37265 - DE GREEF, J., VERBELEN, J.P., CAUBERGS, R., MOEREELS, E., SPRUYT, E. : Comparative study of photosynthetic efficiency and leaf architecture during
leaf development. - In : MARCELLE, R., CLIJSTERS, H., VAN POUCKE, M. (ed.) :
Photosynthesis and Plant Development. Pp. 49 - 56. Dr.W.Junk bv. Publ.,
The Hague - Boston - London 1979.

37266 - DEINUM, B., KNOPPERS, J. : The growth of maize in the cool temperate climate
of the Netherlands : Effect of grain filling on production of dry matter and
on chemical composition and nutritive value. - Neth. J. agr. Sci. 27 : 116 -
130, 1979.

37267 - DE JONG, D.W., WOODLIEF, W.G. : Some factors influencing tobacco leaf senescence. - Beitr. Tabakforsch. 10 : 48 - 56, 1979. [Chl.]

37268 - DE JONG, T.M., BARBOUR, M.G. : Contributions to the biology of Atriplex leu-
cophylla, a C_4 Californian beach plant. - Bull. Torrey bot. Club 106 : 9 - 19,
1979. [Ps.]

37269 - DeJONG, T.M., DRAKE, B.G. : Comparative laboratory and field gas exchange
responses of C_3 and C_4 tidal marsh species. - Plant Physiol. 63 (Suppl.) :
63, 1979.

37270 - DEKOV, I., KUDREV, T., PETROVA, L. : Influence of magnesium deficiency and
water stress on ultrastructural changes of chloroplasts and some parameters
of water regime in maize plants. - In : Mineral Nutrition of Plants. Vol. I.
Pp. 133 - 138. Publ. House Central Cooperative Union, Sofia 1979.

37271 - DÉ LA HARPE, A.C., VISSER, J.H., GROBBELAAR, N. : The chlorophyll concentration and photosynthetic activity of some parasitic flowering plants. - Z.
Pflanzenphysiol. 93 : 83 - 87, 1979.

*37272 - DE LA TORRE,A.,CHUECA,A.,LÓPEZ GORGÉ,J.: Isolation and properties of crystal-
line ferredoxin from Lactuca sativa. - Phytochemistry 17 : 35 - 39, 1978.

37273 - **DE LA TORRE, A., LARA, C., WOLOSIUK, R.A., BUCHANAN, B.B.** : Ferredoxin-thio-
redoxin reductase : A chromophore-free protein of chloroplasts. - FEBS Lett.
107 : 141 - 145, 1979.

37274 - **DELEENS, E., GARNIER-DARDART, J., QUEIROZ, O.** : Carbon isotope composition
of intermediates of the starch-malate sequence and level of the crassulacean
acid metabolism in leaves of *Kalanchoe blossfeldiana* TOM THUMB. - Planta
146 : 441 - 449, 1979.

37275 - **DELEPELAIRE, P., CHUA, N.-H.** :   Lithium dodecyl sulfate/polyacrylamide gel
electrophoresis of thylakoid membranes at 4 °C : Characterizations of two
additional chlorophyll *a*-protein complexes. - Proc.nat. Acad. Sci. USA *76* :
111 - 115, 1979.

*37276 - **DEL RIO, L.A., GOMEZ, M., YAÑEZ, J., LEAL, A., LOPEZ GORGE, J.** : Iron deficiency in pea plants. Effect on catalase, peroxidase, chlorophyll and proteins of leaves. - Plant Soil *49* : 343 - 353, 1978.

37277 - **DELROT, S., BONNEMAIN, J.-L.** : Le transport latéral du carbone assimilé. -
Physiol. vég. *17* : 247 - 270, 1979.

37278 - **DEMERS, S., LAFLEUR, P.E., LEGENDRE, L., TRUMP, C.K.** : Short-term covaria-
bility of chlorophyll and temperature in the St. Lawrence Estuary. - J. Fish.
Res. Board Can. *36* : 568 - 573, 1979.

37279 - **DEMERS, S., LEGENDRE, L.** : Effets des marées sur la variation circadienne
de la capacité photosynthétique du phytoplancton de l'estuaire du Saint-Lau-
rent. - J. exp. mar. Biol. Ecol. *39* : 87 - 99, 1979.

37280 - **DEMETER, S., HERCZEG, T., DROPPA, M., HORVATH, G.** : Thermoluminescence characteristics of granal and agranal chloroplasts of maize. - FEBS Lett. *100* :
321 - 324, 1979.

37281 - **D'EMILIO, M.A., HOURSIANGOU-NEUBRUN, D., BAUGNET-MAHIEU, L., GILLES, J.,
NUYTS, G., BOSSUS, A., MAZZA, A., BONOTTO, S.** : Apicobasal gradient of pro-
tein synthesis in *Acetabularia*. - In : BONOTTO, S., KEFELI, V., PUISEUX-DAO,
S. (ed.) : Developmental Biology of *Acetabularia*. Pp. 269 - 282. Elsevier/
North-Holland Biomedical Press, Amsterdam - New York - Oxford 1979. [Chlo-
roplast.]

37282 - **DEMMIG, B., GIMMLER, H.** : Effect of divalent cations on cation fluxes across
chloroplast envelope and on photosynthesis of intact chloroplasts. - Z. Naturforsch. *34 C* : 233 - 241, 1979.

37283 - **DENCHER, N.A., HEYN, M.P.** : Bacteriorhodopsin monomers pump protons. - FEBS
Lett. *108* : 307 - 310, 1979.

37284 - **DENCHER, N.A., HILDEBRAND, E.** : Sensory transduction in *Halobacterium halobium* : Retinal protein pigment controls UV-induced behavioral response. - Z.
Naturforsch. *34 C* : 841 - 847, 1979.

37285 - **DENESH, M., ANDRIANOV, V.K., BULYCHEV, A.A., KURELLA, G.A.** : Fotoindutsirovannyĭ transport H+ v kletkakh *Nitellopsis obtusa*. [Photoinduced H+-trans-
port in cells of *Nitellopsis obtusa*.] - Biofizika 24 : 657 - 662, 1979. [In
R, ab : E.]

37286 - **DeNIRO, M.J., EPSTEIN, S.** : Relationship between the oxygen isotope ratios
of terrestrial plant ćellulose, carbon dioxide, and water. - Science *204* :
51 - 53, 1979.

37287 - **DEPUIT, E.J.** : Photosynthesis and respiration of plants in the arid ecosystem. -
In : PERRY, R.A., GOODALL, D.W. (ed.) : Arid-land Ecosystems : Structure,
Functioning and Management. Vol. 1. Pp. 509 - 536. Cambridge University
Press, Cambridge 1979.

37288 - **DERYABKIN, V.N., KIREEV, V.B., SKACHKOV, M.P., TRUKHAN, É.M.** : SVCH-foto-
provodimost' v list'yakh rasteniĭ. [Microwave photoconductance in plant
leaves.] - Biofizika 24 : 1026 - 1029, 1979. [Ps; in R, ab : E.]

37289 - **DETLING, J.K.** : Processes controlling blue grama production on the shortgrass prairie. - In : FRENCH, N.R. (ed.) : Perspectives in Grassland Ecology. Pp. 25 - 42. Springer-Verlag, New York - Heidelberg - Berlin 1979.

37290 - DETLING, J.K., DYER, M.I., WINN, D.T. : Net photosynthesis, root respiration, and regrowth of *Bouteloua gracilis* following simulated grazing. - Oecologia *41* : 127 - 134, 1979.

37291 - DETLING, J.K., DYER, M.I., WINN, D.T. : Effect of simulated grasshopper grazing on CO_2 exchange rates of western wheatgrass leaves. - J. econ. Entomol. *72* : 403 - 406, 1979.

37292 - DETLING, J.K., PARTON, W.J., HUNT, H.W. : A simulation model of *Bouteloua gracilis* biomass dynamics on the North American shortgrass prairie. - Oecologia *38* : 167 - 191, 1979.

37293 - DE VILLIERS, O.T. : The metabolism of sorbitol and fructose in isolated chloroplasts of Santa Rosa plum leaves. - Agroplantae *11* : 25 - 26, 1979.

37294 - DEVLIN, R.M., KISIEL, M.J., KOSTUSIAK, A.S. : Influence of R-40244 on pigment content of wheat and corn. - Weed Res. *19* : 59 - 61, 1979.

37295 - DEVLIN, R.M., KISIEL, M.J., KOSTUSIAK, A.S. : The blocking of carotenoid synthesis in .corn with the experimental herbicide R-40244. - Proc. northeast. Weed Sci. Soc. *33* : 95 - 99, 1979.

*37296 - DE VOS, N.M. : Cultivar differences in plant and crop photosynthesis. - In : Crop Physiology and Cereal Breeding. (Proc. Eucarpia Workshop.) Pp. 71 - 74. Wageningen 1978.

37297 - DEWDNEY, S.J., McWHA, J.A. : Abscisic acid and the movement of photosynthetic assimilates towards developing wheat (*Triticum aestivum* L.) grains. - Z. Pflanzenphysiol. *92* : 183 - 186, 1979.

37298 - DE WIT, C.T., VAN LAAR, H.H., VAN KEULEN, H. : Physiological potential of crop production. - In : SNEEP, J., HENDRIKSEN, A.J.T. (ed.) : Plant Breeding Perspectives. Pp. 47 - 82. Pudoc, Wageningen 1979. [Ps.]

*B37299 - DE WIT, C.T. *et al.* : Simulation of Assimilation, Respiration and Transpiration of Crops. - PUDOC, Wageningen 1978.

37300 - DEY, R., VAN ALFEN, N.K. : Influence of *Corynebacterium insidiosum* on water relations of alfalfa. - Phytopathology *69* : 942 - 946, 1979. [Stomatal resistance.]

37301 - DE YOE, D.R., BROWN, G.N. : Glycerolipid and fatty acid changes in eastern white pine chloroplast lamellae during the onset of winter. - Plant Physiol. *64* : 924 - 929, 1979.

37302 - DHILLON, G.S., SINGH, B., KLER, D.S. : Efficient use of solar energy for crop production I. Effect of row-direction on wheat yield with different sowing dates, plant populations and fertilizer levels. - Indian J. Agron. *24* : 322 - 325, 1979. [Growth analysis.]

37303 - DiCAMELLI, C.A., OUTLAW, W.H. : Isolation and characterization of guard cell protoplast phosphoenolpyruvate carboxylase. - Plant Physiol. *63* (Suppl.) : 60, 1979.

37304 - DICKSON, R.E. : Analytical procedures for the sequential extraction of ^{14}C-labeled constituents from leaves, bark and wood of cottonwood plants. - Physiol. Plant. *45* : 480 - 488, 1979.

37305 - DIJKMANS, H., AGHION, J. : Fixation of the chloroplast coupling factor (CF_1) to lipid vesicles. Role of a subunit of CF_0. - Arch. int. Physiol. Biochim. *87* : 1021, 1979.

37306 - DIJKMANS, H., COGNIAUX, F., AGHION, J. : β-carotene or chlorophyll *a* incorporated in lecithin liposomes. Analysis by ultracentrifugation. - Biochem. biophys. Res. Commun. *89* : 1141 - 1145, 1979.

37307 - DIJKMANS, H., LEBLANC, R.M., COGNIAUX, F., AGHION, J. : Properties of chlorophyll-lecithin vesicles : Ultracentrifugation, absorbance, emission and photobleaching. - Photochem. Photobiol. *29* : 367 - 372, 1979.

37308 - DILKS, T.J.K., PROCTOR, M.C.F. : Photosynthesis, respiration and water content in bryophytes. - New Phytol. *82* : 97 - 114, 1979.

37309 - **DILOVA, S.** : Resistance of carotenoids after repeated darkening of post-etio-
lated barley seedlings. - Photosynthetica *13* : 163 - 166, 1979.

37310 - **DILWORTH, M.F., GANTT, E.** : Isolation of active phycobilisome-thylakoid vesicles from *Porphyridium cruentum*. - Plant Physiol. *63* (Suppl.) : 29, 1979.

37311 - **DIMITROVA, O., POPOVA-STAEVSKA, L.** : Vliyanie na nyakoi metaboliti v"rkhu
aktivnostta na ribulozodifosfat karboksilazata v C_3 i C_4 tip rasteniya. [In-
fluence of some metabolites on ribulose bisphosphate carboxylase activity in
C_3 and C_4 plants.] - Fiziol. Rast. (Sofia) *5* (4) : 24 - 29, 1979. [In Bulg.,
ab : E, R.]

37312 - **DIMON, B., GERSTER, R.** : Incorporation d'oxygène dans le glycolate excrété
à la lumière par *Euglena gracilis*. - Compt. rend. Acad. Sci. Paris, Sér. D
283 : 507 - 510, 1976.

37313 - **DINAR, M., RUDICH, J.** : Partitioning of photosynthates in tomatoes grown
under high temperatures. - Plant Physiol. *63* (Suppl.) : 103, 1979.

37314 - **DINER, B.A.** : Energy transfer from the phycobilisomes to photosystem II reaction centers in wild type *Cyanidium caldarium*. - Plant Physiol. *63* : 30 -
- 34, 1979.

37315 - **DINER, B.A., WOLLMAN, F.-A.** : Functional comparison of the photosystem II
center-antenna complex of a phycocyanin-less mutant of *Cyanidium caldarium*
with that of *Chlorella pyrenoidosa*. - Plant Physiol. *63* : 20 - 25, 1979.

37316 - **DINER, B.A., WOLLMAN, F.-A.** : Biosynthesis of photosystem II reaction centers, antenna and plastoquinone pool in greening cells of *Cyanidium caldarium* mutant III-C. - Plant Physiol. *63* : 26 - 29, 1979.

37317 - **DITTRICH, P.** : Der Weg des Kohlenstoffs im Crassulaceen-Säurestoffwechsel. -
Ber. deut. bot. Ges. *92* : 109 - 116, 1979.

37318 - **DITTRICH, P.** : Enzymes of Crassulacean acid metabolism. - In : GIBBS, M.,
LATZKO, E. (ed.) : Photosynthesis II. (Encycl. Plant Physiol. N.S. Vol. 6.)
Pp. 263 - 270. Springer-Verlag, Berlin - Heidelberg - New York 1979.

37319 - **DITTRICH, P., MAYER, M., MEUSEL, M.** : Proton-stimulated opening of stomata
in relation to chloride uptake by guard cells. - Planta *144* : 305 - 309,
1979. [PEPC.]

*37320 - **DMITRIEV, A.P., GLUSHCHA, N.I., GRODZINSKIĬ, D.M.** : Vnutrikletochnyĭ kislo-
rod i radiosensibilizatsiya kletok sineazeleno̐ĭ vodorosli *Anacystis nidulans*.
[Intracellular oxygen and radiosensitization of the blue-green alga *Anacystis nidulans*.] - Mikrobiologiya *47* : 557 - 560, 1978. [Ps.]

*37321 - **DMITRIEV, V.P., NYUPPIEVA, K.A.** : Vliyanie retardantov na soderzhanie khlorofilla v list'yakh kartofelya i ikh zamorozkousto̐ĭchivost'. [Effect of retardants on chlorophyll content in potato leaves and their resistance to low
temperature.] - In : Ekologo-Fiziologicheskie Mekhanizmy Usto̐ĭchivosti Rasteniĭ k De̐ĭstviyu Ekstremal'nykh Temperatur. Pp. 120 - 128, 167. Karel'skiĭ
Filial Akad. Nauk SSSR, Petrozavodsk 1978. [In R.]

37322 - **DOCKERTY, A., MERRETT, M.J.** : Isolation and enzymic characterization of
Euglena proplastids. - Plant Physiol. *63* : 468 - 473, 1979. [Chl.]

37323 - **DODD, J.L., LAUENROTH, W.K.** : Analysis of the response of a grassland ecosystem to stress. - In : FRENCH, N.R. (ed.) : Perspectives in Grassland Ecology. Pp. 43 - 58. Springer-Verlag, New York - Heidelberg - Berlin 1979.

37324 - **DODELET, J.P., BRECH, J. Le, LEBLANC, R.M.** : Photovoltaic efficiencies of
microcrystalline and anhydrous chlorophyll *a*. - Photochem. Photobiol. *29* :
1135 - 1145, 1979.

37325 - **DOEHLERT, D.C., KU, M.S.B., EDWARDS, G.E.** : Dependence of the post-illumi-
nation burst of CO_2 on temperature, light, CO_2 and O_2 concentration in wheat
(*Triticum aestivum*). - Physiol. Plant. *46* : 299 - 306, 1979.

37326 - **DOHERTY, A., GRAY, J.C.** : Synthesis of cytochrome *f* by isolated pea chloroplasts. - Europe. J. Biochem. *98* : 87 - 92, 1979.

37327 - **DÖHLER, G., ROSSLENBROICH, H.-J.** : Diurnal variation in photosynthetic CO_2 fixation of marine phytoplankton populations. - Z. Pflanzenphysiol. *94* : 417 - 425, 1979.

37328 - **DOKULIL, M.** : Optical properties, colour and turbidity. - In : LÖFFLER, H. (ed.) : Neusiedlersee : The Limnology of a Shallow Lake in Central Europe. Pp. 151 - 167. Dr.W.Junk bv. Publ., The Hague - Boston - London 1979. [Chl.]

37329 - **DOKULIL, M.** : Seasonal pattern of phytoplankton. - In : LÖFFLER, H. (ed.) : Neusiedlersee : The Limnology of a Shallow Lake in Central Europe. Pp. 203 - 231. Dr.W.Junk bv. Publ., The Hague - Boston - London 1979.

37330 - **DOKULIL, M.** : Phytoplankton primary production. - In : LÖFFLER, H. (ed.) : Neusiedlersee : The Limnology of a Shalow Lake in Central Europe. Pp. 233 - 234. Dr.W.Junk bv. Publishers, The Hague - Boston - London 1979.

37331 - **DOLINER, L.H., JOLLIFFE, P.A.** : Ecological evidence concerning the adaptive significance of the C_4 dicarboxylic acid pathway of photosynthesis. - Oecologia *38* : 23 - 34, 1979.

37332 - **DOMBROVSKIĬ, Yu.A., MARKMAN, G.S.** : Prostranstvenno-neodnorodnoe raspredelenie fitoplanktona v modeli reguliruemoĭ populyatsii. [Patchness of phytoplankton distribution in a model of regulated population.] - Biofizika *24* : 893 - 896, 1979. [In R, ab : E.]

37333 - **DOMINY, P.J., BAKER, N.R.** : Effect of salinity on PSII primary photochemistry of pea and sea beet. - Plant Physiol. *63* (Suppl.) : 89, 1979.

37334 - **DORAISWAMY, P.C., HODGES, T., PHINNEY, D.E.** : Crop yield literature review for AgRISTARS crops. Corn, soybeans, wheat, barley, sorghum, rice, cotton, and sunflowers. - In : Tech. Rep. No. SR-L9-00405; JSC- 6320. Lockheed Eng. Manage. Serv. Co., Houston, Texas 1979. [Ps.]

37335 - **DOUCE, R., JOYARD, J.** : Isolation and properties of the envelope of spinach chloroplasts. - In : REID, E. (ed.) : Plant Organelles. Pp. 47 - 59. Ellis Horwood, Chichester 1979.

37336 - **DOUCE, R., JOYARD, J.** : Structure and function of the plastid envelope. - Adv. bot. Res. *7* : 1 - 116, 1979.

37337 - **DOUCE, R., JOYARD, J.** : The chloroplast envelope: an unusual cell membrane system. - In : APPELQVIST, L.-Å., LILJENBERG, C. (ed.) : Advances in the Biochemistry and Physiology of Plant Lipids. Pp. 79 - 98. Elsevier/North-Holland Biomedical Press, Amsterdam 1979.

37338 - **DOUGHERTY, P.M., TESKEY, R.O., PHELPS, J.E., HINCKLEY, T.M.** : Net photosynthesis and early growth trends of a dominant white oak (*Quercus alba* L.). - Plant Physiol. *64* : 930 - 935, 1979.

37339 - **DOUILLARD, R., BERGERON, É.** : Activité lipogénasique foliaire : localisation chloroplastique et variations en fonction de l'âge des jeunes plantes. - Physiol. vég. *17* : 457 - 476, 1979.

37340 - **DOUILLARD, R., BERGERON, E.** : Quelques caractéristiques fonctionnelles des activités lipoxygénasiques foliaires solubles ou lamellaire. Conséquences structurales et fonctionnelles. - Physiol. vég. *17* : 749 - 768, 1979. [Chloroplast.]

37341 - **DOUILLARD, R., BERGERON, E.** : Lipoxygenase activity distribution in young wheat chloroplast lamellae. - In : APPELQVIST, L.-Å., LILJENBERG, C. (ed.) : Advances in the Biochemistry and Physiology of Plant Lipids. Pp. 159 - 164. Elsevier/North-Holland Biomedical Press, Amsterdam 1979.

37342 - **DOVNAR, V.S.** : K metodike izmereniya ploshchadi list'ev u zlakovykh kul'tur. [Methods of measuring the leaf surface in cereal crops.] - Sel'skokhoz. Biol. *14* : 235 - 237, 1979. [In R, ab : E.]

37343 - **DOYLE, A.D., FISCHER, R.A.** : Dry matter accumulation and water use relationship in wheat crops. - Aust. J. agr. Res. *30* : 815 - 829, 1979.

37344 - **DRABER, W.** : Neuere Ergebnisse zur Struktur-Aktivitäts-Beziehung. - Z. Naturforsch. *34 C* : 973 - 978, 1979. [Ps.]

37345 - DRACHEV, L.A., KAULEN, A.D., SAMUILOV, V.D., SEVERINA, I.I., SEMENOV, A.Yu.,
         SKULACHEV, V.P., CHEKULAEVA, L.N. : Vstraivanie proteoliposom i khromatoforov
         v membrany na osnove fil'trov. [Incorporation of proteoliposomes and chroma-
         tophores into membranes based on filters.] - Biofizika 24 : 1035 - 1042, 1979.
         [In R, ab : E.]

37346 - DRASKOVITS, R.M. : Light and pigments investigations on species in a Hunga-
         rian beechwood. - Acta bot. Acad. Sci. hung. 25 : 309 - 324, 1979.

37347 - DREW, A.P., BAZZAZ, F.A. : Response of stomatal resistance and photosynthe-
         sis to night temperature in Populus deltoides. - Oecologia 41 : 89 - 98,
         1979.

37348 - DREW, E.A. : Physiological aspects of primary production in seagrasses. -
         Aquat. Bot. 7 : 139 - 150, 1979.

37349 - DREW, M.C., SISWORO, E.J., SAKER, L.R. : Alleviation of waterlogging damage
         to young barley plants by application of nitrate and a synthetic cytokinin,
         and comparison between the effects of waterlogging, nitrogen deficiency and
         root excision. - New Phytol. 82 : 315 - 329, 1979. [Chl.]

37350 - DREWS, G., WEVERS, P., DIERSTEIN, R. : Studies on the photosynthetically
         inactive mutant Ala+pho- of Rhodopseudomonas capsulata which synthesizes
         B870. - FEMS Microbiol. Lett. 5 : 139 - 142, 1979.

37351 - DROMGOOLE, F.I. : Photosynthetic and respiratory transients of oxygen exchan-
         ge in Carpophyllum species (Fucales, Phaeophyceae). - Aquat. Bot. 6 : 133 - 147,
         1979.

37352 - DROZDOV, S.N., KURETS, V.K., POPOV, É.G. : Mnogofaktornyĭ metod modelirovani-
         ya produktivnosti rasteniĭ. [Multiple-factor method for modelling plant pro-
         ductivity.] - Fiziol. Biokhim. kul't. Rast. 11 : 164 - 168, 1979. [In R,
         ab : E.]

37353 - DRUCKMANN, S., SAMUNI, A., OTTOLENGHI, M. : Dynamics of pH-induced spectral
         changes in bacteriorhodopsin. - Biophys. J. 26 : 143 - 145, 1979.

37354 - DUBAY, C.I., SIMMONS, G.M.Jr. : The contribution of macrophytes to the meta-
         limnetic oxygen maximum in a montane, oligotrophic lake. - Amer. Midland Na-
         turalist 101 : 108 - 117, 1979. [Ps.]

37355 - DUBINSKY, Z., BERMAN, T. : Seasonal changes in the spectral composition of
         downwelling irradiance in Lake Kinneret (Israel). - Limnol. Oceanogr. 24 :
         652 - 663, 1979. [Chl.]

37356 - DUGGAN, J.X., GASSMAN, M.L. : Oxidation and metabolism of 5-aminolevulinic
         acid, levulinic acid, and porphobilinogen by shoots of Hordeum vulgare. -
         Plant Physiol. 63 (Suppl.) : 161, 1979.

37357 - DUGGAN, J.X., GASSMAN, M.L. : Oxidation of ^{14}C-porphobilinogen to $^{14}CO_2$ by
         extracts of etiolated shoots of Hordeum vulgare. - Plant Physiol. 63 (Suppl.)
         : 161, 1979.

37358 - DUJARDIN, E., CORREIA, M. : Long-wavelength absorbing pigment-protein com-
         plexes as fluorescence quenchers in etiolated leaves illuminated in liquid
         nitrogen. - Photobiochem. Photobiophys. 1 : 25 - 32, 1979.

37359 - DUKE, S.H., DUKE, S.O. : The photosynthetic independence of light-induced
         nitrate reductase activity. - Plant Physiol. 63 (Suppl.) : 45, 1979.

37360 - DUKE, S.H., SCHRADER, L.E., HENSON, C.A., SERVAITES, J.C., VOGELZANG, R.D.,
         PENDLETON, J.W. : Low root temperature effects on soybean nitrogen metabolism
         and photosynthesis. - Plant Physiol. 63 : 956 - 962, 1979.

37361 - DUKE, S.O., DUKE, S.H. : Photosynthetic independence of initial light-caused
         increase in extractable nitrate reductase activity from maize seedlings. -
         Plant Cell Physiol. 20 : 1371 - 1380, 1979.

37362 - DUKE, S.O., PAUL, R.N., WICKLIFF, J.L. : Tentoxin effects on greening of ivy-
         leaf morningglory [Ipomoea hederacea (L.) JACQ. var. hederacea] cotyledons. -
         Plant Physiol. 63 (Suppl.) : 106, 1979.

37363 - **DUNAEVA, S.E.** : Khloroplasty list'ev raznykh yarusov v ontogeneze pshenitsy
(ultrastruktura). [Chloroplasts in leaves of different insertion levels during wheat ontogeny (ultrastructure).] - Byull. vsesoyuz. nauch.-issled. Inst.
Rastenievod. Im. N.I. Vavilova *87* : 27 - 33, 1979. [In R.]

37364 - **DUNAEVA, S.E.** : Ul'trastruktura khloroplastov pshenitsy v svyazi s vozrastom
lista. [Ultrastructure of wheat chloroplasts in connection with leaf age.] -
Tsitologiya *21* : 5 - 11, 1979. [In R, ab : E.]

37365 - **DUNIEC, J.T., SCULLEY, M.J., THORNE, S.W.** : An analysis of the effect of mo-
no- and di-valent cations on the forces between charged lipid membranes with
special reference to the grana thylakoids of chloroplasts. - J. theor. Biol.
79 : 473 - 484, 1979.

37366 - **DUNKLEY, P.R., ANDERSON, J.M.** : Isolation of the light-harvesting chlorophyll
a/b-protein complex from thylakoid membranes of barley by adsorption chromatography on controlled-pore glass. - Arch. Biochem. Biophys. *193* : 469 - 477,
1979.

37367 - **DUNKLEY, P.R., ANDERSON, J.M.** : The light-harvesting chlorophyll *a/b*-protein
complex from barley thylakoid membranes. Polypeptide composition and characterization of an oligomer. - Biochim. biophys. Acta *545* : 175 - 187, 1979.

37368 - **DÜRING, H.** : Wirkungen der Luft- und Bodenfeuchtigkeit auf das vegetative
Wachstum und den Wasserhaushalt bei Reben. - Vitis *18* : 211 - 220, 1979. [Sto-
matal resistance.]

37369 - **DUSTAN, P.** : Distribution of zooxanthellae and photosynthetic chloroplast
pigments of the reef-building coral *Montastrea annularis* ELLIS and SOLANDER
in relation to depth on a West Indian coral reef. - Bull. mar. Sci. *29* : 79 -
- 95, 1979.

37370 - **DUTTON, P.L., LEIGH, J.S.Jr., PRINCE, R.C., TIEDE, D.M.** : The photochemical
reaction center of photosynthetic bacteria as a model for studying biological
charge separation and electron transfer. - In : **GERISCHER, H., KATZ, J.J.**
(ed.) : Light-Induced Charge Separation in Biology and Chemistry. Pp. 411 -
- 448. Verlag Chemie, Weinheim - New York 1979.

37371 - **DUVAL, J.C., TREMOLIERES, A., DUBACQ, J.P.** : The possible role of transhexa-
decenoic acid and phosphatidylglycerol in light reactions of photosynthesis:
The photochemistry and fluorescence properties of young pea leaf chloroplasts
treated by phospholipase A_2. - FEBS Lett. *106* : 414 - 418, 1979.

37372 - **DUYSEN, M.E., FREEMAN, T.P.** : Wheat chloroplast development and coupled electron transport as indicated by photosynthetic control (PC). - Plant Physiol.
63 (Suppl.) : 30, 1979.

37373 - **DUYSENS, L.N.M.** : Transfer and trapping of excitation energy in photosystem
II. - In : Chlorophyll Organization and Energy Transfer in Photosynthesis.
Pp. 323 - 340. Excerpta Medica, Amsterdam - Oxford - New York 1979.

37374 - **DVORNIKOV, S.S., KNYUKSHTO, V.N., SOLOV'EV, K.N., TSVIRKO, M.P.** : Fosfores-
tsentsiya khlorofillov *a* i *b* i ikh feofitinov. [Phosphorescence of chlorophylls *a* and *b* and their pheophytins.] - Optika Spektroskop. *46* : 689 - 695,
1979. [In R.]

37375 - **DVORNIKOV, S.S., KNYUKSHTO, V.N., SOLOVYOV, K.N., TSVIRKO, M.P.** : Phospho-
rescence polarization and the electronic structure of the lowest triplet
state of photosynthetic pigments. - J. Luminescence *18/19* : 491 - 494, 1979.

37376 - **DVORNIKOV, S.S., SOLOV'EV, K.N., TSVIRKO, M.P.** : Vliyanie dopolnitel'nogo
kompleksoobrazovaniya na spektral'no-lyuminestsentnye svoĭstva Mg-porfirinov.
[Effect of further complexing on the spectral-luminescent properties of magnesium porphyrins.] - Biofizika *24* : 791 - 796, 1979. [In R, ab : E.]

37377 - **DWIVEDI, S., KAR, M., MISHRA, D.** : Biochemical changes in excised leaves of
Oryza sativa subjected to water stress. - Physiol. Plant. *45* : 35 - 40, 1979.
[Chl.]

37378 - **DWIVEDI, S., KAR, M., MISHRA, D.** : Inorganic pyrophosphatase activity in water stressed excised rice leaves. - Irrig. Sci. *1* : 119 - 124, 1979. [Chl.]

37379 - DYKES, M.G., DYCK, L.A. : Light induced chloroplast migrations in *Eremosphaera viridis (Chlorophyceae)*. - J. Phycol. *15* (Suppl.) : 25, 1979.

37380 - DYUTIN, K.E. : Nasledovanie priznaka zhelto-zelenaya okraska molodykh list'ev dyni. [Inheritance of yellow-green coloration character of new leaves in melon.] - Tsitol. Genet. *13* : 407 - 408, 1979. [In R, ab : E.]

37381 - DZEVYATAŬ, A.S., GORNY, A.U. : Uplyŭ aryentatsyi radoŭ sadu s ploskasnymi kronami na intênsiŭnasts' fotasintêzu listsyaŭ i têmperaturny rêzhym krony yablyni. [Effect of row orientation in an hedgerow orchard on leaf photosynthesis and temperature regime of apple-tree crown.] - Vestsi Akad. Navuk belorus. SSR, Ser. biyal. Navuk *1979* (4) : 16 - 21, 137- 138, 1979. [In Belorus., ab : E, R.]

37382 - DZYBOV, D.S. : Metod opredeleniya ploshchadi osnovaniĭ rasteniĭ. [A method for measuring the basal area of plants.] - Bot. Zh. *64* : 1762 - 1768, 1979. [In R.]

*B37383 - EASTIN, J.D., HASKINS, F.A., SULLIVAN, C.Y., VAN BAVEL, C.H.M. (ed.) : Physiological Aspects of Crop Yield. - Amer. Soc. Agron. and Crop Sci. Soc. Amer., Madison, Wisc. 1969. [Ps.]

37384 - ECK, H.V., MUSICK, J.T. : Plant water stress effects on irrigated grain sorghum. II. Effects on nutrients in plant tissues. - Crop Sci. *19* : 592 - 598, 1979. [Dry matter production.]

37385 - ECKERT, H.J., BUCHWALD, H.E., RENGER, G. : Investigation of the reactions of chlorophyll-a_{II} in class II chloroplasts under repetitive double flash group excitation. - FEBS Lett. *103* : 291 - 295, 1979.

37386 - ÉCOCHARD, R., GALLAIS, A., PAUL, M.H., PLANCHON, C. : Héritabilité et réponse à la sélection de caractères physiologiques liés au rendement chez le Soja. - Ann. Amélior. Plant. *29* : 493 - 514, 1979. [Biomass, foliage area.]

37387 - EDER, A., STICHLER, W., ZIEGLER, H. : Typen der photosynthetischen CO_2-Fixierung bei mitteleuropäischen *Euphorbia*-Arten. - Flora *168* : 227 - 240, 1979.

37388 - EDER, F.A., Pyridazinones, their influence on the biosynthesis of carotenoids and the metabolism of lipids in plants (survey of literature). - Z. Naturforsch. *34 C* : 1052 - 1054, 1979.

37389 - EDMEADES, G.O., DAYNARD, T.B. : The relationship between final yield and photosynthesis at flowering in individual maize plants. - Can. J. Plant Sci. *59* : 585 - 601, 1979.

37390 - EDMEADES, G.O., FAIREY, N.A., DAYNARD, T.B. : Influence of plant density on the distribution of ^{14}C-labelled assimilate in maize at flowering. - Can. J. Plant Sci. *59* : 577 - 584, 1979.

37391 - EDWARDS, G.E., HUBER, S.C. : C_4 metabolism in isolated cells and protoplasts. - GIBBS, M., LATZKO, E. (ed.) : Photosynthesis II. (Encycl. Plant Physiol. N.S. Vol. 6.)Pp. 102 - 112. Springer-Verlag, Berlin - Heidelberg - New York 1979.

37392 - EDWARDS, G.E., KU, S.B., HATCH, M.D. : Isolation of bundle sheath protoplasts from C4 species. - Plant Physiol. *63* (Suppl.) : 63, 1979.

37393 - EDWARDS, G.E., LILLEY, R.McC., CRAIG, S., HATCH, M.D. : Isolation of intact and functional chloroplasts from mesophyll and bundle sheath protoplasts of the C4 plant *Panicum miliaceum*. - Plant Physiol. *63* : 821 - 827, 1979.

37394 - EDWARDS, M.B., McNAB, W.H. : Biomass prediction for young southern pines. - J. Forest.*77* : 291 - 292, 1979.

37395 - EGED, Š., JEŠKO, T. : Svetelné krivky fotosyntézy listov druhu *Pulmonaria officinalis* L. [Light curves of photosynthesis of leaves of *Pulmonaria officinalis* L.] - Biológia (Bratislava) *34* : 541 - 546, 1979. [In Slov., ab : E, R.]

37396 - EGUCHI, H., HAMAKOGA, M., MATSUI, T. : Computer control of plant growth by image processing. IV. Digital image processing of reflectance in different wave length regions of light for evaluating vigor of plants. - Environ. Control Biol. *17* (2) : 67 - 77, 1979.

37397 - EHLERINGER, J., MOONEY, H.A., BERRY, J.A. : Photosynthesis and microclimate of *Camissonia claviformis*, a desert winter annual. - Ecology *60* : 280 - 286, 1979.

37398 - EHLERINGER, J.R. : Photosynthesis and photorespiration : Biochemistry, physiology, and ecological implications. - HortScience *14* : 217 - 222, 1979.

37399 - EICKMEIER, W.G. : Eco-physiological differences between high and low elevation CAM species in Big Bend National Park, Texas. - Amer. Midl. Natur. *101* : 118 - 126, 1979.

37400 - EICKMEIER, W.G. : Photosynthetic recovery in the resurrection plant *Selaginella lepidophylla* after wetting. - Oecologia *39* : 93 - 106, 1979.

37401 - EINHELLIG, F.A., RASMUSSEN, J.A. : Effects of three phenolic acids on chlorophyll content and growth of soybean and grain sorghum seedlings. - J. chem. Ecol. *5* : 815 - 824, 1979.

37402 - EISBRENNER, G., BOTHE, H. : Modes of electron transfer from molecular hydrogen in *Anabaena cylindrica*. - Arch. Microbiol. *123* : 37 - 45, 1979.

37403 - EISENBACH, L., EISENBACH, M. : Electrophoretic mobility of membrane fragments on a sucrose gradient. Application to isolated purple membrane fragments from *Halobacterium halobium*. - Anal. Biochem. *92* : 228 - 232, 1979.

37404 - EISENBACH, M., CAPLAN, S.R. : Interaction of purple membrane with solvents II. Mode of interaction. - Biochim. biophys. Acta *554* : 281 - 292, 1979.

37405 - EISENBACH, M., CAPLAN, S.R., TANNY, G. : Interaction of purple membrane with solvents I. Applicability of solubility parameter mapping. - Biochim. biophys. Acta *554* : 269 - 280, 1979.

37406 - EK, A.R. : A model for estimating branch weight and branch leaf weight in biomass studies. - Forest Sci. *25* : 303 - 306, 1979.

37407 - EK, A.R., MONSERUD, R.A. : Performance and comparison of stand growth models based on individual tree and diameter-class growth. - Can. J. Forest Res. *9* : 231 - 244, 1979.

37408 - EKÉS, M. : Luminal chloroplast connections with endoplasmic reticulum-like cisterns and cell wall in *Dryopteris filix-mas* gametophytes. - Acta biol. Acad. Sci. hung. *30* : 201 - 207, 1979.

37409 - ELFERINK, M.G.L., HELLINGWERF, K.J., MICHELS, P.A.M., SEŸEN, H.G., KONINGS, W.N. : Immunochemical analysis of membrane vesicles and chromatophores of *Rhodopseudomonas sphaeroides* by crossed immunoelectrophoresis. - FEBS Lett. *107* : 300 - 307, 1979.

37410 - EL HAMOURI, B., SIRONVAL, C. : A new non-photoreducible protochlorophyll(ide)-protein : P-649-642 from cucumber cotyledons. NADPH mediation of its transformation to photoreducible P-657-650. - FEBS Lett. *103* : 345 - 347, 1979.

37411 - ELIAS, B.A., GIVAN, C.V. : Localization of pyruvate dehydrogenase complex in *Pisum sativum* chloroplasts. - Plant Sci. Lett. *17* : 115 - 122, 1979.

37412 - ELIAS, J.E., GAGIANAS, A.A., GERAKIS, P.A. : Interrelationships and plasticity of growth parameters in *Zea mays* L. populations as influenced by density and nitrogen. - Oecol. Plant. *14* : 159 - 168, 1979. [Growth analysis.]

37413 - ELIÁŠ, P. : Leaf diffusion resistance pattern in an oak-hornbeam forest. - Biol. Plant. *21* : 1 - 8, 1979.

37414 - ELIÁŠ, P. : Stomatal oscillations in adult forest trees in natural environment. - Biol. Plant. *21* : 71 - 74, 1979.

37415 - ELIÁŠ, P. : Stomatal activity within the crowns of tall deciduous trees under forest conditions. - Biol. Plant. *21* : 266 - 274, 1979.

37416 - **ELIÁŠ, P.** : Contribution to the ecophysiological study of the water relations of forest shrubs. - Preslia (Praha) *51* : 77 - 90, 1979. [Stomatal resistance.]

*37417 - **ELIZAROVA, V.A., PYRINA, I.L., GETSEN, M.V.** : Soderzhanie pigmentov fitoplanktona v vodach Kharbeĭskikh ozer. [Content of phytoplankton pigments in Kharbeĭsk lakes.] - In : Produktivnost' Ozer Vostochnoĭ Chasti Bol'shezemel'skoĭ Tundry. Pp. 55 - 63. Nauka, Leningrad 1976. [In R.]

37418 - **ELKIEY, T., ORMROD, D.P.** : Leaf diffusion resistance responses of three petunia cultivars to ozone and/or sulfur dioxide. - Air Pollut. Contr. Assoc. J. *29* : 622 - 625, 1979.

37419 - **ELLEFSON, W.L., KROGMANN, D.W.** : Studies of the multiple forms of ferredoxin-NADP oxidoreductase from spinach. - Arch. Biochem. Biophys. *194* : 593 - 599, 1979.

37420 - **ELLIS, R., TIMSON, C.** : The absence of protochlorophyll(-ide) accumulation in algal cells with inhibited chlorophyll synthesis. - Plant Physiol. *63* (Suppl.) : 96, 1979.

37421 - **ELLIS, R.J.** : The most abundant protein in the world. - Trends biochem. Sci. *4* : 241 - 244, 1979. [RuBPC.]

*37422 - **ELLSWORTH, P.A., STORM, C.B.** : Methyl 10-epipheophorbide a: an unusual epimeric stability relative to chlorophyll a or a'. - J. org. Chem. *43* : 281 - - 283, 1978.

37423 - **ELLSWORTH, R.K., MURPHY, S.J.** : Biosynthesis of protochlorophyllide a from Mg-protoporphyrin IX monomethyl ester *in vivo*. - Photosynthetica *13* : 392 - - 400, 1979.

37424 - **EL-SAYED, M.A., TERNER, J.** : Power- and time-resolved resonance Raman studies and conformational changes in bacteriorhodopsin. - Photochem. Photobiol. *30* : 125 - 132, 1979.

37425 - **ELSTNER, E.F.** : Oxygen activation and superoxide dismutase in chloroplasts. - In : GIBBS, M., LATZKO, E. (ed.) : Photosynthesis II. (Encycl. Plant Physiol. N.S. Vol.6.) Pp. 410 - 415. Springer-Verlag, Berlin - Heidelberg - New York 1979.

37426 - **ELSTNER, E.F., PILS, I.** : Ethane formation and chlorophyll bleaching in DCMU-treated *Euglena gracilis* cells and isolated spinach chloroplast lamellae. - Z. Naturforsch. *34 C* : 1040 - 1043, 1979.

*37427 - **ELSTNER, E.F., YOUNGMAN, R.** : Oxygen activation in chloroplasts : Models for *"in vivo"* observations. - Ber. deut. bot. Ges. *91* : 569 - 577, 1978.

* 37428 - **EL-ZEFTAWI, B.M.** : Chemical and temperature control of rind pigment of citrus fruits. - Proc. int. Soc. Citricult. *1978* : 33 - 36, 1978. [Chl.]

37429 - **EMIGH, V.L.D.** : Metabolic measurements of *Protosiphon botryoides* KLEBS. zoospores. - J. Phycol. *15* (Suppl.) : 13, 1979. [Ps.]

37430 - **ENDRESS, A.G.** : Plastid ultrastructure in the avocado nucellus. - Ann. Bot. *44* : 511 - 512, 1979. [Chloroplast.]

37431 - **ENGLER, D.E., MEEUSE, B.J.D.** : Photorespiration in the protonemata of *Funaria hygrometrica* HEDW. - Acta bot. neerl. *28* : 205 - 212, 1979.

37432 - **ENGLERT, G.** : Homonuclear *Overhauser* ^{1}H-NMR. Experiments on the carotenoid pigments lycopene and prolycopene. - Helv. chim. Acta *62* : 1497 - 1500, 1979.

37433 - **ENGLERT, G., BROWN, B.O., MOSS, G.P., WEEDON, B.C.L., BRITTON, G., GOODWIN, T.W., SIMPSON, K.L., WILLIAMS, R.J.H.** : Prolycopene, a tetra-*cis* carotene with two hindered *cis* double bonds. - J. chem. Soc., chem. Commun. *12* : 545 - - 547, 1979.

37434 - **ENGLISH, S.D., McWILLIAM, J.R., SMITH, R.C.G., DAVIDSON, J.L.** : Photosynthesis and partitioning of dry matter in sunflower. - Aust. J. Plant Physiol. *6* : 149 - 164, 1979.

37435 - **EPPLEY, R.W., PETERSON, B.J.** : Particulate organic matter flux and planktonic new production in the deep ocean. - Nature *282* : 677 - 680, 1979.

37436 - EPPLEY, R.W., RENGER, E.H., HARRISON, W.G. : Nitrate and phytoplankton production in southern California coastal waters. - Limnol. Oceanogr. 24 : 483 -
- 494, 1979.

37437 - EPPLEY, R.W., RENGER, E.H., HARRISON, W.G., CULLEN, J.J. : Ammonium distribution in southern California coastal waters and its role in the growth of
phytoplankton. - Limnol. Oceanogr. 24 : 495 - 509, 1979. [Chl.]

37438 - ERABI, T., YAMASHITA, J., TANAKA, M., HORIO, T. : [Application of polarogra-
phy to studies on redox systems in bio-membranes : especially on photosynthetic electron transport system in chromatophore membrane from photosynthetic
bacterium.] - Tampakushitsu Kakusan Koso [Protein, nucleic Acid, Enzyme] 24 :
696 - 708, 1979. [In Jap.]

37439 - ERICKSON, P.I., KIRKHAM, M.B. : Growth and water relations of wheat plants
with roots split between soil and nutrient solution. - Agron. J. 71 : 361 -
- 364, 1979. [Stomatal resistance.]

37440 - ERICKSON, P.I., KIRKHAM, M.B., ADJEI, G.B. : Water relations, growth and
yield of tall and short wheat cultivars irradiated with X-rays. - Environ.
exp. Bot. 19 : 349 - 356, 1979. [Stomatal resistance.]

37441 - ERICKSON, P.I., KIRKHAM, M.B., STONE, J.F. : Growth, water relations, and
yield of wheat planted in four row directions. - Soil Sci. Soc. Amer. J. 43 :
570 - 574, 1979.

*37442 - ERKENBRECHER, C.W. Jr., STEVENSON, L.H. : The transport of microbial biomass
and suspended material in a high-marsh creek. - Can. J. Microbiol. 24 : 839 -
- 846, 1978. [Chl.]

37443 - ERLENKEUSER, H., WILLKOMM, H. : ^{13}C- und ^{14}C-Untersuchungen an Sedimenten des
Großen Plöner Sees. - Arch. Hydrobiol. 85 : 1 - 29, 1979.

37444 - ERNST, A., KERFIN, W., SPILLER, H., BÖGER, P. : External factors influencing
light-induced hydrogen evolution by the blue-green alga, Nostoc muscorum. -
Z. Naturforsch. 34 C : 820 - 825, 1979.

*B37445 - ERNSTER, L., ESTABROOK, R.W., SLATER, E.C. (ed.) : Dynamics of Energy-Trans-
ducing Membranes. - Elsevier Scientific Publishing Company, Amsterdam - London
- New York 1974. [Ps.]

*37446 - EROKHIN, Yu.E., CHUGUNOV, V.A., MOSKALENKO, A.A., DEMINA, L.P., MAKHNEVA, Z.K.:
Obshchie zakonomernosti organizatsii pigmentnoĭ sistemy purpurnykh fotosinteziruyushchikh bakteriĭ. [General features of organization of pigment system
of purple photosynthetic bacteria.] - In : Itogi Issledovaniya Mekhanizma Fotosinteza. Pp. 148 - 161. Institut Fotosinteza Akad. Nauk SSSR, Pushchino
1974. [In R.]

*37447 - EROKHIN, Yu.E., MOSKALENKO, A.A., GANAGO, A.O. : Nekotorye dannye o roli ka-
rotinoidov v strukture i funktsiyakh pigmentnoĭ sistemy purpurnoĭ bakterii
Chromatium minutissimum. [Some data on the role of carotenoids in structure
and functions of the pigment system in the purple bacterium Chromatium minutissimum.] - In : Itogi Issledovaniya Mekhanizma Fotosinteza. Pp. 162 - 171.
Institut Fotosinteza Akad. Nauk SSSR, Pushchino 1974. [In R.]

37448 - EROKHINA, L.G., KRASNOVSKIĬ, A.A. : Issledovanie spektral'nykh i fotokhimicheskikh éffektov denaturatsii fikobilinovykh pigmentov vodoroslei. [Study
of spectral and photochemical effects of denaturation of phycobilin pigments
of algae.] - In : Itogi Issledovaniya Mekhanizma Fotosinteza. Pp. 61 - 70.
Institut Fotosinteza Akad. Nauk SSSR, Pushchino 1974. [In R.]

37449 - ESKINS, K., BANKS, D.J. : The relationship of accessory pigments to chlorophyll a content in chlorophyll-deficient peanut and soybean varieties. - Photochem. Photobiol. 30 : 585 - 588, 1979.

37450 - ESTRADA, M. : Distribución de las reductasas de nitrato en la región de aflo-
ramiento del noroeste de África. Noviembre 1975. [Nitrate reductase distribution in the upwelling region of northwest Africa, November 1975.] - Re-
sult. Exped. cient. Buque oceanogr. "Cornide de Saavedra" 8 : 153 - 159,
1979. [Chl; in Span., ab : E.]

37451 - **ESYUNINA, A.I.** : Izmenenie aktivnosti glikolatoksidazy v list'yakh pshenitsy za vegetatsionnyĭ period. [Changes in activity of glycolate-oxidase in wheat leaves during the vegetation period.] - Byull. vsesoyuz. nauch.-issled. Inst. Rastenievodstva Im. N.I.Vavilova *87* : 33 - 35, 1979. [In R.]

37452 - **EVANS, E.H., CARR, N.G.** : The interaction of respiration and photosynthesis in microalgae. - In : GIBBS, M., LATZKO, E. (ed.) : Photosynthesis II. (Encycl. Plant Physiol. N.S. Vol. 6.)Pp. 163 - 173. Springer-Verlag, Berlin - Heidelberg - New York 1979.

*37453 - **EVANS, E.H., FOULDS, I., CARR, N.G.** : Environmental conditions and morphological variation in the blue-green alga *Chlorogloea fritschii*. - J. gen. Microbiol. *92* : 147 - 155, 1976. [Ps.]

37454 - **EVANS, E.H., RUSH, J.D., JOHNSON, C.E., EVANS, M.C.W.** : Mössbauer spectra of Photosystem-I reaction centers from the blue-green alga *Chlorogloea fritschii*. - Biochem. J. *182* : 861 - 865, 1979.

37455 - **EVANS, P.K., KROGMANN, D.W.** : Isolation and partial characterization of two cytochromes from *Porphyridium cruentum*. - Plant Physiol. *63* (Suppl.) : 55, 1979.

37456 - **EVENSON, P.D.** : Optimum crown temperatures for maximum alfalfa growth. - Agron. J. *71* : 798 - 800, 1979.

*37457 - **EVERS, A., ERNST-FONBERG, M.L.** : Differential responses of two carboxylases from *Euglena* to the state of chloroplast development. - FEBS Lett. *46* : 233 - 235, 1974.

*B37458 - **EVSTIGNEEV, V.B.** (ed.) : Metody Vydeleniya i Issledovaniya Belkov-Komponentov Fotosinteticheskogo Apparata. [Methods of Isolation and Investigation of the Protein Constituents of the Photosynthetic Apparatus.] - Inst. Fotosinteza Akad. Nauk SSSR, Pushchino-na-Oke 1973. [In R.]

37459 - **EVSTIGNEEV, V.B., GAVRILOVA, V.A.** : O fotokhimicheskom vzaimodeĭstvii khlorofilla *a* i feofitina *a* s malymi kontsentratsiyami gidrokhinona. [Photochemical interaction of chlorophyll *a* and pheophytin *a* with small concentrations of hydroquinone.·] - Biofizika *24* : 797 - 800, 1979. [In R, ab : E.]

*37460 - **EVSTIGNEEV, V.B., STOLOVITSKIĬ, Yu.M., SHKUROPATOV, A.Ya., KADOSHNIKOV, S.I.** : Perenos êlektrona pri osveshchenii i razdelenie zaryadov v sloyakh khlorofila. [Electron transport during irradiation and charge separation in chlorophyll layers.] - In : Itogi Issledovaniya Mekhanizma Fotosinteza. Pp. 3 - 22. Institut Fotosinteza Akad.Nauk SSSR, Pushchino 1974. [In R.]

37461 - **EZHOVA, T.A., GOSTIMSKIĬ, S.A.** : Geneticheskiĭ analiz khlorofil'nykh mutantov gorokha. [Genetic analysis of pea chlorophyll mutants.] - Genetika *15* : 691 - 700, 1979. [In R, ab : E.]

37462 - **FADEEVA, L.G.** : Vliyanie sukhoveya na razvivayushchiĭsya kolos vlagoobespechennoĭ pshenitsy. [Dry wind effect on the developing spike of wheat sufficiently supplied with water.] - Izv. sib. Otd. Akad. Nauk SSSR, Ser. biol. Nauk *1979* (1) : 108 - 115, 1979. [Primary production; in R, ab : E.]

37463 - **FADEEVA, L.G.** : Vliyanie vysokoĭ temperatury vozdukha na fosfornyĭ obmen razvivayushchegosya kolosa pshenitsy. [Influence of high air temperature on phosphorus exchange of developing wheat spike.] - Izv. sib. Otd. Akad. Nauk SSSR, Ser. biol. Nauk *1979* (3) : 96 - 100, 1979. [Dry-matter accumulation; in R, ab : E.]

*37464 - **FAIREY, N.A., DAYNARD, T.B.** : Quantitative distribution of assimilates in component organs of maize during reproductive growth. - Can. J. Plant Sci. *58* : 709 - 717, 1978.

*37465 - **FAIREY, N.A., DAYNARD, T.B.** : Assimilate distribution and utilization of maize. - Can. J. Plant Sci. *58* : 719 - 730, 1978.

37466 - **FALLON, R.D., BROCK, T.D.** : Lytic organisms and photooxidative effects: Influence on blue-green algae (cyanobacteria) in Lake Mendota, Wisconsin. - Appl. environ. Microbiol. *38* : 499 - 505, 1979. [Chl.]

*37467 - **FAN, I-ji, CHIEN, Yue-chin, CHIANG, I-hwa** : The inorganic photoreduction of NADP to NADPH and photophosphorylation of ADP to ATP in visible light. - Sci. sin. *21* : 663 - 668, 1978.

37468 - **FARARD, D.M., HULTIN, H.O.** : Extraction of protein from chloroplasts isolated from alfalfa leaf. - J. Food Biochem. *3* : 151 - 162, 1979.

37469 - **FARKAS, D.L., MALKIN, S.** : Cold storage of isolated class C chloroplasts. Optimal conditions for stabilization of photosynthetic activities. - Plant Physiol. *64* : 942 - 947, 1979.

37470 - **FARQUHAR, G.D.** : Carbon assimilation in relation to transpiration and fluxes of ammonia. - In : **MARCELLE, R., CLIJSTERS, H., VAN POUCKE, M.** (ed.) : Photosynthesis and Plant Development. Pp. 321 - 328. Dr. W. Junk bv. Publ., The Hague - Boston - London 1979.

37471 - **FARQUHAR, G.D.** : Models describing the kinetics of ribulose biphosphate carboxylase-oxygenase. - Arch. Biochem. Biophys. *193*: 456 - 468, 1979.

37472 - **FARQUHAR, G.D., WETSELAAR, R., FIRTH, P.M.** : Ammonia volatilization from senescing leaves of maize. - Science *203*: 1257 - 1258, 1979. [Ps.]

37473 - **FASEHUN, F.E.** : Effect of soil matric potential on leaf water potential, diffusive resistance, growth and development of *Gmelina arborea* L. seedlings. - Biol. Plant. *21* : 100 - 104, 1979. [Ps.]

37474 - **FAWZI, A.F.A., EL-FOULY, M.M.** : Amylase and invertase activities and carbohydrate contents in relation to physiological sink in carnation. - Physiol. Plant. *47* : 245 - 249, 1979. [Photosynthates.]

*37475 - **FAZYLOVA, S.** : Vliyanie intensivnosti osveshcheniya na fotosinteticheskuyu sposobnost' nekotorykh pustynnykh èfemerov. [Effect of irradiance on photosynthetic activity of some desert ephemers.] - In : Fiziologiya i Biokhimiya Dikorastushchikh Kormovykh RasteniĬ Uzbekistana. Pp. 107 - 118. Fan, Tashkent 1975. [In R.]

*37476 - **FAZYLOVA, S.** : Vliyanie faktora vlazhnosti na fotosinteticheskuyu sposobnost' nekotorykh pustynnykh vidov kustarnikov i polukustarnikov. [Effect of humidity factor on photosynthetic activity of some desert shrub species.] - In : Fiziologiya i Biokhimiya Dikorastushchikh Kormovykh RasteniĬ Uzbekistana. Pp. 119 - 126. Fan, Tashkent 1975. [In R.]

*37477 - **FAZYLOVA, S.F.** : Vliyanie rezhima vlazhnosti pochvy na velichinu fotosinteza u pustynnykh èfemerov. [Effect of soil moisture regime on photosynthetic rate in desert ephemers.] - Probl. Osvoeniya Pustyn' *1975* (3) : 77 - 80, 1975. [In R.]

37478 - **FEDERLE, T.W., VESTAL, J.R., HATER, G.R., MILLER, M.C.** : Effects of Prudhoe Bay crude oil on primary production and zooplankton in arctic tundra thaw ponds. - Mar. environ. Res. 2 : 3 - 18, 1979.

37479 - **FEDINA, I., VASILEVA, V., MIRCHEVA, V., VAKLINOVA, S.** : Vliyanie na atrazina v''rkhu rastezha, fotosintezata i aktivnostta na nyakoi enzimi pri mladi tsarevichni rasteniya. [Atrazine effects on growth, photosynthesis and enzyme activity of young maize plants.] - Fiziol. Rast. (Sofia) *5* (3) : 19 - 23, 1979. [In Bulg., ab : E, R.]

37480 - **FEDTKE, C.** : Plant physiological adaptations induced by low rates of photosynthesis. - Z. Naturforsch. *34 C* : 932 - 935, 1979.

*37481 - **FEDULOVA, A.N., KHROMOV, V.M., MAKSIMOV, V.N.** : Vliyanie nekotorykh detergentov, primenyaemykh dlya bor'by s neftyanym zagryazneniem, na protokokkovye vodorosli. [Effect of some detergents used against oil pollution on protococcous algae.] - Nauch. Dokl. vyssh. Shkoly, biol. Nauki *19* (5) : 90 - 95, 1976. [Ps, Chl, Car; in R.]

37482 - **FEDYK, Ya.D.** : Lystovydni plastynky protonemy *Tetraphis pellucida* HEDW. [Leaf-like protonema blades of *Tetraphis pellucida* HEDW.] - Ukr. bot. Zh. *36* : 565 - 569, 624, 1979. [Chl; in Ukr., ab : E, R.]

37483 - **FEE, E.J.** : A relation between lake morphometry and primary productivity and its use in interpreting whole-lake eutrophication experiments. - Limnol. Oceanogr. *24* : 401 - 416, 1979. [Chl.]

*37484 - **FEE, E.J., SHEARER, J.A., DeCLERCQ, D.R.** : *In vivo* chlorophyll profiles from lakes in the Experimental Lakes Area, Northwestern Ontario - 1976 data. - Fish. mar. Serv. Data Rep. *45* : 1 - 104. 1978.

37485 - **FEICK, R., DREWS, G.** : Protein subunits of bacteriochlorophylls B 802 and B 855 of the light-harvesting complex II of *Rhodopseudomonas capsulata*. - Z. Naturforsch. *34 C* : 196 - 199, 1979.

37486 - **FEIERABEND, J., SCHULZ, U., KEMMERICH, P., LOWITZ, T.** : On the action of chlorosis-inducing herbicides in leaves. - Z. Naturforsch. *34 C* : 1036 - 1039, 1979.

*37487 - **FEIGE, G.B.** : Physiologische Charakteristika der Heterocystendifferenzierung in verschiedenen Blaualgenflechten. - Ber. deut. bot. Ges. *91* : 595 - 602, 1978. [Photosynthates.]

*37488 - **FEJER, S.O., SPANGELO, L.P.S.** : Growth and yield of McIntosh, Lawfam and Sandow apples on Anis rootstocks and different stembuilders over 30 years. - Gartenbauwissenschaft *42* : 151 - 154, 1977. [Primary production.]

*37489 - **FEJER, S.O., SPANGELO, L.P.S.** : Morphological yield components in apple and raspberry. - Z. Pflanzenzücht. *79* : 26 - 39, 1977. [Dry-matter distribution.]

37490 - **FELLER, U.** : Effect of changed source/sink relations on proteolytic activities and on nitrogen mobilization in field-grown wheat (*Triticum aestivum* L.). - Plant Cell Physiol. *20* : 1577 - 1583, 1979. [Chl.]

37491 - **FELLOWS, R.J., EGLI, D.B., LEGGETT, J.E.** : Rapid changes in translocation patterns in soybeans following source-sink alterations. - Plant Physiol. *64* : 652 - 655, 1979. [Photosynthates.]

37492 - **FELLOWS, R.J., PATTERSON, R.P., GROSS, H.D., HARRIS, D.** : Recovery from water stress in soybeans: Interaction of net photosynthesis, N-fixation, and dry matter partitioning. - Plant Physiol. *63* (Suppl.) : 139, 1979.

B37493 - **FENCHEL, T., BLACKBURN, T.H.** : Bacteria and Mineral Cycling. - Academic Press, London - New York - San Francisco 1979. [Ps, Chl.]

37494 - **FENTON, J.M., PELLIN, M.J., GOVINDJEE, KAUFMANN, K.J.** : Primary photochemistry of the reaction center of photosystem I. - FEBS Lett. *100* : 1 - 4, 1979.

37495 - **FEOLI, E., LAUSI, D.** : Attività fotosintetica di *Gracilaria verrucosa* (HUDS.) PAPENFUSS. [Photosynthetic activity of *Gracilaria verrucosa* (HUDS.) PAPENFUSS.] - Boll. Soc. adriat. Sci. *63* : 73 - 81, 1979. [In Ital., ab : E.]

37496 - **FERERES, E., CRUZ-ROMERO, G., HOFFMAN, G.J., RAWLINS, S.L.** : Recovery of orange trees following severe water stress. - J. appl. Ecol. *16* : 833 - 842, 1979. [Stomatal resistance.]

37497 - **FERGUSON, P., LEE, J.A.** : The effects of bisulphite and sulphate upon photosynthesis in *Sphagnum*. - New Phytol. *82* : 703 - 712, 1979.

37498 - **FERREE, D.C.** : Influence of pesticides on photosynthesis of crop plants. - In : MARCELLE, R., CLIJSTERS, H., VAN POUCKE, M.(ed.):Photosynthesis and Plant Development. Pp. 331 - 341. Dr. W. Junk bv. Publ., The Hague - Boston - London 1979.

37499 - **FERREIRA, L.G.R., DE SOUZA, J.G., PRISCO, J.T.** : Effects of water deficit on proline accumulation and growth of two cotton genotypes of different drought resistances. - Z. Pflanzenphysiol. *93* : 189 - 199, 1979. [Ps.]

37500 - **FIALA, K.** : Primary production of the Kameničky grassland - underground. - In : RYCHNOVSKÁ, M. (ed.) : Function of Grasslands in Spring Region - Kameničky Project. Pp. 87 - 92. Botanical Institute, Czechoslovak Academy of Sciences, Brno 1979.

37501 - FIALA, K., JAKRLOVÁ, J. : Seasonal dynamics of the growth of *Nardus stricta* L. - In : RYCHNOVSKÁ, M. (ed.) : Function of Grasslands in Spring Region - Kameničky Project. Pp. 115 - 119. Botanical Institute, Czechoslovak Academy of Sciences, Brno 1979.

37502 - FILBIN, G.J., HOUGH, R.A. : The effects of excess copper sulfate on the metabolism of the duckweed *Lemna minor*. - Aquat. Bot. *7* : 79 - 86, 1979. [Ps, Chl.]

37503 - FINDENEGG, G.R. : Inorganic carbon transport in microalgae. I. Location of carbonic anhydrase and HCO_3^-/OH^- exchange. - Plant Sci. Lett. *17* : 101 - 108, 1979.

37504 - FINNEY, M.E. : The influence of infection by *Erysiphe graminis* DC. on the senescence of the first leaf of barley. - Physiol. Plant Pathol. *14* : 31 - 36, 1979. [Chl.]

*37505 - FISCHER, K., LÜTTGE, U. : Light-dependent net production of carbon monoxide by plants. - Nature *275* : 740 - 741, 1978. [Ps.]

37506 - FISCHER, K., LÜTTGE, U. : Lichtabhängige CO-Bildung grüner Pflanzen und ihre Bedeutung für den CO-Haushalt der Atmosphäre. - Flora *168* : 121 - 137, 1979. [Ps.]

37507 - FISCHER, K.H., LATZKO, E. : Chloroplast ribulose-5-phosphate kinase : Light-mediated activation, and detection of both soluble and membrane-associated activity. - Biochem. biophys. Res. Commun. *89* : 300 - 306, 1979.

37508 - FISCHER, U., OESTERHELT, D. : Chromophore equilibria in bacteriorhodopsin. - Biophys. J. *28* : 211 - 230, 1979.

37509 - FISH, L., FRANCESCHI, V.R., STOCKING, C.R. : Effects of pronase on isolated chloroplasts. - Plant Physiol. *64* : 1012 - 1014, 1979.

37510 - FISHER, K.J. : Source sink relationships in the young fruiting greenhouse tomato plant. - Gartenbauwissenschaft *44* : 118 - 120, 1979.

*37511 - FIUSSELLO, N. : Lead pollution : effects on chlorophyll. - Inform. bot. ital. *5* : 107 - 108, 1973.

*37512 - FIUSSELLO, N., MOLINARI, M.T. : Azione del piombo sull'accrescimento dei vegetali. [Effect of lead on plant growth.] - Allionia *19* : 89 - 96, 1973. [Chl;in Ital., ab : E.]

37513 - FLECK, J., DURR, A., LETT, M.C., HIRTH, L. : Changes in protein synthesis during the initial stage of life of tobacco protoplasts. - Planta *145* :279 - 285, 1979. [Fraction I protein.]

37514 - FLÜCKIGER, W., OERTLI, J.J., FLÜCKIGER, H. : Relationship between stomatal diffusive resistance and various applied particle sizes on leaf surfaces. - Z. Pflanzenphysiol. *91* : 173 - 175, 1979. [Stomatal resistance.]

37515 - FOCK, H., KLUG, K., CANVIN, D.T. : Effect of carbon dioxide and temperature on photosynthetic CO_2 uptake and photorespiratory CO_2 evolution in sunflower leaves. - Planta *145* : 219 - 223, 1979.

37516 - FOCK, H., LAWLOR, D.W. : Der Einfluß von Wassermangel auf den Gaswechsel und den primären C-Stoffwechsel von *Helianthus annuus* und *Zea mays*. - Ber. deut. bot. Ges. *92* : 145 - 152, 1979.

37517 - FOKKEMA, N.J., KASTELEIN, P., POST, B.J. : No evidence for acceleration of leaf senescence by phyllosphere saprophytes of wheat. - Trans. brit. mycol. Soc. *72* : 312 - 315, 1979. [Chl.]

*37518 - FOLLETT, R.F., BENZ, L.C., DOERING, E.J., REICHMAN, G.A. : Yield response of corn to irrigation on sandy soils. - Agron. J. *70* : 823 - 828, 1978.

37519 - FOMENKO, N.N., DRYAGINA, I.V., VESELOVA, T.V., KALINICHENKO, I.M. : Issledovanie fiziologicheskogo sostoyaniya obrabotannykh mutagenami semyan yabloni vo vremya ikh stratifikatsii. [Physiological state of apple tree seeds treated with mutagens during stratification.]-Vestn. mosk. gos. Univ., Ser. 16-Biol. *1979* (2) : 51 - 56, 1979. [Chl; in R, ab : E.]

37520 - **FOMISHYNA, R.M.** : Vplyv raptovogo zasolennya na rozklad khlorofilu v izol'o-
vanykh lystkakh *Beta vulgaris* L. [Effect of sudden salting on chlorophyll
decomposition in isolated leaves of *Beta vulgaris* L.] - Ukr. bot. Zh. *36* :
10 - 13, 94, 1979. [In Ukr., ab : E, R.]

37521 - **FONDY, B.R., GEIGER, D.R.** : A method for continuous measurement of export
from a leaf. - Plant Physiol. *63* (Suppl.) : 34, 1979.

37522 - **FONG, F., SCHIFF, J.A.** : Blue-light-induced absorbance changes associated
with carotenoids in *Euglena*. - Planta *146* : 119 - 127, 1979.

37523 - **FORD, M.J., ALHADEFF, M., CHAPMAN, J.M., BLACK, M.** : A rapid and selective
action of 6-benzylaminopurine on 5-aminolevulinate production in excised
sunflower cotyledons. - Plant Sci. Lett. *16* : 397 - 402, 1979. [Chl.]

37524 - **FORK, D.C.** : The influence of changes in the physical phase of thylakoid
membrane lipids on photosynthetic activity. - In : LYONS, J.M., GRAHAM, D.,
RAISON, J.K. (ed.) : Low Temperature Stress in Crop Plants : The Role of the
Membrane. Pp. 215 - 229. Academic Press, New York - San Francisco - London
1979.

37525 - **FORK, D.C., FORD, G.A., CATANZARO, B.** : Measurements with a microprocessor-
-based fluorescence spectrophotometer made on the blue-green alga *Anacystis
nidulans* above and below the phase transition temperature. - Carnegie Inst.
Year Book *78* : 196 - 199, 1979.

37526 - **FORK, D.C., MURATA, N., SATO, N.** : Effect of growth temperature on the lipid
and fatty acid composition, and the dependence on temperature of light-induced redox reactions of cytochrome f and of light energy redistribution in
the thermophilic blue-green alga *Synechococcus lividus*. - Plant Physiol. *63* :
524 - 530, 1979. [Chl.]

37527 - **FORK, D.C., VAN GINKEL, G.** : Phase transitions in thylakoid membranes and
vesicles prepared from lipids separated from *Synechococcus lividus* grown at
55 ° and 38 °C. - Carnegie Inst. Year Book *78* : 178 - 180, 1979.

*B37528 - Formirovanie Pigmentnogo Apparata Fotosinteza. [Formation of Pigment Appara-
tus of Photosynthesis.] - Nauka i Tekhnika, Minsk 1973. [In R.]

37529 - **FORREST, G., VILCINS, G.** : Determination of tobacco carotenoids by resonance
Raman spectroscopy. - J. agr. Food Chem. *27* : 609 - 612, 1979.

*B37530 - **FORTI, G., AVRON, M., MELANDRI, A.** (ed.) : Photosynthesis, Two Centuries
after Its Discovery by Joseph Priestley. Vol. 1, 2, 3. Dr.W.Junk N.V. Publ.,
The Hague 1972.

*B37531 - Fotoregulyatsiya Metabolizma i Morfogeneza Rasteniĭ. [Photoregulation of Me-
tabolism and Morphogenesis of Plants.] - Nauka, Moskva 1975. [In R.]

*B37532 - Fotosintez i Ustoĭchivost' Rasteniĭ. [Photosynthesis and Plant Resistance.] -
Nauka i Tekhnika, Minsk 1973. [In R.]

*37533 - **FOX, S.W., ADACHI, T., STILLWELL, W., ISHIMA, Y., BAUMANN, G.** : Photochemical synthesis of ATP : Protomembranes and protometabolism. - In : DEAMER, D.W.
(ed.) : Light Transducing Membranes : Structure, Function, Evolution. Pp.
61 - 75. Academic Press, San Francisco - London - New York 1978.

37534 - **FOYER, C.H., HALL, D.O.** : A rapid procedure for the preparation of light harvesting chlorophyll *a/b* protein complex. An assessment of its manganese content. - FEBS Lett. *101* : 324 - 328, 1979.

37535 - **FRĄCKOWIAK, D.** : Excitation energy transfer between phycoerythrin and phycocyanin in stretched polyvinyl alcohol film. - Bull. Acad. pol. Sci., Sér.
Sci. biol., Cl. II *27* : 161 - 167, 1979.

37536 - **FRĄCKOWIAK, D., DUDKIEWICZ, J., GRABOWSKI, J., FIKSIŃSKI, K., MANIKOWSKI, H.**
: Spectral properties of phycoerythrin. - Photosynthetica *13* : 21 - 28, 1979.

37537 - **FRĄCKOWIAK, D., EROKHINA, L.G., FIKSIŃSKI, K.** : The influence of aggregation on the excitation energy transfer between phycobiliproteins and chlo-
rophyllin. - Photosynthetica *13* : 245 - 253, 1979.

37538 - FRĄCKOWIAK, D., JADŻYN, C. : Photovoltaic effect of purple membrane. - Bull. Acad. pol. Sci., Sér. Sci. biol. , Cl. II, *27* : 523 - 525, 1979.

37539 - FRADKIN, L.I., NIKALAEVA, G.M., KUPERMAN, N.I. : Uplyŭ umoŭ asvyatlennya i stupeni dyferêntsyyatsyi khlaraplastaŭ na èlektrafarêtychnae fraktsyyaniravanne submembrannykh chastsinak khlaraplastaŭ. [Effect of irradiance and degree of chloroplast differentiation on electrophoretic fractionation of submembrane chloroplast fragments.] - Vestsi Akad. Navuk belarus. SSR, Ser. biyal. Navuk *1979* (5) : 33 - 37, 139, 1979. [In Belorus., ab : E, R.]

*37540 - FRAGA, F. : Fotosíntesis en la ría de Vigo. [Primary production in Ria de Vigo.] - Invest. pesquera *40* : 151 - 167, 1976. [In Span., ab : E.]

*37541 - FRANCIS, C.A., TEMPLE, S.R., FLOR, C.A., GROGAN, C.O. : Effects of competition on yield and dry matter distribution in maize. - Field Crops Res. *1* : 51 - 63, 1978.

37542 - FRANCIS, K. : Photosynthesis by isolated chloroplasts of *Sorghum vulgare*. - Experientia *35* : 1324 - 1326, 1979.

37543 - FRANÇOIS, J. : La jacinthe d'eau [*Eichhornia crassipes* (MART.)SOLMS] fléau aquatique ou hydrophyte d'avenir ? - Ann. Gembloux *85* : 73 - 81, 1979. [Growth analysis.]

37544 - FRANÇOIS, J., RENARD, C. : Étude en milieu contrôlé du comportement d'un tapis de *Festuca arundinacea* SCHREB en régime d'assèchement. - Oecol. Plant. *14* : 417 - 433, 1979. [Growth analysis.]

37545 - FRANK, H.A., BOLT, J., FRIESNER, R., SAUER, K. : Magnetophotoselection of triplet state of reaction centers from *Rhodopseudomonas sphaeroides* R-26. - Biochim. biophys. Acta *547* : 502 - 511, 1979.

37546 - FRANK, H.A., FRIESNER, R., NAIRN, J.A., DISMUKES, G.C., SAUER, K. : The orientation of the primary donor in bacterial photosynthesis. - Biochim. biophys. Acta *547* : 484 - 501, 1979.

37547 - FRANK, H.A., McLEAN, M.B., SAUER, K. : Triplet states in photosystem I of spinach chloroplasts and subchloroplasts particles. - Proc. nat. Acad. Sci. USA *76* : 5124 - 5128, 1979.

37548 - FRASER, L., MATTHEWS, R.E.F. : Strain-specific pathways of cytological change in individual Chinese cabbage protoplasts infected with turnip yellow mosaic virus. - J. gen. Virol. *45* : 623 - 630, 1979. [Chloroplast.]

37549 - FREDRICK, J.F. : Storage glucan biosynthesis in *Cyanidium, Chlorella* and *Prototheca* : Evidence for endosymbiosis ? - Phytochemistry *18* : 1823 - 1826, 1979. [Chloroplast.]

37550 - FRENCH, C.S. : Fifty years of photosynthesis. - Annu. Rev. Plant Physiol. *30* : 1 - 26, 1979.

B37551 - FRENCH, N.R. (ed.) : Perspectives in Grassland Ecology. Ecological Studies 32. Springer-Verlag, New York - Heidelberg - Berlin 1979. [Ps.]

37552 - FRENCH, N.R., STEINHORST, R.K., SWIFT, D.M. : Grassland biomass trophic pyramids. - In : FRENCH, N.R. (ed.) : Perspectives in Grassland Ecology. Pp. 59 - 88. Springer-Verlag, New York - Heidelberg - Berlin 1979.

*37553 - FRENSKA, M.Y., KAFALIEVA-BOEVA, D.N. : Variations in the photo-induced proton transfer depending on the intactness of the chloroplast membrane. - Dokl. bolg. Akad. Nauk *31* : 465 - 468, 1978.

37554 - FREY, M.A., ALBERTE, R.S., SCHIFF, J.A. : Studies by fluorescence of protochlorophyll(ide) and its phototransformation in dark-grown *Euglena gracilis* var. *bacillaris*. - Biol. Bull. *157* : 368 - 369, 1979.

37555 - FREYMAN, S., KEMP, G.A., WILSON, D.B. : Growth of bean accessions at various temperatures. - Can. J. Plant Sci. *59* : 81 - 85, 1979. [Ps.]

37556 - FREYSSINET, G., BELANGER, F., REBEIZ, C.A. : Description of novel chlorophyll chromophores associated with the pigment protein complexes of higher plants. - Plant Physiol. *63* (Suppl.) : 97, 1979.

37557 - FREYSSINET, G., EICHHOLZ, R.L., BUETOW, D.E. : Synthesis of RuBPCase during
light-induced chloroplast development in *Euglena*. - Plant Physiol. *63* (Suppl.)
: 27, 1979.

37558 - FREYSSINET, G., HARRIS, G.C., NASATIR, M., SCHIFF, J.A. : Events surrounding
the early development of *Euglena* chloroplasts. 14. Biosynthesis of cytochrome c-552 in wild type and mutant cells. - Plant Physiol. *63* : 908 - 915,
1979.

37559 - FREYTAG, H.E., JÄGER, R. : Zur berührungslosen Massebestimmung von Pflanzen-
beständen durch Absorption von Gammastrahlen. - Arch. Acker-Pflanzenbau Bo-
denk. *23* : 757 - 764, 1979.

37560 - FRIBOURG, H.A., OVERTON, J.R. : Persistence and productivity of tall fescue
in Bermudagrass sods subjected to different clipping managements. - Agron.
J. *71* : 620 - 624, 1979.

37561 - FRIDLYAND, L.E., KALER, V.L. : Modelirovanie êffekta usileniya i khromatiches-
kikh perekhodov fotosinteza. [Modelling of enhancement and chromatic transient effects of photosynthesis.] - Biofizika *24* : 1016 - 1021, 1979. [In R,
ab : E.]

37562 - FRIEDBERG, D., FINE, M., OREN, A. : Effect of oxygen on the cyanobacterium
Oscillatoria limnetica. - Arch. Microbiol. *123* : 311 - 313, 1979. [Bil.]

37563 - FRIEND, D.J.C., DEPUTY, J., QUEDADO, R. : Photosynthetic and photomorphoge-
netic effects of high photon flux densities on the flowering of two-long-day
plants, *Anagallis arvensis* and *Brassica campestris*. - In : MARCELLE, R.,
CLIJSTERS, H., VAN POUCKE, M. (ed.) : Photosynthesis and Plant Development.
Pp. 59 - 72. Dr.W.Junk bv. Publ., The Hague - Boston - London 1979.

37564 - FRIEND, D.J.C., LYDON, J. : Effects of daylength on flowering, growth, and
CAM of pineapple (*Ananas comosus* (L.) MERILL). - Bot.Gaz. *140* : 280 - 283,
1979. [Ps.]

37565 - FRIESNER, R., DISMUKES, G.C., SAUER, K. : Development of electron spin polarization in photosynthetic electron transfer by the radical pair mechanism. -
Biophys. J. *25* : 277 - 294, 1979.

37566 - FRISCHKNECHT, K., SCHNEIDER, K. : Physiological performances of the blue-
-green alga *Anacystis nidulans* in continuous turbidostat fermenter culture. -
Arch. Microbiol. *120* : 215 - 221, 1979.

B37567 - FRITSCHEN, L.J., GAY, L.W. : Environmental Instrumentation. - Springer-Verlag, New York - Heidelberg - Berlin 1979. [Radiation measurement.]

*37568 - FROLOV, A.K. : Vliyanie usloviĭ osveshchennosti v lesostepnoĭ dubrave na
assimilyatsionnyĭ apparat snyti (*Aegopodium podagraria* L.). [Effect of irradiance in steppe oak forest on the assimilatory apparatus of *Aegopodium podagraria* L.] - Vestn. leningrad. Univ. *1977* (3) : 60 - 65, 1977. [In R, ab : E.]

*37569 - FROLOV, A.K. : Assimilyatsionnyĭ apparat kustarnikov pod pologom lesostep-
noĭ dubravy. [Assimilatory apparatus of shrubs- in the forest-steppe oak forest.] - Tr. petergof. biol. Inst. leningrad. gos. Univ. *27* (Voprosy Êkolo-
gicheskoĭ Anatomii i Fiziologii Rasteniĭ) : 91 - 100, 1978. [In R.]

37570 - FRUGE, D.R., FONG, G.D., FONG, F.K. : Photosynthesis of polyatomic organic
molecules from carbon dioxide and water by the photocatalytic action of vi-
sible-light-illuminated platinized chlorophyll a dihydrate polycrystals. -
J. amer. chem. Soc. *101* : 3694 - 3697, 1979.

37571 - FRY, B., PARKER, P.L. : Animal diet in Texas seagrass meadows : $\delta^{13}C$ evidence for the importance of benthic plants. - Estuarine coastal mar. Sci.
8 : 499 - 509, 1979.

37572 - FRYDMAN, R.B., FRYDMAN, B. : Disappearance of porphobilinogen deaminase activity in leaves before the onset of senescence. - Plant Physiol. *63* : 1154 -
1157, 1979.

37573 - FUJI, A. : Phosphorus budget in natural population of *Corbicula japonica*
prime in Poikilohaline lagoon, Zyusan-ko. - Bull. Fac. Fish. Hokkaido Univ.
30 : 34 - 49, 1979. [Chl.]

37574 - **FUJII, Y.** : [Studies on the mode of herbicidal activity of methoxyphenone.] -
J. Pestic. Sci. *4* : 391 - 399, 1979. [Ps, Chl; in Jap., ab : E.]

*37575 - **FUJITA, K.** : [Photosynthesis and its model reactions.] - Kagaku (Kyoto) *26* :
372 - 381, 1971. [In Jap.]

37576 - **FURUHASHI, K., USUI, H., YATAZAWA, M.** : Unusuall features of respiration in
mixotrophic green tobacco callus tissues. - Plant Cell Physiol. *20* : 363 -
367, 1979. [Photorespiration.]

37577 - **FURUKAWA, A., TOTSUKA, T.** : Effects of NO_2, SO_2 and O_3 alone and in combinations on net photosynthesis in sunflower. - Environ. Control Biol. *17* : 161 -
166, 1979.

37578 - **FUTAMI, A., HAUSKA, G.** : Vectorial redox reactions on physiological quinones.
II. A study of transient semiquinone formation. - Biochim. biophys. Acta *547* :
597 - 608, 1979.

37579 - **FUTAMI, A., HURT, E., HAUSKA, G.** : Vectorial redox reactions of physiological
quinones. I. Requirement of a minimum length of the isoprenoid side chain. -
Biochim. biophys. Acta *547* : 583 - 596, 1979.

*37580 - **GABIDZASHVILI, M.A., CHRELASHVILI, M.N., KACHARAVA, N.F.** : [Dependence of
photosynthesis on temperature in some plants.] - Soobshch. Akad. Nauk gruz.
SSR *73*(1):157- 160, 1974. [In Georg., ab : E, R.]

37581 - **GÄCHTER, R., MARÈS, A.** : Comments to the acidification and bubbling method
for determining phytoplankton production. - Oikos *33* : 69 - 73, 1979.

37582 - **GAFF, D.F., McGREGOR, G.R.** : The effect of dehydration and rehydration on
the nitrogen content of various fractions from resurrection plants. - Biol.
Plant. *21* : 92 - 99, 1979. [Chl.]

37583 - **GAFNI, A., WERBER, M.M.** : Ferredoxin from *Halobacterium* of the Dead Sea.
Structural properties revealed by fluorescence techniques. - Arch. Biochem.
Biophys. *196* : 363 - 370, 1979.

37584 - **GAGLIANO, A.G., GEACINTOV, N.E., BRETON, J., ACKER, S., REMY, R.** : Electric
linear dichroism of P700 chlorophyll *a*-protein complexes. - Photochem. Photobiol. *29* : 415 - 418, 1979.

*37585 - **GAILHOFER, M., THALER, I.** : "Stromazentrum" in Leukoplasten der Epidermis
von *Asphodelus microcarpus*. - Phyton (Austria) *19* : 97 - 102, 1978. [Chloro-
plast.]

37586 - **GALE, J., EASTON, J.** : The effect of limestone dust on vegetation in an area
with a Mediterranean climate. - Environ. Pollut. *19* (2) : 89 - 102, 1979.
[Ps.]

37587 - **GALITSKIĬ, V.V., KOMAROV, A.S.** : O modelirovanii rosta rasteniĬ. [The plant
growth modelling.] - Izv. Akad. Nauk SSSR, Ser. biol. *1979* (5) : 714 - 723,
1979. [In R.]

37588 - **GALKIN, V.I.** : Primenenie rostovogo analiza dlya polucheniya ontogenetiches-
koĬ formuly produktivnosti genotipa zernovykh zlakov. [Use of growth analysis for getting the ontogenetic formula of productivity of cereal genotype.]
- Byull. vses. nauch.-issled. Inst. Rasteniev. Im. N.I.Vavilova *87* : 18 - 27,
1979. [In R.]

37589 - **GALLAGHER, J.N.** : Field studies of cereal leaf growth I. Initiation and expansion in relation to temperature and ontogeny. - J. exp. Bot. *30* : 625 -
- 636, 1979.

37590 - **GALLAGHER, J.N.** : Field studies of cereal leaf growth II. The relation between auxanometer and dissection.measurements of leaf extension and their relation to crop leaf area expansion. - J. exp. Bot. *30* : 637 - 646, 1979.

37591 - GALLAGHER, J.N., BISCOE, P.V. : Field studies of cereal leaf growth III. Barley leaf extension in relation to temperature, irradiance, and water potential. - J. exp. Bot. *30* : 645 - 655, 1979.

37592 - GALLAGHER, J.N., BISCOE, P.V., WALLACE, J.S. : Field studies of cereal leaf
growth IV. Winter wheat leaf extension in relation to temperature and leaf
water status. - J. exp. Bot. *30* : 657 - 668, 1979.

37593 - GALLOWAY, L., FRUGE, D.R., FONG, F.K. : Gaseous.evolution of molecular hydrogen and oxygen in photochemical splitting of water by platinized chlorophyll
a dihydrate polycrystals. Laboratory simulation of the primary light reaction
in plant photosynthesis. - In : KING, R.B. (ed.) : Advances in Chemistry
Series. No. 173. Inorganic Compounds with Unusual Properties.-II. Pp. 210 -
- 224. Amer. chem. Soc. 1979.

37594 - GALLOWAY, L., FRUGE, D.R., HALEY, G.M., CODDINGTON, A.B., FONG, F.K. : Rela-
tively low-temperature thermochemical   generation of molecular hydrogen from
decomposition of water by platinum. Red-light photochemical origin of the
chlorophyll *a* water splitting reaction. - J. amer. chem. Soc. *101* : 229 - 231,
1979.

37595 - GALUTVA, O.A., KUZNETSOVA, T.A., NEKRASOV, L.I. : Fotosensibilizirovannoe
khlorofillom vosstanovlenie metilovogo krasnogo v rastvorakh detergenta.
[Chlorophyll-photosensitized reduction of methylene red in detergent solutions.] - Biofizika *24* : 5 - 8, 1979. [In R, ab : E.]

37596 - GALUTVA, O.A., NEKRASOV, L.I. : Fluorestsentsiya khlorofilla, adsorbirovannogo sovmestno s tritonom X-100 na silokhrome. [Fluorescence of chlorophyll
adsorbed with *Triton X-100* on silochrome.] - Biofizika *24* : 362, 1979. [In R.]

37597 - GALUTVA, O.A., NEKRASOV, L.I. : Adsorbtsiya khlorofilla *a* i *b* iz vodnykh mi-
tsellyarnykh rastvorov tritona X-100 na silokhrome. [Adsorption of chlorophyll *a* and *b* from aqueous micellar solutions of *Triton X-100* on silochrome.]
- Biofizika *24* : 564, 1979. [In R.]

37598 - GANAGO, A.O., EROKHIN, Yu.E., SOLOV'EV, A.A. : Linear dichroism of light-induced absorbance changes in oriented reaction centers from *Rhodopseudomonas
sphaeroides* R-26. - Stud. biophys. *77* : 5 - 12, 1979.

37599 - GANTT, E. : Phycobiliproteins of *Cryptophyceae*. - In : LEVANDOWSKY, M., HUT-
NER, S.H. (ed.) : Biochemistry and Physiology of *Protozoa*. Vol.1. 2nd Ed.
Pp. 121 - 137. Academic Press, New York - San Francisco - London 1979.

37600 - GANTT, E., LIPSCHULTZ, C.A. : Structure and composition of phycobilisomes
from *Griffithsia pacifica (Rhodophyceae)*. - J. Phycol. *15* (Suppl.) : 17,
1979.

37601 - GANTT, E., LIPSCHULTZ, C.A., GRABOWSKI, J., ZIMMERMAN, B.K. : Phycobilisomes
from blue-green and red algae. Isolation criteria and dissociation characteristics. - Plant Physiol. *63* : 615 - 620, 1979.

37602 - GAPONENKA, V.I., BALEVA, E.F., SHAŬCHUK, S.M. : Fotasintez i asimilyatsyĭny-
ya liki raslin kukuruzy, vyrashchanykh na zyalěnym i belym svyatle. [Photo-
synthesis and assimilation numbers of maize plants grown in green and white
light.] - Vestsi Akad. Navuk belarus. SSR, Ser. biyal. Navuk *1979* (4) : 52 -
- 55, 139, 1979. [In Belorus., ab : E, R.]

37603 - GAPONENKO, V.I., SHEVCHUK, S.N. : Skorost' obnovleniya khlorofilla i assimi-
lyatsionnye chisla u zolotistoĭ i zelenolistnoĭ form klena yasenelistnogo.
[Chlorophyll turnover rate and assimilation numbers in golden and green-leaf
forms of *Acer negundo*.] - Dokl. Akad. Nauk belorus. SSR *23* : 756 - 759, 1979.
[In R, ab : E.]

37604 - GARAB, G.I., PAILLOTIN, G., JOLIOT, P. : Flash-induced scattering transient
in the 10 μs - 5 s time range between 450 and 540 nm with *Chlorella* cells. -
Biochim. biophys. Acta *545* : 445 - 453, 1979. [Chl.]

37605 - GARDINER, T.R., VIETOR, D.M., CRAKER, L.E. : Growth habit and row width effects on leaf area development and light interception of field beans. - Can.
J. Plant Sci. *59* : 191 - 199, 1979.

37606 - GARGAS, E., HARE, I., MARTENS, P., EDLER, L. : Diel changes in phytoplankton photosynthetic efficiency in brackish waters. - Mar. Biol. *52* : 113 - 122, 1979.

37607 - GARNIER, J., GUYON, D., PICAUD, A. : Characterization of new strains of non-photosynthetic mutants of *Chlamydomonas reinhardtii* I. Fluorescence, photo-chemical activities, chlorophyll-protein complexes. - Plant Cell Physiol. *20* : 1013 - 1027, 1979.

37608 - GARTY, H., EISENBACH, M., SHULDMAN, R., CAPLAN, S.R. : Light-induced pH changes in sub-bacterial particles of *Halobacterium halobium*. Effect of ionophores. - Biochim. biophys. Acta *545* : 365 - 375, 1979.

37609 - GARWOOD, E.A., TYSON, K.G., SINCLAIR, J. : Use of water by six grass species 1. Dry-matter yields and response to irrigation. - J. agr. Sci. *93* : 13 - 24, 1979.

37610 - GASANOV, R., ABILOV, Z.K., GAZANCHYAN, R.M., KURBONOVA, U.M., KHANNA, R., GOVINDJEE : Excitation energy transfer in photosystems I and II from grana and in photosystem I from stroma lamellae, and identification of emission bands with pigment-protein complexes at 77 K. - Z. Pflanzenphysiol. *95* : 149 - 169, 1979.

37611 - GAST, P., HOFF, A.J. : Transfer of light-induced electron-spin polarization from the intermediary acceptor to the prereduced primary acceptor in the reaction center of photosynthetic bacteria. - Biochim. biophys. Acta *548* : 520 - 535, 1979.

37612 - GAUDILLÈRE, J.P. : Caractéristiques photosynthétiques d'espèces appartenant aux genres *Aegilops et Triticum*. - Ann. Amélior. Plant. *29* : 523 - 533, 1979.

37613 - GAUDILLÈRE, J.-P. : Effet des caroténoïdes sur le rendement quantique de la photosynthèse chez *Triticum aestivum*. - Physiol. vég. *17* : 777 - 787, 1979.

37614 - GAUHL, E. : Sun and shade ecotypes of *Solanum dulcamara* L. : Photosynthetic light dependence characteristics in relation to mild water stress. - Oecologia *39* : 61 - 70, 1979.

*37615 - GAVRILENKO, V.F., ZHIGALOVA, T.V., RUBIN, B.A. : Indutsiruemoe svetom pogloshchenie protonov khloroplastami pshenitsy razlichnykh po produktivnosti sortov v zavisimosti ot pH reaktsionnoĭ smesi i intensivnosti osveshcheniya. [Light-induced proton absorption by chloroplasts of wheat cultivars of different productivity in relation to pH and irradiance.] - Nauch. Dokl. vyssh. Shkoly, biol. Nauki *19* (4) : 97 - 103, 1976. [In R.]

37616 - GEETHA, V., GNANAM, A. : Identification of RuBPCase as the product of *in vitro* protein synthesis in the isolated chloroplasts of *Sorghum vulgare*. - Plant Physiol. *63* (Suppl.) : 36, 1979.

37617 - GEIGER, D.R. : Control of partitioning and export of carbon in leaves of higher plants. - Bot. Gaz. *140* : 241 - 248, 1979.

37618 - GEIGER, D.R., FONDY, B.R. : A method for continuous measurement of export from a leaf. - Plant Physiol. *64* : 361 - 365, 1979.

*37619 - GELIN, C. : The restoration of freshwater ecosystems in Sweden. - In : HOLDGATE, M.W., WOODMAN, M.J. (ed.) : The Breakdown and Restoration of Ecosystems. Pp. 323 - 338. Plenum Press, New York - London 1978. [Primary production.]

37620 - GENCHEV, S., K"DREV, T., GEORGIEVA, V., RANKOV, V., DIMITROV, G. : Izmeneniya v s"d"rzhanieto na plastidnite pigmenti pod vliyanie na razlichni s"otnosheniya na khranitelnite elementi pri domatite. [Changes in plastid pigment content of tomatoes under the influence of various nutrient element ratios.] - Fiziol. Rast. (Sofia) *5* (4) : 67 - 74, 1979. [In Bulg., ab : E, R.]

37621 - GENCHEV, S., RANKOV, V. : Changes in plastid pigments content as influenced by different ratios of nutrient elements in eggplant. - In : Mineral Nutrition of Plants. Vol. II. Pp. 45 - 51. Publ. House Central Cooperative Union, Sofia 1979.

37622 - GENDEL, S., OHAD, I., BOGORAD, L. : Control of phycoerythrin synthesis during chromatic adaptation. - Plant Physiol. *64* : 786 - 790, 1979.

37623 - GERAKIS, P.A., PAPAKOSTA-TASOPOULOU, D. : Growth dynamics of *Zea mays* L. populations differing in genotype and density and grown under illuminance stress. - Oecol. Plant. *14* : 13 - 26, 1979.

37624 - GERBAUD, A., ANDRE, M. : Photosynthesis and photorespiration in whole plants of wheat. - Plant Physiol. *64* : 735 - 738, 1979.

37625 - GERBER, G.E., ANDEREGG, R.J., HERLIHY, W.C., GRAY, C.P., BIEMANN, K., KHORANA, H.G. : Partial primary structure of bacteriorhodopsin : Sequencing methods for membrane proteins. - Proc. nat. Acad. Sci. USA *76* : 227 - 231, 1979.

37626 - GEROLA, P.D., JENNINGS, R.C., FORTI, G., GARLASCHI, F.M. : Influence of protons on thylakoid membrane stacking. - Plant Sci. Lett. *16* : 249 - 254, 1979.

37627 - GERWICK, B.C., BLACK, C.C. Jr. : Sulfur assimilation in C_4 plants. Intercellular compartmentation of adenosine 5'-triphosphate sulfurylase in crabgrass leaves. - Plant Physiol. *64* : 590 - 593, 1979. [Ps.]

37628 - GERWICK, B.C., WILLIAMS, G.J. III. : Effects of growth temperature on carboxylase enzyme activity in *Opuntia polyacantha*. - Photosynthetica *13* : 254 - 259, 1979.

37629 - GIAQUINTA, R.T. : Phloem loading of sucrose. Involvement of membrane ATPase and proton transport. - Plant Physiol. *63* : 744 - 748, 1979.

37630 - GIBBS, M., BAMBERGER, E.S. : Photoevolution of H_2 and CO_2 in anaerobically adapted *Chlamydomonas reinhardi* F-60 and *Scenedesmus obliquus*. - Plant Physiol. *63* (Suppl.) : 40, 1979.

37631 - GIBBS, M., LATZKO, E. : Introduction. - In : GIBBS, M., LATZKO, E. (ed.) : Photosynthesis II. (Encycl. Plant Physiol. N.S. Vol. 6.) Pp. 1 - 5. Springer--Verlag, Berlin - Heidelberg - New York 1979. [Ps.]

37632 - GIBSON, J., STACKEBRANDT, E., ZABLEN, L.B., GUPTA, R., WOESE, C.R. : A phylogenetic analysis of the purple photosynthetic bacteria. - Curr. Microbiol. *3* : 59 - 64, 1979. [Cyt.]

37633 - GIDDINGS, T.H. Jr., STAEHELIN, L.A. : Changes in thylakoid structure associated with the differentiation of heterocysts in the cyanobacterium, *Anabaena cylindrica*. - Biochim. biophys. Acta *546* : 373 - 382, 1979.

37634 - GIERSCH, C. : Quantitative high-performance liquid chromatographic analysis of ^{14}C-labelled photosynthetic intermediates in isolated intact chloroplasts. - J. Chromatogr. *172* : 153 - 161, 1979.

37635 - GIESKES, W.W.C., KRAAY, G.W., BAARS, M.A. : Current ^{14}C methods for measuring primary production : gross underestimates in oceanic waters. - Neth. J. Sea Res. *13* : 58 - 78, 1979.

37636 - GIFFORD, R.M. : Carbon dioxide and plant growth under water and light stress: Implications for balancing the global carbon budget. - Search *10* : 316 - 318, 1979. [Ps.]

37637 - GIFFORD, R.M. : Growth and yield of CO_2-enriched wheat under water-limited conditions. - Aust. J. Plant Physiol. *6* : 367 - 378, 1979.

37638 - GIGON, A. : CO_2-gas exchange, water relations and convergence of mediterranean shrub-types from California and Chile. - Oecol. Plant. *14* : 129 - 150, 1979.

37639 - GILBOA, A., BEN-AMOTZ, A. : An improved method for rapid assaying of viability of cryopreserved unicellular algae. - Plant Sci. Lett. *14* : 317 - 320, 1979. [Ps.]

*37640 - GILLER, Yu.E., SHCHERBAKOVA, I.Yu., MAĬSTER, A., SHUTILOVA, N.I., KADOSHNIKOVA, I.G. : O svyazi obrazovaniya nativnykh form khlorofilla s biosintezom RNK i belka. [Relation of the formation of native forms of chlorophyll with RNA and protein biosynthesis.] - Dokl. Akad. Nauk tadzh. SSR *21* (9) : 52 - 56, 1978. [In R, ab : Tajik.]

37641 - GILLER, Yu.E., SHCHERBAKOVA, I.Yu., SHUTILOVA, N.I., KADOSHNIKOVA, I.G. :
O svyazi obrazovaniya nativnykh form khlorofilla s protsessami translyatsii
v khloroplastakh i tsitoplazme. [Relation between the formation of native
forms of chlorophyll and translation processes in chloroplasts and cytoplasm.] -
Dokl. Akad. Nauk SSSR *244* : 739 - 742, 1979. [In R.]

37642 - GIMENEZ-GALLEGO, G., RAMIREZ-PONCE, M.P., RAMIREZ, J.M. : Regulation of cyclic photophosphorylation in *Rhodospirillum rubrum* by the redox state of ni-
cotinamide-adenine dinucleotide. - Biochim. biophys. Acta *547* : 211 - 217,
1979.

*37643 - GINS, V.K. : Issledovanie konformatsii ferredoksinov gorokha i kukuruzy pri
deĭstvii spetsificheskikh reagentov na negeminovoe zhelezo i tiolovye gruppy.
[Conformation of pea and maize ferredoxins affected with specific reagents
on non-heme iron and thiol groups.] - In : Biologiya i Nauchno-tekhnicheskiĭ
Progress. Pp. 78 - 81. Pushchino 1974. [In R.]

37644 - GINS, V.K., CHERMNYKH, R.M., MUKHIN, E.N. : Reaktsionnaya sposobnost' nege-
movogo zheleza i tiolovykh grupp ferredoksinov gorokha i kukuruzy pri deĭstvii spetsificheskikh reagentov i mocheviny. [The effect of specific reagents
and urea on the reactivity of non-haem iron and thiol groups of pea and maize
ferredoxins.] - Biokhimiya *44* : 1184 - 1191, 1979. [In R, ab : E.]

*37645 - GIOVANNOZZI-SERMANNI, G., EVANGELISTI, L., VERI, G., PIETROSANTI, T., TRICO-
LI, M. : Metabolismo di *Spinacia oleracea* in relazione al microclima. I. -
Andamento diurno di alcuni componenti organici ed inorganici. [Metabolism
of *Spinacia oleracea* related to the microclimate. I. - Diurnal behavior of
organic and inorganic compounds.] - Atti Simp. int. Agrochim. *9* (IX Simposio
Internazionale di Agrochimica su "La Fitonutrizione Oligominerale" : 309 -
318, 1972. [Chl; in Ital., ab : E.]

*37646 - GIRS, G.I., ZUBAREVA, O.N. : Izmenenie zelenykh pigmentov sosny obyknovennoĭ
pod deĭstviem vysokikh temperatur. [Changes in green pigments of *Pinus nigra*
under high temperatures.] - In : Fiziologo-Biokhimicheskie Protsessy u Khvoĭnykh Rasteniĭ. Pp. 34 - 47, 142. Akad. Nauk SSSR, Krasnoyarsk 1978. [In R.]

*37647 - GIRS, G.I., ZUBAREVA, O.N. : Reaktsiyą karotinoidov khvoi sosny obyknovennoĭ
na deĭstvie vysokikh temperatur. [Response of carotenoids of *Pinus nigra* to
high temperatures.] - In : GIRS, G.I., SUDACHKOVA, N.E. (ed.) : Fiziologo-
-Biokhimicheskie Mekhanizmy Rosta Khvoĭnykh. Pp. 96 - 109. Nauka, Sibir.
Otd., Novosibirsk 1978. [In R.]

37648 - GITEL'ZON, I.I., SHEVYRNOGOV, A.P., MOLVINSKIKH, S.L., CHEPILOV, V.V.,
KARAEV, N.D., PSAKHIS, M.B. : Polevoĭ mnogokanal'nyĭ spektrofotometr MKS-12.
[A field multichannel spectrophotometer MK-12.] - Okeanologiya *19* : 911 -
914, 1979. [In R, ab : E.]

37649 - GIURGEVICH, J.R., DUNN, E.L. : Seasonal patterns of CO_2 and water vapor exchange of the tall and short height forms of *Spartina alterniflora* LOISEL in
a Georgia salt marsh. - Oecologia *43* : 139 - 156, 1979.

37650 - GIVAN, A.L. : Ribulose bisphosphate carboxylase from a mutant strain of *Chlamydomonas reinhardii* deficient in chloroplast ribosomes. The absence of both
subunits and their pattern of synthesis during enzyme recovery. - Planta *144* :
271 - 276, 1979.

37651 - GLAZER, A.N., HIXSON, C.S., DeLANGE, R.J. : Determination of the number of
thioether-linked cysteine residues in cytochromes *c* and phycobiliproteins. -
Anal. Biochem. *92* : 489 - 496, 1979.

37652 - GLAZER, A.N., WILLIAMS, R.C., YAMANAKA, G., SCHACHMAN, H.K. : Characterization of cyanobacterial phycobilisomes in zwitterionic detergents. - Proc.
nat. Acad. Sci. USA *76* : 6162 - 6166, 1979.

37653 - GLIDEWELL, S.M., RAVEN, J.A. : $^{18}O_2$ and $^{13}O_2$ studies of photorespiratory metabolism and oxalate synthesis in *Spinacia oleracea* leaves. - Plant Physiol.
63 (Suppl.) : 111, 1979.

37654 - GLIME, J.M., ACTON, D.W. : Temperature effects on assimilation and respiration in the *Fontinalis duriaei*-periphyton association. - Bryologist *82* : 382 -
392, 1979.

*37655 - GLIWICZ, Z.M., HILLBRICHT-ILKOWSKA, A. : Ecosystem of the Mikołajskie Lake. Elimination of phytoplankton biomass and its subsequent fate in lake through the year. - Pol. Arch. Hydrobiol. 22 : 39 - 52, 1975.

37656 - GLORY, M., VANDEN DRIESCHE, T. : The polysaccharide synthesis circadian rhythm and the hours of changing responsiveness to morphactins in *Acetabularia*. - In : BONOTTO, S., KEFELI, V., PUISEUX-DAO, S. (ed.) : Developmental Biology of *Acetabularia*. Pp. 205 - 219. Elsevier/North-Holland Biomedical Press, Amsterdam - New York - Oxford 1979.

37657 - GLOSER, J. : Photosynthetic characteristics of several important plant species. - In : RYCHNOVSKÁ, M. (ed.) : Function of Glasslands in Spring Region - Kameničky Project. Pp. 147 - 152. Botanical Institute, Czechoslovak Academy of Sciences, Brno 1979.

37658 - GLOSER, J. : A versatile chamber for gas exchange measurements of grassland swards. - In : RYCHNOVSKÁ, M. (ed.) : Function of Grasslands in Spring Region - Kameničky Project. Pp. 161 - 164. Botanical Institute, Czechoslovak Academy of Sciences, Brno 1979.

37659 - GLOVER, H.E., MORRIS, I. : Photosynthetic carboxylating enzymes in marine phytoplankton. - Limnol. Oceanogr. 24 : 510 - 519, 1979.

37660 - GOCHEV, A.D. : Quantum-mechanical study of the primary and secondary electron transfer reactions in bacterial photosynthesis. - Izv. Khim. 12 : 608 - 616, 1979.

37661 - GOCHEV, A.D., CHRISTOV, S.G. : Quantum-mechanical study of electron transfer rate from primary to secondary acceptors in bacterial photosynthesis. - Dokl. bolg. Akad. Nauk 32 : 321 - 324, 1979.

37662 - GODIK, V.I., BORISOV, A.Yu. : Short-lived delayed luminescence of photosynthetic organisms. I. Nanosecond afterglows in purple bacteria at low redox potentials. - Biochim. biophys. Acta 548 : 296 - 308, 1979.

37663 - GODIK, V.I., KOTOVA, E.A., BORISOV, A.Yu. : Nanosekundnaya rekombinatsionnaya lyuminestsentsiya purpurnykh bakteriĭ. [Nanosecond luminescence of purple bacteria.] - Dokl. Akad. Nauk SSSR 249 : 990 - 993, 1979. [In R.]

37664 - GOLBECK, J.H., KOK, B. : Redox titration of electron acceptor Q and the plastoquinone pool in Photosystem II. - Biochim. biophys. Acta 547 : 347 - 360, 1979.

37665 - GOL'D, V.M., GAEVSKIĬ, N.A., BELONOG, N.P., GRIGOR'EV, Yu.S. : Deĭstvie Cd^{2+} i Mg^{2+} na fotosinteticheskuyu aktivnost' khloroplastov, ikh strukturu i vyvod fluorestsentsii. [The effects of Cd^{2+} and Mg^{2+} on the structure, photosynthetic activity and fluorescence efficiency of chloroplasts.] - Biofizika 24 : 778, 1979. [In R.]

37666 - GOL'DFEL'D, M.G., DMITROVSKIĬ, L.G., BLYUMENFEL'D, L.A. : Vliyanie intensivnosti sveta i pronikayushchikh ionov na èffektivnost' fotofosforilirovaniya v khloroplastakh pri impul'snom osveshchenii. [Irradiance and penetrating ions effects on photophosphorylation yield in chloroplasts under pulse excitation.] - Biofizika 24 : 1106 - 1108, 1979. [In R, ab : E.]

37667 - GOL'DFEL'D, M.G., KHALILOV, R.I. : Lokalizatsiya medi v fotosinteticheskom apparate khloroplasta. [Localization of copper in the photosynthetic apparatus of chloroplast.] - Biofizika 24 : 762 - 765, 1979. [In R, ab : E.]

37668 - GOL'DFEL'D, M.G., KHALILOV, R.I., KHANGULOV, S.V. : Svetozavisimyĭ paramagnitnyĭ tsentr v fotosisteme 2 vysshikh rasteniĭ. [Light-dependent paramagnetic centre in photosystem 2 of higher plant chloroplasts.] - Mol. Biol. (Moskva) 13 : 324 - 336, 1979. [In R, ab : E.]

37669 - GOL'DFEL'D, M.G., KONONENKO, A.A., NOKS, P.P., KHALILOV, R.I., KHANGULOV, S.B.: Molyarnaya èkstinktsiya pigmenta v reaktsionnom tsentre fotosistemy II khloroplastov. [Molar extinction of pigment in the reaction centre of chloroplast photosystem II.] - Biofizika 24 : 162 - 163, 1979. [In R, ab : E.]

37670 - GOL'DFEL'D, M.G., TIMOFEEV, V.P., KHALILOV, R.I. : Vliyanie orientatsii v magnitnom pole na formu signala è.p.r. II v fotosinteticheskikh sistemakh.

[Effect of orientation in a magnetic field on the shape of the EPR II signal
in photosynthetic systems.] - Dokl. Akad. Nauk SSSR *247* : 235 - 237, 1979.
[In R.]

37671 - GOL'DFEL'D, M.G., VOZVYSHAEVA, L.V., YUSHMANOV, V.E. : Magnitnaya relaksa-
tsiya protonov vody i sostoyanie sistemy fotorazlozheniya vody v khloroplastakh. [Magnetic relaxation of water protons and the state of water photodis-
sociation system   in chloroplasts.]   - Biofizika *24* : 264 - 269, 1979.
[In R, ab : E.]

37672 - GOLDMAN, J.C. : Outdoor algal mass cultures - I. Applications. - Water Res.
13 : 1 - 19, 1979. [Ps.]

37673 - GOLDMAN, J.C. : Outdoor algal mass cultures - II. Photosynthetic yield limitations. - Water Res. *13* : 119 - 136, 1979.

37674 - GOLDSCHMIDT, E.E. : The fate of chlorophyll-protein complexes during senescence of detached parsley leaves. - Plant Physiol. *63* (Suppl.) : 74, 1979.

37675 - GOLECKI, J., DREWS, G., BÜHLER, R. : The size and number of intramembrane
particles in cells of the photosynthetic bacterium *Rhodopseudomonas capsulata* studied by freeze-fracture electron microscopy. - Cytobiologie *18* : 381 -
- 389, 1979.

37676 - GOLECKI, J.R. : Ultrastructure of cell wall and thylakoid membranes of the
thermophilic cyanobacterium *Synechococcus lividus* under the influence of temperature shifts. - Arch. Microbiol. *120* : 125 - 133, 1979.

*37677 - GOLOMAZOVA, G.M. : Optimal'nye usloviya fotosinteza listvennitsy sibirskoĭ.
[Optimum conditions for photosynthesis of *Larix sibirica*.] - In : Fiziologo-
-Biokhimicheskie Protsessy u Khvoĭnykh Rasteniĭ. Pp. 24 - 34, 142. Akad. Nauk
SSSR, Krasnoyarsk 1978. [In R.]

37678 - GOLUBKOVA, B.M., KISLYAKOVA, T.E., SHAKHOV, A.A. : K voprosu o biogeneze membran khloroplastov u prorostkov *Magnolia grandiflora* na svetu. [Biogenesis
of chloroplast membranes in sprouts of *Magnolia grandiflora* in the light.] -
Dokl. Akad. Nauk SSSR *249* : 1274 - 1276, 1979. [In R.]

37679 - GOMÓŁKA, B. : Odkrycie fotosyntezy. (W 200 rocznicę ogłoszenia dzieła Jana
Ingen-Housza, 1779.) Cz.I.Od starożytności do końca XVII wieku. [Discovery of
photosynthesis. (At the 200 years anniversary of announcement of Jan Ingen-
-Housz work, 1779.) Part I. From ancient times to the end of XVIIth century.]
- Wiadom. bot. *23* : 155 - 168, 1979. [In Pol.]

37680 - GOMÓŁKA, B. : Odkrycie fotosyntezy. (Część II. Wiek XVIII.) [Discovery of
photosynthesis. (Part II. 18th century.)] - Wiadom. bot. *23* : 271 - 281,
1979. [In Pol.]

37681 - GONCHAROVA, É.A., UDOVENKO, G.V., YAKOVLEV, A.F. : Osobennosti transporta
veshchestv u rasteniĭ pri raznoĭ vodoobespechennosti. [Peculiarities of substance transport in plants with different degree of water supply.] - Fiziol.
Biokhim. kul't. Rast. *11* : 358 - 364, 1979. [Photosynthates; in R, ab : E.]

37682 - GONCHAROVA, N.V., EVSTIGNEEV, V.B. : Élektronnoe stroenie fosfatov v svyazi
s problemoĭ sinteza i ispol'zovaniya ATF. [Electron structure of phosphates
in relation with synthesis and utilization of ATP.] - Biofizika *24* : 9 - 14,
1979. [In R, ab : E.]

37683 - GOODWIN, T.W. : Biosynthesis of terpenoids. - Annu. Rev. Plant Physiol. *30* :
369 - 404, 1979. [Car.]

37684 - GOODWIN, T.W. : Isoprenoid distribution and biosynthesis in flagellates. -
In : LEVANDOWSKY, M., HUTNER, S.H. (ed.) : Biochemistry and Physiology of
Protozoa. 2nd Ed. Pp. 91 - 120. Academic Press, New York - San Francisco -
- London 1979. [Car.]

37685 - GOODWIN, T.W. : Thirty years of biosynthesis. - Pure appl.Chem. *51* : 593 -
- 596, 1979. [Car.]

37686 - GOPALAM, A., GOPALACHARI, N.C. : Biochemical changes during maturation of
flue-cured tobacco. I-Changes in leaf pigments. - Tobacco Res. *5* : 113 - 117,
1979.

37687 - **GORDON, A.J., RYLE, G.J.A., POWELL, C.E.** : The strategy of carbon utilization in uniculm barley II. The effect of continuous light and continuous dark treatments. - J. exp. Bot. *30* : 589 - 599, 1979.

*37688 - **GORDON, J.C., WHEELER, C.T.** : Whole plant studies on photosynthesis and acetylene reduction in *Alnus glutinosa*. - New Phytol. *80* : 179 - 186, 1978.

37689 - **GORHAM, E.** : Shoot height, weight and standing crop in relation to density of monospecific plant stands. - Nature *279* : 148 -   150, 1979.

*37690 - **GORLENKO, V.M.** : Kharakteristika nitchatykh fototrofnykh bakteriĭ iz presnykh ozer. [Characteristics of filamentous phototrophic bacteria from fresh-water lakes.] - Mikrobiologiya *44* : 756 - 758, 1975. [ChI; in R, ab : E.]

37691 - **GORTON, H.L., BRIGGS, W.R.** : Phytochrome responses in light-grown green and achlorophyllous corn seedlings. - Carnegie Inst. Year Book *78* : 134 - 137, 1979. [ChI.]

37692 - **GORTON, H.L., BRIGGS, W.R.** : Shibata shift in corn seedlings grown with and without Sandoz. - Carnegie Inst. Year Book *78* : 138 - 139, 1979.

37693 - **GORYSHINA, T.K.** : O nekotorykh strukturno-funktsional'nykh kharakteristikakh assimilyatsionnogo apparata lista u rasteniĭ lesostepnoĭ dubravy. I. Osobennosti plastidnogo apparata u rasteniĭ raznykh yarusov. [Some structural and functional features of leaf assimilatory apparatus in plants of the forest-steppe oakwood. I. Leaf plastid apparatus in plants of various forest layers.] - Bot. Zh. *64* : 331 - 340, 1979. [In R, ab : E.]

37694 - **GORYSHINA, T.K.** : O nekotorykh strukturno-funktsional'nykh kharakteristikakh assimilyatsionnogo apparata lista u rasteniĭ lesostepnoĭ dubravy. II. Sezonnaya dinamika plastidnogo apparata v travyannom pokrove. [Some structural and functional features of leaf assimilatory apparatus in plants of the forest-steppe oakwood. II. Seasonal dynamics of plastid apparatus in the herbaceous cover.] - Bot. Zh. *64* : 469 - 478, 1979. [In R, ab : E.]

37695 - **GORYSHINA, T.K., ZABOTINA, L.N., PRUZHINA, E.G.** : Osobennosti assimilyatsionnykh tkaneĭ i plastidnogo apparata lista v raznykh chastyakh krony drevesnykh porod v lesostepnoĭ dubrave. [Characteristics of assimilatory tissues and plastid apparatus in leaves of various parts of crown in some tree species of the oakwood.] - Vest. leningr. Univ. *1979* (3) : 67 - 76, 125, 1979. [In R, ab : E.]

37696 - **GOSSE, G., MONTENY, B., PERRIER, A.** : Détermination de la photosynthèse d'une culture de *Panicum maximum* par la méthode aérodynamique. - Photosynthetica *13* : 186 - 197, 1979.

37697 - **GOTO, K.** : Modes of control by the circadian oscillator and the hourglass mechanism of the activities of cytoplasmic and chloroplast glyceraldehyde 3-phosphate dehydrogenases in Lemna gibba G3. - Plant Cell Physiol. *20* : 513 - - 521, 1979.

37698 - **GOTO, K.** : Mechanism of control by a circadian oscillator of chloroplast NADP-linked glyceraldehyde 3-phosphate dehydrogenase in *Lemna gibba* G3. - Plant Cell Physiol. *20* : 523 - 532, 1979.

37699 - **GOTTSCHALK, K.W., DICKMANN, D.I.** : Environmental effects on photosynthesis and stomatal conductance of four, two-year-old *Populus* clones grown in the field. - Plant Physiol. *63* (Suppl.) : 127, 1979.

37700 - **GOUDRIAAN, J.** : A family of saturation type curves, especially in relation to photosynthesis. - Ann. Bot. *43* : 783 - 785, 1979.

37701 - **GOUDRIAAN, J., AJTAY, G.L.** : The possible effects of increased CO_2 on photosynthesis. - In : BOLIN, B., DEGENS, E.T., KEMPE, S., KETNER, P. (ed.) : The Global Carbon Cycle. Pp. 237 - 249. John Wiley & Sons, Chichester 1979.

37702 - **GOUDRIAAN, J., VAN KEULEN, H.** : The direct and indirect effects of nitrogen shortage on photosynthesis and transpiration in maize and sunflower. - Neth. J. agr. Sci. *27* : 227 - 234, 1979.

37703 - **GOUGH, S.P., KANNANGARA, C.G.** : Biosynthesis of Δ-aminolevulinate in greening barley leaves. III: The formation of Δ-aminolevulinate in *tigrina* mutants of barley. - Carlsberg Res. Commun. *44* : 403 - 416, 1979.

37704 - GOULD, J.M., PATTERSON, L.K., LING, E., WINGET, G.D. : Phosphorylation in a simple system of lipids and chloroplast ATP synthetase driven by pulsed ionising radiation. - Nature *280* : 607 - 609, 1979.

37705 - GOVINDJEE, JURSINIC, P.A. : Photosynthesis and fast changes in light emission by green plants. - In : SMITH, K.C. (ed.) : Photochemical and Photobiological Reviews. Vol.4. Pp. 125 - 205. Plenum Press, New York - London 1979.

37706 - GOVINDJEE, MATHIS, P., VERNOTTE, C., WONG, D., SAPHON, S., WYDRZYNSKI, T., BRIANTAIS, J.-M. : Cation effects on system II reactions in thylakoids : Measurements on oxygen evolution, the electrochromic change at 515 nanometers, the primary acceptor and the primary donor. - Z. Naturforsch. *34 C* : 826 - - 830, 1979.

37707 - GOVINDJEE, WONG, D., PRÉZELIN, B.B., SWEENEY, B.M. : Chlorophyll *a* fluorescence of *Gonyaulax polyedra* grown on a light-dark cycle and after transfer to constant light. - Photochem. Photobiol. *30* : 405 - 411, 1979.

37708 - GOWER, R.A., POSNER, H.B. : Effects of light and 3-(3,4-dichlorophenyl)-1,1- -dimethylurea on levels of ATP in *Lemna paucicostata* 6746 and a photosynthetic mutant with abnormal flowering responses. - Plant Physiol. *63* : 548 - 551, 1979.

*37709 - GRACE, J., MARKS, T.C. : Physiological aspects of bog production at Moor House. - In : HEAL, O.W., PERKINS, D.F. (ed.) : Production Ecology of British Moors and Montane Grasslands. Pp. 38 - 51. Springer-Verlag, Berlin - Heidelberg - New York 1978. [Ps.]

37710 - GRADINARSKI, L. : Razvitie assimilyatsionnoĭ poverkhnosti i produktivnost' sakharnoĭ svekly. [Development of the assimilation surface and productivity of sugar beet.] - In : VAKLINOVA, S.G., VANKOVA-RADEVA, R., VASILEVA, V.S. (ed.) : Fotosinteticheskaya Assimilyatsiya CO$_2$ i Fotodykhanie. Pp. 148 - 153. Izdatel'stvo bolgarskoĭ Akademii Nauk, Sofia 1979. [In R.]

37711 - GRADINARSKY, L. : Effect of magnesium deficiency on the photosynthesis of *Zea mays*. - In : Mineral Nutrition of Plants. Vol. II. Pp 153 - 156. Publ. House Central Cooperative Union, Sofia 1979.

37712 - GRAHAM, D., CHAPMAN, E.A. : Interactions between photosynthesis and respiration in higher plants. - In : GIBBS, M., LATZKO, E. (ed.) : Photosynthesis II. (Encycl. Plant Physiol. N.S. Vol.6.) Pp. 150 - 162. Springer-Verlag, Berlin - Heidelberg - New York 1979.

37713 - GRAHAM, D., HOCKLEY, D.G., PATTERSON, B.D. : Temperature effects on phosphoenol pyruvate carboxylase from chilling sensitive and chilling resistant plants. - In : LYONS, J.M., GRAHAM, D., RAISON, J.K. (ed.) : Low Temperature Stress in Crop Plants: The Role of the Membrane. Pp. 453 - 461. Academic Press, New York - San Francisco - London 1979.

37714 - GRAHAM, D., HOCKLEY, D.G., PATTERSON, B.D. : Effect of temperature on PEP- -carboxylase from chilling-sensitive and -resistant plants. - Plant Physiol. *63* (Suppl.) : 77, 1979.

37715 - GRAVES, D.A., GIBSON, M.A., BLEAKNEY, J.S. : The digestive diverticula of *Alderia modesta* and *Elysia chlorotica (Opisthobranchia : Sacoglossa)*. - Veliger *21* : 415 - 422, 1979. [Ps.]

37716 - GREBANIER, A.E., STEINBACK, K.E., BOGORAD, L. : Comparison of the molecular weights of proteins synthesized by isolated chloroplasts with those which appear during greening in *Zea mays*. - Plant Physiol. *63* : 436 - 439, 1979.

37717 - GREEN, B.R. : Five unique minor chlorophyll-protein complexes from *Acetabularia* thylakoid membranes. - Plant Physiol. *63* (Suppl.) : 29, 1979.

37718 - GREENBAUM, E. : The turnover times and pool sizes of photosynthetic hydrogen production by green algae. - Sol. Energy *23* : 315 - 320, 1979.

37719 - GREENE, D.M., SUTHERLAND, S.M., KIRKHAM, M.B. : Influence of area on winter wheat climatic models. - Climatic Change *2* : 21 - 32, 1979.

37720 - GREENE, R.V., LANYI, J.K. : Proton movements in response to a light-driven electrogenic pump for sodium ions in *Halobacterium halobium* membranes. - J. biol. Chem. *254* : 10986 - 10994, 1979.

37721 - GREGORY, P.J., SQUIRE, G.R. : Irrigation effects on roots and shoots of pearl
millet (*Pennisetum typhoides*). - Exp. Agr. *15* : 161 - 168, 1979. [Growth analysis.]

*B37722 - GREGORY, R.P.F. : Biochemistry of Photosynthesis. 2nd Edition. - J.Willey &
Sons Ltd., London - New York - Sydney - Toronto 1977.

37723 - GREGORY, R.P.F., DROPPA, M., HORVÁTH, G., EVANS, E.H. : A comparison based
on delayed light emission and fluorescence induction of intact chloroplasts
isolated from mesophyll protoplasts and bundle-sheath cells of maize. - Biochem. J. *180* : 253 - 256, 1979.

37724 - GRESSEL, J. : A review of the place of *in vitro* cell culture systems in studies of action, metabolism and resistance of biocides affecting photosynthesis. - Z. Naturforsch. *34 C* : 905 - 913, 1979.

37725 - GRIERSON, D. : Light-stimulated accumulation of 14S RNA during chloroplast
development in spinach cotyledons. - Z. Pflanzenphysiol. *95* : 171 - 177,
1979.

*37726 - GRIFFITHS, W.T. : Source of reducing equivalents for the *in vitro* synthesis
of chlorophyll from protochlorophyll. - FEBS Lett. *46* : 301 - 304, 1974.

B37727 - GRIME, J.P. : Plant Strategies and Vegetation Processes. - John Wiley & Sons,
Chichester - New York - Brisbane - Toronto 1979. [Primary production.]

37728 - GRODEN, D., BECK, E. : H_2O_2 destruction by ascorbate-dependent systems from
chloroplasts. - Biochim. biophys. Acta *546* : 426 - 435, 1979.

37729 - GRODZINSKI, B. : A study of formate production and oxidation in leaf peroxi-
somes during photorespiration. - Plant Physiol. *63* : 289 - 293, 1979.

37730 - GROSS, E.L. : Cation-induced increases in the rate of P700 recovery in Photosystem I particles. - Arch. Biochem. Biophys. *195* : 198 - 204, 1979.

37731 - GROSS, E.L., ABDELLA, P.M., BURKEY, K. : The effect of cations, chemical modification and immobilization on the rate of P700 recovery using plastocyanin
as an electron donor. - Plant Physiol. *63* (Suppl.) : 54, 1979.

37732 - GROSS, J. : The pigments of three hybrid varieties of broccoli (*Brassica oleracea* var. *italica*). - Gartenbauwissenschaft *44* : 213 - 216, 1979.

37733 - GROSS, J., AYADI, A., MARMÉ, D. : Protochlorophyll(ide)$_{630}$ photosensitizes
active Ca^{2+} accumulation in microsomal and mitochondrial fractions isolated
from plants. - Photochem. Photobiol. *30* : 615 - 621, 1979.

37734 - GROSS, J., LENZ, F. : Violaxanthingehalt als Indikator zur Erntezeitbestim-
mung bei "Golden Delicious" Äpfeln. - Gartenbauwissenschaft *44* : 134 - 135,
1979.

37735 - GROSS, J., LENZ, F. : The effect of an induced water stress on pigment changes in broccoli (*Brassica oleracea* var. *italica*). - Gartenbauwissenschaft *44* :
159 - 161, 1979.

37736 - GROSS, L.J., CHABOT, B.F. : Time course of photosynthetic response to changes
in incident light energy. - Plant Physiol. *63* : 1033 - 1038, 1979.

37737 - GROSSMAN, A., TOGASAKI, R.K. : Temperature sensitive mutations in *Chlamydomonas reinhardi* affecting photophosphorylation. - Plant Physiol. *63* (Suppl.):
145, 1979.

37738 - GROVER, I.S., TYAGI, P.S. : Induction of chlorophyll mutants by some common
pesticides. - Indian J. exp. Biol. *17* : 609 - 611, 1979.

37739 - GRUMBACH, K.H. : Evidence for the existence of two β-carotene pools and two
biosynthetic β-carotene pathways in the chloroplast. - Z. Naturforsch. *34 C* :
1205 - 1208, 1979.

37740 - GRUMBACH, K.H., BACH, T.J. : The effect of PS II herbicides, Amitrol and
SAN 6706 on the activity of 3-hydroxy-3-methylglutaryl-coenzyme-A-reductase
and the incorporation of [2-^{14}C]acetate and [2-^{3}H]mevalonate into chloroplast pigments of radish seedlings. - Z. Naturforsch. *34 C* : 941 - 943, 1979.

*37741 - **GRUMBACH, K.H., BRITTON, G., GOODWIN, T.W.** : Incorporátion von [D-4,5-³H]-leu-
cine und [2-¹⁴C]-Acetat in die photosynthetischen Pigmente und Chinone von
Chlorella pyrenoidosa. - Ber. deut. bot. Ges. *91* : 495 - 508, 1978.

37742 - **GRZESIAK, S.** : Influence of sulphur dioxide on the relative rate of photosynthesis in four species of cultivated plants under optimum soil moisture
and drought conditions. - Bull. Acad. pol. Sci., Sér. Sci. biol. *27* : 309 -
- 321, 1979.

37743 - **GUDKOV, N.D., STOLOVITSKIĬ, Yu.M., EVSTIGNEEV, V.B.** : O roli singletno-voz-
buzhdennogo i tripletnogo sostoyaniĭ pri fotookislenii khlorofilla. [Role of
singlet-excited and triplet states in chlorophyll photooxidation.] - Biofizika
24 : 197 - 201, 1979. [In R, ab : E.]

37744 - **GUERS, J., MOUSSEAU, M.** : Influence de la température sur l'activité photosynthétique du Cacaoyer (*Theobroma cacao* L.). - Compt. rend. Acad. Sci. Paris,
Sér. D *289* : 797 - 800, 1979.

37745 - **GUIKEMA, J.A., YOCUM, C.F.** : Steady-state kinetic analyses of Photosystem II
activity catalyzed by lipophilic electron acceptors. - Biochim. biophys. Acta
547 : 241 - 251, 1979.

37746 - **GUILLOT, F.S., DROSTE, T., HART, W.** : Egg attachment of citrus blackfly *Aleu-
rocanthus woglumi* ASHBY to citrus leaves. - Southwest. Entomol. *4* : 167 -
- 169, 1979. [Ps.]

37747 - **GUILLOT-SALOMON, T., TUQUET, C., LUBAC, M. de, HALLAIS, M.F., SIGNOL, M.** :
Ultrastructure and lipid composition of chloroplasts of shade and sun plants.
- In : APPELQVIST, L.-Å., LILJENBERG, C. (ed.) : Advances in the Biochemistry
and Physiology of Plant Lipids. Pp. 169 - 174. Elsevier/North-Holland Biomedical Press, Amsterdam - New York - Shannon 1979.

37748 - **GULYA, T. Jr., DUNLEAVY, J.M.** : Inhibition of chlorophyll synthesis by *Pseudomonas glycinea.* - Crop Sci. *19* : 261 - 264, 1979.

37749 - **GULYAEV, B.A., KUKARSKIKH, G.P., TIMOFEEV, K.N., KRENDELEVA, T.E.** : Fotokhi-
micheskie i spektral'nye svoĭstva chastits, obogashchennykh reaktsionnymi
tsentrami fotosistemy I. [Photochemical and spectral properties of particles
enriched with photosystem I reaction centres.] - Biokhimiya *44* : 564 - 569,
1979. [In R, ab : E.]

37750 - **GULYAEV, B.A., TETEN'KIN, V.L., POMERANTSEVA, O.M.** : Svetosobirayushchiĭ pig-
ment-belkovyĭ kompleks vysshikh rasteniĭ. [Light-gathering pigment-protein
complex of higher plants.] - Dokl. Akad. Nauk SSSR *248* : 752 - 755, 1979.
[In R.]

37751 - **GULYAEV, B.I.** : Fotosintez i potentsial'naya produktivnost' sel'skokhozyaĭstvennykh kul'tur. [Photosynthesis and potential productivity of agricultural crops.] - Fiziol. Biokhim. kul't. Rast. *11* : 527 - 536, 1979. [In R, ab :
E.]

37752 - **GULYAEV, B.I.** : Reaktsiya ust'its na izmenenie intensivnosti sveta i kontsentratsii CO₂. [Stomata response to changes in irradiance and CO₂ concentration.]- Fiziol. Biokhim. kul't. Rast. *11* : 593 - 600, 1979. [Ps ; in R, ab :
E.]

*37753 - **GULYAEV, B.I., SLUKHAĬ, S.I., LIKHOLAT, D.A., PETRENKO, N.I.** : Vliyanie urovnya kaliĭnogo pitaniya na fotosintez, dykhanie i diffuzionnoe soprotivlenie
list'ev kukuruzy. [Effect of potassium nutrition level on photosynthesis,
respiration and diffusion resistance of maize leaves.] - Dokl. Akad. Nauk
Ukr. SSR, Ser.B *1978* : 1045 - 1048, 1978. [In R, ab : E.]

37754 - **GÜNTHER, G., BAUMANN, G., KLOS, J., BALFANZ, J.** : Nettophotosynthese und Photorespiration bei Zuckerrübe (*Beta vulgaris*) und Weißem Gänsefuß (*Chenopodium
album*). - Biochem. Physiol. Pflanzen *174* : 616 - 628, 1979.

37755 - **GUSHCHINA, L.M., VAKLINOVA, S.G.** : Raspredelenie i funktsiya karboangidrazy,
FEP- i RDF-karboksilaz i vozdeĭstvie diamoksa na ikh aktivnost' v list'yakh
kukuruzy. [Distribution and function of carbonic anhydrase, PEP carboxylase
and RuBP carboxylase and the diamox influence on their activity in maize
leaves.] - In : VAKLINOVA, S.G., VANKOVA-RADEVA, R., VASILEVA, V.S. (ed.) :

Fotosinteticheskaya Assimilyatsiya CO_2 i Fotodykhanie. Pp. 57 - 61. Izdatel'-stvo bolgarskoĭ Akademii Nauk, Sofiya 1979. [In R.]

*37756 - GYURJÁN, I., H.NAGY, A., KERESZTES, Á. : Defektusos kloroplasztiszok szerke-zete és makromolekuláris összetétele variegált *Tradescantia albiflora* leve-lekben. [The structure and macromolecular composition of defective chloro-plasts in the variegated leaves of *Tradescantia albiflora*.] - Biológia (Bu-dapest) *23* : 175 - 184, 1975. [In Hung., ab : E.]

37757 - GYURJÁN, I., YURINA, N.P., THURISHCHEVA, M.S., ODINTSOVA, M.S. : Altered chloroplast ribosomal proteins in a yellow mutant of *Chlamydomonas reinhard-tii*. - Mol. gen. Genet. *170*: 203 - 211, 1979.

37758 - HAAS, R., HEINZ, E., POPOVICI, G., WEISSENBÖCK, G. : Protoplasts from oat primary leaves as tools for experiments on the compartmentation in lipid and flavonoid metabolism. - Z. Naturforsch. *34 C* : 854 - 864, 1979. [Ps.]

37759 - HABERKORN, R., MICHEL-BEYERLE, M.E. : On the mechanism of magnetic field effects in bacterial photosynthesis. - Biophys. J. *26* : 489 - 498, 1979.

37760 - HABERKORN, R., MICHEL-BEYERLE, M.E., MARCUS, R.A. : On spin-exchange and electron-transfer rates in bacterial photosynthesis. - Proc. nat. Acad. Sci. USA *76* : 4185 - 4188, 1979.

37761 - HABESHAW, D. : The effect of foliar pathogens on the leaf photosynthetic car-bon dioxide uptake of barley. - In : MARCELLE, R., CLIJSTERS, H., VAN POUCKE, M. (ed.) : Photosynthesis and Plant Development. Pp. 355 - 373. Dr.W.Junk bv. Publ., The Hague - Boston - London 1979.

37762 - HACHE, A., SEITZ, H.-P., GÜNTHER, K. : Some theoretical aspects of electron transport chain. - Stud. biophys. *75* : 175 - 182, 1979.

37763 - HACKNEY, D.D., ROSEN, G., BOYER, P.D. : Subunit interaction during catalysis: Alternating site cooperativity in photophosphorylation shown by substrate modulation of $[^{18}O]$ATP species formation. - Proc. nat. Acad. Sci. USA *76* : 3646 - 3650, 1979.

37764 - HÄDER, D.-P. : Photomovement. - In : HAUPT, W., FEINLEIB, M.E. (ed.) : Phy-siology of Movements. (Encycl. Plant Physiol. N.S. Vol.7.) Pp. 268 - 309. Springer-Verlag, Berlin - Heidelberg - New York 1979. [Chl.]

37765 - HAENHEL, W., HOCHHEIMER, H.J. : On the current generated by a galvanic cell driven by photosynthetic electron transport. - Bioelectrochem. Bioenerg. *6* : 563 - 574, 1979. J. Electroanal. Chem. *104* : 563 - 574, 1979.

37766 - HAGAR, W.G., FREEBERG, J.A. : Photosynthetic rates of fern sporophytes and gametophytes. - Plant Physiol. *63* (Suppl.) : 65, 1979.

37767 - HAGAR, W.G., HIYAMA, T. : Characterization of the light-induced transient states of the chlorophyll proteins 668 and 743 from *Atriplex rosea*. - Plant Physiol. *63* : 1182 - 1186, 1979.

37768 - HAGEMAN, R.H. : Integration of nitrogen assimilation in relation to yield. - In : HEWITT, E.J., CUTTING, C.V. (ed.) : Nitrogen Assimilation of Plants. Pp. 591 - 611. Academic Press, London - New York - San Francisco 1979.

*37769 - HAGEMANN, R. : Hipotezy na temat ewolucji chloroplastów i mitochondriów. Po-chodzenie endosymbiontyczne czy endogenna kompartmentacja ? [Hypotheses on the evolution of chloroplasts and mitochondria. Are they of endosymbiotic or endogenous compartmentation origin ?] - Kosmos (Warszawa), Ser.A *27* : 477 - 491, 1978. [In Pol.]

37770 - HAGEMANN, R. : Genetics and molecular biology of plastids of higher plants. - Stadler Symp. *11* : 91 - 115, 1979.

37771 - HAGEMANN, R., BÖRNER, T. : Genetische, cytologische und molekularbiologische Analyse der Plastiden höherer Pflanzen. - Wiss. Z. Martin Luther Univ. Halle--Wittenberg, Math.-naturwiss. Reihe *28* (3) : 49 - 60, 137 - 141, 1979.

☆37772 - HAGIHARA, A., HOZUMI, K. : Studies on photosynthetic production and its sea-
         sonal change in a *Chamaecyparis obtusa* plantation. - J. jap. Forest. Soc. *59* :
         327 - 337, 1977.

37773 - HALES, B.J., GUPTA, A.D. : Orientation of the bacteriochlorophyll triplet
         and the primary ubiquinone acceptor of *Rhodospirillum rubrum* in membrane mul-
         tilayers determined by ESR spectroscopy. - Biochim. biophys. Acta *548* : 276 -
         286, 1979.

37774 - HALL, A.E. : A model of leaf photosynthesis and respiration for predicting
         carbon dioxide assimilation in different environments. - Oecologia *143* :
         299 - 316, 1979.

☆37775 - HALL, A.E., COGGINS, C.W. Jr. : Chlorophyll destruction in lemon fruit by
         light and 2', 4'-dichloro-1-cyanoethanesulphonanilide. - Physiol. Plant. *44* :
         221 - 223, 1978.

37776 - HALL, D.O. : "Fortunately for us, plants are very adaptable and exist in
         great diversity ... they could thus continue indefinitely to supply us with
         renewable quantities of food, fibre, fuel and chemicals." - Nature *278* :
         114 - 117, 1979. [Ps.]

37777 - HALL, D.O. : Photosynthesis research in the USSR. - Nature *281* : 255 - 256, 1979.

37778 - HALL, D.O. : Solar energy use through biology - past , present and future. -
         Solar Energy *22* : 307 - 328, 1979. [Ps.]

37779 - HALL, D.O. : The plant chloroplast : The renewable source of food, chemicals
         and fuel. - Proc. royal Inst. Great Britain (Speaking of Science Series) *51* :
         1 - 20, 1979.

37780 - HALL, D.O. : The status of solar energy as fuel. - Impact Sci. Soc. *29* :
         307 - 317, 1979. [Ps.]

37781 - HALL, D.O. : World biomass : An overview. - In : Biomass for Energy. Pp. 1 -
         14. UK-ISES, London 1979. [Productivity.]

☆B37782 - HALL, D.O., RAO, K.K. : Photosynthesis. Second Edition. - Edward Arnold,
         London 1977.

37783 - HALL, N.P., McCURRY, S.D., TOLBERT, N.E. : Storage and stabilizing ribulose-
         -P_2 carboxylase/oxygenase preparations. - Plant Physiol. *63* (Suppl.) : 64,
         1979.

37784 - HALLDAL, P. : Biological energy capture : Special features of marine plants.
         - In : SIMONETT, D.S. (ed.) : Marine Sciences and Ocean Policy Symposium.
         Pp. 21 - 31. Univ. California, Santa Barbara 1979. [Ps.]

37785 - HALLDAL, P. : Effects of changing levels of ultraviolet radiation on phyto-
         plankton. - In : BISWAS, A.K. (ed.) : The Ozone Layer. Pp. 21 - 34. Perga-
         mon Press, Oxford - New York - Toronto - Sydney - Paris - Frankfurt 1979.
         [Ps.]

37786 - HALLDAL, P., HOLMEN, A.T. : The effect of DCMU (3-(,4-dichlorophenyl)-1,1-
         -dimethylurea) on photosynthesis, glycolate excretion ("photorespiration")
         and amino acid metabolism in the blue-green alga *Anacystis*. - Physiol. Plant.
         47 : 195 - 199, 1979.

37787 - HALLDAL, P., HOLMEN, A.T. : The interrelationship between photosynthetic
         electron transport, glycolate excretion and amino acid metabolism in the
         blue-green alga *Anacystis nidulans*. - Plant Cell Physiol. *20* : 757 - 763,
         1979.

37788 - HALLENBECK, P.C., BENEMANN, J.R. : Hydrogen from algae. - In : BARBER, J.
         (ed.) : Photosynthesis in Relation to Model Systems. Pp. 331 - 364. Elsevier,
         Amsterdam - New York - Oxford 1979.

37789 - HALLENSTVET, M., LIAAEN-JENSEN, S., SKULBERG, O.M. : Carotenoids of *Oscilla-
         toria bornetii* f. *tenuis*. - Biochem. Syst. Ecol. *7* : 1 - 2, 1979.

37790 - HAMILTON, R., MAGUIRE, D., McCABE, M. : A versatile microstirrer and oxygen
         electrode system for a spectrophotometer cuvette. - Anal. Biochem. *93* : 386 -
         389, 1979.

37791 - HAMMEL, K.E., CORNWELL, K.L., BASSHAM, J.A. : Stimulation of dark CO_2 fixation by ammonia in isolated mesophyll cells of *Papaver somniferum* L. - Plant Cell Physiol. *20* : 1523 - 1529, 1979.

*37792 - HAMMERSCHLAG, R.S., HILTON, J.L., BARTELS, P.G., MORELAND, D.E. : Contribution of side chains to karbutilate mode of action. - Weed Sci. *23* : 425 - 427, 1975. [Ps.]

37793 - HAMPP, R. : Kinetics of mitochondrial phosphate transport and rates of respiration and phosphorylation during greening of etiolated *Avena* leaves. - Planta *144* : 325 - 332, 1979. [RuBPC, Cyt.]

37794 - HAMPP, R., WELLBURN, A.R. : Control of mitochondrial activities by phytochrome during greening. - Planta *147* : 229 - 235, 1979. [Ps, Chl.]

37795 - HANIFFA, M.A. : Studies of monthly variation of macrophytic biomass in pond Idumban, Palni. - Comp. Physiol. Ecol. *4* : 64 - 67, 1979. [Growth analysis.]

37796 - HANISAK, M.D. : Growth patterns of *Codium fragile* ssp. *tomentosoides* in response to temperature, irradiance, salinity, and nitrogen source. - Mar. Biol. *50* : 319 - 332, 1979.

*37797 - HANNAN, P.J., LAMONTAGNE, R.A., SWINNERTON, J.W., PATOUILLET, C. : Algae, ultraviolet light, and the production of trace gases. - NRL Memorandum Rep. (Washington) *3664* : 1 - 19, 1977. [Production of CO and C_1-C_4 hydrocarbons by irradiated algae.]

37798 - HANNAN, P.J., PATOUILLET, C. : An algal toxicity test and evaluation of adsorption effect. - J. Water Pollut. Control Fed. *51* : 834 - 840, 1979. [Chl.]

37799 - HANSCOM, Z. III., JOHNSON, H.B., HOFFMAN, G.J. : The effect of environmental stress on the salt tolerance and gas exchange of pinto bean. - Plant Physiol. *63* (Suppl.) : 60, 1979.

37800 - HANSELMANN, K.W., BEYELER, W., PFLUGSHAUPT, C., BACHOFEN, R. : Photophosphorylation with chromatophore membranes from *Rhodospirillum rubrum*. - In : CARAFOLI, E., SEMENZA, G. (ed.) : Membrane Biochemistry. A Laboratory Manual on Transport and Bioenergetics. Pp. 120 - 143. Springer-Verlag, Berlin - Heidelberg - New York 1979.

37801 - HANSEN, G.K. : Influence of nitrogen form and nitrogen absence on utilization of assimilates for growth and maintenance in tops of *Lolium multiflorum*. - Physiol. Plant. *46* : 165 - 168, 1979. [Ps.]

37802 - HANSEN, P. : ^{14}C-studies on apple trees. IX. Seasonal changes in the formation of fruit constituents and their subsequent conversions. - Physiol. Plant. *47* : 190 - 194, 1979.

37803 - HANSON, A.D., TULLY, R.E. : Amino acids translocated from turgid and water-stressed barley leaves. II. Studies with ^{13}N and ^{14}C. - Plant Physiol. *64* : 467 - 471, 1979.

37804 - HANSON, W.D., YEH, R.Y. : Genotypic differences for reduction in carbon exchange rates as associated with assimilate accumulation in soybean leaves. - Crop Sci. *19* : 54 - 58, 1979.

37805 - HARAUX, F., KOUCHKOVSKY, Y. de : About the use of 9-aminoacridine for measuring photosynthetic protons movements. - Plant Physiol. *63* (Suppl.) : 42, 1979.

37806 - HARAUX, F., KOUCHKOVSKY, Y. de : Quantitative estimation of the photosynthetic proton binding inside the thylakoids by correlating internal acidification to external alkalinisation and to oxygen evolution in chloroplasts. - Biochim. biophys. Acta *546* : 455 - 471, 1979.

37807 - HARDY, R.W.F. : Chemical plant growth regulation in world agriculture. - In : SCOTT, T.K. (ed.) : Plant Regulation and World Agriculture. Pp. 165 - 206. Plenum Press, New York - London 1979. [Ps.]

37808 - HARI, P., KANNINEN, M., KELLOMÄKI, S., LUUKKANEN, O., PELKONEN, P., SALMINEN, R., SMOLANDER, H. : An automatic system for measurements of gas exchange and environmental factors in a forest stand, with special reference to measuring principles. - Silva fenn. *13* : 94 - 100, 1979.

37809 - **HARJULA, H.** : Analysis of errors in estimating phytoplankton primary productivity and chlorophyll *a* with special reference to Lake Päijänne. - Ann. bot. fenn. *16* : 307 - 337, 1979.

37810 - **HARJULA, H., ROOS, A., GRANBERG, K., KAATRA, K.** : On phytoplankton counting. - Ann. bot. fenn. *16* : 76 - 78, 1979. [Algae biomass estimates.]

37811 - **HARNISCHFEGER, G.** : Connection between the rate of cooling and fluorescence properties at 77 K of isolated chloroplasts. - Biochim. biophys. Acta *546* : 348 - 355, 1979.

37812 - **HARPER, L.A., PALLAS, J.E. Jr., BRUCE, R.R., JONES, J.B. Jr.** : Greenhouse microclimate for tomatoes in the southeast. - J. amer. Soc. Hort. Sci. *104* : 659 - 663, 1979. [Ps.]

37813 - **HARRIMAN, A., BARBER, J.** : Photosynthetic water-splitting process and artificial chemical systems. - In : BARBER, J. (ed.) : Photosynthesis in Relation to Model Systems. Pp. 243 - 280. Elsevier, Amsterdam - New York - Oxford 1979.

37814 - **HARRINGTON, G.** : Estimation of above-ground biomass of trees and shrubs in a *Eucalyptus populnea* F. MUELL. woodland by regression of mass on trunk diameter and plant height. - Aust. J. Bot. *27* : 135 - 143, 1979.

37815 - **HARRIS, D.A., BALTSCHEFFSKY, M.** : Bound nucleotides and phosphorylation in *Rhodospirillum rubrum*. - Biochem. biophys. Res. Commun. *86* : 1248 - 1255, 1979.

37816 - **HARRIS, G., RENTHAL, R., TULEY, J., ROBINSON, N.** : Dansylation of bacteriorhodopsin near the retinal attachment site. - Biochem. biophys. Res. Commun. *91* : 926 - 931, 1979.

37817 - **HARRIS, G.C., O'BRIEN, R.J., MARVEL, D.J.** : Photosynthetic carbon metabolism in *Typha latifolia*. - Plant Physiol. *63* (Suppl.) : 2, 1979.

37818 - **HARRIS, G.P., HEANEY, S.I., TALLING, J.F.** : Physiological and environmental constraints in the ecology of the planktonic dinoflagellate *Ceratium hirundinella*. - Freshw. Biol. *9* : 413 - 428, 1979. [Ps.]

37819 - **HART, F.X., SCHOTTENFELD, R.S.** : Evaporation and plant damage in electric fields. - Int. J. Biometeorol. *23* : 63 - 68, 1979. [Stomatal resistance.]

37820 - **HARVEY, G.W.** : Alteration of photosynthetic unit number and size of mature leaves in response to the lower radient flux. - Plant Physiol. *63* (Suppl.) : 126, 1979.

37821 - **HARVEY, G.W.** : Analysis of photosynthesis cytochrome content with a spectrophotometer-computer interfaced system. - Carnegie Inst. Year Book *78* : 194 - 196, 1979.

37822 - **HASHIMOTO, K., NISHIMURA, M.** : Regulation of electron transfer in *Chromatium vinosum* chromatophores by intravesicular H^+ concentration. - J. Biochem. (Tokyo) *85* : 57 - 64, 1979.

37823 - **HASUMI, H., NAKAMURA, S., KOGA, K., YOSHIZUMI, H.** : Effects of neutral salts on thermal stability of spinach ferredoxin. - Biochem. biophys. Res. Commun. *87* : 1095 - 1101, 1979.

37824 - **HATCH, A., JENSEN, R.G.** : Activation of RuBP carboxylase by intermediates. - Plant Physiol. *63* (Suppl.) : 64, 1979.

*37825 - **HATCH, C.R., GERRARD, D.J., TAPPEINER, J.C. II** : Exposed crown surface area : a mathematical index of individual tree growth potential. - Can. J. Forest Res. *5* : 224 - 228, 1975.

37826 - **HATCH, M.D.** : Mechanism of C_4 photosynthesis in *Chloris gayana*: Pool sizes and kinetics of $^{14}CO_2$ incorporation into 4-carbon and 3-carbon intermediates. - Arch. Biochem. Biophys. *194* : 117 - 127, 1979.

37827 - **HATCH, M.D.** : Regulation of C_4 photosynthesis : factors affecting cold-mediated inactivation and reactivation of pyruvate, P_i dikinase. - Aust. J. Plant Physiol. *6* : 607 - 619, 1979. '

37828 - HATFIELD, J.L. : Canopy temperatures : the usefulness and reliability of remote measurements. - Agron. J. *71* : 889 - 892, 1979.

37829 - HATFIELD, J.L., CARLSON, R.E. : Light quality distributions and spectral albedo of three maize canopies. - Agr. Meteorol. *20* : 215 - 226, 1979.

37830 - HATZIOS, K.K., PENNER, D., BELL, D. : Inhibition of photosynthetic electron
transport in isolated spinach chloroplasts by two 1,3,4-thiadiazolyl derivatives. - Plant Physiol. *63* (Suppl.) : 41, 1979.

B37831 - HAUPT, W., FEINLEIB, M.E. (ed.) : Physiology of Movements. Encyclopedia of
Plant Physiology, New Series, Vol. 7. - Springer-Verlag, Berlin - Heidelberg
- New York 1979. [Chloroplast.]

37832 - HAUPT, W., POLACCO, E. : Phytochrome-mediated response in *Mougeotia* to very
short laser flashes. - Plant Sci. Lett. *17* : 67 - 73, 1979. [Chloroplast.]

37833 - HAUPT, W., REIF, G. : Short-term reactions in phytochrome-regulation of chloroplast orientations. - Acta protozool. *18* : 145, 1979.

*37834 - HAYASHIYA, K., NISHIDA, J., UCHIDA, Y. : [The mechanism of formation of the
red fluorescent protein in the digestive juice of silkworm larvae - The formation of chlorophyllide-α.] - Jap. J. appl. Entomol. Zool. *20* : 37 - 43,
1976. [In Jap., ab : E.]

37835 - HAYHOME, B.A., SCHIFF, J.A., GUILLARD, R.R.L., ALBERTE, R.S. : Absorption
properties of *Platymonas* sp. Rey 2 containing a high proportion of chlorophyll *b*. - Biol. Bull. *157* : 370, 1979.

37836 - HAYSTEAD, A., KING, J., LAMB, W.I.C. : Photosynthesis, respiration and nitrogen fixation in white clover. - Grass Forage Sci. *34* : 125 - 130, 1979.

37837 - HEAGLE, A.S., PHILBECK, R.B., ROGERS, H.H., LETCHWORTH, M.B. : Dispensing
and monitoring ozone in open-top field chambers for plant-effects studies. -
Phytopathology *69* : 15 - 20, 1979.

37838 - HEARN, A.B. : Water relationships in cotton. - Outlook Agr. *10* : 159 - 166,
1979. [Stomatal resistance.]

37839 - HEATHCOTE, L., BAMBRIDGE, K.R., McLAREN, J.S. : Specially constructed growth
cabinets for simulation of the spectral photon distributions found under natural vegetation canopies. - J. exp. Bot. *30* : 347 - 353, 1979.

37840 - HEATHCOTE, P., TIMOFEEV, K.N., EVANS, M.C.W. : Detection by EPR spectrometry
of a new intermediate in the primary photochemistry of photosystem I particles isolated using Triton X-100. - FEBS Lett. *101* : 105 - 109, 1979.

37841 - HEBER, U., ENSER, U., WEIS, E., ZIEM, U., GIERSCH, C. : Regulation of the
photosynthetic carbon cycle, phosphorylation and electron transport in illuminated intact chloroplasts. - In : ATKINSON, D.E., FOX, C.F. (ed.) : Modulation of Protein Function. Pp. 113 - 138. Academic Press, New York - San
Francisco - London 1979.

37842 - HEBER, U., KIRK, M.R., BOARDMAN, N.K. : Photoreactions of cytochrome *b*-559
and cyclic electron flow in Photosystem II of intact chloroplasts. - Biochim.
biophys. Acta *546* : 292 - 306, 1979.

37843 - HEBER, U., VOLGER, H., OVERBECK, V., SANTARIUS, K.A. : Membrane damage and
protection during freezing. - In : FEENEMA, O. (ed.) : Proteins at Low Temperatures. (Advances in Chemistry Ser. Vol. 180.) Pp. 159 - 189. Amer. Chem.
Soc. 1979. [Ps.]

37844 - HEBER, U., WALKER, D.A. : The chloroplast envelope - barrier or bridge ? -
Trends biochem. Sci. *4* : 252 - 256, 1979.

37845 - HEDLEY, C.L., AMBROSE, M.J. : The effects of shading on the yield components
of six "leafless" pea genotypes. - Ann. Bot. *44* : 469 - 478, 1979.

37846 - HEELIS, D., KERNICK, W., PHILLIPS, G.O., DAVIES, K. : Separation and identification of the carotenoid pigments of stigmata isolated from light grown
cells of *Euglena gracilis* strain Z. - Arch. Microbiol. *121* : 207 - 211,
1979.

37847 - HEIM, G., LANDSBERG, J.J., WATSON, R.L., BRAIN, P. : Eco-physiology of apple
trees : dry matter production and partitioning by young Golden Delicious
trees in France and England. - J. appl. Ecol. *16* : 179 - 194, 1979.

37848 - HEIN, M.B., BRENNER, M.L., BRUN, W.A. : Source/sink interactions in soybeans.
II. A possible role of IAA. - Plant Physiol. *63* (Suppl.) : 43, 1979.

37849 - HEINZ, E., SIEBERTZ, H.P., LINSCHEID, M., JOYARD, J., DOUCE, R. : Investigations on the origin of diglyceride diversity in leaf lipids. - In : APPEL-
QVIST, L.-Å., LILJENBERG, C. (ed.) : Advances in the Biochemistry and Physiology of Plant Lipids. Pp. 99 - 120. Elsevier/North-Holland Biomedical Press,
Amsterdam 1979. [Chloroplast.]

37850 - HEISE, K.-P., HARNISCHFEGER, G. : Correlation between photosynthesis and
plant lipid composition. - In : APPELQVIST, L.-Å., LILJENBERG, C. (ed.) :
Advances in the Biochemistry and Physiology of Plant Lipids. Pp. 175 - 180.
Elsevier/North-Holland Biomedical Press, Amsterdam 1979.

*37851 - HEISE, K.-P., STOTTMEISTER, A. : Die diurnale Veränderung des Lipidmusters
in Blättern von *Spinacia oleracea*. - Ber. deut. bot. Ges. *89* : 677 - 694,
1976. [Ps.]

37852 - HELDT, H.W. : Light-dependent changes of stromal H^+ and Mg^{2+} concentrations
controlling CO_2 fixation. - In : GIBBS, M., LATZKO, E. (ed.) : Photosynthesis
II. (Encycl. Plant Physiol. N.S. Vol. 6.) Pp. 202 - 207. Springer-Verlag,
Berlin - Heidelberg - New York 1979.

37853 - HELLINGWERF, K.J., ARENTS, J.C., SCHOLTE, B.J., WESTERHOFF, H.V. : Bacteriorhodopsin in liposomes. II. Experimental evidence in support of a theoretical model. - Biochim. biophys. Acta *547* : 561 - 582, 1979.

37854 - HENDERSON, R. : The structure of bacteriorhodopsin and its relevance to other membrane proteins. - In : CONE, R.A., DOWLING, J.E. (ed.) : Membrane
Transduction Mechanisms. Pp. 3 - 15. Raven Press, New York 1979.

37855 - HENDLER, R.W. : Limitations of the use of and interpretation of data from
the Aminco-Chance dual-wavelength split-beam recording spectrophotometer and
related instruments. - Anal. Biochem. *94* : 450 - 464, 1979.

37856 - HENRY, E.W., RICHARD, L.B. : A study of the effects of applied ethephon on
enzyme (polyphenol oxidase, peroxidase, catalase, ATPase) activity in tobacco
(*Nicotiana tabacum*) apical tissue chloroplasts. - Z. Pflanzenphysiol. *92* :
11 - 22, 1979.

37857 - HERATH, H.M.W., ORMROD, D.P. : Photosynthetic rates of citronella and lemon-
grass. - Plant Physiol. *63* : 406 - 408, 1979.

37858 - HERBERT, M., BURKHARD, C., SCHNARRENBERGER, C. : A survey for isoenzymes
of glucosephosphate isomerase, phosphoglucomutase, glucose-6-phosphate dehydrogenase and 6-phosphogluconate dehydrogenase in C_3-, C_4- and Crassulacean-acid-metabolism plants, and green algae. - Planta *145* : 95 - 104, 1979.

37859 - HERBLAND, A., VOITURIEZ, B. : Hydrological structure analysis for estimating
the primary production in the tropical Atlantic Ocean. - J. mar. Res. *37* :
87 - 101, 1979.

37860 - HERCZEG, T., LEHOCZKI, E., FARKAS, T., ROJIK, I., SZALAY, L. : Cerulenin-in-
duced modification of structural and photosynthetic characteristics in *Chlorella*. - Z. Pflanzenphysiol. *94* : 55 - 64, 1979.

37861 - HERCZEG, T., LEHOCZKI, E., SZALAY, L. : The prompt effect of pyridazinone
herbicides on the primary processes of photosynthesis. - FEBS Lett. *108* :
226 - 228, 1979.

37862 - HEROLD, A., McNEIL, P.H. : Restoration of photosynthesis in pot-bound tobacco plants. - J. exp. Bot. *30* : 1187 - 1194, 1979.

37863 - HEROLD, A., WALKER, D.A. : Transport across chloroplast envelopes. The role
of phosphate. - In : GIEBISCH, G., TOSTESON, D.C., USSING, H.H. (ed.) :
Membrane Transport in Biology. Vol. 2. Pp. 411 - 439. Springer-Verlag, Berlin - Heidelberg - New York 1979.

37864 - HERTZBERG, S., BORCH, G., LIAAEN-JENSEN, S. : Circular dichroic spectra of mono-*cis* carotenoids. - Acta chem. scand. *33* : 42 - 46, 1979.

37865 - HESS, B., KUSCHMITZ, D. : Kinetic interaction between aromatic residues and the retinal chromophore of bacteriorhodopsin during the photocycle. - FEBS Lett. *100* : 334 - 340, 1979.

37866 - HEWITT, E.J., HUCKLESBY, D.P., MANN, A.F., NOTTON, B.A., RUCKLIDGE, G.J. : Regulation of nitrate assimilation in plants. - In : HEWITT, E.J., CUTTING, C.V. (ed.) : Nitrogen Assimilation of Plants. Pp. 255 - 287. Academic Press, London - New York - San Francisco 1979. [Ps.]

37867 - HEYLAND, K.-U., SOLANSKY, S., BECKER, F.A. : Die Assimilatspeicherung in der Sommerweizenähre unter dem Einfluß von Mehltaubefall (*Erysiphe graminis*) auf verschiedenen Assimilationsorganen. - Z. Pflanzenkrank. Pflanzensch. *86* : 513 - 532, 1979.

37868 - HICKMAN, M. : Seasonal succession, standing crop and determinants of primary productivity of the phytoplankton of Ministik Lake, Alberta, Canada. - Hydrobiologia *64* : 105 - 121, 1979.

37869 - HIEKE, B., HOFFMANN, P., LEUPOLD, D., MORY, S., SCHOTTE, J. : Das Absorptionsverhalten von Chlorophyll *in vivo* und *in vitro* bei Anregung mit Laserimpulsen von 694,3 nm. - Photosynthetica *13* : 37 - 44, 1979.

37870 - HIEKE, B., KÖCKRITZ, A. : The DCPIP reduction of isolated chloroplasts from selected species of *Triticum* und *Aegilops* under the influence of the plastoquinone antagonist DBMIB. - Biochem. Physiol. Pflanzen *174* : 517 - 522, 1979.

37871 - HILL, J., ROBSON, A.D., LONERAGAN, J.F. : The effect of copper supply on the senescence and the retranslocation of nutrients of the oldest leaf of wheat. - Ann. Bot. *44* : 279 - 287, 1979. [Chl.]

37872 - HILLER, R.G., PILGER, T.B.G., CAMPBELL, D. : Extraction and stabilisation of the dimer of chlorophyll-protein complex II. - Plant Sci. Lett. *14* : 7 - 11, 1979.

37873 - HILLMER, P., FAHLBUSCH, K. : Evidence for an involvement of glutamine synthetase in regulation of nitrogenase activity in *Rhodopseudomonas capsulata*. - Arch. Microbiol. *122* : 213 - 218, 1979. [H_2 production.]

37874 - HINCKLEY, T.M., DOUGHERTY, P.M., LASSOIE, J.P., ROBERTS, J.E., TESKEY, R.O. : A severe drought : impact on tree growth, phenology, net photosynthetic rate and water relations. - Amer. Midl. Natur. *102* : 307 - 316, 1979.

37875 - HIND, G., SLOVACEK, R.E. : Inhibition and uncoupling of cyclic electron transport. - Plant Physiol. *63* (Suppl.) : 55, 1979.

*37876 - HINDE, R. : The metabolism of photosynthetically fixed carbon by isolated chloroplasts from *Codium fragile* (*Chlorophyta* : *Siphonales*) and by *Elysia viridis* (*Mollusca* : *Sacoglossa*). - Biol. J. Linn. Soc. *10* : 329 - 342, 1978.

37877 - HIPKINS, M.F. : The primary processes of photosynthesis. - Phys. Bull. *30* : 210 - 212, 1979.

37878 - HIRAYAMA, O., MATSUDA, H., SENZAKI, K., MASUDA, T.´: Extraction and reconstitution of lyophilized photosystem I subchloroplast fragments from spinach chloroplasts. - Agr. biol. Chem. *43* : 1205 - 1210, 1979.

37879 - HIRSCH, R.E., BRODY, S.S. : Spectral properties of chlorophyll *a* monolayers : Monolayers of chlorophyll *a* and pheophytin at a gas-water interface. - Photochem. Photobiol. *29* : 589 - 596, 1979.

37880 - HISCOX, J.D., ISRAELSTAM, G.F. : A method for the extraction of chlorophyll from leaf tissue without maceration. - Can. J. Bot. *57* : 1332 - 1334, 1979.

37881 - HITZ, W.D., STEWART, C.R. : Pool sizes of photorespiratory intermediates : Effects of O_2, CO_2 and darkening. - Plant Physiol. *63* (Suppl.) : 153, 1979.

37882 - HIYAMA, T., FORK, D.C. : Kinetic identification of component X as P430 : acceptor of photosystem I. - Carnegie Inst. Year Book *78* : 180 - 182, 1979.

37883 - HIYAMA, T., FORK, D.C. : Kinetic identification of component X as P430 :
A primary electron acceptor of Photosystem I. - Plant Physiol. 63 (Suppl.) :
53, 1979.

37884 - HIYAMA, T., TSUJIMOTO, H.Y., ARNON, D.I. : Photoreduction of membrane-bound
paramagnetic component X by water as electron donor. - FEBS Lett. 98 : 381 -
385, 1979.

37885 - HO, K.K., ULRICH, E.L., KROGMANN, D.W., GOMEZ-LOJERO, C. : Isolation of photosynthetic catalysts from cyanobacteria. - Biochim. biophys. Acta 545 :
236 - 248, 1979.

37886 - HO, L.C. : Regulation of assimilate translocation between leaves and fruits
in the tomato. - Ann. Bot. 43 : 437 - 448, 1979.

37887 - HO, L.C. : Partitioning of ^{14}C-assimilate within individual tomato leaves
in relation to the rate of export. - In : MARCELLE, R., CLIJSTERS, H., VAN
POUCKE, M. (ed.) : Photosynthesis and Plant Development. Pp. 243 - 250.
Dr.W.Junk bv. Publ., The Hague - Boston - London 1979.

37888 - HO, L.C., SHAW, A.F. : Net accumulation of minerals and water and the carbon
budget in an expanding leaf of tomato. - Ann. Bot. 43 : 45 - 54, 1979.

37889 - HO, Y.B. : Chemical composition studies on some aquatic macrophytes in three
Scottish lochs.I. Chlorophyll, ash, carbon, nitrogen and phosphorus. - Hydrobiologia 63 : 161 - 166, 1979.

37890 - HO, Y.B. : Shoot development and production studies of *Phragmites australis*
(CAV.) TRIN. *ex* STEUDEL in Scottish lochs. - Hydrobiologia 64 : 215 - 222,
1979. [Growth analysis.]

37891 - HO, Y.-K., LIU, C.J., SAUNDERS, D.R., WANG, J.H. : Light dependence of the
decay of the proton gradient in broken chloroplasts. - Biochim. biophys.
Acta 547 : 149 - 160, 1979.

37892 - HO, Y.-K., WANG, J.H. : Regulation of proton leakage from broken chloroplasts
by CF_o. - Biochem. biophys. Res. Commun. 89 : 294 - 299, 1979.

37893 - HOBSON, L.A., HARTLEY, F.A., KETCHAM, D.E. : Effects of variations in daylength and temperature on net rates of photosynthesis, dark respiration,
and excretion by *Isochrysis galbana* PARKE. - Plant Physiol. 63 : 947 - 951,
1979.

37894 - HOCHMAN, Y., LANIR, A., WERBER, M.M., CARMELI, C. : The effect of binding
of cobalt(III)-nucleotide complexes on the kinetic properties of adenosine
triphosphatase activity in coupling factor 1 from chloroplasts. - Arch.
Biochem. Biophys. 192 : 138 - 147, 1979.

37895 - HODÁŇOVÁ, D. : Sugar beet canopy photosynthesis as limited by leaf age and
irradiance. Estimation by models. - Photosynthetica 13 : 376 - 385, 1979.

37896 - HODDINOTT, J. : The influence of light quality on carbohydrate translocation
within corn leaf strips. - Plant Physiol. 63 (Suppl.) : 34, 1979.

37897 - HODDINOTT, J., EHRET, D.L., GORHAM, P.R. : Rapid influence of water stress
on photosynthesis and translocation in *Phaseolus vulgaris*. - Can. J. Bot.
57 : 768 - 776, 1979.

37898 - HODGES, T., DORAISWAMY, P.C. : Crop phenology literature review for corn,
soybean, wheat, barley, sorghum, rice, cotton, and sunflower. - Tech. Rep.
No. SR-L9-00409; JSC-16088. Lockhead Electronics Co., Houston, Texas 1979.
[Modelling productivity.]

37899 - HODGES, T., KANEMASU, E.T., TEARE, I.D. : Modeling dry matter accumulation
and yield of grain sorghum. - Can. J. Plant Sci. 59 : 803 - 818, 1979.

37900 - HOFF, A.J., GAST, P. : Transfer of light-induced electron spin polarization
in bacterial photosynthetic reaction centers. - J. phys. Chem. 83 : 3355 -
3358, 1979.

37901 - HOFFMANN, P., HIEKE, B., KÖCKRITZ, A. : On the problems of alternative ways
in photosynthetic electron transport of higher plants. - Biochem. Physiol.
Pflanz. 174 : 579 - 596, 1979.

37902 - HOGETSU, D., MIYACHI, S. : Role of carbonic anhydrase in photosynthetic CO_2 fixation in *Chlorella*. - Plant Cell Physiol. *20* : 747 - 756, 1979.

37903 - HOGETSU, D., MIYACHI, S. : Operation of the reductive pentose phosphate cycle during the induction period of photosynthesis in *Chlorella*. - Plant Cell Physiol. *20* : 1427 - 1432, 1979.

37904 - HÖHLER, T., SCHAUB, H. : Effect of low oxygen partial pressure during growth on the metabolism of aspartic acid and 3-phosphoglyceric acid in leaves of the C_4 plant *Amaranthus paniculatus*. - Biochem. Physiol. Pflanzen *174* : 58 - 67, 1979.

37905 - HOLADAY, S., BROWN, R.H., BLACK, C.C. : Characterization of PEP carboxylase in *Panicum* species. - Plant Physiol. *63* (Suppl.) : 38, 1979.

B37906 - HOLBEN, B.N., TUCKER, C.J., FAN, C. : Assessing Soybean Leaf Area and Leaf Biomass by Spectral Measurements. - NASA technical Memorandum 80312. Pp. 1 - 17. Goddard Space Flight Center, Greenbelt, Md. 1979.

37907 - HOMANN, P.H. : The light dependent quenching of chloroplast fluorescence by cofactors of cyclic electron flow in Photosystem I. - Photochem. Photobiol. *29* : 815 - 822, 1979.

37908 - HOMMERTZHEIM, D.L. : Analytical description of a soybean canopy. - Agron. J. *71* : 405 - 409, 1979.

*37909 - HONDA, H., FISHER, J.B. : Tree branch angle : maximizing effective leaf area. - Science *199* : 888 - 890, 1978.

37910 - HONG, F.T., MONTAL, M. : Bacteriorhodopsin in model membranes. A new component of the displacement photocurrent in the microsecond time scale. - Biophys. J. *25* : 465 - 472, 1979.

37911 - HONIG, B., EBREY, T., CALLENDER, R.H., DINUR, U., OTTOLENGHI, M. : Photoisomerization, energy storage, and charge separation : A model for light energy transduction in visual pigments and bacteriorhodopsin. - Proc. nat. Acad. Sci. USA *76* : 2503 - 2507, 1979.

37912 - HOOPER, A.W., SMITH, R.A., BOWMAN, G.E. : A near-infrared diffuse reflectance spectrophotometer. - J. agr. eng. Res. *24* : 79 - 85, 1979.

37913 - HOPE, A.B., MORLAND, A. : Proton translocation in isolated spinach chloroplasts after single-turnover actinic flashes. - Aust. J. Plant Physiol. *6* : 289 - 304, 1979.

37914 - HOPFIELD, J.J. : Nonadiabatic electron tunneling : implications for bacterial photosynthesis and for critical physical tests of the mechanism. - In : CHANCE, B., FRAUENFELDER, H., MARCUS, R.A., SUTIN, N., DEVAULT, D.C., SCHREIFFER, J.R. (ed.) : Tunneling in Biological Systems. Pp. 417 - 432. Academic Press, New York - San Francisco - London 1979.

37915 - HORIE, T. : Studies on photosynthesis and primary production of rice plants in relation to meteorological environments. II. Gaseous diffusive resistances, photosynthesis and transpiration in the leaves as influenced by atmospheric humidity, and air and soil temperatures. - J. agr. Meteorol. *35* : 1 - 12, 1979.

37916 - HORRUM, M.A., SCHWARTZBACH, S.D. : Glyoxysomal and mitochondrial enzyme activity during light induced chloroplast development in *Euglena*. - Plant Physiol. *63* (Suppl.) : 160, 1979.

37917 - HORTON, P., CROZE, E. : Characterization of two quenchers of chlorophyll fluorescence with different midpoint oxidation-reduction potentials in chloroplasts. - Biochim. biophys. Acta *545* : 188 - 201, 1979.

37918 - HORTON, P., NAYLOR, B. : The influence of chloroplast membrane stacking on the redox properties of the fluorescence quencher. - Photobiochem. Photobiophys. *1* : 17 - 23, 1979.

37919 - HORVÁTH, G., NIEMI, H.A., DROPPA, M., FALUDI-DANIEL, Á. : Characteristics of the flash-induced 515 nanometer absorbance change of intact isolated chloroplasts. - Plant Physiol. *63* : 778 - 782, 1979.

37920 - HOSHINA, S. : Inhibition of electron transport and uncoupling of photophos-
        phorylation in spinach chloroplasts by lysolecithin. - Sci. Rep. Kanazawa
        Univ. *24* : 45 - 53, 1979.

37921 - HOSHINA, S. : Replacement of galactolipids in chloroplasts by lecithin in
        the presence of lysolecithin. - Sci. Rep. Kanazawa Univ. *24* : 81 - 89, 1979.

37922 - HOSHINA, S. : Restoration of lysolecithin-induced inhibition of the Hill re-
        action in spinach chloroplasts by the addition of lecithin. - Plant Cell Phy-
        siol. *20* : 1107 - 1116, 1979.

37923 - HOUCHINS, J.P. : Physiological characterization of a reversible hydrogenase
        from *Anabaena* 7120. - Plant Physiol. *63* (Suppl.) : 30, 1979.

37924 - HOUGH, R.A. : Photosynthesis, respiration, and organic carbon release in
        Elodea canadensis MICHX. - Aquat. Bot. *7* : 1 - 11, 1979.

37925 - HOURSIANGOU-NEUBRUN, D., DUBACQ, J.-P., BONOTTO, S., PUISEUX-DAO, S. : Chlo-
        roplast evolution in anucleate *Acetabularia*. - In : BONOTTO, S., KEFELI, V.,
        PUISEUX-DAO, S. (ed.) : Developmental Biology of *Acetabularia*. Pp. 131 - 139.
        Elsevier/North-Holland Biomedical Press, Amsterdam - New York - Oxford 1979.

37926 - HOUSLEY, T.L., SCHRADER, L.E., MILLER, M., SETTER, T.L. : Partitioning of
        ^{14}C-photosynthate, and long distance translocation of amino acids in preflo-
        wering and flowering, nodulated and nonnodulated soybeans. - Plant Physiol.
        64 : 94 - 98, 1979.

37927 - HOWSLEY, R., PEARSON, H.W. : pH dependent sulphide toxicity to oxygenic pho-
        tosynthesis in cyanobacteria. - FEMS Microbiol. Lett. *6* : 287 - 292, 1979.

37928 - HUBER, S.C. : Effect of pH on chloroplast photosynthesis. Inhibition of O_2
        evolution by inorganic phosphate and magnesium. - Biochim. biophys. Acta
        545 : 131 - 140, 1979.

37929 - HUBER, S.C. : Control of phosphate exchange across the chloroplast envelope
        by stromal pH. - Plant Physiol. *63* (Suppl.) : 3, 1979.

37930 - HUBER, S.C. : Effect of photosynthetic intermediates on the magnesium inhi-
        bition of oxygen evolution by barley chloroplasts. - Plant Physiol. *63* :
        754 - 757, 1979.

37931 - HUBER, S.C. : Orthophosphate control of glucose-6-phosphate dehydrogenase
        light modulation in relation to the induction phase of chloroplast photo-
        synthesis. - Plant Physiol. *64* : 846 - 851, 1979.

37932 - HUBER, S.C., MAURY, W. : Magnesium-activated $(Na^+)K^+/H^+$ exchange across the
        chloroplast envelope. - Plant Physiol. *63* (Suppl.) : 3, 1979.

37933 - HUBER, W. : Die Rolle von Abscisinsäure und Cytokininen in Pflanzen unter
        Stresseinwirkungen. - Ber. deut. bot. Ges. *92* : 193 - 207, 1979. [Ps.]

37934 - HUBER, W., SANKHLA, N. : Effect of sodium chloride on photosynthesis of
        Lemna minor L. - Z. Pflanzenphysiol. *91* : 147 - 156, 1979.

37935 - HUBÍK, E. : Potenciální výnosy obilnin v ČSSR. [Potential yield of cereal
        crops in ČSSR.] - In : Bilancia Energie a Vody v Pol'ných a Lesných Ekosys-
        témoch. Zborník Referátov z Odborného Seminára. Pp. 16 - 24. Vysoká Škola
        pol'nohospodárska, Nitra 1979. [Ps; in Czech.]

37936 - HÜCKEL, D., BEYERSMANN, D. : Rapid purification and direct spectrophotomet-
        ric assay for 5-aminolevulinic acid dehydratase. - Anal. Biochem. *97* : 277 -
        281, 1979.

37937 - HUDOCK, M.O., TOGASAKI, R.K., LIEN, S., HOSEK, M., SAN PIETRO, A. : A uni-
        parently inherited mutation affecting photophosphorylation in *Chlamydomonas
        reinhardi*. - Biochem. biophys. Res. Commun. *87* : 66 - 71, 1979.

37938 - HUEBNER, J.S. : Apparatus for recording light flash induced membrane voltage
        transients with 10ns resolution. - Photochem. Photobiol. *30* : 233 - 241,
        1979. [Chl.]

37939 - HUET, R. : Extraction, dosage et stabilisation des caroténoïdes d'agrumes. -
        Fruits *34* : 479 - 488, 1979.

37940 - HUGHES, G., KEATINGE, J.D.H., SCOTT, S.P. : Leaf area estimation by non-destructive methods in pigeon pea *Cajanus cajan* (L.) MILLSP. - Trop. Agr. *56* :
371 - 374, 1979.

37941 - HUISINGA, B. : Control of loading and unloading by turgor regulation in long
distance transport. - Acta bot. neerl. *28* : 67 - 72, 1979. [Photosynthates.]

37942 - HUISMAN, J.G., TOUW, I., LIEBREGTS, P., BERNARDS, A. : Biosynthesis of ferredoxin in *Chlamydomonas reinhardii*. - Planta *145* : 351 - 356, 1979.

37943 - HUMPHREY, G.F. : Photosynthetic characteristics of algae grown under constant
illumination and light-dark regimes. - J. exp. mar. Biol. Ecol. *40* : 63 - 70,
1979.

*37944 - HUMPHREYS, T. : A model for sucrose transport in the maize scutellum. - Phy-
tochemistry *17* : 679 - 684, 1978. [ATPase.]

37945 - HUNDING, C. : The oxygen balance of Lake Mývatn, Iceland. - Oikos *32* : 139 -
150, 1979. [Primary production.]

37946 - HUNER, N.P.A., MACDOWALL, F.D.H. : Changes in the net charge and subunit properties of ribulose bisphosphate carboxylase-oxygenase during cold hardening
of Puma rye. - Can. J. Biochem. *57* : 155 - 164, 1979.

37947 - HUNER, N.P.A., MACDOWALL, F.D.H. : The effects of low temperature acclimation of winter rye on catalytic properties of its ribulose bisphosphate carboxylase-oxygenase. - Can. J. Biochem. *57* : 1036 - 1041, 1979.

37948 - HUNER, N.P.A., MILLER, R.W. : Spin-labelling of RuBP carboxylase from cold-
-hardened and unhardened Puma rye. - Plant Physiol. *63* (Suppl.) : 109, 1979.

37949 - HUNER, N.P.A., PARSONS, L.R., CARTER, J.V. : The structure and activity of
RuBP carboxylase from rye exposed to low leaf water potentials. - Plant Physiol. *63* (Suppl.) : 139, 1979.

37950 - HUNT, L.D., OGREN, W.L. : Glyoxylate stimulation of photosynthesis in isolated soybean leaf mesophyll cells. - Plant Physiol. *63* (Suppl.) : 38, 1979.

*37951 - HUNT, R. : Demography versus plant growth analysis. - New Phytol. *80* : 269 -
272, 1978.

37952 - HUNT, R. : Plant growth analysis : The rationale behind the use of the fitted
mathematical function. - Ann. Bot. *43* : 245 - 249, 1979.

37953 - HUNT, W.F., LOOMIS, R.S. : Respiration modelling and hypothesis testing with
a dynamic model of sugar beet growth. - Ann. Bot. *44* : 5 - 17, 1979. [Ps.]

37954 - HUNTER, C.N., HOLMES, N.G., JONES, O.T.G., NIEDERMAN, R.A. : Membranes of
Rhodopseudomonas sphaeroides. VII. Photochemical properties of a fraction
enriched in newly synthesized bacteriochlorophyll *a*-protein complexes. -
Biochim. biophys. Acta *548* : 253 - 266, 1979.

37955 - HUNTER, C.N., JONES, O.T.G. : The incorporation of reaction centres into
membranes from a bacteriochlorophyll-less mutant of *Rhodopseudomonas sphaeroides*. - Biochim. biophys. Acta *545* : 325 - 338, 1979.

37956 - HUNTER, C.N., JONES, O.T.G. : The kinetics of flash-induced electron flow
in bacteriochlorophyll-less membranes of *Rhodopseudomonas sphaeroides* reconstituted with reaction centres. - Biochim. biophys. Acta *545* : 339 - 351,
1979.

37957 - HUNTER, C.N., VAN GRONDELLE, R., HOLMES, N.G., JONES, O.T.G. : The reconstitution of energy transfer in membranes from a bacteriochlorophyll-less mutant of *Rhodopseudomonas sphaeroides* by addition of light-harvesting and
reaction centre pigment-protein complexes. - Biochim. biophys. Acta *548* :
458 - 470, 1979.

37958 - HUNTER, C.N., VAN GRONDELLE, R., HOLMES, N.G., JONES, O.T.G., NIEDERMAN, R.A.
: Fluorescence yield properties of a fraction enriched in newly synthesized
bacteriochlorophyll *a*-protein complexes from *Rhodopseudomonas sphaeroides*. -
Photochem. Photobiol. *30* : 313 - 316, 1979.

37959 - HUNTJENS, J.L.M. : A sensitive method for continuous measurement of the carbon dioxide evolution rate of soil samples. - Plant Soil *53* : 529 - 534, 1979. [Also for Ps.]

37960 - HURKMAN, W.J. : Ultrastructural changes of chloroplasts in attached and detached, aging primary wheat leaves. - Amer. J. Bot. *66* : 64 - 70, 1979.

37961 - HURKMAN, W.J., MORRÉ, D.J., BRACKER, C.E., MOLLENHAUER, H.H. : Identification of etioplast membranes in fractions from soybean hypocotyls. - Plant Physiol. *64* : 398 - 403, 1979.

37962 - HÜSEMANN, W., PLOHR, A., BARZ, W. : Photosynthetic characteristics of photomixotrophic and photoautotrophic cell suspension cultures of *Chenopodium rubrum*. - Protoplasma *101* : 101 - 112, 1979.

37963 - HUSZÁR, J. : Genetic analysis of the different chlorophyll types of tobacco.- Biológia (Bratislava) *34* : 219 - 225, 1979.

*37964 - HUZISIGE, H. : [One of aspects of studying photosynthesis I.]-Kagaku *45* : 43 - 49, 1975. [In Jap.]

*37965 - HUZISIGE, H. : [One of aspects of studying photosynthesis II.] - Kagaku *45* : 110 - 117, 1975. [In Jap.]

37966 - HUZISIGE, H. : [Development of photosynthesis research.] - Tanpakushitsu Kakusan Koso, Bessatsu *21* : 4 - 12, 1979. [In Jap.]

37967 - HUZISIGE, H., DOI, M. : A new, light-induced absorbance change related to the reaction-center of photosystem II. - Plant Cell Physiol. *20* : 925 - 933, 1979.

37968 - HUZISIGE, H., DOI, M., NATUGA, T. : Relationships among the three light-induced absorbance changes related to the reaction-center of photosystem II. - Plant Cell Physiol. *20* : 935 - 946, 1979.

37969 - HYNNINEN, P.H. : Application of elution analysis to the study of chlorophyll transformations by column chromatography on sucrose. - J. Chromatogr. *175* : 75 - 88, 1979.

37970 - HYNNINEN, P.H. : Reduction of chlorophyll a, a', and b by sodium borohydride: separation of diastereoisomeric desoxo-chlorophyll alcohols on a sucrose column. - J. Chromatogr. *175* : 89 - 103, 1979.

37971 - HYNNINEN, P.H., WASIELEWSKI, M.R., KATZ, J.J. : Chlorophylls. VI. Epimerization and enolization of chlorophyll a and its magnesium-free derivatives. - Acta chem. scand. *B 33* : 637 - 648, 1979.

37972 - IBENTHAL, W.-D., HEITEFUSS, R. : Nebenwirkungen herbizider Harnstoff- und Triazinderivate auf den Befall von Weizen mit *Erysiphe graminis* f.sp. *tritici*. II. Physiologische und biochemische Untersuchungen über die Ursachen der indirekten Nebenwirkungen der Herbizide auf den Mehltaubefall. - Phytopathol. Z. *95* : 193 - 209, 1979. [Chl.]

37973 - IBRAGIMOVA, G.B., ALIJEV, K.A., MACHMADBEKOVA, L.M. : Effect of transcription and translation inhibitors on the synthesis and activity of RuDP carboxylase in *Acetabularia*. - In : BONOTTO, S., KEFELI, V., PUISEUX-DAO, S. (ed.) : Developmental Biology of *Acetabularia*. Pp. 255 - 258. Elsevier/North-Holland Biomedical Press, Amsterdam - New York - Oxford 1979.

37974 - IDSO, S.B., PINTER, P.J. Jr., HATFIELD, J.L., JACKSON, R.D., REGINATO, R.J. : A remote sensing model for the prediction of wheat yields prior to harvest. - J. theor. Biol. *77* : 217 - 228, 1979.

37975 - IGNAT'EV, A.R., POLEVAYA, V.S., SHABAEVA, Ė.V., MARKOVA, E.B. : Fotosinteticheskiĭ i temnovoĭ metabolizm ugleroda kul'tury tkani *Ruta graveolens* raznogo vozrasta. [Photosynthetic and dark metabolism of carbon in *Ruta graveolens* tissue culture of various age.]-Fiziol. Rast. *26* : 5 - 13, 1979. [In R, ab : E.]

37976 - **IGNATOV, G., KUDREV, T.** : Changes in the ultrastructure and the function of cell organelles from tissues of magnesium ion-deficient maize plants. - In : Mineral Nutrition of Plants. Vol. I. Pp. 119 - 132. Publ. House Central Cooperative Union, Sofia 1979. [Chloroplast.]

*37977 - **IKAN, R., AIZENSHTAT, Z., BAEDECKER, M.J., KAPLAN, I.R.** : Thermal alteration experiments on organic matter in recent marine sediment - I. Pigments. - Geochim. cosmochim. Acta *39* : 173 - 185, 1975.

37978 - **IKEDA, T.** : Electron microscopic evidence for the reversible transformation of *Euonymus* plastids. - Bot. Mag. (Tokyo) *92* : 23 - 30, 1979.

37979 - **IL'INA, M.D., BORISOV, A.Yu.** : Izmeneniya vykhoda i vremeni zhizni fluorestsentsii khloroplastov pod deĭstviem kationov. [The effect of cations on the quantum yield and the lifetime of the chloroplast fluorescence.] - Biokhimiya *44* : 40 - 49, 1979. [In R, ab : E.]

37980 - **IL'INA, M.D., DOMNINSKIĬ, D.A., BORISOV, A.Yu.** : Razdelenie fotosistem rasteniĭ metodom sitosorbtsionnoĭ khromatografii. [Separation of plant photosystems by sieve-sorptive chromatography.] - Biokhimiya *44* : 1994 - 2004, 1979. [In R, ab : E.]

37981 - **IMAI, K., MURATA, Y.** : Changes in apparent photosynthesis, CO_2 compensation point and dark respiration of leaves of some *Poaceae* and *Cyperaceae* species with senescence. - Plant Cell Physiol. *20* : 1653 - 1658, 1979.

37982 - **IMAI, K., MURATA, Y.** : Effect of carbon dioxide concentration on growth and dry matter production of crop plants. VI. Effect of oxygen concentration on the carbon dioxide-dry matter production relationship in some C_3- and C_4-crop species. - Jap. J. Crop Sci. *48* : 58 - 65, 1979.

37983 - **IMAI, K., MURATA, Y.** : Effect of carbon dioxide concentration on growth and dry matter production of crop plants. VII. Influence of light intensity and temperature on the effect of carbon dioxide-enrichment in some C_3- and C_4-species. - Jap. J. Crop Sci. *48* : 409 - 417, 1979.

37984 - **IMBAMBA, S.K., PAPA, G.** : Distribution of the Kranz type anatomy in some dicotyledonous families of Kenya. - Photosynthetica *13* : 315 - 322, 1979.

*37985 - **IMEVBORE, A.M.A., MESZES, G., BÖSZÖRMENYI, Z.** : The primary productivity of a fish-pond at Ile-Ife, Nigeria. - In : **KAJAK, Z., HILLBRICHT-ILKOWSKA, A.** (ed.) : Productivity Problems of Freshwaters. Pp. 715 - 723. Polish Sci. Publ., Warszawa - Kraków 1972.

37986 - **IMHOFF, H., VOIGTLÄNDER, G.** : Bewegungsrichtung und Verteilung von ^{14}C-Assimilation in Sproß und Wurzeln von *Rumex obtusifolius* L. und *Polygonum bistorta* L. als Indikatoren für eine termingerechte Herbizidanwendung. - Z. Acker-Pflanzenbau *148* : 418 - 429, 1979.

37987 - **INANAGA, S., KUMURA, A., MURATA, Y.** : [Studies on matter production of rape plant (*Brassica napus* L.) II. Photosynthesis and matter production of pods.]- Jap. J. Crop Sci. *48* : 260 - 264, 1979. [In Jap., ab : E.]

37988 - **INANAGA, S., KUMURA, A., MURATA, Y.** : [Studies on matter production of rape plant (*Brassica napus* L.) III. Photosynthesis, respiration and carbon balance sheet of the single pod.]- Jap. J. Crop Sci. *48*: 265-271, 1979. [In Jap.,ab:E.]

37989 - **INANAGA, S., KUMURA, A., MURATA, Y.** : Photosynthesis and yield of rapeseed. - Jap. agr. Res. quart. *13* : 169 - 173, 1979.

37990 - **INCROPERA, F.P., PRIVOZNIK, K.G.** : Radiative property measurements for selected water suspensions. - Water Resour. Res. *15* : 85 - 89, 1979.

*37991 - **INDENKO, I.F., RASULOV, A.R.** : Usovershenstvovanie metoda uskorennogo opredeleniya obshcheĭ ploshchadi list'ev na dereve. [Improvement of the method of rapid determination of total leaf area in a tree.] - Sbornik nauch. Trudov vsesoyuz. nauch.-issled. Inst. Im. I.V. Michurina *27* (Sovershenstvovanie Sortimenta i Agrotekhnicheskikh Priemov v Sadovodstve) : 66 - 70, 128, 1978. [In R.]

37992 - INDIATI, R., FAVOLA, G., MORETTI, R., TOMBESI, L. : Profili di CO_2 e produt-
tività. [Profiles of CO_2 and productivity.] - Ann. Inst. sper. Nutr. Piante
9(2): 1 - 16,1978-1979. [In Ital., ab : E.]

37993 - INDREBØ, G., PENGERUD, B., DUNDAS, I. : Microbial activities in a permanently
stratified estuary. I. Primary production and sulfate reduction. - Mar. Biol.
51 : 295 - 304, 1979.

37994 - INDREBØ, G., PENGERUD, B., DUNDAS, I. : Microbial activities in a permanently
stratified estuary. II. Microbial activities at the oxic-anoxic interface. -
Mar. Biol. *51* : 305 - 309, 1979. [Ps.]

37995 - INOUE, K., UCHIJIMA, Z. : Experimental study of microstructure of wind turbulence in rice and maize canopies. - Bull. nat. Inst. agr. Sci. Ser. A *26* :
1 - 88, 1979.

37996 - INOUE, Y., SHIBATA, K. : Thermoluminescence from the oxygen-evolving system
in photosynthesis. - Trends biochem. Sci. *4* : 182 - 184, 1979.

37997 - INOUE, Y., WATANABE, A., SHIBATA, K. : Transient variation of photoacoustic
signal from leaves accompanying photosynthesis. - FEBS Lett. *101* : 321 - 323,
1979.

37998 - ĬORDANOV, I., DILOVA, S., PETKOVA, R., ZEĬNALOV, Yu. : Posledeĭstvie na viso-
kata temperatura v"rkhu formiraneto i aktivnostta na fotosintetichniya aparat
v etiolirani fasulevi lista. [Post-effect of high temperature on photosynthetic apparatus formation in etiolated bean leaves.] - Fiziol. Rast. (Sofiya)
5 (2) : 9 - 18, 1979. [In Bulg., ab : E, R.]

37999 - IORDANOV, I.T. : Fotosinteticheskaya aktivnost' raznykh list'ev rasteniĭ fasoli. [Photosynthetic activity in various leaves of the bean plants.] - Sel'-
skokhoz. Biol. *14* : 655 - 660, 1979. [In R, ab : E.]

38000 - IPATOV, V.S., KIRIKOVA, L.A., BIBIKOV, V.P. : Skvozistost' drevostoev (izmerenie i vozmozhnosti ispol'zovaniya v kachestve pokazatelya mikroklimatiches-
kikh usloviĭ pod pologom lesa). [The value of transparence of a woodstand
(its measurement and the possibility of its usage as an index of microclima-
tic conditions under forest canopy).] - Bot. Zh. *64* : 1615 - 1624, 1979.
[In R.]

38001 - IRELAND, C., SCHEUER, P.J. : Photosynthetic marine mollusks : *In vivo* ^{14}C
incorporation into metabolites of the sacoglossan *Placobranchus ocellatus*. -
Science *205* : 922 - 923, 1979.

38002 - IRELAND, C.R., GOLDWIN, G.K. : Are the *in vitro* effects of disalicylidenepro-
panediamine mediated by salicylaldehyde ? - Plant Physiol. *63* : 1210 - 1211,
1979. [Ps.]

38003 - IRELAND, R.J., DELUCA, V., DENNIS, D.T. : Isoenzymes of pyruvate kinase in
etioplasts and chloroplasts. - Plant Physiol. *63* : 903 - 907, 1979.

38004 - IRIYAMA, K. : [Biomimetic chlorophyll science and method for the preparation of electrodes coated with chlorophyll multilayers.] - Bull. Res. Inst.
Polymer Textiles *120* : 161 - 175, 1979. [In Jap., ab : E.]

38005 - IRIYAMA, K. : Chemical stability of chlorophyll *a* in benzene solutions and
in monolayers at a water-air interface. - J. Colloid Interface Sci. *68* :
391 - 392, 1979.

38006 - IRIYAMA, K. : Methods of preparing chlorophyll *a* multilayers on glass plates.
- Photochem. Photobiol. *29* : 633 - 636, 1979.

38007 - IRIYAMA, K. : [On the enolic form of chlorophyll *a*.] - Bull. Res. Inst. Polymer Textiles *121* : 31 - 35, 1979. [In Jap., ab : E.]

38008 - IRIYAMA, K., SHIRAKI, M., YOSHIURA, M. : An improved method for extraction,
partial purification, separation and isolation of chlorophyll from spinach
leaves. - J. Liq. Chromatogr. *2* : 255 - 276, 1979.

38009 - IRIYAMA, K., YOSHIURA, M. : Separation of chlorophyll *a* and chlorophyll *b*
by column chromatography with Sephadex LH-20 or powdered sugar. - J. Chromatogr. *177* : 154 - 156, 1979.

38010 - IRIYAMA, K., YOSHIURA, M., AMADA, M. : [Methods for the preparation of chlo-
rophyll-a', chlorophyll-b' pheophytin-a and pheophytin-b.] - Bull. Res. Inst.
Polymer Textiles 121 : 37 - 43, 1979. [In Jap., ab : E.]

38011 - IRIYAMA, K., YOSHIURA, M., SHIRAKI, M. : [Method for extraction, partial purification, separation and isolation of chlorophylls.] - Bull. Res. Inst.
Polymer Textiles 119 : 13 - 21, 1979. [In Jap., ab : E.]

38012 - IRIYAMA, K., YOSHIURA, M., SHIRAKI, M. : [Micro-method for the qualitative
analysis of photosynthetic pigments and their degradation products.] - Bull.
Res. Inst. Polymer Textiles 119 : 23 - 28, 1979. [In Jap., ab : E.]

38013 - IRIYAMA, K., YOSHIURA, M., SHIRAKI, M. : Preparation of chlorophyll-a and
chlorophyll-b by means of column chromatography with Sephasorb HP Ultrafine.
- J. chem. Soc., chem. Commun. 1979 : 406 - 407, 1979.

38014 - IRIYAMA, K., YOSHIURA, M., SHIRAKI, M., OKADA, A. : [A rapid and convenient
method for purification and isolation of chlorophyll a from Porphyra yezoensis.] - Bull. Res. Inst. Polymer Textiles 120 : 177 - 181, 1979. [In·Jap.,
ab : E.]

38015 - ISAAKIDOU, J., PAPAGEORGIOU, G.C. : Functional effects of chemical modification of unstacked and stacked chloroplasts with glutaraldehyde. - Arch. Biochem. Biophys. 195 : 280 - 287, 1979.

38016 - ISHIHARA, K. : Diurnal course of stomatal aperture of leaf blades in rice
plants. - Jap. agr. Res. quart. 13 : 85 - 89, 1979.

38017 - ISHIHARA, K., HIRASAWA, T., IIDA, O., OGURA, T. : [An improved infiltration
method for measuring the narrow stomatal aperture of leaf blades in rice
plants.] - Jap. J. Crop Sci. 48 : 319 - 320, 1979. [In Jap.]

38018 - ISHIHARA, K., IIDA, O., HIRASAWA, T., OGURA, T. : [Relationship between nitrogen content in leaf blades and photosynthetic rate of rice plants with reference to stomatal aperture and conductance.] - Jap. J. Crop Sci. 48 : 543 -
550, 1979. [In Jap., ab : E.]

38019 - ISHIHARA, K., KURODA, E., ISHII, R., OGURA, T. : [Relationship between nitrogen content in leaf blades and photosynthetic rate in rice plants measured
with an infrared gas analyzer and an oxygen electrode.] - Jap. J. Crop Sci.
48 : 551 - 556, 1979. [In Jap., ab : E.]

38020 - ISHII, R., SHIBAYAMA, M., MURATA, Y. : Effect of light on the CO_2 evolution
of C_3 and C_4 plant in relation to the Kok effect. - Jap. J. Crop Sci. 48 :
52 - 57, 1979.

*B38021 - ISLER, O. : Carotenoids. - Birkhäuser Verlag, Basel - Stuttgart 1971.

38022 - ISRAELSTAM, G.F. : Chloroplastic activity in response to gibberellic acid
treatment of dwarf and normal cultivars of pea (Pisum sativum L.). - Biol.
Plant. 21:468 - 471, 1979. [Ps.]

38023 - ITOH, S. : Surface potential and reaction of the membrane-bound electron
transfer components. I. Reaction of P-700 in sonicated chloroplasts with
redox reagents. - Biochim. biophys. Acta 548 : 579 - 595, 1979.

38024 - ITOH, S. : Surface potential and reaction of the membrane-bound electron-
-transfer components, II. Integrity of the chloroplast membrane and reaction
of P-700. - Biochim. biophys. Acta 548 : 596 - 607, 1079.

38025 - ITOH, S., MATSUURA, K., MASAMOTO, K., NISHIMURA, M. : Estimation of the surface potential in photosynthetic membranes. - In : MUKOHATA, Y., PACKER, L.
(ed.) : Cation Flux across Biomembranes. Pp. 229 - 242. Academic Press,
New York - San Francisco - London 1979.

38026 - IVANCHENKO, V.M. : Regulyatsiya funktsiĭ membran sformirovannykh khloroplastov i intensivnost' fotosinteza. [Regulation of functions of membranes of
developed chloroplasts and photosynthetic rate.] - In : GONCHARIK, M.N.
(ed.) : Regulyatsiya Funktsiĭ Membran Rastitel'nykh Kletok. Pp. 147 - 197.
Nauka i Tekhnika, Minsk 1979. [In R.]

38027 - IVANCHENKO, V.M., LEGENCHENKO, B.I., URBANOVICH, T.A., KRUCHININA, S.S., MARSHAKOVA, M.I. : O vremennoĭ izmenchivosti strukturno-funktsional'nykh kharakteristik fotosinteticheskogo apparata. [Time-dependent changes in structural and functional parameters of the photosynthetic apparatus.] - Fiziol. Rast. *26* : 28 - 34, 1979. [In R, ab : E.]

38028 - IVANOV, A.F., FILIN, V.I. : Teoreticheskie osnovy programmirovaniya urozhaev. [Theoretical bases of the yield programming.] - Sel'skokhoz. Biol. *14* : 323 - 330, 1979. [In R, ab : E.]

38029 - IVANOV, B.N., POVALYAEVA, T.V. : Psevdotsiklicheskiĭ transport èlektronov pri fotovosstanovlenii NADF⁺ izolirovannymi khloroplastami gorokha. [Pseudocyclic electron transport in photoreduction of NADP⁺ by isolated pea chloroplasts.] - Fiziol. Rast. *26* : 276 - 282, 1979. [In R, ab : E.]

38030 - IVANOV, B.N., TIKHONOVA, L.N. : Pogloshchenie protonov izolirovannymi khloroplastami pri tsiklicheskom i netsiklicheskom transportakh èlektronov, kataliziruemykh fotosistemoĭ I v prisutstvii ferredoksina. [Proton uptake by isolated chloroplasts during cyclic and non-cyclic electron transport catalyzed by photosystem I in the presence of ferredoxin.] - Biokhimiya *44* : 983 - 989, 1979. [In R, ab : E.]

38031 - IVERSON, R.L., WHITLEDGE, T.E., GOERING, J.J. : Chlorophyll and nitrate fine structure in the southeastern Bering Sea shelf break front. - Nature *281* : 664 - 666, 1979.

38032 - IWASA, T., TOKUNAGA, F., YOSHIZAWA, T. : Photoreaction of *trans*-bacteriorhodopsin at liquid helium temperature. - FEBS Lett. *101* : 121 - 124, 1979.

38033 - IZHAR, S., DAVEY, M.R., GATENBY, A.A. : A light-sensitive mutant in *Petunia*. Growth, ultrastructure and regulation of chlorophyll and ribulose-1,5-bisphosphate carboxylase content in the leaves. - Plant Sci. Lett. *15* : 75 - 82, 1979.

38034 - JACKSON, J.E., PALMER, J.W. : A simple model of light transmission and interception by discontinuous canopies. - Ann. Bot. *44* : 381 - 383, 1979.

38035 - JACKSON, M.B. : Rapid injury to peas by soil waterlogging. - J. Sci. Food Agr. *30* : 143 - 152, 1979. [Chl.]

38036 - JACOB, J.-L., PRIMOT, L., PRÉVÔT, J.-C. : Purification et étude de la phosphoénolpyruvate carboxylase du latex d'*Hevea brasiliensis*. - Physiol. vég. *17* : 501 - 516, 1979.

*B38037 - JACOBI, G. (ed.) : Biochemische Cytologie der Pflanzenzelle. Ein Praktikum. - G. Thieme Verlag, Stuttgart 1974. [Ps.]

38038 - JACOBS, E., FISHER, R.R. : Resolution and reconstitution of *Rhodospirillum rubrum* pyridine dinucleotide transhydrogenase : Chemical modification with *N*-ethylmaleimide and 2,4-pentanedione. - Biochemistry *18* : 4315 - 4322, 1979.

38039 - JACOBS, N.J., JACOBS, J.M. : Microbial oxidation of protoporphyrinogen, an intermediate in heme and chlorophyll biosynthesis. - Arch. Biochem. Biophys. *197* : 396 - 403, 1979.

38040 - JACOBS, R.P.W.M. : Distribution and aspects of the production and biomass of eelgrass, *Zostera marina* L., at Roscoff, France. - Aquat. Bot. *7* : 151 - 172, 1979.

38041 - JACQUOT, J.P., MAUDINAS, B., GADAL, P. : Occurence of thioredoxin *m* activity in the photosynthetic bacteria *Rhodopseudomonas capsulata*. - Biochem. biophys. Res. Commun. *91* : 1371 - 1376, 1979.

38042 - JAGOW, G. von : Zum Prinzip der Photophosphorylierung und oxidativen Phosphorylierung. - Naturwissenschaften *66* : 539 - 546, 1979.

38043 - **JAHN, O.L.** : Penetration of photosynthetically active radiation as a measurement of canopy density of citrus trees. - J. amer. Soc. hort. Sci. *104* :
557 - 560, 1979.

38044 - **JAKRLOVÁ, J.** : Primary production of the Kameničky grasslands - aboveground.
- In : **RYCHNOVSKÁ, M.** (ed.) : Function of Grasslands in Spring Region -
Kameničky Project. Pp. 77 - 86. Botanical Institute, Czechoslovak Academy of
Sciences, Brno 1979.

38045 - **JAKRLOVÁ, J.** : Vertical structure of the leaf canopy in the Kameničky grassland. - In : **RYCHNOVSKÁ, M.** (ed.) : Function of Grasslands in Spring Region
- Kameničky Project. Pp. 93 - 101. Botanical Institute, Czechoslovak Academy
of Sciences, Brno 1979.

38046 - **JAMIESON, P.D.** : Relationship between net and solar radiation received by
pastures at Lincoln, New Zealand. - New Zeal. J. Sci. *22* : 245 - 247, 1979.

38047 - **JANARDHAN, K., RAO, P.H.** : Estimation of leaf area in *Capsicum frutescens* L.
by rapid method. - Indian J. Bot. *2* : 76 - 79, 1979.

38048 - **JANSON, T.R., KATZ, J.J.** : Nuclear magnetic resonance spectroscopy of dia-
magnetic porphyrins. - In : **DOLPHIN, D.** (ed.) : The Porphyrins. Vol. IV.
Pp. 1 - 59. Academic Press, New York - San Francisco - London 1979. [Chl.]

38049 - **JANSSON, C., ANDERSSON, B., ÅKERLUND, H.-E.** : Trypsination of inside-out
chloroplast thylakoid vesicles for localization of the water-splitting site.
- FEBS Lett. *105* : 177 - 180, 1979.

38050 - **JANUS, L.L., DUTHIE, H.C.** : Phytoplankton and primary production of lakes
in the Matamek Watershed, Quebec. - Int. Rev. ges. Hydrobiol. *64* : 89 - 98,
1979.

38051 - **JANZEN, A.F., BOLTON, J.R.** : Photochemical electron transfer in monolayer
assemblies. 2. Photoelectric behavior in chlorophyll *a*/acceptor systems. -
J. amer. chem. Soc. *101* : 6342 - 6348, 1979.

38052 - **JANZEN, A.F., BOLTON, J.R., STILLMAN, M.J.** : Photochemical electron transfer
in monolayer assemblies. 1. Spectroscopic study of radicals produced in chlorophyll *a*/acceptor systems. - J. amer. chem. Soc. *101* : 6337 - 6341, 1979.

38053 - **JENKINS, G.I., WOOLHOUSE, H.W.** : A study of photosynthetic electron transport during senescence of the primary leaves of *Phaseolus vulgaris*. - Plant
Physiol. *63* (Suppl.) : 74, 1979.

38054 - **JENNINGS, R.C., GARLASCHI, F.M., GEROLA, P.D., FORTI, G.** : Partition zone
penetration by chymotrypsin, and the localization of the chloroplast flavo-
protein and photosystem II. - Biochim. biophys. Acta *546* : 207 - 219, 1979.

38055 - **JENNINGS, R.C., GEROLA, P.D., FORTI, G., GARLASCHI, F.M.** : The influence of
proton-induced grana formation on partial electron-transport reactions in
chloroplasts. - FEBS Lett. *106* : 247 - 250, 1979.

38056 - **JENSEN, R.G.** : The isolation of intact leaf cells, protoplasts and chloroplasts. - In : **GIBBS, M., LATZKO, E.** (ed.) : Photosynthesis II. (Encycl.
Plant Physiol. N.S. Vol. 6.) Pp. 31 - 40. Springer-Verlag, Berlin - Heidelberg - New York 1979.

38057 - **JERNIGAN, R.W., TSOKOS, C.P.** : Phytoplankton modeling involving random rate
constants. Part 1 : Deterministic setting. - Int. J. environ. Stud. *14* :
97 - 105, 1979.

38058 - **JERNSTEDT, J.A., CLARK, C.** : Stomata on the fruits and seeds of *Eschschol-
tzia* (*Papaveraceae*). - Amer. J. Bot. *66* : 586 - 590, 1979. [Ps.]

38059 - **JESKE, C., SENGER, H.** : Synthese von Pigmenten und Entwicklung des Photosyntheseapparates in einigen Höheren Pflanzen. - Ber. deut. bot. Ges. *92* :
609 - 617, 1979.

*38060 - **JIMENEZ, S.R.** : Mise en évidence de l'upwelling équatorial a l'est des Ga-
lapagos. - Cah. ORSTOM, Sér. Oceanogr. *16* (2) : 137 - 155, 1978. [Chl.]

38061 - JOHAL, S., BOURQUE, D.P. : Crystalline ribulose 1,5-bisphosphate carboxylase-oxygenase from spinach. - Science *204* : 75 - 77, 1979.

38062 - JOHAL, S., BOURQUE, D.P. : Crystallization of RuBP carboxylase/oxygenase from
higher plants. - Plant Physiol. *63* (Suppl.) : 153, 1979.

38063 - JOHANSSON, B.C. : A reaction center mutant of *Rhodospirillum rubrum*. - Acta
chem. scand. *B 33* : 605 - 606, 1979.

*38064 - JOHNSON, C.E., CLARK, J.A. : A comparison between diffusivities measured by
"ring diffusers" and an aerodynamic method. - Agr. Meteorol. *19* : 363 - 377,
1978.

38065 - JOHNSON, C.R., KRANTZ, J.K., JOINER, J.N., CONOVER, C.A. : Light compensation
point and leaf distribution of *Ficus benjamina* as affected by light intensity
and nitrogen-potassium nutrition. - J. amer. Soc. hort. Sci. *104* : 335 - 338,
1979.

38066 - JOHNSON, H.B., ROWLANDS, P.G., TING, I.P. : Tritium and carbon-14 double
isotope porometer for simultaneous measurements of transpiration and photosynthesis. - Photosynthetica *13* : 409 - 418, 1979.

38067 - JOHNSON, K.S., PYTKOWICZ, R.M., WONG, C.S. : Biological production and the
exchange of oxygen and carbon dioxide across the sea surface in Stuart Channel, British Columbia. - Limnol. Oceanogr. *24* : 474 - 482, 1979.

38068 - JOHNSTONE, I.M. : Papua New Guinea seagrasses and aspects of the biology and
growth of *Enhalus acoroides*. - Aquat.Bot. *7* : 197 - 208, 1979. [Ps.]

38069 - JOLIOT, P., JOLIOT, A. : Comparative study of the fluorescence yield and of
the C550 absorption change at room temperature. - Biochim. biophys. Acta
546 : 93 - 105, 1979.

38070 - JÓNASSON, P.M., ADALSTEINSSON, H. : Phytoplankton production in shallow eutrophic Lake Mývatn, Iceland. - Oikos *32* : 113 - 138, 1979. [Chl.]

38071 - JONES, C.A., ZIMMERMANN, F.J.P., DALL'ACQUA, F.M. : Light penetration in
wide-row upland rice. - Trop. Agr. *56* : 367 - 369, 1979.

38072 - JONES, C.E., JONES, C.A., MACKAY, R.A. : Reactions in microemulsions. 4.
Kinetics of chlorophyll sensitized photoreduction of methyl red and crystal
violet by ascorbate. - J. phys. Chem. *83* : 805 - 810, 1979.

38073 - JONES, C.E., MACKAY, R.A. : Chlorophyll mediated photoreactions in micro-
emulsion media. - In : LONGO, F.R. (ed.) : Porphyrin Chemistry Advances.
Pp. 71 - 88. Ann Arbor Science, Ann Arbor 1979.

38074 - JONES, H.G. : Screening for tolerance of photosynthesis to osmotic and sali-
ne stress using rice leaf-slices. - Photosynthetica *13* : 1 - 8, 1979.

38075 - JONES, H.G. : Stomatal behavior and breeding for drought resistance. - In :
MUSSELL, H., STAPLES, R. (ed.) : Stress Physiology in Crop Plants. Pp. 407 -
428. John Wiley & Sons, Inc., New York 1979. [Ps.]

38076 - JONES, J.G., SIMON, B.M. : The measurement of electron transport system
activity in freshwater benthic and planktonic samples. - J. appl. Bacteriol.
46 : 305 - 315, 1979.

38077 - JONES, M.M., RAWSON, H.M. : Influence of rate of development of leaf water
deficit upon photosynthesis, leaf conductance, water use efficiency, and
osmotic potential in sorghum. - Physiol. Plant. *45* : 103 - 111, 1979.

38078 - JONES, R.A., RAST, W., LEE, G.F. : Relationship between summer mean and maximum chlorophyll a concentrations in lakes. - Environ. Sci. Technol. *13* :
869 - 870, 1979.

38079 - JONES, R.C., ADAMS, M.S. : Productivity of algal epiphytes of *Myriophyllum
spicatum* in lake Wingra, Wisconsin. - J. Phycol. *15* (Suppl.) : 21, 1979.

38080 - JORDAN, B.R., GIVAN, C.V. : Effects of light and inhibitors on glutamate metabolism in leaf discs of *Vicia faba* L. Sources of ATP for glutamine synthesis and photoregulation of tricarboxylic acid cycle metabolism. - Plant Physiol. *64* : 1043 - 1047, 1979. [Ps.]

38081 - JORDAN, D.B., OGREN, W.L. : Purification and properties of soybean RuBP carboxylase. - Plant Physiol. *63* (Suppl.) : 154, 1979.

38082 - JORDAN, L.S., SHANER, D.L. : Weed control. - In : HALL, A.E., CANNELL, G.H.,
LAWTON, H.W. (ed.) : Agriculture in Semi-Arid Environments. Pp. 266 - 296.
Springer-Verlag, Berlin - Heidelberg - New York 1979. [Ps.]

38083 - JORDAN, P.W., NOBEL, P.S. : Infrequent establishment of seedlings of *Agave
deserti* (*Agavaceae*) in the northwestern Sonoran desert. - Amer. J. Bot. *66* :
1079 - 1084, 1979. [Ps.]

38084 - JORDAN, W.R., McCRARY, M., MILLER, F.R. : Compensatory growth in the crown
root system of sorghum. - Agron. J. *71* : 803 - 806, 1979. [Growth analysis.]

*38085 - JORGA, W., WEISE, G. : Beziehungen zwischen Kohlendioxidgasstoffwechsel sub-
merser Makrophyten (Typ : Hydrogencarbonatspalter) und Sauerstoffproduktion
in langsam fließenden Gewässern. - Acta hydrochim. hydrobiol. *6* : 199 - 226,
1978.

38086 - JØRGENSEN, B.B., REVSBECH, N.P., BLACKBURN, T.H., COHEN, Y. : Diurnal cycle
of oxygen and sulfide microgradients and microbial photosynthesis in a cyanobacterial mat sediment. - Appl. environ. Microbiol. *38* : 46 - 58, 1979.

38087 - JOSHI, G.V., KARADGE, B.A. : Effect of sodium chloride on photosynthetic
$^{14}CO_2$ assimilation in *Portulaca oleracea* LINN. - Indian J. exp. Biol. *17* :
167 - 170, 1979.

*38088 - JOSHI, G.V., KARADGE, B.A., BARTAKKE, S.P. : Photosynthetic carbon metabolism
in succulents. - In : SEN, D.N., BANSAL, R.P. (ed.) : Environmental Physiology and Ecology of Plants. Pp. 87 - 96. Bishen Singh Mahendra Pal Singh, Dehra Dun 1978.

38089 - JOUSSAUME, M. : Action du blanc d'oeuf sur la survie de chloroplastes isolés
de feuilles de Pois. - Plant Sci. Lett. *16* : 219 - 224, 1979.

38090 - JOUSSOT-DUBIEN, J., ALBRECHT, A.C., GERISCHER, H., KNOX, R.S., MARCUS, R.A.,
SCHOTT, M., WELLER, A., WILLIG, F. : Mechanisms of charge separation and sub-
sequent processes. Group report. - In : GERISCHER, H., KATZ, J.J. (ed.) :
Light-Induced Charge Separation in Biology and Chemistry. Pp. 129 - 149.
Verlag Chemie, Weinheim - New York 1979. [Chl.]

38091 - JOUY, M., SIRONVAL, C. : Quenching of the fluorescence emitted by $P_{695-682}$
at room temperature in etiolated illuminated leaves. - Planta *147* : 127 -
133, 1979.

38092 - JOYARD, J., DOUCE, R. : Characterization of phosphatidate phosphohydrolase
activity associated with chloroplast envelope membranes. - FEBS Lett. *102* :
147 - 150, 1979.

38093 - JOYARD, J., DOUCE, R. : The chloroplast envelope is the site of galactolipid
synthesis. - Plant Physiol. *63* (Suppl.) : 120, 1979.

38094 - JUNGE, W., AUSLÄNDER, A., McGEER, A.J., RUNGE, T. : The buffering capacity of
the internal phase of thylakoids and the magnitude of the pH changes inside
under flashing light. - Biochim. biophys. Acta *546* : 121 - 141, 1979.

38095 - JUNGE, W., BARBER, J., BOLTON, J.R., CROFTS, A.R., DUTTON, P.L., FAJER, J.,
JONES, O.T., KATZ, J.J., MICHEL-BEYERLE, M.E., REICH, R., RENGER, G., WITT,
H.T. : Chlorophyll mediated processes. Group report. - In : GERISCHER, H.,
KATZ, J.J. (ed.) : Light-Induced Charge Separation in Biology and Chemistry.
Pp. 449 - 470. Verlag Chemie, Weinheim - New York 1979.

38096 - JUNGE, W., SCHAFFERNICHT, H. : The field of possible structures for the
chlorophyll a dimer in photosystem I of green plants delineated by polarized photochemistry. - In : Chlorophyll Organization and Energy Transfer in
Photosynthesis. Pp. 127 - 146. Excerpta Medica, Amsterdam - Oxford - New
York 1979.

38097 - JUNGES, W. : Vergleichende lichtökologische Untersuchungen an Pflanzen aus
verschiedenen Klimagebieten der Erde unter Farbfolien. - Arch. Gartenbau
27 : 453 - 469, 1979. [Growth analysis.]

38098 - **JURIK, T.W., CHABOT, J.F., CHABOT, B.F.** : Ontogeny of photosynthetic performance in *Fragaria virginiana* under changing light regimes. - Plant Physiol. *63* : 542 - 547, 1979.

38099 - **JURSINIC, P.** : Flash-yield pattern for photosynthetic oxygen evolution in *Chlorella* and chloroplasts as a function of excitation intensity. - Arch. Biochem. Biophys. *196* : 484 - 492, 1979.

38100 - **JURY, W.A.** : Water transport through soil, plant, and atmosphere. - In : HALL, A.E., CANNELL, G.H., LAWTON, H.W. (ed.) : Agriculture in Semi-Arid Environments. Pp. 180 - 199. Springer-Verlag, Berlin - Heidelberg - New York 1979. [Canopy.]

38101 - **JUTTE, S.M., DURBIN, R.D.** : Ultrastructural effects in zinnia leaves of a chlorosis-inducing toxin from *Pseudomonas tagetis*. - Phytopathology *69* : 839 - 842, 1979. [Chloroplast.]

38102 - **JÜTTNER, F.** : The algal excretion product, geranylacetone : a potent inhibitor of carotene biosynthesis in *Synechococcus*. - Z. Naturforsch. *34 C* : 957 - 960, 1979.

38103 - **KABAKI, N., SAKA, H., AKITA, S.**: [Effects of nitrogen, phosphorus and potassium deficiencies on photosynthesis and RuBP carboxylase-oxygenase activities in rice plants.] - Jap. J. Crop Sci. *48* : 378 - 384, 1979. [In Jap., ab : E.]

38104 - **KABI, T., NANDA, H.P., HALDAR, A.M.** : Chlorophyll degradation and starch metabolism in bean (*Dolichos lablab* L.) leaves intected with yellow bean mosaic virus (YBMV). - J. indian bot. Soc. *58* : 25 - 30, 1979.

38105 - **KACHAN, A.A., LITSOV, N.I., NIKOLAEVSKAYA, V.I.** : O roli vtorichnykh protsessov pri okislenii vody v model'nykh fotosinteticheskikh sistemakh. [Role of secondary processes of water oxidation in model photosynthetic systems.] - Dokl. Akad. Nauk ukr. SSR, Ser. B : geol. khim. biol. Nauki *1979* : 637 - 640, 1979. [In R , ab : E.]

38106 - **KACHRU, D.N., KRISHNAN, P.S.** : Chlorophyll and enzymes of photorespiration in *Dendrophthoe falcata* seeds. - Plant Sci. Lett. *16* : 165 - 170, 1979.

38107 - **KADOSHNIKOVA, I.G., KISELEV, B.A.** : Kolloidnye rastvory khlorofilla. Ėlektricheskiĭ zaryad chastits. [Colloid solutions of chlorophyll. Electrical charge of particles.] - Biofizika *24* : 811 - 814, 1979. [In R, ab : E.]

38108 - **KADOSHNIKOVA, I.G., KISELEV, B.A., EVSTIGNEEV, V.B.** : Kolloidnye rastvory khlorofilla. II. O fluorestsentsii i fotosensibiliziruyushchem deĭstvii kolloidnykh rastvorov khlorofilla. [Colloid solutions of chlorophyll. II. Fluorescence and photosensitizing action of colloid chlorophyll solutions.] - Biofizika *24* : 770, 1979. [In R.]

38109 - **KAFALIEVA, D.N., BUSHEVA, M.K.** : Vliyanie temperatury na fotoindutsirovannyĭ gradient pH v izolirovannykh khloroplastakh. [Effect of temperature on photoinduced pH gradient in isolated chloroplasts.] - Biofizika *24* : 676 - 680, 1979. [In R, ab : E.]

38110 - **KAFALIEVA-BOEVA, D.N., VAKLINOVA, S.G.** : Soderzhanie pigmentov i izmenenie reaktsii Khilla v mutantakh gorokha. [Pigment content and Hill reaction in pea mutants.] - In : VAKLINOVA, S.G., VANKOVA-RADEVA, R., VASILEVA, V.S. (ed.) : Fotosinteticheskaya Assimilyatsiya CO_2 i Fotodykhanie. Pp. 104 - 108. Izdatel'stvo bolgarskoĭ Akademii Nauk, Sofiya 1979. [In R.]

38111 - **KAGAN, N.E.** : Strati-bisporphyrins : A promising gauge for the study of heme and chlorophyll interactions. - In : LONGO, F.R. (ed.) : Porphyrin, Chemical Advances. Pp. 43 - 50. Ann Arbor Sci., Ann Arbor 1979.

38112 - **KAGEYAMA, A., YOKOHAMA, Y., NISIZAWA, K.** : Diurnal rhythm of apparent photosynthesis of a brown alga, *Spatoglossum pacificum*. - Bot. mar. 22 : 199 - 201, 1979.

38113 - **KAISER, W .M.** : Reversible inhibition of the Calvin cycle and activation of oxidative pentose phosphate cycle in isolated intact chloroplasts by hydrogen peroxide. - Planta *145* : 377 - 382, 1979.

38114 - **KAISER, W.M., BASSHAM, J.A.** : Carbon metabolism of chloroplasts in the dark : oxidative pentose phosphate cycle versus glycolytic pathway. - Planta *144* : 193 - 200, 1979.

38115 - **KAISER, W.M., BASSHAM, J.A.** : Light-dark regulation of starch metabolism in chloroplasts. I. Levels of metabolites in chloroplasts and medium during light-dark transition. - Plant Physiol. *63* : 105 - 108, 1979.

38116 - **KAISER, W.M., BASSHAM, J.A.** : Light-dark regulation of starch metabolism in chloroplasts. II. Effect of chloroplastic metabolite levels on the formation of ADP-glucose by chloroplast extracts. - Plant Physiol. *63* : 109 - 113, 1979.

38117 - **KAISER, W.M., PAUL, J.S., BASSHAM, J.A.** : Release of photosynthates from mesophyll cells *in vitro* and *in vivo*. - Z. Pflanzenphysiol. *94* : 377 - 385, 1979.

*B38118 - **KALER, V.L.** : Avtoregulyatsiya Obrazovaniya Khlorofilla v Vysshikh Rasteniyakh. [Autoregulation of Chlorophyll Formation in Higher Plants.] - Nauka i Tekhnika, Minsk 1976. [In R.]

38119 - **KALININA, L.M., FRADKIN, L.I., SHLYK, A.A.** : Razvitie dlinnovolnovoĭ fluorestsentsii v postétiolirovannykh list'yakh ne soderzhavshchego khlorofilla *b* mutanta yachmenya. [Development of long-wavelength fluorescence in postetiolated leaves of a barley mutants not containing chlorophyll *b*.] - Vestsi Akad.Navuk belarus. SSR, Ser. biyal. Navuk *1979* (4) : 121 - 123, 143, 1979. [In R, ab : E.]

38120 - **KALTOFEN, H.** : Die mathematische Behandlung des Pflanzenwachstums und der Ertragsbildung - Rückblick und Ausblick. - Biol. Rundschau *17* : 229 - 248, 1979. [Ps.]

38121 - **KAMÍNEK, M., PAČES, V., CORSE, J., CHALLICE, J.S.** : Effect of stereospecific hydroxylation of N^6-(Δ^2-isopentenyl) adenosine on cytokinin activity. - Planta *145* : 239 - 243, 1979. [Chl.]

38122 - **KAMINSKI, Z., GUTKOWSKI, R., MALESZEWSKI, S.** : Photosynthesis in bean leaves treated with α-hydroxy-2-pyridine-methanesulfonic acid (α-HPMS) the glycolic acid oxidase inhibitor. - Z. Pflanzenphysiol. *91* : 17 - 24, 1979.

38123 - **KAMPRATH, E.J., CASSEL, D.K., GROSS, H.D., DIBB, D.W.** : Tillage effects on biomass production and moisture utilization by soybeans on coastal plain soils. - Agron. J. *71* : 1001 - 1005, 1979. [Growth analysis.]

38124 - **KAMYKOWSKI, D.** : The growth response of a model *Gymnodinium splendens* in stationary and wavy water columns. - Mar. Biol. *50* : 289 - 303, 1979. [Growth model.]

38125 - **KANAI, R.** : [Biochemistry of C_4-pathway of photosynthesis.] - Tanpakushitsu Kakusan Koso, Bessatsu *21* : 173 - 185, 1979. [In Jap.]

38126 - **KANDELER, R., HELDWEIN, R.** : Significance of photosynthesis, N-deficiency, ABA and pH for synthesis of malate in *Lemna*. - In : MARCELLE, R., CLIJSTERS, H., VAN POUCKE, M. (ed.) : Photosynthesis and Plant Development. Pp. 103 - 110. Dr.W.Junk bv.Publ., The Hague - Boston - London 1979.

38127 - **KANDIAH, S.** : Turnover of carbohydrates in relation to growth in apple trees. I. Seasonal variation of growth and carbohydrate reserves. - Ann. Bot. *44* : 175 - 183, 1979.

38128 - **KANDIAH, S.** : Turnover of carbohydrates in relation to growth of apple trees. II. Distribution of ^{14}C assimilates labelled in autumn, spring and summer. - Ann. Bot. *44* : 185 - 195, 1979.

38129 - **KANIUGA, Z., ZĄBEK, J., MICHALSKI, W.P.** : Photosynthetic apparatus in chilling-sensitive plants. VI. Cold and dark-induced changes in chloroplast superoxide dismutase activity in relation to loosely-bound manganèse content. - Planta *145* : 145 - 150, 1979.

38130 - **KANNANGARA, C.G., GOUGH, S.P.** : Biosynthesis of Δ-aminolevulinate in greening
barley leaves II: Induction of enzyme synthesis by light. - Carlsberg Res.
Commun. *44* : 11 - 20, 1979.

38131 - **KAO, C.H.** : The inter-organ control of leaf senescence of rice seedlings. -
Plant Physiol. *63* (Suppl.) : 75, 1979. [Chl.]

38132 - **KAPPEN, L., LANGE,O.L., SCHULZE, E.-D., EVENARI, M., BUSCHBOM, U.** : Ecophysiological investigations on lichens of the Negev Desert VI. Annual course of
the photosynthetic production of *Ramalina maciformis* (DEL.) BORY. - Flora
168 : 85 - 108, 1979.

38133 - **KAR, R.K., NANDA, H.P., KABI, T.** : Metabolic changes in papaya infected with
papaya mosaic virus. - Geobios (Jodhpur) *6* (2) : 49 - 52, 1979. [Ps, Chl.]

38134 - **KARAMANOS, A.J.** : Water stress : A challenge for the future of agriculture. -
In : SCOTT, T.K. (ed.) : Plant Regulation and World Agriculture. Pp. 415 -
455. Plenum Press, New York - London 1979. [Ps.]

38135 - **KARAPETYAN, N.V.** : Spectroscopy of biological samples at room and low temperatures. - Zagad. Biofiz. wspó/czes. *4* : 39 - 48, 1979. [Ps.]

38136 - **KARAPETYAN, N.V., BUKHOV, N.G.** : Vliyanie degidratatsii na funktsionirovanie
fotosistem vysshikh rasteniĭ. [The effect of dehydration on functioning of
photosystems of higher plants.]-Mol.Biol.(Moskva)*13* : 947 - 954, 1979. [In R,
ab : E.]

38137 - **KARAVAEV, V.A., KUKUSHKIN, A.K.** : Teoreticheskoe issledovanie kinetiki okis-
litel'no-vosstanovitel'nykh prevrashcheniĭ v tsepi ėlektronnogo transporta.
[Theoretical study of redox transformations in the electron transport chain.]
- Biofizika *24* : 92 - 95, 1979. [Ps, in R., ab : E.]

38138 - **KARNAUKHOV, V.N., MENDGUL, M.I., MARTSENYUK, P.P.** : Izmenenie funktsional'-
nogo sostoyaniya kletok sinezelenoĭ vodorosli v protsesse razvitiya virusnoĭ infektsii. [Changes in the functional state of blue-green alga cells in
the course of the development of virus infection.] - Fiziol. Rast. *26* : 190 -
192, 1979. [Chl; in R.]

*B38139 - **KARPILOV, Yu.S.**(ed.) : Fotosintez Kukuruzy. [Photosynthesis in Maize.] -
Akad. Nauk SSSR, Pushchino na Oke 1974. [In R, ab : E.]

*38140 - **KARTUSCH, B.** : Gaswechselmessungen an stadtbewohnenden Pflanzen I. Im schwach
immittierten Raum.   -   Sitzungsber. österr. Akad. Wiss., math.-nat. Kl.,
Abt. I, *185* : 239 - 247, 1976.

38141 - **KARTUSCH, B.** : Freilandmessungen der Gaswechselverläufe bei den Frühjahrs-
geophyten *Galanthus nivalis* und *Ficaria verna*. - Oecol. Plant. *14* : 177 -
185, 1979. [Ps.]

38142 - **KARUBE, I., AIZAWA, K., IKEDA, S., SUZUKI, S.** : Carbon dioxide fixation by
immobilized chloroplasts. - Biotechnol. Bioeng. *21* : 253 - 260, 1979.

38143 - **KARUNEN, P.** : Effect of light intensity on growth, CO_2 fixation, and chlorophyll and polar lipid production in germinating *Polytrichum commune* spo-
res. - Physiol. Plant. *45* : 197 - 200, 1979.

38144 - **KARUNEN, P., MIKOLA, H., LINKO, R., EURANTO, E.K.** : Lipids in *Sphagnum* mosses of various ages. - Can. J. Bot. *57* : 1335 - 1339, 1979. [Xanthophylls.]

38145 - **KARYDIS, M.** : Short term effects of hydrocarbons on the photosynthesis and
respiration of some phytoplankton species. - Bot. Mar. *22* : 281 - 285, 1979.

38146 - **KÄSEMIR, H.** : Mini review. Control of chloroplast formation by light. -
Cell Biol. internat. Rep. *3* : 197 - 214, 1979.

*38147 - **KASHIRO, Yu.P.** : Metody i apparatura dlya izucheniya vneshneĭ sredy drevesnykh rasteniĭ na ikh nachal'nykh ėtapakh ontogeneza. [Methods and equipment
for studying environment of woody plants during initial stages of ontogeny.]
- Tr. Inst. Ėkol. Rast. Zhiv. (Sverdlovsk) *100* (Ėkologo-fiziologicheskie Issledovaniya Khvoĭnykh Drevešnykh Vidov na Urale) : 54 - 94, 98, 1976.
[In R.]

38148 - KATO, S., HOZYO, Y., SHIMOTSUBO, K. : [Translocation of ^{14}C-photosynthates from the leaves at different stages of development in *Ipomoea* grafts.] - Jap. J. Crop Sci. *38* : 254 - 259, 1979. [In Jap., ab : E.]

38149 - KATOH, K., OHTA, Y., HIROSE, Y., IWAMURA, T. : Photoautotrophic growth of *Marchantia polymorpha* L. cells in suspension culture. - Planta *144* : 509 - 510, 1979. [Chl.]

38150 - KATOH, T., GANTT, E. : Photosynthetic vesicles with bound phycobilisomes from *Anabaena variabilis*. - Biochim. biophys. Acta *546* : 383 - 393, 1979.

38151 - KATS, E.Yu., KOZLOV, Yu. N., KISELEV, B.A. : Fotosensibilizirovannoe vydelenie vodoroda v fotokhimicheskikh sistemakh s ispol'zovaniem khlorofilla. [Photosensitized relase of hydrogen in photochemical systems using chlorophyll.] - Biofizika *24* : 801 - 805, 1979. [In R, ab : E.]

38152 - KATS, E.Yu., KOZLOV, Yu.N., KISELEV , B.A. : Izuchenie fotokhimicheskikh reaktsiĭ khlorofilla metodom fotopolyarografii. II. Opredelenie konstant skorosteĭ fotosensibilizirovannogo khlorofillom vosstanovleniya krasiteleĭ metodom konkuriruyushchego aktseptora. [Study of photochemical reactions of chlorophyll by the method of photopolarography. II. Determination of kinetic constants of photosensibilized by chlorophyll reduction of dyes by the method of competitive acceptors.] - Biofizika *24* : 946, 1979. [In R.]

38153 - KATZ, J.J. : Charge separation in synthetic photoreaction centers. - In : GERISCHER, H., KATZ, J.J. (ed.) : Light-induced Charge Separation in Biology and Chemistry. Pp. 331 - 359. Verlag Chemie, Weinheim - New York 1979. [Chl.]

38154 - KATZ, J.J., SHIPMAN, L.L., NORRIS, J.R. : Structure and function of photoreaction-centre chlorophyll. - In : Chlorophyll Organization and Energy Transfer in Photosynthesis. Pp. 1 - 40. Excerpta Medica, Amsterdam - Oxford - New York 1979.

38155 - KATZ, J.J., WASIELEWSKI, M.R. : Biomimetic approaches to artificial photosynthesis.- Biotechnol. Bioeng. Symp. *8* (Biotechnol. Energy Prod. Conserv.) : 423 - 452, 1979.

38156 - KATZFUSS, M. : ^{14}C-Verteilung im Herbst und -Mobilisierung in jungen Apfelbäumen. - Arch. Gartenbau *27* (3) : 119 - 123, 1979.

38157 - KAVON, D.L., ZEEVAART, J.A.D. : Simultaneous inhibition of translocation of photosynthate and of the floral stimulus by localized low-temperature treatment in the short-day plant *Pharbitis nil*. - Planta *144* : 201 - 204, 1979.

38158 - KAWAKUBO, K., SHINDO, M., KONOTSUNE, T. : A mechanism of chlorosis caused by 1,3-dimethyl-4-(2,4-dichlorobenzoyl)-5-hydroxypyrazole, a herbicidal compound. - Plant Physiol. *64* : 774 - 779, 1979.

38159 - KAWAMURA, M., FUJITA, Y. : A probable lack of function of *c*-type cytochrome in the photosynthetic system of the blue-green alga *Anabaena variabilis*. - Plant Cell Physiol. *20* : 331 - 339, 1979.

38160 - KAWAMURA, M., MIMURO, M., FUJITA, Y. : Quantitative relationship between two reaction centers in the photosynthetic system of blue-green algae. - Plant Cell Physiol. *20* : 697 - 705, 1979.

38161 - KAZAKOV, E.A., OKANENKO, A.S. : Potentsial'nye vozmozhnosti intensivnosti fotosinteza i produktivnosti sakharnoĭ svekly v razlichnykh usloviyakh vodnogo rezhima pochvy. [Capacity of photosynthetic rate and productivity of sugar beet under different soil water regime.] - Fiziol. Biokhim. kul't. Rast. *11* : 574 - 582, 1979. [In R, ab : E.]

38162 - KAZAKOVA, A.S., VASIN, Yu.A., VERKHOTUROV, V.N. : Izmeneniya lineĭnogo dikhroizma i polyarizatsii flyuorestsentsii kletok *Chlorella vulgaris* pri teplovom vozdeĭstvii. [Changes in linear dichroism and fluorescence polarization of *Chlorella vulgaris* cells during heat treatment.] - Nauch. Dokl. vyssh. Shkoly, biol. Nauki *1979* (6) : 24 - 29, 1979. [In R.]

*38163 - KAZARYAN, V.O., AKOPOVA, Zh.M. : O vliyanii vitaminov i gibberellina na obra-
zovanie khlorofilla, aktivnost' fotosinteza i prodolzhitel'nost' zhizni izolirovannykh list'ev. [Effect of vitamins and gibberellin on chlorophyll formation, photosynthetic activity and life span of isolated leaves.] - Dokl.
Akad. Nauk arm. SSR 67 : 237 - 242, 1978. [In R, ab : Arm.]

*38164 - KAZARYAN, V.V., ZAKARYAN, S.O. : K voprosu o reaktsii khvoi nekotorykh intro-
dutsentov k zimnim usloviyam erevanskogo botanicheskogo sada. [Reaction of
needles of some introduced trees to winter conditions in the Erevan botanical garden.] - Dokl. Akad. Nauk arm. SSR 67 : 56 - 60, 1978. [In R, ab : Arm.]

38165 - KE, B., DEMETER, S., ZAMARAEV, K.I., KHAIRUTDINOV, R.F. : Charge recombinat-
ion in photosystem I at low temperatures. Kinetics of electron tunneling. -
Biochim. biophys. Acta 545 : 265 - 284, 1979.

38166 - KECK, R.W. : Carbon dioxide exchange rate inhibition in salt stressed soybean.
- Plant Physiol. 63 (Suppl.) : 149, 1979.

38167 - KEELING, C.D., MOOK, W.G., TANS, P.P. : Recent trends in the $^{13}C/^{12}C$ ratio
of atmospheric carbon dioxide. - Nature 277 : 121 - 123, 1979.

38168 - KEENER, M.E., DeMICHELE, D.W., SHARPE, P.J.H. : Sink metabolism: A conceptu-
al framework for analysis. - Ann. Bot. 44 : 659 - 669, 1979. [Photosynthates.]

38169 - KEIFER, D.W., SPANSWICK, R.M. : Correlation of adenosine triphosphate levels
in Chara corallina with the activity of the electrogenic pump. - Plant Physiol. 64 : 165 - 168, 1979.

38170 - KEITH, B. : Mass transfer and ^{14}C translocation in detached maize leaves. -
Can. J. Bot. 57 : 657 - 665, 1979.

38171 - KEL'BALIKHANOV, B.F., RUSANOV, S.Yu., SHNYREV, G.D. : Metody opredeleniya
pokazatelei pogloshcheniya sveta morskoi vodoi.[Methods for the determination
of light absorption index in the sea water.] - Okeanologiya 19 : 168 - 174,
1979. [In R, ab : E.]

38172 - KELL, D.B. : On the functional proton current pathway of electron transport
phosphorylation. An electrodic view. - Biochim. biophys. Acta 549 : 55 - 99,
1979.

38173 - KELLEY, B.C., JOUANNEAU, Y., VIGNAIS, P.M. : Nitrogenase activity in Rhodopseudomonas sulfidophila. - Arch. Microbiol. 122 : 145 - 152, 1979. [Ps.]

38174 - KELLOMÄKI, S., SALMINEN, R., HARI, P., VENTILÄ, M., KANNINEN, M., KAUPPI, P.,
SMOLANDER, H. : A method for approximating the photosynthetic production of
stand members inside the canopy. - J. appl. Ecol. 16 : 243 - 252, 1979.

38175 - KENT, S.S. : Autotrophic citrate synthesis from C_2 and C_4 units of malate. -
Plant Physiol. 63 (Suppl.) : 4, 1979. [Ps.]

38176 - KENT, S.S. : Photosynthesis in the higher plant Vicia faba. 5. Role of malate as a precursor of the tricarboxylic acid cycle. - Plant Physiol. 64 : 159 -
- 161, 1979.

38177 - KENYON, W.H., BLACK, C.C., Jr. : The role of vacuoles in metabolite compartmentation in CAM plants. - Plant Physiol. 63 (Suppl.) : 37, 1979.

*38178 - KERBY, N.W., EVANS, L.V. : Isolation and partial characterization of pyreno-
ids from the brown alga Pilayella littoralis (L.) KJELLM. - Planta 142 : 91 -
- 95, 1978.

38179 - KERFIN, W., BÖGER, P. : Light-induced hydrogen evolution in blue-green algae.
- In : DELLWEG, H. (ed.) : 4. Symposium Technische Mikrobiologie. Pp. 313 -
- 324. Verlag Versuchs- u. Lehranstalt f.Spiritusfabrikation u. Fermentations-
technologie i. Institut f. Gärungsgewerbe u. Biotechnologie, Berlin 1979.

38180 - KERSHANSKAYA, O.I., URAZALIEV, R.A., BEDENKO,V.P. : Pokazateli fotosinteza
kak test na vysokuyu produktivnost' geterozisnykh gibridov pshenitsy. [Pho-
tosynthesis indices as a test on high productivity of wheat heterotic hybrids.]
- Sel'skokhoz. Biol. 14 : 593 - 596, 1979. [In R, ab : E.]

38181 - KERSHAW, K.A., MORRIS, T., TYSIACZNY, M.J., MacFARLANE, J.D. : Physiological-
environmental interactions in lichens. VIII. The environmental control of .

dark CO_2 fixation in *Parmelia caperata* (L.) ACH. and *Peltigera canina* var. *praetextata* MUE. - New Phytol. *83* : 433 - 444, 1979. [Ps.]

38182 - KESSELL, S.R. : Adaptation and dimorphism in eastern hemlock, *Tsuga canadensis* (L.) CARR. - Amer. Natur. *113* : 333 - 350, 1979. [Growth analysis.]

38183 - KESSELMEIER, J., RUPPEL, H.G. : Relations between saponin concentration and prolamellar body structure in etioplasts of *Avena sativa* during greening and reetiolating and in etioplasts of *Hordeum vulgare* and *Pisum sativum*. - Z. Pflanzenphysiol. *93* : 171 - 184, 1979.

38184 - KESZTHELYI, L., ORMOS, P. : Electric signals associated with the photocycle of bacteriorhodopsin. - FEBS Lett. *109* : 189 - 193, 1979.

38185 - KEYS, A.J. : Mechanisms of carbon dioxide assimilation in photosynthesis. - School Sci. Rev. *60* : 670 - 677, 1979.

38186 - KHAN, M.-U., LEM, N.W., CHANDORKAR, K.R., WILLIAMS, J.P. : Effects of substituted pyridazinones (San 6706, San 9774, San 9785) and glycerolipids and their associated fatty acids in the leaves of *Vicia faba* and *Hordeum vulgare*. - Plant Physiol. *64* : 300 - 305, 1979. [Chl, chloroplast.]

38187 - KHANOVA, L.A., TARASEVICH, M.R. : Electrochemical study of chlorophyll adsorption layers. - Biochem. Bioenerg. *6* (J. Electroanal. Chem. *104*) : 155 - 163, 1979.

38188 - KHANOVA, L.A., TARASEVICH, M.R., ZAKHARKIN, G.I. : Spektral'nye issledovaniya adsorbirovannogo na èlektrode khlorofilla. [Spectral studies of chlorophyll adsorbed on an electrode.] - Élektrokhimiya *15* : 1377 - 1380, 1979. [In R.]

38189 - KHARKYANEN, V.N., KHRISTOFOROV, L.N., KUKHTIN, V.V., PETROV, E.G. : Mechanisms of charge separation in bacterial photosynthesis. - Int. J. Quantum Chem. *16* : 877 - 882, 1979.

38190 - KHATYLEVA, L.U., DYLYANOK, L.A., YATSÈVICH, A.P. : Manasomny analiz èlementaŭ praduktsyĭnastsi ŭ yaravoĭ pshanitsy sortu Pityk-62. [Monosomic analysis of productivity elements in spring wheat Pitic-62.] - Vestsi Akad. Navuk belarus. SSR, Ser. biyal. Navuk *1979* (3) : 29 - 31, 139, 1979. [In Belorus., ab, : E, R.]

38191 - KHITROV, Yu.A., KAUROV, B.S., GAVRILOV, A.G., RUBIN, L.B. : Deĭstvie moshchnogo izlucheniya rubinovogo lazera na transport èlektronov i sopryazhennye s nim protsessy v khloroplastakh gorokha (*Pisum sativum*). [Effect of power laser radiation on electron transport and coupled processes in pea (*Pisum sativum*) chloroplasts.]- Fiziol.Rast. *26* : 808 - 814, 1979. [In R, ab : E.]

*38192 - KHLYASTIKOV, G.P. : Fotosintez i metabolizm fosfora u fasoli pri razlichnykh urovnyakh azotnogo i fosfornogo pitaniya. [Photosynthesis and phosphorus metabolism in *Phaseolus* under different levels of nitrogen and phosphorus nutrition.] - In : Produktivnosf' Nazemnykh Fotosinteziruyushchikh Sistem v Èkstremal'nykh Usloviyakh. Pp. 91 - 99, 188. Sib. Otd. Akad. Nauk SSSR, Buryat. Filial, Ulan Udè 1977. [In R.]

*38193 - KHMARA, L.A., SEMICHAEVSKIĬ, V.D. : Vliyanie nedostatka margantsa na sostoyanie khlorofilla v khloroplastakh i subkhloroplastnykh fraktsiyakh rastenyĭ gorokha. [Effect of manganese deficiency on chlorophyll state in chloroplasts and subchloroplast fractions of pea plants.] - In : Biologiya i Nauchno-Tekhnicheskiĭ Progress. Pp. 53 - 57. Pushchino 1974. [In R.]

38194 - KHODASEVICH, È.V., ARNAUTOVA, A.I., GVARDIYAN, V.N., MYSHKOVETS, E.N. : Strukturnaya organizatsiya khloroplasta i fotosinteticheskaya funktsiya pri dlitel'noĭ vegetatsii lista. [Structural organization of the chloroplast and photosynthetic function during prolonged leaf life span.] - Zh. obshch. Biol. *40* : 603 - 609, 1979. [In R, ab : E.]

*38195 - KHOLUPENKO, I.P., MEDYANNIKOV, V.M. : Nekotorye voprosy raspredeleniya produktov fotosinteza u soi v reproduktivnyĭ period. [Distribution of photosynthates in soybean during the reproduction period.] - In : Pogloshchenie i Peredvizhenie Veshchestv u Rastenyĭ. Pp. 56 - 67, 81. Akad. Nauk SSSR, Vladivostok 1978. [In R.]

38196 - KHOR, H.T. : Removal of chlorophyll pigments from plant neutral lipids. -
J.Chromatogr. *179* : 225 - 226, 1979.

38197 - KHORANA, H.G., GERBER, G.E., HERLIHY, W.C., GRAY, C.P., ANDEREGG, R.J.,
NIHEI, K., BIEMANN, K. : Amino acid sequence of bacteriorhodopsin. - Proc.
nat. Acad. Sci. USA *76* : 5046 - 5050, 1979.

38198 - KHOTYLEVA, L.V., SHEVELUKHA, T.A., POLCHANINOVA, T.V., STAL'MAKOVA, R.N. :
Geneticheskiĭ kontrol' intensivnosti dykhaniya v ontogeneze u yarovoĭ pshenitsy. [Genetic control of respiration rate during ontogenesis of spring
wheat.] - Dokl. Akad. Nauk belorus. SSR *23* : 1045 - 1047, 1979. [In R,
ab : E.]

*38199 - KHRISTIN, M.S., AKULOVA, E.A. : Obnaruzhenie dvukh funktsional'no aktivnykh
ferredoksinov v gorokhe. [Finding of two functionally active ferredoxins in
pea.] - In : Itogi Issledovaniya Mekhanizma Fotosinteza. Pp. 141 - 147.
Pushchino 1974. [In R.]

38200 - KIEFER, D.A., OLSON, R.J., WILSON, W.H. : Reflectance spectroscopy of marine
phytoplankton. Part I. Optical properties as related to age and growth rate.
- Limnol. Oceanogr. *24* : 664 - 672, 1979.

38201 - KIKUCHI, R., ASHIDA, K., HIRAO, S. : Phycobilins in different color types
of *Porphyra yezoensis* UEDA. - Bull. jap. Soc. sci. Fish. *45* : 1461 - 1464,
1979.

38202 - KIKUYAMA, M., HAYAMA, T., FUJII, S., TAZAWA, M. : Relationship between light-
-induced potential change and internal ATP concentration in tonoplast-free
Chara cells. - Plant Cell Physiol. *20* : 993 - 1002, 1979. [Ps.]

*38203 - KIM, V.A. : Mekhanizm fotookisleniya khlorofilla nitrosoedineniyami. [Mecha-
nism of chlorophyll photooxidation by nitrous compounds.] - In : Biologiya
i Nauchno-Tekhnicheskiĭ Progress. Pp. 51 - 53. Pushchino 1974. [In R.]

*38204 - KIM, V.A., VOZNYAK, V.M., EVSTIGNEEV, V.B. : Izuchenie metodom ÉPR fotokhi-
micheskoĭ generatsii kation-radikalov bakteriokhlorofilla i khlorofilla.
[Studying of photochemical generation of cation-radicals of bacteriochlorophyll and chlorophyll by means of EPR.] - In : Itogi Issledovaniya Mekhanizma Fotosinteza. Pp. 46 - 53. Pushchino 1974. [In R.]

38205 - KIMBALL, B.A., MITCHELL, S.T. : Low-cost carbon dioxide analyzer for green-
houses. - HortScience *14* : 180 - 182, 1979.

38206 - KIMENOV, G.P., MINKOV, I.N. : Vliyanie vodnogo defitsita na raspredelenie
^{14}C sredi produktov fotosinteza v list'yakh *Haberlea rhodopensis* FRIV. i
Ramonda serbica PANC. [Effect of water deficit on the ^{14}C-incorporation
into photosynthates in the leaves of *Haberlea rhodopensis* FRIV. and *Ramonda
serbica* PANC.] - In : VAKLINOVA, S.G., VANKOVA-RADEVA, R., VASILEVA, V.S.
(ed.) : Fotosinteticheskaya Assimilyatsiya CO_2 i Fotodykhanie. Pp. 92 - 97.
Izdat. bolg. Akad. Nauk, Sofiya 1979. [In R.]

38207 - KIMES, D.S., SMITH, J.A., BERRY, J.K. : Extension of the optical diffraction
analysis technique for estimating forest canopy geometry. - Aust. J. Bot.
27 : 575 - 578, 1979.

38208 - KIMPEL, D.L., VAUGHN, K.C., WILSON, K.G. : Investigations of the plastome
of *Chlorophytum*. - Plant Physiol. *63* (Suppl.) : 28, 1979. [Ps, Chl.]

38209 - KINERSON, R.S. : Studies of photosynthesis and diffusion resistance in paper birch (*Betula papyrifera* MARSH.) with synthesis through computer simulation. - Oecologia *39* : 37 - 49, 1979.

38210 - KING, J., LAMB, W.I.C., McGREGOR, M.T. : Regrowth of ryegrass swards subject
to different cutting regimes and stocking densities. - Grass Forage Sci.
34 : 107 - 118, 1979. [Ps.]

*B38211 - KING, K.M.(ed.) : Measurement and Modelling of Photosynthesis in Relation
to Productivity. - Can. Committee IBP, Guelph 1972.

38212 - KIPE-NOLT, J.A., STEVENS, S.E. Jr. : Effect of levulinic acid on pigment
biosynthesis in *Agmenellum quadruplicatum*. - J. Bacteriol. *137* : 146 - 152,
1979.

38213 - KIPE-NOLT, J.A., STEVENS, S.E. Jr. : Synthesis of δ-aminolevulinic acid by
         blue-green algae. - Plant Physiol. *63* (Suppl.) : 27, 1979.

38214 - KIRICHENKO, E.B., KUZNETSOVA, L.G., SMOLYGINA, L.D., SERDYUK, O.P. : Vklyu-
         chenie ^{14}C-aminokislot v polipeptidy lamell ètiokhloroplastov mezofilla i
         obkladki *Zea mays* L. [Incorporation of ^{14}C-amino acids into polypeptides
         from etiochloroplast lamellae of *Zea mays* L. mesophyll and bundle sheath.] -
         Fiziol. Rast. *26* : 14 - 19, 1979. [In R, ab : E.]

*38215 - KIRITA, H., HOZUMI, K. : Estimation of the total chlorophyll amount and its
         seasonal change in a warm-temperate evergreen oak forest at Minamata, Japan.
         - Jap. J. Ecol. *23* : 195 - 200, 1973.

38216 - KIRK, J.T.O. : Spectral distribution of photosynthetically active radiation
         in some south-eastern Australian waters. - Aust. J. mar. Freshw. Res. *30* :
         81 - 91, 1979.

38217 - KIRKHAM, M.B. : Water relations of wheat alternated between two root tempe-
         ratures. - New Phytol. *82* : 89 - 96, 1979. [Stomatal resistance.]

38218 - KIRST, G.O., BISSON, M.A. : Regulation of turgor pressure in marine algae:
         ions and low-molecular-weight organic compounds. - Aust. J. Plant Physiol.
         6 : 539 - 556, 1979. [Ps.]

38219 - KIRYAKOV, K., PRESOLSKA, P. : Vliyanie na g"stotata na poseva i toreneto
         v"rkhu nyakoi fotosintetichni pokazateli i dobiva pri samooprasheni linii
         tsarevitsa. [Effect of planting density and fertilizer application on some
         photosynthetic features and yield of self-pollinated maize lines.] - Raste-
         niev. Nauki *16* (2) : 5 - 12, 1979. [Growth analysis; in Bulg., ab : E, R.]

38220 - KIS, P. : Preparation of chlorophyll *a* by a nonchromatographic method. -
         Anal. Biochem. *96* : 126 - 129, 1979.

38221 - KISELEVA, T.M. : Sezonnaya dinamika plastidnogo apparata i pigmentov u neko-
         torykh rastenii elovogo lesa. [Seasonal dynamics of plastid apparatus and
         pigments in some plants of spruce forest.] - Vestnik leningrad. Univ. *1979*
         [9 (Biol.2)]: 28 - 33, 124, 1979. [In R, ab : E.]

38222 - KISLYUK, I.M. : Protecting and injurious effects of light on photosynthetic
         apparatus during and after heat treatment of leaves. - Photosynthetica *13* :
         386 - 391, 1979.

38223 - KLEIN, G., RÜDIGER, W. : Thioether formation of phycocyanobilin: A model re-
         action of phycocyanin biosynthesis. - Z. Naturforsch. *34C* : 192 - 195, 1979.

38224 - KLEIN, J.J. : An apparent violation of second law of thermodynamics in bio-
         logical systems. - J. chem. Educ. *56* : 314, 1979. [Ps.]

38225 - KLEIN, O., DÖRNEMANN, D., SENGER, H. : Zwei Biosynthesewege für δ-Aminolävu-
         linsäure in der Pigmentmutante C-2A' der einzelligen Grünalge *Scenedesmus
         obliquus*. - Ber. deut. bot. Ges. *92* : 619 - 628, 1979.

38226 - KLEINIG,H., LIEDVOGEL, B. : On the energy requirements of fatty acid synthe-
         sis in spinach chloroplasts in the light and in the dark. - FEBS Lett. *101* :
         339 - 342, 1979.

38227 - KLENOVSKÁ, S. : Morphogenetical and physiological processes in tobacco ex-
         plants by cultivating in water potential decreased conditions. - Acta Fac.
         Rerum nat. Univ. comenianae, Physiol. Plant. *16* : 37 - 44, 1979. [Chl.]

38228 - KLEUDGEN, H.K. : Changes in composition of chlorophylls, carotenoids, and
         prenylquinones in green seedlings of *Hordeum* and *Raphanus* induced by the
         herbicide San 6706 - an effect possibly antagonistic to phytochrome action. -
         Pestic. Biochem. Physiol. *12* : 231 - 238, 1979.

38229 - KLEUDGEN, H.K. : Veränderungen der Pigment- und Prenylchinongehalte in Chlo-
         roplasten von Gerstenkeimlingen nach Applikation des Wuchsstoffherbizids
         MCPA (4-Chlor-2-methylphenoxyessigsäure). - Z. Naturforsch. *34C* : 106 - 109,
         1979.

38230 - KLEUDGEN, H.K. : Die Wirkung von Simazin auf die Bildung der plastidären
         Prenyllipide in Keimlingen von *Hordeum vulgare* L. - Z. Naturforsch. *34C* :
         110 - 113, 1979. [Chl.]

38231 - KLIMOV, S.V., DZHANUMOV, D.A., BOCHAROV, E.A. : O sootnoshenii avto- i geterotrofnogo pitaniya u prorostkov ozimoĭ pshenitsy *Triticum aestivum* L. [Re-
lationship between autotrophic and heterotrophic nutrition in winter wheat
Triticum aestivum L. seedlings.] - Fiziol. Rast. *26* : 1143 - 1149, 1979. [Ps;
in R, ab : E.]

38232 - KLIMOV, V.V., ALLAKHVERDIEV, S.I., DEMETER, Sh., KRASNOVSKIĬ, A.A. : Fotovos-
stanovlenie feofitina v fotosisteme 2 khloroplastov v zavisimosti ot okislitel'no-vosstanovitel'nogo potentsiala sredy. [Photoreduction of pheophytin
in the photosystem 2 of chloroplasts with respect to the redox potential of
the medium.] - Dokl. Akad. Nauk SSSR *249* : 227 - 230, 1979. [In R.]

38233 - KLIMOV, V.V., ALLAKHVERDIEV, S.I., KRASNOVSKIĬ, A.A. : Signal É.P.R. pri fotovosstanovlenii feofitina v reaktsionnykh tsentrakh fotosistemy 2 khloroplastov. [EPR signal in the photoreduction of pheophytin in reaction centres
of the photosystem 2 of chloroplasts.] - Dokl. Akad. Nauk SSSR *249* : 485 -
488, 1979. [In R.]

38234 - KLOCKARE, B., MELO, T.B., JOHNSSON, A. : A system approach to fluorescence
induction in green leaves. - Physiol. Plant. *46* : 101 - 108, 1979. [Chl.]

38235 - KLUGE, M. : Crassulacean Acid Metabolism (CAM) : Einige stoffwechselphysio-
logische und ökologische Aspekte. - Ber. deut. bot. Ges. *92* : 95 - 107,
1979.

38236 - KLUGE, M. : The flow of carbon in Crassulacean Acid Metabolism (CAM). - In :
GIBBS, M., LATZKO, E. (ed.) : Photosynthesis II.(Encycl. Plant Physiol. N.S.
Vol.6.) Pp. 113 - 125. Springer-Verlag, Berlin - Heidelberg - New York 1979.

38237 - KLUGE, M., KNAPP, I., KRAMER, D., SCHWERDTNER, I., RITTER, H. : Crassulacean
acid metabolism (CAM) in leaves of *Aloe arborescens* MILL. Comparative studies of the carbon metabolism of chlorenchym and central hydrenchym. - Planta
145 : 357 - 363, 1979.

B38238 - KLUGE, M., LORENZEN, H. (ed.) : Biochemische Grundlagen ökologischer Anpas-
sungen bei Pflanzen. - Gustav Fischer Verlag, Stuttgart - New York 1979.
[Ps.]

38239 - KNAFF, D.B., CARR, J.W. : The energy-linked carotenoid band-shift in *Chromatium vinosum*. - Arch. Biochem. Biophys. *193* : 379 - 384, 1979.

38240 - KNAFF, D.B., OLSON, J.M., PRINCE, R.C. : The light-reaction of the green
photosynthetic bacterium *Chlorobium limicola* f. *thiosulfatophilum* at cryogenic temperatures. - FEBS Lett. *98* : 285 - 289, 1979.

38241 - KNAFF, D.B., WHETSTONE, R., CARR, J.W. : The role of the membrane potential
in active transport by the photosynthetic bacterium *Chromatium vinosum*. -
FEBS Lett. *99* : 283 - 286, 1979.

38242 - KNEE, M., HATFIELD, S.G.S., RATNAYAKE, M. : Acceleration of some ripening
processes by light treatment of stored apples. - J. exp. Bot. *30* : 1013 -
1020, 1979. [Ps, Chl.]

38243 - KNOBLOCH, K., BALTSCHEFFSKY, M. : Photophosphorylation and pyrophosphate-
-driven ATP generation in chromatophores from *Rhodopseudomonas palustris*. -
Hoppe Seyler's Z. physiol. Chem. *360* : 1165 - 1166, 1979.

38244 - KNOX, R.S. : Conversion of light into free energy. - In : GERISCHER, H.,
KATZ, J.J. (ed.) : Light-Induced Charge Separation in Biology and Chemistry.
Pp. 45 - 59. Verlag Chemie, Weinheim - New York 1979.[Ps, Chl.]

38245 - KNOX, R.S., VAN METTER, R.L. : Fluorescence of light-harvesting chlorophyll
a/b-protein complexes: implications for the photosynthetic unit. - In : Chlorophyll Organization and Energy Transfer in Photosynthesis. Pp. 177 - 190.
Excerpta Medica, Amsterdam - Oxford - New York 1979.

38246 - KOBATA, T., TAKAMI, S. : [The effects of water stress on the grain-filling
in rice.] - Jap. J. Crop Sci. *48* : 75 - 81, 1979. [Photosynthates; in Jap.,
ab : E.]

38247 - KOBAYASHI, H., TAKABE, T., NISHIMURA, M., AKAZAWA, T. : Roles of the large
and small subunits of ribulose-1,5-bisphosphate carboxylase in the activation
by CO_2 and Mg^{2+}. - J. Biochem. (Tokyo) *85* : 923 - 930, 1979.

38248 - KOBAYASHI, T., DEGENKOLB, E.O., BERSOHN, R., RENTZEPIS, P.M., MacCOLL, R.,
        BERNS, D.S. : Energy transfer among the chromophores in phycocyanins measu-
        red by picosecond kinetics. - Biochemistry 18 : 5073 - 5078, 1979.

38249 - KOBAYASHI, Y., INOUE, Y., FURUYA, F., SHIBATA, K., HEBER, U. : Regulation
        of adenylate levels in intact spinach chloroplasts. - Planta 147 : 69 - 75,
        1979.

38250 - KOBAYASHI, Y., INOUE, Y., SHIBATA, K., HEBER, U. : Control of electron flow
        in intact chloroplasts by the intrathylakoid pH, not by the phosphorylation
        potential. - Planta 146 : 481 - 486, 1979.

38251 - KOBLENTS-MISHKE, O.I. : Fotosintez morskogo fitoplanktona v zavisimosti ot
        podvodnoĭ obluchennosti. [Photosynthesis of marine phytoplankton in relation
        to light conditions.] - Fiziol. Rast. 26 : 908 - 920, 1979. [In R, ab : E.]

38252 - KOCH, J., BERGMANN, H. : Einfluss von Phytohormonen auf die Wasserausnutzung
        (WUE) von Hordeum vulgare (L.) und Triticum aestivum (L.). - Biochem. Phy-
        siol. Pflanzen 174 : 486 - 490, 1979. [Ps.]

38253 - KOCH, K.E., KENNEDY, R.A. : Crassulacean acid metabolism (CAM) in Portulaca
        oleracea L. under natural environmental conditions. - Plant Physiol. 63
        (Suppl.) : 37, 1979.

38254 - KOCHUBEĬ, S.M. : O prirode dlinnovolnovoĭ fluorestsentsii fotosistemy I vys-
        shikh rasteniĭ. [Nature of long-wave fluorescence of higher plant photosys-
        tem I.] - Fiziol. Biokhim. kul't. Rast. 11 : 563 - 573, 1979. [In R, ab : E.]

38255 - KOCHUBEĬ, S.M., GULIEV, F.A., SMIRNOV, A.A. : Primenenie lazernoĭ spektro-
        fluorimetrii dlya issledovaniya fotosinteziruyushchikh chastits. [Use of la-
        ser spectro-fluorimetry for studying photosynthesizing particles.] - Dokl.
        Akad. Nauk SSSR 244 : 743 - 746, 1979. [In R.]

38256 - KOEDZHIKOV, Kh., GOTEVA, M. : Izyasnyavane s pomoshchta na izotopniya metod
        polyata na korenovata sistema za obmyana na asimilati mezhdu rasteniyata v
        proizvodstven posev. [The role of the root system in the exchange of assimi-
        lates between crop plants studied by an isotope method.] - Rasteniev. Nauki
        16 (2) : 25 - 28, 1979. [In Bulg., ab : E, R.]

38257 - KOEDZHIKOV, Kh., NANCHEVA, R., GENCHEV, D. : Vliyanie na g"stotata na poseva
        v"rkhu s"d"rzhanieto na maznini i proteini v semkata i kalorichniya efekt
        na fotosintezata pri sl"nchogleda. [Effect of planting density on oil and
        protein levels of sunflower seeds and the caloric effect of photosynthesis
        in sunflower.] - Rasteniev. Nauki 16 (4) : 3 - 12, 1979. [In Bulg., ab : E.]

*38258 - KOGURE(TAMAKI), K., NAKA, J., ASANUMA, K. : Behavior of ^{14}C photosynthetic
        products during the reproductive growth in broad bean plant. - Tech. Bull.
        Fac. Agr. Kagawa Univ. 30 (63) : 1 - 8, 1978.

38259 - KOHL, H.C. Jr., THIGPEN, S.P. : Rate of dry weight gain of chrysanthemum as
        a function of leaf area index and night temperature. - J. amer. Soc. hort.
        Sci. 104 : 300 - 303, 1979. [Growth analysis.]

*38260 - KÖHLER, K.-H. : Licht und Leben. - Greifswalder Universitätsreden, neue Folge
        38 : 1 - 37, 1977. [Chl.]

*38261 - KÖHLER, K.-H. : Zu einigen aktuellen Problemen der lichtabhängigen Pigment-
        synthesen bei Pflanzen. - Wiss. Z. E.-M.-Arndt-Univ. Greifswald, math.-natur-
        wiss. Reihe 26 : 103 - 111, 1977.

38262 - KOHOUT, V. : Rozdíly v tvorbě hmoty plevelů a cukrovky v regulovaných bio-
        energetických podmínkách. [Differences in the production of weed and sugar
        beet biomass under controlled bio-energetic conditions.] - Rostl. Výroba
        (Praha) 25 : 1081 - 1089, 1979. [Growth analysis; in Czech, ab : E, G, R.]

38263 - KOIKE, H., KATOH, S. : Heat-stabilities of cytochromes and ferredoxin isola-
        ted from a thermophilic blue-green alga. - Plant Cell Physiol. 20 : 1157 -
        1161, 1979.

38264 - KOIZUMI, J., YABE, I., AIBA, S. : [Light absorption rate of photosynthetic
        bacterium in fermentor - Analysis by Monte Carlo method.] - Kagaku Kogaku
        Ronbunshu 5 : 644 - 649, 1979. [In Jap., ab : E.]

*38265 - **KOKSHAROV, V.P., KLYUKINA, E.M.** : Matematicheskiĭ metod opredeleniya ploshchadi list'ev kartofelya. [Mathematical method for leaf area determination in potato.] - Tr. ural'. nauch.-issled. Inst. sel'. Khoz. (Sverdlovsk) *23* (Ovoshchevodstvo Zashchishchennogo i Otkrytogo Grunta na Urale) : 135 - 136, 1978. [In R.]

38266 - **KOLESNIKOV, M.P., EGOROV, I.A.** : Metalloporfiriny v otlozheniyakh dokembriya kak veroyatnye svidetel'stva drevnego fotosinteza. [Metalloporphyrins in precambrian formations as possible evidence of ancient photosynthesis.] - Dokl. Akad. Nauk SSSR *244* : 470 - 473, 1979. [Models; in R.]

38267 - **KOLESNIKOV, M.P., EGOROV, I.A.** : Metalloporphyrins and molecular complexes of amino acids with porphyrins in juvenile volcanic ash. - Origins Life *9* : 267 - 277, 1979. [Chl derivatives.]

38268 - **KOLESNIKOV, P.A., ZORĖ, S.V., PETROCHENKO, E.I., SHUMOVA, T.A.** : Ingibirovanie perekis'yu vodoroda metabolizma ribozo-5-fosfata i triozofosfat degidrogenazy v ėkstraktakh iz khloroplastov. [Hydrogen peroxide inhibition of the metabolism of ribose-5-phosphate and triose phosphate dehydrogenase in extracts from chloroplasts.] - Dokl. Akad. Nauk SSSR *247* : 1502 - 1505, 1979. [In R.]

38269 - **KOLESNIKOV, P.A., ZORĖ, S.V., PETROCHENKO, E.I., SHUMOVA, T.A., PSHENOVA, K.V., MUTUSKIN, A.A.** : Ferredoksin-NADF reduktaza v metabolizme ribozo-5-fosfata v ėkstraktakh iz khloroplastov. [Ferredoxin-NADP reductase in the metabolism of ribose-5-phosphate in chloroplast extracts.] - Dokl. Akad. Nauk SSSR *247* : 499 - 502, 1979. [In R.]

*38270 - **KOLLER, K.-P., WEHRMEYER, W., MÖRSCHEL, E.** : Biliprotein assembly in the disc-shaped phycobilisomes of *Rhodella violacea*. On the molecular composition of energy-transfering complexes (tripartite units) forming the periphery of the phycobilisome. - Europe. J. Biochem. *91* : 57 - 63, 1978.

38271 - **KOLLMAN, V.H., HANNERS, J.L., LONDON, R.E., ADAME, E.G., WALKER, T.E.** : Photosynthetic preparation and characterization of ^{13}C-labeled carbohydrates in *Agmenellum quadruplicatum*. - Carbohydrate Res. *73* : 193 - 202, 1979.

38272 - **KOLOMEĬCHENKO, V.V.** : Povyshenie produktivnosti i KPD ispol'zovaniya FAR ovrazhno-balochnykh zemel'. [Increasing productivity and efficiency of PhAR utilization on ravine-gorge lands.] - Vest. sel'skokhoz. Nauki *1979* (3) : 61 - 64, 1979. [Car; in R, ab : E.]

*38273 - **KOLTUNOVA, A.G., LEBEDEVA, G.P., NASYROV, Yu.S.** : Rol' fitokhroma v biogeneze khloroplastov. [Role of phytochrome in chloroplast biogenesis.] - Izv. Akad. Nauk tadzh. SSR, Otd. biol. Nauk *1978* (1) : 46 - 52, 1978. [In R, ab : Tajik.]

*38274 - **KOMÁRKOVÁ, J.** : Circadian periodicity of algal photosynthetic activity and chlorophyll *a* content in laboratory experiments. - Arch. Hydrobiol., Suppl. *51* (algol. Stud. 21) : 434 - 443, 1978.

38275 - **KOMBRINK, E., WÖBER, G., WALKER, D.A.** : Einfluß von DEAE-Dextran auf den Elektronentransport in Chloroplasten von *Dunaliella marina*. - Ber. deut. bot. Ges. *92* : 379 - 392, 1979.

38276 - **KONDO, T., NAKASHIMA, H.** : Content of adenosine phosphate compounds in a long-day duckweed, *Lemna gibba* G_3, under different light and nutritional conditions. - Physiol. Plant. *45* : 357 - 362, 1979.

38277 - **KONDRAT'EVA, T.M.** : Pervichnaya produktsiya v vodakh tropicheskoĭ Atlantiki i ee sutochnye izmeneniya. [Primary production in the tropical Atlantic and its diurnal changes.] - Okeanologiya *19* : 869 - 877, 1979. [In R, ab : E.]

38278 - **KONISHI, T., TRISTRAM, S., PACKER, L.** : The effect of cross-linking on photocycling activity of bacteriorhodopsin. - Photochem. Photobiol. *29* : 353 - 358, 1979.

*38279 - **KONONCHUK, G.L., PRIKHOD'KO, S.I., TROFIMCHUK, G.E.** : Mnogokanal'nyĭ kvantometr dlya gidrobiologicheskikh issledovaniĭ. [Multichannel quantometer for hydrobiological research.] - Gidrobiol. Zh. *13* (6) : 83 - 87, 1977. [In R.]

*38280 - KONONCHUK, G.L., PRIKHOD'KO, S.I., TROFIMCHUK, G.E. : Izmerenie fotosinteti-
cheski aktivnoĭ radiatsii. [Measurement of photosynthetically active radiation.] - Gidrobiol. Zh. *13* (3) : 116 - 119, 1977. [In R.]

38281 - KOOYMAN, R.P.H., SCHAAFSMA, T.J., JANSEN, G., CLARKE, R.H., HOBART, D.R.,
LEENSTRA, W .R. : A comparative study of dimerization of chlorophylls and
pheophytins by fluorescence and ODMR. - Chem. Phys. Lett. *68* : 65 - 70, 1979.

*38282 - KORENSTEIN, R., HESS, B. : Immobilization of bacteriorhodopsin and orienta-
tion of its transition moment in purple membrane. - FEBS Lett. *89* : 15 - 20,
1978.

38283 - KORENSTEIN, R., HESS, B., MARKUS, M. : Cooperativity in the photocycle of
purple membrane of *Halobacterium halobium* with a mechanism of free energy
transduction. - FEBS Lett. *102* : 155 - 161, 1979.

*38284 - KÖRNER, C., HILSCHER, H. : Wachstumsdynamik von Grünerlen auf ehemaligen Alm-
flächen an der zentralalpinen Waldgrenze. - In : CERNUSCA, A. (ed.) : Ökologische Analysen von Almflächen im Gasteiner Tal. Veröffentlichungen des Öster-
reichischen MaB-Hochgebirgsprogramms Hohe Tauern. Vol.2. Pp. 187 - 193. Uni-
versitätsverlag Wagner, Innsbruck 1978.

38285 - KÖRNER, C., SCHEEL, J.A., BAUER, H. : Maximum leaf diffusive conductance in
vascular plants. - Photosynthetica *13* : 45 - 82, 1979.

38286 - KORZH, B.V. : Vliyanie razvivayushchikhsya geterozisnykh semyan na fotosintez i dykhanie materinskikh rasteniĭ kukuruzy. [Influence of developing heterotic seeds on photosynthesis and respiration of mother plants of maize.] -
Byull. vsesoyuz. nauch.-issled. Inst. Rastenievod. N.I.Vavilova *87* : 45 - 49,
1979. [In R.]

38287 - KORZH, B. V. : Fotosintez i fotodykhanie vidov-istochnikov genomov pshenitsy-
pri razlichnykh temperaturakh. [Photosynthesis and photorespiration of species - sources of wheat genome - under various temperatures.] - Byull. vsesoyuz. nauch.-issled. Inst. Rastenievod. N.I.Vavilova *87* : 50 - 54, 1979. [In
R.]

38288 - KOSAKOVSKAYA, I.V., KOMARNITSKIĬ, I.K., GLEBA,Yu.Yu., SYTNIK, K.M. : Ribulo-
zodifosfatkarboksilaza roda *Nicotiana*. [Ribulosebisphosphate carboxylase in
the genus *Nicotiana*.] - In : VAKLINOVA, S.G., VANKOVA-RADEVA, R., VASILEVA,
V.S. (ed.) : Fotosinteticheskaya Assimilyatsiya CO_2 i Fotodykhanie. Pp. 44 -
49. Izdatel'stvo bolgarskoĭ Akademii Nauk, Sofiya 1979. [In R.]

38289 - KOSITSYN, A.V., IGOSHINA, T.I. : Deĭstvie kompleksoobrazovateleĭ i tsinka
na karboangidrazu tradeskantsii. [Effect of chelating agents and zinc on
Tradescantia carbonic anhydrase.] - Fiziol. Rast. *26* : 81 - 85, 1979. [In R,
ab : E.]

38290 - KOSOVEL, V., TALARICO, L. : Seasonal variations of photosynthetic pigments
in *Gracilaria verrucosa* (HUDS.) PAPENFUSS (*Florideophyceae-Gigartinales*). -
Boll. Soc. adriat. Sci. *63* : 5 - 15, 1979.

38291 - KOSTIKOV, A.P., LADYGIN, V.G., MEZENTSEV, V.V., IL'IN, Yu.N. : Ustanovlenie
mesta narusheniya élektrotransportnoĭ tsepi khloroplastov mutantov *Chlamydomonas* s neaktivnoĭ fotosistemoĭ 2. [Discovery of damage localization in electron transport chain of chloroplasts in *Chlamydomonas* mutants with inactive
photosystem 2.] - Biofizika *24* : 925 - 927, 1979. [In R, ab : E.]

38292 - KOSTLAN, N.V., CHERNYA, V.F. : Vplyv umov osvitlennya na pigmentnyĭ sklad
riznykh shtamiv khloreli. [Effect of illumination on pigment composition of
Chlorella strains.] - Ukr. bot. Zh. *36* : 411 - 414, 508, 1979. [In Ukr.,
ab : E, R.]

38293 - KOSTOV, K.D., NAĬDENOVA, Ts.: Izmenenie transpiratsii i fotosinteza u molodykh rasteniĭ duba krasnogo v zavisimosti ot vlazhnosti i kislotnosti pochvy.
[Change in young red oak photosynthesis and transpiration in relation to
soil humidity and acidity.] - In : VAKLINOVA, S.G., VANKOVA-RADEVA, R.,
VASILEVA, V.S. (ed.) : Fotosinteticheskaya Assimilyatsiya CO_2 i Fotodykhanie. Pp. 160 - 165.Izdatel'stvo bolgarskoĭ Akademii Nauk, Sofiya 1979. [In R.]

38294 - **KOSTREJ, A.** : Vstup energie do pol'nohospodárskych produkčných systemov.
[Energy input to agricultural production ecosystems.] - In : Bilançia Energie a Vody v Pol'ných a Lesných Ekosystémoch. Pp. 25 - 31. Vysoká Škola pol'-
nohospodárská, Nitra 1979. [In Slovak.]

38295 - **KÖST-REYES, E., KÖST, H.-P.** : The protein-chromophore bond in B phycoerythrin from *Porphyridium cruentum*. Radiosulfur labeling experiments. - Europe.
J. Biochem. *102* : 83 - 91, 1979.

38296 - **KOSTYAEV, V.Ya.** : Fiksatsiya molekulyarnogo azota i fotosintez u sinezelenoĭ
vodorosli *Anabaena spiroides* v dlinnovolnovykh ul'trafioletovykh luchakh.
[Fixation of molecular nitrogen and photosynthesis in the blue-green alga
Anabaena spiroides in long-wave ultraviolet rays.] - Dokl. Akad. Nauk SSSR
248 : 1018 - 1020, 1979. [In R.]

38297 - **KOVATCHEVA, N., BERGMAN, B.** : Some characteristics of malate dehydrogenase
of the blue-green alga *Nostoc muscorum*. - Plant Sci. Lett. *16* : 189 - 194,
1979.

38298 - **KOW , Y.W., GIBBS, M.** : Oxidation of glyceraldehyde-3-P to glycerate-3-P in
the darkened chloroplast. - Plant Physiol. *63* (Suppl.) : 3, 1979.

38299 - **KOWAL, T., KRUPIŃSKA, A.** : Produktywność qatunku *Thymus pulegioides* L. w wa-
runkach naturalnych. [Productivity of the species *Thymus pulegioides* L. in
natural conditions.] - Acta agrobot. *32* : 81 - 89, 1979. [In Pol., ab : E, R.]

38300 - **KOWAL, T., PIC, S.** : Produktivność gatunku *Achillea millefolium* L. w warun-
kach naturalnych. [Productivity of the species *Achillea millefolium* L. in
natural conditions.] - Acta agrobot. *32* : 91 - 100, 1979. [In Pol., ab :
E, R.]

*38301 - **KOWALLIK, K.V., HERRMANN, R.G.** : Cytology. a) General and molecular cytology:
Chloroplasts. - Progr. Bot. *39* : 1 - 17, 1977.

*38302 - **KOYAMA, T.** : Comparison of two methods for estimating the exchange rate of
carbon dioxide at an air-lake interface. - Jap. J. Limnol. *37* : 118 - 122,
1976.

*38303 - **KOYAMA, T., NISHIMURA, M., MATSUDA, H.** : Estimation of exchange rate of carbon dioxide between air and lake surface. - Jap. J. Limnol. *36* : 111 - 116,
1975.

*38304 - **KOYAMA, T., NISHIMURA, M., MATSUDA, H.** : Diurnal and seasonal variations of
exchange rate of carbon dioxide across an air-lake water interface. - Jap. J.
Limnol. *38* : 131 - 137, 1977.

38305 - **KOYAMA, Y., LONG, R.A., MARTIN, W.G., CAREY, P.R.** : The resonance Raman
spectrum of carotenoids as an intrinsic probe for membrane potential. Oscil-
latory changes in the spectrum of neurosporene in the chromatophores of
Rhodopseudomonas sphaeroides. - Biochim. biophys. Acta *548* : 153 - 160,1979.

*38306 - **KOZHOVA, O.M., BARASHOVA, N.I.** : O vzaimootnosheniyakh fito- i zooplanktona
v Bratskom vodokhranilishche. [Interrelationships of phyto- and zooplankton
in Bratsk water reservoir.] - In : Gidrobiologicheskie i Ikhtiologicheskie
Issledovaniya v Vostochnoĭ Sibiri. Pp. 105 - 110, 214. Irkutsk.gos. Univ.,
Irkutsk 1978. [Primary production; in R.]

*38307 - **KOZHOVA, O.M., ZAGORENKO, G.F., POMAZKOVA, G.I., PUTYATINA, T.N., ERBAEVA,
Ė.A., IZMEST'EVA, L.R.** : Gidrobiologicheskiĭ rezhim ozera Khubsugul. [Hydro-
biological regimen of the lake Khubsugul.] - In : Gidrobiologicheskie i Ikhtiologicheskie Issledovaniya v Vostochnoĭ Sibiri. Pp. 3 - 20, 210. Irkutsk.
gos. Univ., Irkutsk 1978. [Chl; in R.]

38308 - **KOZHOVA, O.M., ZAGORENKO, G.P.** : Peculiarities of the Khubsugul Lake ecosystem (Mongolia). - Pol. Arch. Hydrobiol. *26* : 337 - 350, 1979. [Primary
production.]

38309 - **KOZLOV, Yu.N., KATS, E.Yu., KISELEV, B.A., EVSTIGNEEV, V.B.** : Izuchenie kinetiki fotosensibilizirovannogo khlorofillom vosstanovleniya krasiteleĭ
metodom fotopolyarografii. [Kinetics of chlorophyll photosensitized reduction of dyes by the method of photopolarography.] - Biofizika *24* : 583 -
587, 1979. [In R, ab : E.]

38310 - KOZLOWSKI, T.T., PALLARDY, S.G. : Stomatal responses of *Fraxinus pennsylvanica* seedlings during and after flooding. - Physiol. Plant. *46* : 155 - 158, 1979. [Stomatal resistance.]

38311 - KRALJIČ, I., BARBOY, N., LEICKNAM, J.-P. : Photosensitized formation of singlet oxygen by chlorophyll *a* in neutral aqueous micellar solutions with Triton X-100. - Photochem. Photobiol. *30* : 631 - 633, 1979.

*38312 - KRAMAR, T.I., MALINOVSKIĬ, V.I., ZHURAVLEV, Yu.N. : Vliyanie obezvozhivaniya i vysokoĭ temperatury na intensivnost' fotosinteza zdorovykh i porazhennykh virusami rasteniĭ. [Effect of desiccation and high temperature on photosynthetic rate of healthy and virus-diseased plants.] - In : Fiziologicheskie i Biokhimicheskie Issledovaniya Rasteniĭ na Dal'nem Vostoke. Pp. 51 - 62. Biol.-pochv. Inst. sib. Otd. Akad. Nauk SSSR, Vladivostok 1970. [In R.]

38313 - KRANS, J.V., BEARD, J.B., WILKINSON, J.F. : Classification of C_3 and C_4 turfgrass species based on CO_2 compensation concentration and leaf anatomy. - HortScience *14* : 183 - 185, 1979.

38314 - KRASAVTSEV, O.A. : Osobennosti mekhanizma vymerzaniya drevesiny i tsvetkovykh pochek vishni. [Characteristics of the mechanism of xylem and flower bud frost killing in cherry trees.] - Fiziol. biokhim. kul't. Rast. *11* : 176 - 180, 1979. [Energy content; in R, ab : E.]

38315 - KRASICHKOVA, G.V., BOBODZHANOV, V.A., VAKHIDOVA, L.P., CHANDYLOVA, L.V., GILLER, Yu.E. : Sravnitel'naya kharakteristika pigmentnogo sostava i fotokhimicheskoĭ aktivnosti khloroplastov mutantnykh form gorokha i nekotorye pokazateli produktivnosti. [Comparative characteristics of the pigment composition and photochemical activity of chloroplasts of pea mutants and some indexes of productivity.] - Dokl. Akad. Nauk tadzh. SSR *22* : 446 - 449, 1979. [In R, ab : Tajik.]

38316 - KRASICHKOVA, G.V., GILLER, Yu.E. : Sravnitel'naya kharakteristika fotokhimicheskoĭ aktivnosti khloroplastov nekotorykh sortov i gibridov khlopchatnika. [Comparative characteristics of photochemical activity of chloroplasts in some cotton cultivars and hybrids.] - Fiziol. Rast. *26* : 270 - 275, 1979. [In R, ab : E.]

38317 - KRASLOVÁ, J. : Dynamika cukrů v listech a bulvách cukrové řepy jako měřitelný příznak změn energetických podmínek. [Dynamics of sugars in the leaves and roots of sugar beet as a measurable sign of changes in energetic conditions.] - Rost. Výroba (Praha) *25* : 1057 - 1064, 1979. [In Czech, ab : E, G, R.]

38318 - KRASNA, A.I. : Hydrogenase : properties and applications. - Enzyme Microbiol. Technol. *1* : 165 - 172, 1979. [Ps.]

38319 - KRASNOVSKIĬ, A.A. : Biologicheskoe preobrazovanie solnechnoĭ énergii. [Biological transformation of solar energy.] - Vestnik Akad. Nauk SSSR *1979* (1) : 83 - 96, 1979. [In R.]

*38320 - KRASNOVSKIĬ, A.A., UMRIKHINA, A.V., BUBLICHENKO, N.V. : Svobodnye radikaly pri fotokhimicheskikh reaktsiyakh khlorofilla. [Free radicals in photochemical reactions of chlorophyll.] - In : Spektroskopiya Fotoprevrashcheniĭ v Molekulakh. Pp. 106 - 131. Nauka, Leningrad 1977. [In R.]

38321 - KRASNOVSKY, A.A. : Photoproduction of hydrogen in photosynthetic and artificial systems. - In : BARBER, J. (ed.) : Photosynthesis in Relation to Model Systems. Pp. 281 - 298. Elsevier, Amsterdam - New York - Oxford 1979.

38322 - KRASNOVSKY, A.A. Jr. : Photoluminescence of singlet oxygen in pigment solutions. - Photochem. Photobiol. *29* : 29 - 36, 1979.

*38323 - KREMER, B.P. : Studies on $^{14}CO_2$-assimilation in marine *Rhodophyceae*. - Mar. Biol. *48* : 47 - 54, 1978.

38324 - KREMER, B.P. : Photoassimilatory products and osmoregulation in marine *Rhodophyceae*. - Z. Pflanzenphysiol. *93* : 139 - 147, 1979.

38325 - KREMER, B.P., FEIGE, G.B. : Accumulation of photoassimilatory products by phycobiliprotein-containing algae with special reference to *Cyanidium caldarium*. - Z. Naturforsch. *34 C* : 1209 - 1214, 1979.

38326 - KREMER, B.P., KIES, L., ROSTAMI-RABET, A. : Photosynthetic performance of cyanelles in the endocyanomes *Cyanophora*, *Glaucosphaera*, *Gloeochaete* and *Glaucocystis*. - Z. Pflanzenphysiol. *92* : 303 - 317, 1979.

38327 - KREMER, B.P., MARKHAM, J.W. : Carbon assimilation by different developmental stages of *Laminaria saccharina*. - Planta *144* : 497 - 501, 1979.

38328 - KRESLAVSKIĬ, V.D., OLOVYANISHNIKOVA, G.D., STOLOVITSKIĬ, Yu.M., KUTYSHENKO, V.P., EVSTIGNEEV, V.B . : Kompleksy khlorofilla i ego analogov s tetratsiankhinodimetanom. [The complexes of chlorophyll and its analogs with tetracyanchinodimethane.] - Biofizika *24* : 770, 1979. [In R.]

38329 - KRESLAVSKIĬ, V.D., STOLOVITSKIĬ, Yu.M., GERTS, S.M., EVSTIGNEEV, V.B. : Spetsificheskaya sol'vatatsiya khlorofilla éfirami. [Specific solvation of chlorophyll by ethers.] - Biofizika *24* : 771, 1979. [In R.]

38330 - KRIEBEL, A.N., GILLBRO, T., WILD, U.P. : A low temperature investigation of the intermediates of the photocycle of light-adapted bacteriorhodopsin. Optical absorption and fluorescence measurements. - Biochim. biophys. Acta *546* : 106 - 120, 1979.

38331 - KRIEBEL, K.T. : Albedo of vegetated surfaces : Its variability with differing irradiances. - Remote Sens. Environ. *8* : 283 - 290, 1979.

38332 - KRINSKY, N.I. : Carotenoid pigments : multiple mechanisms for coping with the stress of photosensitized oxidations. - In : SHILO, M. (ed.) : Strategies of Microbial Life in Extreme Environments. Pp. 163 - 177. Verlag Chemie, Weinheim - New York 1979.

38333 - KRINSKY, N.I. : Carotenoid protection against oxidation. - Pure appl. Chem. *51* : 649 - 660, 1979.

38334 - KRISHNAN, M., GNANAM, A. : A basic chlorophyll-protein complex. - FEBS Lett. *97* : 322 - 324, 1979.

38335 - KROCHKO, J.E., WINNER, W.E., BEWLEY, J.D. : Respiration in relation to adenosine triphosphate content during desiccation and rehydration of a desiccation-tolerant and a desiccation-intolerant moss. - Plant Physiol. *64* : 13 - 17, 1979. [Ps.]

38336 - KRUMBEIN, W.E., BUCHHOLZ, H., FRANKE, P., GIANI, D., GIELE, C., WONNEBERGER, K. : O_2 and H_2S coexistence in stromatolites. A model for the origin of mineralogical lamination in stromatolites and banded iron formations. - Naturwissenschaften *66* : 381 - 389, 1979. [Ps.]

38337 - KRUPATKINA, D.K., KUZ'MENKO, L.V. : Sravnenie trekh metodov opredeleniya pervichnoĭ produktsii. [Comparison of three methods for determination of primary production.] - Biol. Morya *49* (Ékosistemy Pelagiali Atlanticheskogo Okeana i Moreĭ Sredizemnomorskogo Basseĭna) : 59 - 65, 118, 1979. [In R, ab : E.]

38338 - KRYSTEVA , N.G. : Issledovaniya glitseral'degid-3-fosfatdegidrogenazy khlorelly. [Study of glyceraldehyde-3-phosphate dehydrogenase in *Chlorella*.] - In : VAKLINOVA, S.G., VANKOVA-RADEVA, R., VASILEVA, V.S. (ed.) : Fotosinteticheskaya Assimilyatsiya CO_2 i Fotodykhanie. Pp. 68 - 73. Izdat. bolg. Akad. Nauk, Sofiya 1979. [Ps; in R.]

*38339 - KSENZHEK, O.S., KOGANOV, M.M., LOBACH, G.A. : Élektrokhimicheskie protsessy na membranakh khloroplastov i ikh issledovanie. [Electrochemical processes on chloroplast membranes and their study.] - In : Élektrodnye Protsessy i Metody ikh Izucheniya. Pp. 286 - 289. Naukova Dumka, Kiev 1978. [In R.]

38340 - KU, M.S.B., SCHMITT, M.R., EDWARDS, G.E. : Quantitative determination of RuBP carboxylase-oxygenase protein in leaves of several C_3 and C_4 plants. - J. exp. Bot. *30* : 89 - 98, 1979.

38341 - KU, S.B., SPALDING, M.H., EDWARDS, G.E. : Intracellular localization of
phosphoenolpyruvate carboxykinase in leaves of C_4 and CAM plants. - Plant
Physiol. *63* (Suppl.)   : 63, 1979.

38342 - KUANG Ting-yun, CHANG Chi-de, HAO Nai-pin, LIN Shih-ching, LOU Shi-quing,
LI Tung-zhu, ZSAO Pao-yu : [Structure and function of chloroplast membrane.
(I) Ultrastructure and constituents of chloroplast membrane in relation to
function of Photosystem II.] - Acta physiol. sin. *5* : 99 - 107, 1979. [In
Chin., ab : E.]

38343 - KUBÍN, Š. : Vývojové světelné zdroje národního podniku Tesla Holešovice z
hlediska fotosynteticky účinného záření. [New light sources of Tesla Holeso-
vice from the point of view of PhAR.] - Světelná Technika (Praha) *1979* :
6 - 9, 1979.[In Czech.]

38344 - KUBÍN, Š., DOUCHA, J. : Využití nových typů výbojových světelných zdrojů
při intenzívní kultivaci řas. [Use of new types of discharge lamps in cultivation of algae.] - Světelná Technika (Praha) *1979* : 59 - 62, 1979. [Chl;
in Czech, ab : E, G, R.]

38345 - KUCHEROVA, T.P., LISHCHUK, A.I., STADNIK, S.A. : Vliyanie khlorkholinkhlorida i lateksa na vodnyǐ rezhim, soderzhanie pigmentov i bioělektricheskuyu
reaktsiyu razlichnykh po zasukhoustoǐchivosti sortov abrikosa. [Effect of
chlorocholine chloride and latex on water relations, content of pigments
and bioelectric reaction of apricot cultivars differing in drought resistance.] - Fiziol. Biokhim. kul't. Rast. *11* : 68 - 72, 1979. [In R, ab : E.]

38346 - KUENEMAN, E.A., WALLACE, D.H., LUDFORD, P.M. : Photosynthetic measurements
of field-grown dry beans and their relation to selection for yield. - J.
amer. Soc. hort. Sci. *104* : 480 - 482, 1979.

38347 - KUFER, W ., SCHEER, H. : Chemical modification of biliprotein chromophores. -
Z. Naturforsch. *34 C* : 776 - 781, 1979.

38348 - KUFER, W., SCHEER, H. : Studies on plant bile pigments, VII. Preparation
and characterization of phycobiliproteins with chromophores chemically modified by reduction. - Hoppe-Seylers Z. physiol. Chem. *360* : 935 - 956,
1979.

38349 - KUKUSHKIN, A.K., TIKHONOV, A.N. : Issledovanie protsessov migratsii ěnergii
i ělektronnogo transporta v fotosinteze vysshikh rasteniǐ : Ěffekty usileniya i spektry deǐstviya. [Energy migration and electron transport in photosynthesis of higher plants : Emerson effect and action spectra.] - Biofizika *24* : 87 - 91, 1979. [In R, ab : E.]

38350 - KULAEVA, O.N., SELIVANKINA, S.Yu., ROMANKO, E.G., NIKOLAEVA, M.K., NICHIPO-
ROVICH, A.A. : Aktivatsiya tsitokininom RNK-polimeraznoǐ aktivnosti v izolirovannykh yadrakh i khloroplastakh. [Activation by cytokinin of RNA-polyme-
rase activity in isolated nuclei and chloroplasts.] - Fiziol. Rast. *26* :
1016 - 1028, 1979. [Ps; in R, ab : E.]

38351 - KULIKOV, A.V., MEL'NIKOV, A.V., BOGATYRENKO, V.R., SYRTSOVA, L.A., LIKHTEN-
SHTEǏN, G.I. : Opredelenie rasstoyaniya mezhdu zaryadami posle ikh fotoraz-
deleniya v khromatoforakh iz *R. rubrum*. [Determination of the distance between charges after their photoseparation in chromatophores from *Rhodospirillum rubrum*.] - Biofizika *24* : 337 - 339, 1979. [In R, ab : E.]

38352 - KUMAZAWA, S., SKJOLDAL, H.R., MITSUI, A. : Hydrogen photoproduction by a
marine blue-green alga, under combined nitrogen limited culture. - Plant
Physiol. *63* (Suppl.) : 85, 1979.

*38353 - KUMMEROVÁ, M., TESAŘOVÁ, Z. : Intenzita fotosyntézy a obsah chlorofylu u
dvou odrůd jarního ječmene. [The photosynthetic rate and chlorophyll content observed in two summer barley varieties.] - Acta Univ. agr. (Brno),
Fac. agron. *24* : 379 - 386, 1976. [In Czech, ab : E, G. R.]

38354 - KUNERT, K.-J., BÖGER, P. : Influence of bleaching herbicides on chlorophyll
and carotenoids. - Z. Naturforsch. *34 C* : 1047 - 1051, 1979.

38355 - KUNERT, K.J., BÖGER, P. : Influence of norflurazon, difunon and fluridone
on the photosynthetic apparatus. - Plant Physiol. *63* (Suppl.) : 42, 1979.

✸38356 - KUNG, S.D., RHODES, P.R. : Interaction of chloroplast and nuclear genomes
in regulating RuBP carboxylase activity. - In : SIEGELMAN, H.W., HIND, G.
(ed.) : Photosynthetic Carbon Assimilation. Pp. 307 - 324. Plenum Press,
New York - London 1978.

38357 - KUNO, H. : [Effects of photochemical oxidant on the growth of poplar cuttings I. Changes with days and years on growth and leaf drop of poplar cuttings by filtered air method.] - Taiki Osen Gakkaishi *14* : 265 - 274, 1979.
[Growth analysis; in Jap., ab : E.]

38358 - KURAMOTO, R.T., BREST, D.E. : Physiological response to salinity by four
salt marsh plants. - Bot. Gaz. *140* : 295 - 298, 1979. [Ps.]

38359 - KUREĬSHEVICH, A.V. : Izmeneniya soderzhaniya khlorofilla v fitoplanktone
v period intensivnogo "tsveteniya" vody sinezelenymi vodoroslyami. [Changes
of chlorophyll content in phytoplankton during the period of intensive
water-blooming caused by blue-green algae.] - Gidrobiol. Zh. *15* (1) : 64 -
69, 1979. [In R, ab : E.]

✸38360 - KURETS, V.K. : Vliyanie faktorov vneshneĭ sredy na produktivnost' rasteniĭ
zashchishchennogo grunta. [Influence of external factors on productivity
of plants from protected grounds.] - In : Ėkologo-Fiziologicheskie Mekhanizmy Ustoĭchivosti Rasteniĭ k Deĭstviyu Ėkstremal'nykh Temperatur. Pp.
29 - 37. Petrozavodsk 1978. [In R.]

38361 - KURIHARA, K., SUKIGARA, M., TOYOSHIMA, Y. : Photoinduced charge separation
in liposomes containing chlorophyll *a*. I. Photoreduction of copper(II) by
potassium ascorbate through liposome bilayer containing purified chlorophyll *a*. - Biochim. biophys. Acta *547* : 117 - 126, 1979.

38362 - KURIHARA, K., TOYOSHIMA, Y., SUKIGARA, M. : Photoinduced charge separation
in liposomes containing chlorophyll *a*. II. The effect of ion transport
across membrane on the photoreduction of $Fe(CN)_6^{3-}$.- Biochem. biophys. Res.
Commun. *88* : 320 - 326, 1979.

✸38363 - KURINNYĬ, F.I., GAVVA, I.A., YAVORSKAYA, T.K. : Transport pitatel'nykh veshchestv i produktivnost' sakharnoĭ svekly v zavisimosti ot soderzhaniya se-
ry v pitatel'noĭ srede. [Transport of nutrients and productivity of sugar
beet in dependence on sulphur content in nutrient medium.] - In : Mineral'-
noe Pitanie i Produktivnost' Rasteniĭ. Pp. 151 - 158, 323. Naukova Dumka,
Kiev 1978. [Chl; in R.]

38364 - KUROSAKI, T., IZUMI, K., MOCHIZUKI, T. : [Studies on the histochemical of
carrot - II. Studies on carotenoid and starch in cultured tissues of carrot.] - Bull. Fac. School Educ., Hiroshima Univ., Part II, *2* : 141 - 151,
1979. [In Jap., ab : E.]

✸38365 - KURSAKOV, G.A., PALFITOV, V.F., KURSAKOVA, L.E., ABRAMOVA, I.V. : Soderzhanie pigmentov v list'yakh spontannykh pochkovykh mutatsiĭ vishne-cheresh-
nevykh gibridov. [Pigment contents in leaves of spontaneous bud mutations
of sour cherry-sweet cherry hybrids.] - Byul. nauch. Inform. tsentr. genet.
Labor. Im. I.V. Michurina *23* : 15 - 18, 59, 1976. [In R.]

38366 - KURSANOV, A.L., NICHIPOROVICH, A.A., PARAMONOVA, N.V., SLOBODSKAYA, G.A.,
CHMORA, S.N. : Ul'trastruktura khloroplastov pri fotosinteze list'ev sakharnoĭ svekly v usloviyakh razlichnoĭ kontsentratsii CO_2 i O_2. [Chloro-
plast ultrastructure of sugar beet leaves photosynthesizing under various
CO_2 and O_2 concentrations.] - Fiziol. Rast. *26* : 250 - 258, 1979. [In R,
ab : E.]

38367 - KURSAR, T.A., ALBERTE, R.S. : Light intensity adaptation in *Gracilaria
tikvahiae*. - Plant Physiol. *63* (Suppl.) : 128, 1979.

38368 - KUSHWAHA, S.C., KATES, M. : Effect of nicotine on carotenogenesis in ex-
tremely halophilic bacteria. - Phytochemistry *18* : 2061 - 2062, 1979.

38369 - KUSHWAHA, S.C., KATES, M. : Effect of glycerol on carotenogenesis in the
extreme halophile, *Halobacterium cutirubrum*. - Can. J. Microbiol. *25* :
1288 - 1291, 1979.

38370 - **KUSHWAHA, S.C., KATES, M.** : Studies of the biosynthesis of C_{50} carotenoids in *Halobacterium cutirubrum*. - Can. J. Microbiol. *25* : 1292 - 1297, 1979.

38371 - **KUTAS, E.N.** : Vliyanie intensivnosti osveshcheniya na soderzhanie khlorofilla v list'yakh oranzhereĭnykh rasteniĭ. [Effect of illuminance on chlorophyll content in leaves of greenhouse plants.]- Bot. Zh. *64* : 420 - 426, 1979. [In R.]

38372 - **KUTAS, E.N.** : Vliyanie intensivnosti osveshcheniya na anatomicheskoe stroenie list'ev nekotorykh oranzhereĭnykh rasteniĭ. [The effect of illuminance on the anatomical structure of leaves of some greenhouse plants.]-Bot. Zh. *64* : 1650 - 1657, 1979. [In R.]

38373 - **KUWABARA, T., MURATA, N.** : Purification and characterization of 33 kilodalton protein of spinach chloroplasts. - Biochim. biophys. Acta *581* : 228 - 236, 1979.

✻38374 - **KUZNETSOV, S.I.** : Trends in the development of ecological microbiology. - In : DROOP, M.R., JANNASCH, H.W. (ed.) : Advances in Aquatic Microbiology. Vol. 1. Pp. 1 - 48. Academic Press, London - New York 1977. [Ps.]

38375 - **KUZNETSOV, S.I., ROMANENKO, V.I., KARPOVA, N.S.** : Kharakteristika chislennosti bakteriĭ i mikrobiologicheskie protsessy krugovorota organicheskogo veshchestva v Rybinskom vodokhranilishche v 1975 g. [Characteristic of bacteria number and microbiological processes of organic matter cycle in Rybinsk reservoir in 1975.] - Tr. Inst. Biol. vnutr. Vod. Akad. Nauk SSSR *37* (Mikrobiologicheskie i Khimicheskie Protsessy Destruktsii Organicheskogo Veshchestva v Vodoemakh) : 5 - 20, 1979. [In R.]

38376 - **KYLE, D.J., ZALIK, S.** : A virescens mutant of barley with a reduced quantum yield. - Plant Physiol. *63* (Suppl.) : 41, 1979. [Ps, Chl.]

38377 - **LAASCH, N., KAISER, W., URBACH, W.** : Effects of disalicylidenepropanediamines on photosynthetic electron transport of isolated spinach chloroplasts. - Plant Physiol. *63* : 605 - 608, 1979.

38378 - **LaBAUGH, J.W.** : Chlorophyll prediction models and changes in assimilation numbers in Spruce Knob Lake, West Virginia. - Arch. Hydrobiol. *87* : 178 - 197, 1979.

38379 - **LADYGIN, V.G.** : Fluorestsentsiya i formy khlorofilla fotosistemy 1 i 2 *Chlamydomonas reinhardii*. [Fluorescence and chlorophyll forms in photosystems 1 and 2 of *Chlamydomonas reinhardii*.] - Dokl. Akad. Nauk SSSR *248* : 984 - 987, 1979. [In R.]

38380 - **LADYGIN, V.G.** : Spektral'nye formy khlorofilla mutantov *Chlamydomonas* s neaktivnymi fotosistemami. [Spectral forms of chlorophyll in *Chlamydomonas* mutants with inactive photosystems.] - Biofizika *24* : 254 - 259, 1979. [In R, ab : E.]

38381 - **LADYGIN, V.G., KHRISTIN, M.S.** : Izuchenie belkovykh perenoschikov êlektrontransportnoĭ tsepi khloroplastov u nefotosinteziruyushchikh mutantov *Chlamydomonas reinhardii*. [Protein carriers of the electron transport chain of non-photosynthesizing mutants of *Chlamydomonas reinhardii*.] - Biokhimiya *44* : 1310 - 1316, 1979. [In R, ab : E.]

38382 - **LADYGIN, V.G., SEMENOVA, G.A., TAGEEVA, S.V.** : Spektral'nye formy khlorofilla i struktura khloroplastov mutantov *Chlamydomonas* s narusheniyami v svetosobirayushchikh pigmentakh. [Spectral forms of chlorophyll and chloroplast structure of *Chlamydomonas* mutants with disturbed light-converging pigments.] - Biofizika *24* : 681 - 687, 1979. [In R, ab : E.]

38383 - **LADYGINA, M.E., GRISHKOVA, V.P.** : Sostav lipidov khloroplastov tabaka pri virusnoĭ infektsii. [Composition of lipids of chloroplasts in tobacco plants as affected by the virus infection.] - Sel'skokhoz. Biol. *14* : 50 - 55, 1979. [In R, ab : E.]

38384 - LADYGINA, M.E., GRISHKOVA, V.P., ALESHINA, N.V. : Membrannye belki khloroplastov zdorovykh i infitsirovannykh VTM rasteniĭ tabaka. [Membrane proteins
of chloroplasts of intact and TMV-infected tobacco plants.] - Biokhimiya
44 : 1635 - 1642, 1979. [In R, ab : E.]

38385 - LAETSCH, W.M. : Reflections on C₄ photosynthesis and plant productivity. -
In : SCOTT, T.K. (ed.) : Plant Regulation and World Agriculture. Pp. 479 -
489. Plenum Press, New York - London 1979.

38386 - LAGARIAS, J.C., GLAZER, A.N., RAPOPORT, H. : Chromopeptides from C-phycocyanin. Structure and linkage of a phycocyanobilin bound to the β-subunit. -
J. amer. chem. Soc. 101 : 5030 - 5037, 1979.

38387 - LAI, C.-N. : Chlorophyll : The active factor in wheat sprout extract inhibiting the metabolic activation of carcinogens in vitro. - Nutrition Cancer
1 (3) : 19 - 21, 1979.

38388 - LAĬSK, A., OYA, V. : Komplekt apparatury dlya issledovaniya kinetiki fotosinteza list'ev. Rostovaya kamera, gazometricheskaya apparatura, uzel po-
dachi ¹⁴CO₂. [Equipment for kinetic studies of photosynthesis in leaves :
growth chamber, gasometric apparatus, unit for ¹⁴CO₂ injection.] - Fiziol.
Rast. 26 : 199 - 206, 1979. [In R, ab : E.]

38389 - LAKSO, A.N. : Seasonal changes in stomatal response to leaf water potential
in apple. - J. amer. Soc. hort. Sci. 104 : 58 - 60, 1979. [Ps.]

*38390 - LAL, B.B., CHAKRAVARTI, B.P. : Infection and development of Physoderma
maydis under various light conditions. - Philippine Agr. 61 : 46 - 54,
1977. [Chloroplast.]

*38391 - LAMB, J.E. : Plant carbonic anhydrase. - Life Sci. 20 : 393 - 406, 1977.

38392 - LAMBERS, H., NOORD, R., POSTHUMUS, F. : Respiration of Senecio shoots :
Inhibition during photosynthesis, resistance to cyanide and relation to
growth and maintenance. - Physiol. Plant. 45 : 351 - 356, 1979.

38393 - LAMBERT, G.R., DADAY, A., SMITH, G.D. : Hydrogen evolution from immobilized
cultures of the cyanobacterium Anabaena.cylindrica B629. - FEBS Lett. 101 :
125 - 128, 1979.

38394 - LAMBERT, R., KUNERT, K.-J., BÖGER, P. : On the phytotoxic mode of action
of nitrofen. - Pesticide Biochem. Physiol. 11 : 267 - 274, 1979. [Ps.]

38395 - LAMPPA, G.K., BENDICH, A.J. : Changes in chloroplast DNA levels during development of pea (Pisum sativum). - Plant Physiol. 64 : 126 - 130, 1979.

38396 - LANDEN, W.O. Jr., EITENMILLER, R.R. : Application of gel permeation chromatography and nonaqueous reverse phase chromatography to high pressure
liquid chromatographic determination of retinyl palmitate and β-carotene
in oil and margarin. - J. Assoc. off. anal. Chem. 62 : 283 - 289, 1979.

38397 - LANDGREN, C.R., BONNETT, H.T. : The culture of albino tobacco protoplasts
treated with polyethylene glycol to induce chloroplast incorporation. -
Plant Sci. Lett. 16 : 15 - 22, 1979.

38398 - LANG, W., WOLF, H.U., ZANDER, R. : A sensitive continuous and discontinuous
photometric determination of oxygen, carbon dioxide, and carbon monoxide
in gases and fluids. - Anal. Biochem. 92 : 255 - 264, 1979.

38399 - LANGE, O.L., LÖSCH, R. : Plant water relations. - Progr. Bot. 41 : 10 - 43,
1979. [Ps, Chl.]

38400 - LANGE, O.L., MEDINA, E. : Stomata of the CAM plant Tillandsia recurvata
respond directly to humidity. - Oecologia 40 : 357 - 363, 1979. [Ps.]

38401 - LANGE, O.L., MEYER, A. : Mittäglicher Stomataschluss bei Aprikose (Prunus
armeniaca) und Wein (Vitis vinifera) im Freiland trotz guter Bodenwasser-
-Versorgung. - Flora 168 : 511 - 528, 1979. [Ps.]

38402 - LÄNNERGREN, C. : Buoyancy of natural populations of marine phytoplankton. -
Mar. Biol. 54 : 1 - 10, 1979. [Chl.]

38403 - LANYI, J.K., HELGERSON, S.L., SILVERMAN, M.P. : Relationship between proton motive force and potassium ion transport in *Halobacterium halobium* envelope vesicles. - Arch. Biochem. Biophys. *193* : 329 - 339, 1979.

38404 - LANYI, J.K., MacDONALD, R.E. : Light-induced transport in *Halobacterium halobium*. - In : COLOWICK, S.P., KAPLAN, N.O. (ed.) : Methods in Enzymology. Vol. 56. Pp. 398 - 407. Academic Press, New York - San Francisco - London 1979.

38405 - LANYI, J.K., SILVERMAN, M.P. : Gating effects in *Halobacterium halobium* membrane transport. - J. biol. Chem. *254* : 4750 - 4755, 1979.

*38406 - LANZA, P., BULDINI, P.L. : An improved conductimetric measurement of carbon dioxide. - Anal. chim. Acta *85* : 61 - 68, 1976.

38407 - LAPČEVIČ, R. : Uticaj broja biljaka i vremena ubiranja na prinos i kvalitet zelene mase nekih hibrida kukuruza. [Effect of plant density and time of harvesting on the yield and quality of the green matter of some maize hybrids.] - Arh. poljopr. Nauke *32* : 45 - 56, 1979. [Dry-matter accumulation; in Croat., ab : E.]

38408 - LARKUM, A.W.D., WYN JONES, R.G. : Carbon dioxide fixation by chloroplasts isolated in glycinebetaine. A putative cytoplasmic osmoticum. - Planta *145* : 393 - 394, 1979.

38409 - LARSEN, D.P., VAN SICKLE, J., MALUEG, K.W., SMITH, P.D. : The effect of wastewater phosphorus removal on Shagawa Lake, Minnesota : phosphorus supplies, lake phosphorus and chlorophyll a . - Water Res. *13* : 1259 - 1272, 1979.

38410 - LARSON, D.W. : Discussion : Whole plant gas exchange and acclimation in lichens. - In : UNDERWOOD, L.S., TIESZEN, L.L., CALLAHAN, A.B., FOLK, G.E. (ed.) : Comparative Mechanisms of Cold Adaptation. Pp. 303 - 310. Academic Press, New York - San Francisco - London 1979. [Ps.]

38411 - LARSON, D.W. : Preliminary studies of the physiological ecology of *Umbilicaria* lichens. - Can. J. Bot. *57* : 1398 - 1406, 1979. [Ps.]

38412 - LARSSON, C. : $^{14}CO_2$ fixation and compartmentation of carbon metabolism in a recombined chloroplast-"cytoplasm" system. - Physiol. Plant. *46* : 221 - 226, 1979.

38413 - LARSSON, C., ALBERTSSON, E. : Enzymes related to serine synthesis in spinach chloroplasts. - Physiol. Plant. *45* : 7 - 10, 1979.

*38414 - LARSSON, P., TANGEN, K. : The input and significance of particulate terrestrial organic carbon in a subalpine freshwater ecosystem. - In : WIELGOLASKI, F.E. (ed.) : Fennoscandian Tundra Ecosystems. Part 1. Pp. 351 - 359. Springer-Verlag, Berlin - Heidelberg - New York 1975. [Ps.]

*B38415 - LASCELLES, J. (ed.) : Microbial Photosynthesis. - Dowden, Hutchinson & Ross, Inc., Stroudsburg, Pennsylvania 1973.

38416 - LASLEY, S.E., GARBER, M.P., HODGES, C.F. : Aftereffects of light and chilling temperatures on photosynthesis in excised cucumber cotyledons. - J. amer. Soc. hort. Sci. *104* : 477 - 480, 1979.

*38417 - LASTEIN, E., GARGAS, E. : Relationship between phytoplankton photosynthesis and light, temperature and nutrients in shallow lakes. - Verh. int. Verein. Limnol. *20* : 678 - 689, 1978.

38418 - LASZLO, J.A., GROSS, E.L. : Polypeptide composition of the Photosystem II core complex isolated from spinach chloroplasts. - Plant Physiol. *63* (Suppl.) : 54, 1979.

38419 - LATZKO, E., KELLY, G.J. : Enzymes of the reductive pentose phosphate cycle. - In : GIBBS, M., LATZKO, E. (ed.) : Photosynthesis II. (Encycl. Plant Physiol. N.S. Vol. 6.) Pp. 239 - 250. Springer-Verlag, Berlin - Heidelberg - New York 1979.

38420 - LATZKO, E., KELLY, G.J., WIRTH, E., MEYER, A.O., SCHMIDT, H.-L., WINKLER, F.J., FISCHBECK, G. : Biochemische Kriterien für Calvin-Cyclus und C_4-Metabolismus bei Getreidearten. - Ber. deut. bot. Ges. *92* : 153 - 156, 1979.

38421 - LÄUCHLI, A., PFLÜGER, R. : Potassium transport through plant cell membranes and metabolic role of potassium in plants. - In : Potassium Research - Review and Trends. Pp. 111 - 163. International Potash Research Institute, Bern 1979. [Ps, Chl, photosynthates.]

38422 - LAUENROTH, W.K. : Grassland primary production : North American grasslands in perspective. - In : FRENCH, N.R. (ed.) : Perspectives in Grassland Ecology. Pp. 3 - 24. Springer-Verlag, New York - Heidelberg - Berlin 1979.

38423 - LAUENROTH, W.K., DODD, J.L. : Response of native grassland legumes to water and nitrogen treatments. - J. Range Manage. 32 : 292 - 294, 1979. [Dry-matter accumulation.]

*38424 - LAURINAVICHENE, T.V., GOGOTOV, I.N. : O mekhanizme vydeleniya molekulyarno-go vodoroda Rhodopseudomonas palustris pri ispol'zovanii piruvata i formia-ta. [Mechanism of molecular hydrogen production by Rhodopseudomonas palustris using pyruvate and formiate.] - In : Biologiya i Nauchno-Tekhnicheskiĭ Progress. Pp. 70 - 72. Akad. Nauk SSSR, Pushchino 1974. [In R.]

38425 - LAURITIS, J.A., PRIOLI, L.M. : Ultrastructural cytochemical localization of photosystem II in dimorphic plastids of Zea mays L. - Rev. brasil. Bot. 2 : 91 - 96, 1979.

38426 - LAVAL-MARTIN, D.L., SHUCH, D.J., EDMUNDS, L.N. Jr. : Cell cycle-related and endogenously controlled circadian photosynthetic rhythms in Euglena. - Plant Physiol. 63 : 495 - 502, 1979.

38427 - LaVELLE, J.M. : Translocation in Calliarthron tuberculosum and its role in the light-enhancement of calcification. - Mar. Biol. 55 : 37 - 44, 1979. [Ps, photosynthates.]

38428 - LAVERGNE, D., BISMUTH, E., SARDA, C., CHAMPIGNY, M.L. : Physiological studies on two cultivars of Pennisetum : P. americanum 23 DB, a cultivated species and P. mollissimum, a wild species. II. Effects of leaf age on biochemical characteristics and activities of the enzymes associated with the photosynthetic carbon metabolism. - Z. Pflanzenphysiol. 93 : 159 - 170, 1979.

38429 - LAWLOR, D.W. : Effects of water and heat stress on carbon metabolism of plants with C₃ and C₄ photosynthesis. - In : MUSSELL, H., STAPLES, R. (ed.) : Stress Physiology in Crop Plants. Pp. 303 - 326. John Wiley & Sons, Inc., New York 1979.

38430 - LAWYER, A.L., ZELITCH, I. : Inhibition of glycine decarboxylation and serine formation in tobacco by glycine hydroxamate and its effect on photorespiratory carbon flow. - Plant Physiol. 64 : 706 - 711, 1979.

38431 - LAYZELL, D.B., RAINBIRD, R.M., ATKINS, C.A., PATE, J.S. : Economy of photosynthate use in nitrogen-fixing legume nodules. Observations on two contrasting symbioses. - Plant Physiol. 64 : 888 - 891, 1979.

38432 - LAZAREV, Yu.A., SHNYROV, V.L. : Issledovanie teplovoĭ denaturatsii bakterio-rodopsina iz Halobacterium halobium. [Heat denaturation of bacteriorhodopsin from Halobacterium halobium.]- Bioorg. Khim. 5 : 105 - 112, 1979. [In R, ab : E.]

38433 - LEA, P.J., MIFLIN, B.J. : Photosynthetic ammonia assimilation. - In : GIBBS, M., LATZKO, E. (ed.) : Photosynthesis II. (Encycl. Plant Physiol. N.S. Vol. 6.) Pp. 445 - 456. Springer-Verlag, Berlin - Heidelberg - New York 1979.

38434 - LEA, P.J., MIFLIN, B.J. : The assimilation of ammonium nitrogen by chlorophyllous tissue. - In : HEWITT, E.J., CUTTING, C.V. (ed.) : Nitrogen Assimilation of Plants. Pp. 475 - 487. Academic Press, London - New York - San Francisco 1979.

38435 - LEACH, J.E. : A field enclosure apparatus for measuring crop photosynthesis. - Ann. appl. Biol. 92 : 125 - 132, 1979.

38436 - LEACH, J.E. : Some effects of air temperature and humidity on crop and leaf photosynthesis, transpiration and resistance to gas transfer. - Ann. appl. Biol. 92 : 287 - 297, 1979.

38437 - **LEAN, D.R.S., BURNISON, B.K.** : An evaluation of errors in the ^{14}C method of primary production measurement. - Limnol. Oceanogr. *24* : 917 - 928, 1979.

38438 - **LEAWITT, J.R.C., DOBRENZ, A.K., STONE, J.E.** : Physiological and morphological characteristics of large and small leaflet alfalfa genotypes. - Agron. J. *71* : 529 - 532, 1979. [Ps, photorespiration.]

*38439 - **LEBEDEV, S.I., ALEĬNIKOV, I.M.** : Obnovlenie fotosinteticheskikh pigmentov v list'yakh podsolnechnika pri raznom urovne fosfornogo pitaniya. [Recovery of photosynthetic pigments in sunflower leaves at various levels of phosphorus nutrition.] - In : Mineral'noe Pitanie i Produktivnost' Rasteniĭ. Pp. 167 - 170, 323. Naukova Dumka, Kiev 1978. [In R.]

38440 - **LEBEDEV, S.I., LYTVYNENKO, L.G., BORSHCH, T.M.** : Okyslyuval'no-vidnovyĭ rezhim i nagromadzhennya khlorofilu v lystkakh kukurudzy zalezhno vid umov korenevogo zhyvlennya. [Redox regime and accumulation of chlorophyll in maize leaves depending on conditions of root nutrition.] - Ukr. bot. Zh. *36* : 248 - 250, 287, 1979. [In Ukr., ab : E, R.]

38441 - **LEBEDEV, S.I., NAGORNAYA, R.V., SAVCHENKO, N.P.** : Photosynthesis and optimum level of mineral nutrition of maize. - In : Mineral Nutrition of Plants. Vol. II. Pp. 41 - 44. Publ. House Central Cooperative Union, Sofia 1979.

38442 - **LEBEDEV, V.V.** : Opredelenie potokov CO_2 v lesnykh fitotsenozakh. [Determination of carbon dioxide flows in forest phytocenoses.] - Lesovedenie *1979* (5) : 26 - 32, 1979. [In R.]

38443 - **LEBEDEVA, L.A., SMETANINA, N.A., RUMYANTSEVA, V.I.** : Izmeneniya v pigmentnom komplekse list'ev ogurtsov pod vliyaniem predposevnogo vozdeĭstviya na semena khimicheskimi i termicheskimi reagentami. [Changes in pigment complex of cucumber leaves induced by pre-sowing treatment with chemical and thermical agents.] - Uch. Zapiski gos. pedag. Inst. (Kazan') *195* : 3 - 17, 1979. [In R.]

38444 - **LECHOWICZ, M.J., ADAMS, M.S.** : Net CO_2 exchange in *Cladonia* lichen species endemic to southeastern North America. - Photosynthetica *13* : 155 - 162, 1979.

38445 - **LECHOWSKI, Z.** : Mechanizmy translokacji chloroplastów. [Mechanisms of chloroplast translocation.] - Zesz. nauk. Uniw. jagielloń. *549* (Prace Biol. mol. 6) : 111 - 120, 1979. [In Pol., ab : E.]

*38446 - **LECLERC, J.-C., HOARAU, J.** : Analyse spectroscopique de biliprotéines et d'holochromes chlorophylliens : application à quelques algues rouges marines. - Bull. Union océanogr. France *6* (4) : 50 - 52, 1974.

38447 - **LECLERC, J.C., HOARAU, J., REMY, R.** : Analysis of absorption spectra changes induced by temperature lowering on phycobilisomes, thylakoids and chlorophyll-protein complexes. - Biochim. biophys. Acta *547* : 398 - 409, 1979.

*38448 - **LEDIG, F.T., CLARK, J.G.** : Photosynthesis in a half-sib family experiment in pitch pine. - Can. J. Forest Res. *7* : 510 - 514, 1977.

38449 - **LEDOIGT, G., LOUVEL, C.** : Etude de la régulation du développement plastidial chez *Euglena gracilis*. II. Localisation fonctionnelle et synthèse des particules ribosomiques chloroplastiques.- Biochim. biophys. Acta *563* : 432 - 444, 1979.

38450 - **LEE, H.J., McKEE, G.W., KNIEVEL, D.P.** : Determination of physiological maturity in oat. - Agron. J. *71* : 931 - 935, 1979. [Photosynthates.]

38451 - **LEE, K., NALEWAJKO, C., JACK, T.R.** : Effects of vanadium on freshwater algae. - Fish. mar. Serv. tech. Rep. *862* : 297 - 310, 1979. [Ps.]

38452 - **LEEGOOD, R.C., WALKER, D.A.** : Isolation of protoplasts and chloroplasts from flag leaves of *Triticum aestivum* L. - Plant Physiol. *63* : 1212 - 1214, 1979.

38453 - **LEE-KADEN, J., SIMONIS, W.** : Photosystem I and photophosphorylation-dependent leucine incorporation in the blue-green alga *Anacystis nidulans*. - Plant Cell Physiol. *20* : 1179 - 1190, 1979.

38454 - LEGENDRE, L., SIMARD, Y. : Océanographie biologique estivale et phytoplancton dans le sud-est de la baie d'Hudson. - Mar. Biol. 52 : 11 - 22, 1979.
[Chl.]

38455 - LEGG, B.J., DAY, W., LAWLOR, D.W., PARKINSON, K.J. : The effects of drought
on barley growth : models and measurements showing the relative importance
of leaf area and photosynthetic rate. - J. agr. Sci. 92 : 703 - 716, 1979.

38456 - LEHMAN, J.L., VASCONCELOS, A.C. : Physiology of copper and mercury stress
in the marine diatom Cylindrotheca closterium. - J. Phycol. 15 (Suppl.) :
18, 1979. [Ps.]

38457 - LEHMAN, J.L., VASCONCELOS, A.C. : Physiology of zinc and cadmium stress in
the marine diatom Cylindrotheca closterium. - J. Phycol. 15 (Suppl.) : 19,
1979.

38458 - LEHMANN-KIRK, U., BADER, K.P., SCHMID, G.H., RADUNZ, A. : Inhibition of
photosynthetic electron transport in tobacco chloroplasts and thylakoids
of the blue green alga Oscillatoria chalybea by an antiserum to synthetic
zeaxanthin. - Z. Naturforsch. 34 C : 1218 - 1221, 1979.

38459 - LEHMANN-KIRK, U., SCHMID, G.H., RADUNZ, A. : The effect of antibodies to
violaxanthin on photosynthetic electron transport. - Z. Naturforsch. 34 C :
427 - 430, 1979.

38460 - LEHOCZKI, E., HERCZEG, T., SZALAY, L. : Dichlorophenylurea-resistant oxygen
evolution in Chlorella after cerulenin treatment. - Biochim. biophys. Acta
545 : 376 - 380, 1979.

*38461 - LEKVEISHVILI, N.I. : Primenenie nekotorykh fiziologicheskikh pokazateleĭ v
kachestve bioindikatora bezvrednosti pri ispytanii fosfororganicheskikh
preparatov na tsitrusovykh. [Use of physiological indexes as bioindicators
of non-toxicity in tests of organophosphorus preparations on Aurantiaceae.]
- Tr. Inst. Zashch. Rast. gruz. SSR (Tbilisi) 23 : 150 - 156, 1971. [Ps,
Chl.]

38462 - LELETKIN, V.A., ZVALINSKIĬ, V.I. : Izmerenie intensivnosti kislorodnogo obmena morskikh rasteniĭ i zhivotnykh. [Measurement of oxygen exchange rate
in marine plants and animals.] - Biol. Morya 1979 (6) : 80 - 84, 1979.
[In R, ab : E.]

38463 - LEMAIRE, J.M., CARPENTIER, F., DALLE, J.F., DOUSSINAULT, G. : Lutte biologique contre le Piétin-échaudage des céréales. Modifications physiologiques
chez le Blé inoculé par une souche atténuée d'Ophiobolus graminis. 2. Chan-
gement de la teneur en chlorophylle. - Ann. Phytopathol. 11 : 193 - 197,
1979.

38464 - LENZ, F. : Fruit effects on photosynthesis, light- and dark-respiration. -
In : MARCELLE, R., CLIJSTERS, H., VAN POUCKE, M. (ed.) : Photosynthesis and
Plant Development. Pp. 271 - 281. Dr. W. Junk bv. Publ., The Hague - Boston
- London 1979.

38465 - LENZ, F. : Sink-source relationships in fruit trees. - In : SCOTT, T.K.
(ed.) : Plant Regulation and World Agriculture. Pp. 141 - 153. Plenum Press,
New York - London 1979. [Ps.]

38466 - LEONARD, J.M., ROSE, R.J. : Sensitivity of the chloroplast division cycle
to chloramphenicol and cycloheximide in cultured spinach leaves. - Plant
Sci. Lett. 14 : 159 - 167, 1979.

38467 - LEONG, T.-Y., SCHWEIGER, H.-G. : The role of chloroplast-membrane-protein
synthesis in the circadian clock. Purification and partial characterization
of a polypeptide which is suggested to be involved in the clock. - Europe.
J. Biochem. 98 : 187 - 194, 1979.

38468 - LERER, M., BAR-AKIVA, A. : Effect of manganese deficiency on chloroplasts
of lemon leaves. - Physiol. Plant. 47 : 163 - 166, 1979.

38469 - LETO, K., BECKETT, J., ARNTZEN, C. : Effect of nuclear gene dosage on photosynthetic light reactions. - Plant Physiol. 63 (Suppl.) : 160, 1979.

*38470 - LEUPOLD, D., MORY, S., HOFFMANN, O. : Nonlinear absorption of chlorophyll-a.
- Acta phys. chem. *23* : 33 - 35, 1977.

*38471 - LEUPOLD, D., VOIGT, B., MORY, S., KÖNIG, R., HOFFMANN, P. : Stepwise multi-
photon absorption of *in vivo* chlorophyll-a. - In : WEST, M. (ed.) : Lasers
in Chemistry. Pp. 299 - 304. Elsevier, Amsterdam 1977.

38472 - LEVERENZ, J.W., JARVIS, P.G. : Photosynthesis in Sitka spruce. VIII. The
effects of light flux density and direction on the rate of net photosynthesis and the stomatal conductance of needles. - J. appl. Ecol. *16* : 919 -
932, 1979.

38473 - LEWIS, M.A., GERKING, S.D. : Primary productivity in a polluted intermittent
desert stream. - Amer. Midl. Naturalist *102* : 172 - 174, 1979.

38474 - LI Jing-yan : [The endosymbiotic theory on the origin of chloroplasts and
mitochondria.] - Tzu Jan Tsa Chih [J. Nat.] *1* : 192 - 199, 202, 253 - 259,
1978. [In Chin.]

B38475 - LI Jing-yan : [The Origin of Eucaryotic Cells in Evolution.] - Science Press,
Peking 1979. [Chloroplast phylogeny; in Chin.]

38476 - LIAAEN-JENSEN, S. : Carotenoids - a chemosystematic approach. - Pure appl.
Chem. *51* : 661 - 675, 1979.

*B38477 - LICHTENTHALER, H., PFISTER, K. : Praktikum der Photosynthese. - Quelle &
Meyer, Heidelberg 1978.

38478 - LICHTENTHALER, H.K. : Effect of biocides on the development of the photosynthetic apparatus of radish seedlings grown under strong and weak light
conditions. - Z. Naturforsch. *34 C* : 936 - 940, 1979.

38479 - LICHTENTHALER, H.K. : Occurence and function of prenyllipids in the photo-
synthetic membrane. - In : APPELQVIST,L.Å.,LILJENBERG,C.(ed.):Advances in the
Biochemistry and Physiology of Plant Lipids. Pp. 57 - 78. Elsevier/North-
-Holland Biomedical Press, Amsterdam - New York - Oxford 1979.

38480 - LICHTENTHALER, H.K., SUNDQVIST, C. : Association of plastoquinone-9 and
phylloquinone K_1 with photoconvertible protochlorophyllide holochrome. -
Physiol. Plant. *45* : 381 - 386, 1979.

38481 - LIEN, S., SAN PIETRO, A. : Interaction of plastocyanin and P700 in PSI
reaction center particles from *C. reinhardi* and spinach. - Arch. Biochem.
Biophys. *194* : 128 - 137, 1979.

38482 - LIEN, S., SAN PIETRO, A. : Effect of oxygen and pentachlorophenol on the
induction lag of photosynthesis in *Chlamydomonas reinhardi*. - Arch. Biochem.
Biophys. *197* : 178 - 184, 1979.

38483 - LIEN, S., SAN PIETRO, A. : On the reactivity of oxygen with photosystem I
electron acceptors. - FEBS Lett. *99* : 189 - 193, 1979.

38484 - LIEN, T., KNUTSEN, G. : Synchronous growth of *Chlamydomonas reinhardtii*
(*Chlorophyceae*) : a review of optimal conditions. - J. Phycol. *15* : 191 -
200, 1979. [Chl, Car.]

*38485 - LIETH, H., BOX, E. : The gross primary productivity pattern of the land
vegetation : A first attempt.- Trop. Ecol. *18* : 109 - 115, 1977.

*B38486 - LIETH, H.F.H. (ed.) : Patterns of Primary Production in the Biosphere.
(Benchmark Papers in Ecology. Vol.8.) - Dowden,Hutchinson & Ross, Stroudsburg 1978.

38487 - LILLEY, R.McC., WALKER, D.A. : Studies with the reconstituted chloroplast
system. - In : GIBBS, M., LATZKO, E. (ed.) : Photosynthesis II. (Encycl.
Plant Physiol. N.S. Vol. 6.) Pp. 41 - 53. Springer-Verlag, Berlin - Heidelberg - New York 1979.

38488 - LIMAR', R.S., ANPILOGOVA, N.N. : Dinamika transporta ugleroda ^{14}C v rasteniyakh yarovoY pshenitsy. [Dynamics of carbon ^{14}C transport in spring wheat
plants.] - Byull. vsesoyuz. nauch.-issled. Inst. Rastenievodstva Im. N.I.
Vavilova *87* : 55 - 58, 1979. [In R.]

38489 - LIMPINUNTANA, V., GREENWAY, H. : Sugar accumulation in barley and rice
grown in solutions with low concentrations of oxygen. - Ann. Bot. *43* : 373 -
381, 1979.

38490 - LIN De-hui, LI Cun-xin, ZHANG He, LI Heng : [Influence of industrial waste
water on photosynthetic pigments in *Hydrilla verticillata* L.] - Acta bot.
yunnanica *1* (2) : 62 - 65, 1979.[In Chin., ab : E.]

38491 - LIN, M., SELTZER, S. : The consequences of a deuterium exchange test on pro-
posed mechanisms for the purple membrane proton pump. - FEBS Lett. *106* :
135 - 138, 1979.

38492 - LINDEMAN, W. : Inhibition of photosynthesis in *Lemna minor* by illumination
during chilling in the presence of oxygen. - Photosynthetica *13* : 175 - 185,
1979.

38493 - LINDER, S. : Photosynthesis and respiration in conifers. A classified reference list 1891 - 1977. - Stud. forest. suec. *149* : 1 - 71, 1979.

38494 - LINDLEY, E.V., MacDONALD, R.E. : A second mechanism for sodium extrusion
in *Halobacterium halobium* : A light-driven sodium pump. - Biochem. biophys.
Res. Commun. *88* : 491 - 499, 1979.

38495 - LINNETT, P.E., MITCHELL, A.D., PARTIS, M.D., BEECHEY, R.B. : Preparation of
the soluble ATPase from mitochondria, chloroplasts, and bacteria by the
chloroform technique. - In : COLOWICK, S.P., KAPLAN, N.O. (ed.) : Methods
in Enzymology. Vol.55. Pp. 337 - 343. Academic Press, New York - San Francisco - London 1979.

38496 - LIPS, S.H. : Photosynthesis and photorespiration in nitrate metabolism. -
In : HEWITT, E.J., CUTTING, C.V. (ed.) : Nitrogen Assimilation of Plants.
Pp. 445 - 450. Academic Press, London - New York - San Francisco 1979.

38497 - LIPSKAYA, G.A. : Fotokhimicheskaya aktivnost' khloroplastov i nakoplenie
pigmentov v protsesse zeleneniya ètiolirovannykh prorostkov yachmenya pri
neodinakovom soderzhanii kobal'ta v semenakh. [Photochemical activity of
chloroplasts and the accumulation of pigments during the greening of etiolated barley seedlings grown from seeds with different cobalt levels.] -
Vest. belorus. gos. Univ. Ser.2,*1979* (1) : 38 - 41, 1979. [In R.]

38498 - LIS, E.K., ANTOSZEWSKI, R. : Modification of the strawberry receptacle accu-
mulation ability by growth regulators. - In : MARCELLE, R., CLIJSTERS, H.,
VAN POUCKE, M. (ed.) : Photosynthesis and Plant Development. Pp. 263 - 270.
Dr.W.Junk bv. Publ., The Hague - Boston - London 1979. [Photosynthates.]

38499 - LITTLER, M.M., MURRAY, S.N., ARNOLD, K.E. : Seasonal variations in net photosynthetic performance and cover of intertidal macrophytes. - Aquat. Bot.
7 : 35 - 46, 1979.

38500 - LITTLETON, E.J., DENNETT, M.D., ELSTON, J., MONTEITH, J.L. : The growth and
development of cowpeas (*Vigna unguiculata*) under tropical field conditions
1. Leaf area. - J. agr. Sci. *93* : 291 - 307, 1979. [Growth analysis.]

38501 - LITTLETON, E.J., DENNETT, M.D., MONTEITH, J.L., ELSTON, J. : The growth and
development of cowpeas (*Vigna unguiculata*) under tropical field conditions
2. Accumulation and partition of dry weight. - J. agr. Sci. *93* : 309 - 320,
1979. [Dry-matter accumulation.]

38502 - LITVIN, F.F., STADNICHUK, I.N. : O soblyudenii sootnosheniya Stepanova dlya
parametrov spektrov pogloshcheniya i fluorestsentsii khlorofilla *a* v rast-
vore i kletke. [Applicability of Stepanov's relation to absorption and fluorescence spectra of chlorophyll *a in vivo* and *in vitro*.] - Biofizika *24* :
651 - 656, 1979. [In R, ab : E.]

38503 - LĪVANSKÝ, K. : Effect of the nonilluminated part of suspension on biomass
production in an algal reactor. - Folia microbiol. *24* : 339 - 345, 1979.

38504 - LJUBEŠIČ, N. : Chromoplasts of *Forsythia suspensa* (THUNB.) VAHL. I. Ultrastructure and pigment composition. - Acta bot. croat. *38* : 23 - 28, 1979.
[Chloroplast.]

38505 - LJUBEŠIĆ, N., RADIĆ, M. : Chromoplasts of *Forsythia suspensa* (THUNB.)VAHL. II. The effect of isopropyl N-phenylcarbamate. - Acta bot. croat. *38* : 29 - 34, 1979. [Chloroplast.]

38506 - LLAMA, M.J., SERRA, J.L., RAO, K.K., HALL, D.O. : Separation and purification of two hydrogenase activities in *Chromatium* : kinetics and subunit analysis. - Biochem. Soc. Trans. *7* : 223 - 224, 1979.

38507 - LLAMA, M.J., SERRA, J.L., RAO, K.K., HALL, D.O. : Isolation and characterization of the hydrogenase activity from the non-heterocystous cyanobacterium *Spirulina maxima*. - FEBS Lett. *98* : 342 - 346, 1979.

38508 - LLOYD, N.D.H., WOOLHOUSE, H.W. : Comparative aspects of photosynthesis, photorespiration and transpiration in four species of the *Cyperaceae* from the relict flora of Teesdale, Northern England. - New Phytol. *83* : 1 - 7, 1979.

*38509 - LOBANOV, G.A. : Metodika otsenki fotosinteticheskoĭ aktivnosti list'ev u plodovykh rasteniĭ. [Methods of determining photosynthetic activity of leaves of fruit plants.] - In : Programma i Metodika Sortoizucheniya Plodovykh, Yagodnykh i Orekhoplodnykh Kul'tur. Pp. 332 - 339. Vses. nauch.-issl. Inst. Sadov., Michurinsk 1973. [In R.]

38510 - LOCKAU, W. : The inhibition of photosynthetic electron transport in spinach chloroplasts by low osmolarity. - Europe. J. Biochem. *94* : 365 - 373, 1979.

*38511 - LODH, S.B., SELVARAJ, Y., DIVAKAR, N.G. : Changes in pigments & oxidative & cellulytic enzymes in tomato (*Lycopersicon esculentum* MILL) infected with tomato leaf curl virus. - Indian J. exp. Biol. *11* : 210 - 212, 1973.

38512 - LODHI, G.P., SINGH, R.K., SHARMA, S.C. : Production and distribution of dry matter in plant components and its effect on seed yield in brown-seeded Indian colza. - Indian J. agr. Sci. *49* : 463 - 469, 1979.

38513 - LÖFFELHARDT, W., KINDL, H. : Conversion of 4-hydroxyphenylpyruvic acid into homogentisic acid at the thylakoid membrane of *Lemna gibba*. - FEBS Lett. *104* : 332 - 334, 1979.

38514 - LOFTUS, M.E., PLACE, A.R., SELIGER, H.H. : Inorganic carbon requirements of natural populations and laboratory cultures of some Chesapeake Bay phytoplankton. - Estuaries *2* : 236 - 248, 1979. [Ps.]

38515 - LONERGAN, T.A., SARGENT, M.L. : Regulation of the photosynthesis rhythm in *Euglena gracilis* II. Involvement of electron flow through both photosystems. - Plant Physiol. *64* : 99 - 103, 1979.

38516 - LONG, S.P., INCOLL, L.D. : The prediction and measurement of photosynthetic rate of *Spartina townsendii* (*sensu-lato*) in the field. - J. appl. Ecol. *16* : 879 - 891, 1979.

38517 - LONGO, G.P., PEDRETTI, M., ROSSI, G., LONGO, C.P. : Effect of benzyladenine on the development of plastids and microbodies in excised watermelon cotyledons. - Planta *145* : 209 - 217, 1979.

38518 - LONGSTRETH, D.J., NOBEL, P.S. : Nutrient effects on photosynthesis of cotton. - Plant Physiol. *63* (Suppl.) : 39, 1979.

38519 - LONGSTRETH, D.J., NOBEL, P.S. : Salinity effects on leaf anatomy. Consequences for photosynthesis. - Plant Physiol. *63* : 700 - 703, 1979.

38520 - LOOMIS, R.S., RABBINGE, R., NG, E. : Explanatory models in crop physiology. - Annu. Rev. Plant Physiol. *30* : 339 - 367, 1979. [Ps.]

38521 - LOPEZ, E. : Algal chloroplasts in the protoplasm of three species of benthic foraminifera : taxonomic affinity, viability and persistence. - Mar. Biol. *53* : 201 - 211, 1979.

38522 - LÖPPERT, H., KRONBERGER, W. : Control of nitrate uptake by photosynthesis in *Lemna paucicostata* 6746. - In : MARCELLE, R., CLIJSTERS, H., VAN POUCKE, M. (ed.) : Photosynthesis and Plant Development. Pp. 301 - 308. Dr. W. Junk bv. Publ., The Hague - Boston - London 1979.

38523 - LORENC-PLUCIŃSKA, G. : The effect of ozone on photosynthesis and respiration in Scots pines differing in resistance to this gas. - Arboretum kórnickie *24* : 329 - 338, 1979.

38524 - LORENZEN, C.J. : Ultraviolet radiation and phytoplankton photosynthesis. - Limnol. Oceanogr. *24* : 1117 - 1120, 1979.

38525 - LORIMER, G.H. : Evidence for the existence of discrete activator and substrate sites for CO_2 on ribulose-1,5-bisphosphate carboxylase. - J. biol. Chem. *254* : 5599 - 5601, 1979.

38526 - LOSADA, M. : Photoproduction of ammonia and hydrogen peroxide. - Bioelectrochem. Bioenergetics *6* : 205 - 225, 1979. J. electroanal. Chem. *104* : 205 - 225, 1979. [Ps.]

38527 - LOSADA, M., QUERRERO, M.G. : The photosynthetic reduction of nitrate and its regulation. - In : BARBER, J. (ed.) : Photosynthesis in Relation to Model Systems. Pp. 365 - 408. Elsevier, Amsterdam - New York - Oxford 1979.

B38528 - LOSADA VILLASANTE, M. : Reflexiones en Torno a la Transduccion Biologica de la Energia. [Biological Transduction of Energy.] - Real Acad. Med. Sevilla, Sevilla 1979. [Ps; in Span.]

38529 - LÖSCH, R. : Stomatal responses to changes in air humidity. - In : SEN, D.N., CHAWAN, D.D., BANSAL, R.P. (ed.) : Structure, Function and Ecology of Stomata. Pp. 189 - 216. Bishen Singh Mahendra Pal Singh, Dehra Dun 1979. [Ps.]

38530 - LOSEV, A.P., LYAL'KOVA, N.D. : Issledovanie nachal'nykh stadiĭ fotogidrirovaniya protokhlorofillida v ètiolirovannykh rasteniyakh. [Initial steps of protochlorophyllide photoreduction in etiolated plants.]- Mol. Biol. *13* : 837 - 844, 1979. [In R, ab : E.]

38531 - LOTINA, B., ARIAS, C., PARRA, M.C., ALBORES, M., DILLEY, R.A. : Uncoupling of photophosphorylation by amines bound to Sepharose beads. - Plant Physiol. *63* (Suppl.) : 29, 1979.

38532 - LOVE, J.M., BARDEN, J.A. : The effects of ethyl 5-(4-chlorophenyl)-2H-tetrazole-2-acetate on net photosynthesis, stomatal resistance, and transpiration in 'Delicious' and 'Golden Delicious' apple leaves. - HortScience *14* : 515 - 516, 1979.

38533 - LÜDERS, W. : Erste Erfahrungen mit dem Blattflächenmeßgerät Li-Cor. - Nachrichtenbl. deut. Pflanzenschutzdienst. *31* : 28 - 30, 1979.

38534 - LUDLOW, M.M., IBARAKI, K. : Stomatal control of water loss in siratro (*Macroptilium atropurpureum* (DC) URB.), a tropical pasture legume. - Ann. Bot. *43* : 639 - 647, 1979. [Resistances.]

38535 - LUGG, D.G., SINCLAIR, T.R. : A survey of soybean cultivars for variability in specific leaf weight. - Crop Sci. *19* : 887 - 892, 1979.

38536 - LUISETTI, J., MÖHWALD, H., GALLA, H.-J. : Monitoring the location profile of fluorophores in phosphatidyl-choline bilayers by the use of paramagnetic quenching. - Biochim. biophys. Acta *552* : 519 - 530, 1979. [Chl.]

38537 - LUISETTI, J., MÖHWALD, H., GALLA, H.J. : Spectroscopic and thermodynamic studies of chlorophyll containing monolayers and vesicles. Part II : Chlorophyll *a* and pheophytin *a* aggregation on DMPC vesicles. - Z. Naturforsch. *34 C* : 406 - 413, 1979.

38538 - LUKASHEV, E.P., KONONENKO, A.A., TIMOFEEV, K.N., USPENSKAYA, N.Ya., RUBIN, A.B. : Issledovanie aktseptorov èlektrona v fotosinteticheskikh reaktsionnykh tsentrakh iz *Rhodopseudomonas sphaeroides*. [Electron acceptors in photosynthetic reaction centres from *Rhodopseudomonas sphaeroides*.] - Biokhimiya *44* : 1223 - 1233, 1979. [In R, ab : E.]

38539 - LUPTON, F.G.H., OLIVER, R.H., MURTY, K.S., RUWALI, K.N. : The importance of sink/source balance in determining yielding capacity in wheat. - In : RAMANUJAM, S. (ed.) : Proceedings of the Fifth International Wheat Genetics Symposium. Vol. 2. Pp. 891 - 898. Indian Soc. Genetics and Plant Breeding, New Delhi 1979.

38540 - LURIE, S. : Photosynthetic capacity of stomatal chloroplasts and their contribution to stomatal opening. - In : SEN, D.N., CHAWAN, D.D., BANSAL, R.P.
(ed.) : Structure, Function and Ecology of Stomata. Pp. 61 - 76. Bishen
Singh Mahendra Pal Singh, Dehra Dun 1979.

38541 - LURIE, S., HENDRIX, D.L. : Differential ion stimulation of plasmalemma adenosine triphosphatase from leaf epidermis and mesophyll of *Nicotiana rusti-
ca* L. - Plant Physiol. *63* : 936 - 939, 1979,

38542 - LURIE, S., PAZ, N., STRUCH, N., BRAVDO, B.A. : Effect of leaf age on photo-
synthesis and photorespiration. - In : MARCELLE, R., CLIJSTERS, H., VAN
POUCKE, M. (ed.) : Photosynthesis and Plant Development. Pp. 31 - 38.
Dr. W. Junk bv. Publ., The Hague - Boston - London 1979.

38543 - LUSH, W.M., RAWSON, H.M. : Effects of domestication and region of origin
on leaf gas exchange in cowpea (*Vigna unguiculata* (L.) WALP.). - Photosynthetica *13* : 419 - 427, 1979.

B38544 - LÜTTGE, U., HIGINBOTHAM, N. : Transport in Plants. - Springer-Verlag, New
York - Heidelberg - Berlin 1979. [Ps.]

38545 - LÜTTKE, A., NUYTS, G., GILLES, J., BAUGNET-MAHIEU, L., BONOTTO, S. : Effect
of X-rays on the chloroplasts of *Acetabularia mediterranea*. - Arch. int.
Physiol. Biochim. *87* : 826 - 827, 1979.

38546 - LÜTTKE, A., RAHMSDORF, U. : Heterogeneity of chloroplasts from *Acetabularia
mediterranea* : Separation by equilibrium centrifugation. - In : BONOTTO, S.,
KEFELI, V., PUISEUX-DAO, S. (ed.) : Developmental Biology of *Acetabularia*.
Pp. 168 - 177. Elsevier/North-Holland Biomedical Press, Amsterdam - New York
- Oxford 1979.

38547 - LÜTZ, C., KLEIN, S. : Biochemical and cytological observations on chloroplast development. VI. Chlorophylls and saponins in prolamellar bodies and
prothylakoids separated from etioplasts of etiolated *Avena sativa* L. leaves. - Z. Pflanzenphysiol. *95* : 227 - 237, 1979.

38548 - LUTZ, M., BROWN, J.S., RÉMY, R. : Resonance Raman spectroscopy of chlorophyll-protein complexes. - In : Chlorophyll Organization and Energy Transfer
in Photosynthesis. Pp. 105 - 125. Excerpta Medica, Amsterdam - Oxford -
New York 1979.

38549 - LUTZ, M., KLEO, J. : Bacteriochlorophyll *a* cation radical in solution and
in reaction centers of *Rhodopseudomonas sphaeroides*. Resonance Raman scattering. - Biochim. biophys. Acta *546* : 365 - 369, 1979.

38550 - LYLES, L., ALLISON, B.E. : Wind profile parameters and turbulence intensity
over several roughness element geometries. - Trans. ASAE *22* : 334 - 338,
343, 1979. [Equations.]

38551 - LYTLE, R. Jr., HULL, R. : Photoassimilate partitioning in natural stands
of *Spartina alterniflora* LOISEL. - Plant Physiol. *63* (Suppl.) : 34, 1979.

*38552 - LYUBIMOV, V.Yu., KARPILOV, Yu.S. : Vliyanie organicheskikh kislot na sve-
toindutsirovannoe pogloshchenie kisloroda khloroplastami kukuruzy. [The
effect of organic acids on light-induced absorption of oxygen by maize chloroplasts.] - Dokl. Akad. Nauk SSSR *237* : 746 - 748, 1977. [In R.]

38553 - MABESA, L.B., BALDWIN, R.E., GARNER, G.B. : Non-volatile organic acid profiles of peas and carrots cooked by microwaves. - J. Food Protection *42* : 385 -
- 388, 1979. [Chl.]

38554 - MacCOLL, R., BERNS, D.S. : Evolution of the biliproteins. - Trends biochem.
Sci. *4* (2) : 44 - 47, 1979,

38555 - MACHOLD, O., MEISTER, A. : Resolution of the light-harvesting chlorophyll
a/b-protein of *Vicia faba* chloroplasts into two different chlorophyll-protein
complexes. - Biochim. biophys. Acta *546* : 472 - 480, 1979.

38556 - MACHOLD, O., MEISTER, A. : Stability of chlorophyll-protein complexes *in vitro*. - Biochem. Physiol. Pflanzen *174* : 92 - 98, 1979.

38557 - MACHOLD, O., SIMPSON, D.J., MØLLER, B.L. : Chlorophyll-proteins of thylakoids from wild-type and mutants of barley (*Hordeum vulgare* L.). - Carlsberg Res. Commun. *44* : 235 - 254, 1979.

38558 - MACIEJEWSKA, U. : The effect of embryonal axis on the development of photosynthetic activity in apple seedlings. - New Phytol. *82* : 81 - 88, 1979.

38559 - MacISAAC, J.J., DUGDALE, R.C., HUNTSMAN, S.A., CONWAY, H.L. : The effect of sewage on uptake of inorganic nitrogen and carbon by natural populations of marine phytoplankton. - J. mar. Res. *37* : 51 - 66, 1979.

38560 - MacKAY, B., MARSHO, T. : Regulation of oxaloacetate reduction by salts in intact spinach chloroplasts. - Plant Physiol. *63* (Suppl.) : 55, 1979. [Ps.]

38561 - MACKENDER, R.O. : The galactolipid and pigment composition of the thylakoid membranes from naturally differentiating chloroplasts of *Avena sativa* L. - In : APPELQVIST, L.-Å., LILJENBERG, C. (ed.) : Advances in the Biochemistry and Physiology of Plant Lipids. Pp. 205 - 210. Elsevier/North-Holland Biomedical Press, Amsterdam - New York - Shannon 1979.

38562 - MACKENDER, R.O. : Galactolipid and chlorophyll synthesis and changes in fatty acid composition during the greening of etiolated maize leaf segments of different ages. - Plant Sci. Lett. *16* : 101 - 109, 1979.

38563 - MACKINNON, J.C. : Energy allocation during growth by six maize hybrids in Nova Scotia. - Can. J. Plant Sci. *59* : 667 - 677, 1979.

*38654 - MACKINNON, J.C., GARTLEY, C.H., WILKIE, K.I. : Energy efficiency of forage maize production in an atlantic canadian environment. - J. appl. Ecol. *15* : 503 - 514, 1978.

38565 - MACLER, B.A., PELROY, R.A., BASSHAM, J.A. : Hydrogen formation in nearly stoichiometric amounts from glucose by a *Rhodopseudomonas sphaeroides* mutant. - J. Bacteriol. *138* : 446 - 452, 1979.

38566 - MÄDER, M., SCHLOSS, P. : Isolation of malate dehydrogenase from cell walls of *Nicotiana tabacum*. - Plant Sci. Lett. *17* : 75 - 80, 1979.

*38567 - MADGWICK, H.A.I., JACKSON, D.S., KNIGHT, P.J. : Above-ground dry matter, energy, and nutrient contents of trees in an age series of *Pinus radiata* plantations. - N. Zeal. J. Forest Sci. *7* : 445 - 468, 1977.

*38568 - MADGWICK, H.A.I., OLAH, F.D., BURKHART, H.E. : Biomass of open-grown Virginia pine. - Forest Sci. *23* : 89 - 91, 1977. [Dry-matter distribution.]

38569 - MADIGAN, M.T., GEST, H. : Growth of the photosynthetic bacterium *Rhodopseudomonas capsulata* chemoautotrophically in darkness with H_2 as the energy source. - J. Bacteriol. *137* : 524 - 530, 1979.

38570 - MADIGAN, M.T., WALL, J.D., GEST, H. : Dark anaerobic dinitrogen fixation by a photosynthetic microorganism. - Science *204* : 1429 - 1430, 1979.

38571 - MAGGIORA, G.M. : Assessment of reaction center special-pair chlorophyll models. - Int. J. Quantum Chem. *16* : 331 - 352, 1979.

*38572 - MAGGS, D.H., ALEXANDER, D.McE. : The foliage-light product, a measure for assessing orchard canopies, and its relation to the yields of three apple varieties trained to three forms. - J. appl. Ecol. *10* : 501 - 511, 1973. [Canopy structure.]

38573 - MAGNUSSON, A. : Mise en évidence d'un rythme circadien de teneur en Ca^{2+} et Mg^{2+} dans les chloroplastes d'*Acetabularia mediterranea*. - Arch. int. Physiol. Biochim. *87* : 830 - 831, 1979.

38574 - MAHON, J.D. : Selection of field peas for attached leaf photosynthesis in the field. - Plant Physiol. *63* (Suppl.) : 39, 1979.

38575 - MAHON, J.D., DOMEY, J. : A light-weight battery operated infra-red gas analyzer for field measurements of photosynthetic CO_2 exchange. - Photosynthetica *13* : 459 - 466, 1979.

38576 - **MAIER, M., KAPPEN, L.** : Cellular compartmentalization of salt ions and protective agents with respect to freezing tolerance of leaves. Investigations with the halophyte *Halimione portulacoides* (L.) AELLEN. - Oecologia *38* : 303 - 316, 1979. [Chl.]

*38577 - **MAIER, R.** : Aspects of production of *Utricularia vulgaris* L. in some vegetation types in the reed-belt of Lake Neusiedlersee. - Pol. Arch. Hydrobiol. *20* : 169 - 174, 1973. [Ps.]

38578 - **MAIER, R.** : Production of *Utricularia vulgaris* L. - In : LÖFFLER, H. (ed.) : Neusiedlersee : The Limnology of a Shallow Lake in Central Europe. Pp. 273 - 279. Dr. W. Junk bv. Publ., The Hague - Boston - London 1979. [Chl.]

*38579 - **MAIER, R., SIEGHARDT, H.** : Untersuchungen zur Primärptoduktion im Grüngürtel des Neusiedler Sees. Teil II : *Phragmites communis* TRIN. - Pol. Arch. Hydrobiol. *24* : 245 - 257, 1977.

*38580 - **MAKAROV, A.D., STAKHOV, L.F.** : Pterinbelkovyĭ kompleks v protsessakh perenosa ėlektronov i fotofosforilirovaniya. [Pterin-protein complex in processes of electron transport and photophosphorylation.] - In : Itogi Issledovaniya Mekhanizma Fotosinteza. Pp. 80 - 93. Institut Fotosinteza Akad. Nauk SSSR, Pushchino 1974. [In R.]

38581 - **MAKARSKA, E., BUBICZ, M.** : Wpływ NPK, Afalonu i Gesagardu 50 na zawartość karotenowców w peluszce. [The NPK, Afalon and Gesagard 50 effects on the content of carotenoids in field pea.] - Rocz. Nauk rol., Ser. A *104* (2) : 61 - 73, 1979. [In Pol., ab : E, R.]

38582 - **MAKEDONSKA, Ts.** : Stimulirovanie fotosinteticheskoĭ assimilyatsii CO_2 v usloviyakh zagryazneniya vozdukha. [Stimulation of photosynthetic assimilation of CO_2 in air pollution.] - In : VAKLINOVA, S.G., VANKOVA-RADEVA, R., VASILĖVA, V.S. (ed.) : Fotosinteticheskaya Assimilyatsiya CO_2 i Fotodykhanie. Pp. 154 - 159. Izdat. bolg. Akad. Nauk, Sofiya 1979. [In R.]

38583 - **MAKEDONSKA, Ts., ĬORDANOV, Ts.** : Vliyanie na pochvenite usloviya v"rkhu intenzivnostta na fotosintezata pri gornata granitsa na gorata. [Effect of soil conditions on photosynthetic rate at upper boundaries of forests.] - Nauch. Tr. vissh. lesotekh. Inst., Ser. Gorskostopanstvo *24* : 77 - 83, 1979. [In Bulg., ab : G, R.]

38584 - **MAKI, T.** : Vertical profiles of wind turbulence in teosinte and sorgo canopies. - J. agr. Meteorol. *35* : 133 - 143, 1979.

*38585 - **MAKSIMOV, V.N., MAKSIMOVA, Ė.A.** : Geterotrofnaya i fotosinteticheskaya assimilyatsiya karbonatov biotsenozami pelagiali Baĭkala pri dobavkakh stochnykh vod Baĭkal'skogo tsellyulozno-bumazhnogo kombinata. [Heterotrophic and photosynthetic assimilation of carbonates by biocenoses of the pelagic zone of Lake Baikal during introduction of wastewater from the Baikal paper and pulp mill.] - In : Gidrobiologicheskie i Ikhtiologicheskie Issledovaniya v Vostochnoĭ Sibiri. Pp. 179 - 183, 218-219. Irkutsk. gos. Univ., Irkutsk 1978. [In R.]

38586 - **MAKSIMOVA, I.V.** : Intenzivnost' fotosinteza i obrazovanie vnekletochnykh produktov u *Nitzschia ovalis* ARN. pri razlichnykh kontsentratsiyakh kisloroda. [Photosynthetic rate and formation of extracellular products by *Nitzschia ovalis* ARN. at different oxygen concentrations.] - Biol. Nauki *1979*(1): 78 - 84, 1979. [In R.]

38587 - **MAKSIMOVA, I.V.** : Intensivnost' vydeleniya organicheskikh veshchestv suspenziyami kletok zelenykh vodorosleĭ v usloviyakh, obespechivayushchikh vysokuyu skorost' fotosinteza. [Intensity of efflux of organic substances by suspensions of green algae cells under conditions ensuring a high rate of photosynthesis.] - In : Rol' Nizshikh Organizmov v Krugovorote Veshchestv v Zamknutykh Ėkologicheskikh Sistemakh. Pp. 140 - 144. Naukova Dumka, Kiev 1979. [In R.]

*38588 - **MALAKONDAIAH, N., FANG, S.C.** : Influence of monuron on photosystem II and light-dependent $^{14}CO_2$ fixation in isolated cells of C_3 and C_4 plants. - Pestic. Biochem. Physiol. *9* : 33 - 38, 1978.

38589 - MALAKONDAIAH, N., FANG, S.C. : Differential effects of phenoxy herbicides on light-dependent $^{14}CO_2$ fixation and isolated cells of C_3 and C_4 plants. - Pestic. Biochem. Physiol. *10* : 268 - 274, 1979.

38590 - MALET, P. : Liaison statistique entre le développement et la croissance chez les plantes : application d'un modèle simplifié à l'analyse de la croissance. - Ann. agron. *30* : 415 - 430, 1979.

38591 - MALIK, M.N., KHAN, M.A., SIDDIQUE, M. : Control of chlorophyll formation and sprouting of tubers by oil dipping in five potato cultivars of Pakistan. - Sci. Hort. (Amsterdam) *10* : 331 - 336, 1979.

38592 - MALKIN, R., BARBER, J. : On the function of the fluorescence quenchers in chloroplasts and their relation to the primary electron acceptor of photosystem II. - Arch. Biochem. Biophys. *193* : 169 - 178, 1979.

38593 - MALKIN, R., BEARDEN, A.J. : Iron-sulfur centers of the chloroplast membrane. - Coordination Chem. Rev. *28* : 1 - 22, 1979.

*38594 - MALKIN, R., KNAFF, D.B., McSWAIN, B.D. : The effect of oxidation-reduction potential on the fluorescence yield of spinach chloroplasts at 77°K. - FEBS Let. *47* : 140 - 142, 1974.

38595 - MALKIN, S., CAHEN, D. : Photoacoustic spectroscopy and radiant energy conversion : theory of the effect with special emphasis on photosynthesis. - Photochem. Photobiol. *29* : 803 - 813, 1979.

38596 - MALKIN, S., FORD, G.A., FORK, D.C. : Computer-assisted measurements of parameters of the rise curve of fluorescence. - Carnegie Inst. Year Book *78* : 199 - 201, 1979.

38597 - MALKIN, S., WONG, D., GOVINDJEE, MERKELO, H. : Fluorescence studies on leaves during induction : parallel relationship between intensity and life-time during the "P" to "S" decay. - Plant Physiol. *63* (Suppl.) : 41, 1979.

38598 - MALNOË, P., ROCHAIX, J.-D., CHUA, N.H., SPAHR, P.-F. : Characterization of the gene and messenger RNA of the large subunit of ribulose 1,5-diphosphate carboxylase in *Chlamydomonas reinhardii*. - J. mol. Biol. *133* : 417 - 434, 1979.

38599 - MAL'YAN, A.N. : Ob uchastii karboksil'noĭ gruppy v aktivnom tsentre CF_1-ATFazy khloroplastov. [Participation of carboxyl groups in active centre CF_1-ATPase of chloroplasts.] - Dokl. Akad. Nauk SSSR *247* : 993 - 996, 1979. [In R.]

38600 - MAMLEEVA, N.A., KUZNETSOV, S.L., NEKRASOV, L.I. : Deĭstvie kisloroda na signal ÉPR adsorbirovannogo khlorofilla a i feofitina a. [Effect of oxygen on ESR signal of adsorbed chlorophyll a and pheophytin a.] - Biofizika *24* : 943, 1979. [In R.]

38601 - MAMLEEVA, N.A., LUNINA, E.V., NEKRASOV, L.I. : ÉPR adsorbirovannogo khlorofilla a. [ESR spectra of adsorbed chlorophyll a.] - Biofizika *24* : 594 - 597, 1979. [In R, ab : E.]

38602 - MANAKOV, K.N., NIKONOV, V.V. : Pervichnaya biologicheskaya produktivnost' elovykh lesov Kol'skogo poluostrova. [Primary biological productivity of spruce forests of the Koĭa peninsula.] - Bot. Zh. *64* : 232 - 241, 1979. [In R.]

38603 - MANLEY, S.A.M., LEDIG, F.T. : Photosynthesis in black and red spruce and their hybrid derivatives : ecological isolation and hybrid adaptive inferiority. - Can. J. Bot. *57* : 305 - 314, 1979.

38604 - MANLEY, S.L., CHAPMAN, D.J. : Metabolism of L-tyrosine to 4-hydroxybenzaldehyde and 3-bromo-4-hydroxybenzaldehyde by chloroplast-containing fractions of *Odonthalia floccosa* (ESP.) FALK. - Plant Physiol. *64* : 1032 - 1038, 1979. [Chl.]

38605 - MANN, J.E., CURRY, G.L., SHARPE, P.J.H. : Light interception by isolated plants.- Agr. Meteorol. *20* : 205 - 214, 1979.

38606 - **MANN, K., MECKE, D.** : Inhibition of spinach glyceraldehyde-3-phosphate dehydrogenases by pentalenolactone. - Nature *282* : 535 - 536, 1979.

38607 - **MANOLOV, P., BORICHENKO, N., RANGELOV, B.** : Obrazovanie sorbitola v protsesse fotosinteza u 14-ti plodovykh porod sem. rozotsvetnykh. [Sorbitol formation during photosynthesis of fourteen *Rosaceae* species.] - In : VAKLINOVA, S.G., VANKOVA-RADEVA, R., VASILEVA, V.S. (ed.) : Fotosinteticheskaya Assimilyatsiya CO_2 i Fotodykhanie. Pp. 86 - 91. Izdat. bolg. Akad. Nauk, Sofiya 1979. [In R.]

38608 - **MANOLOV, P., RANGELOV, B., BORICHENKO, N.** : ^{14}C-kineticheskie issledovaniya glikolatnogo puti v list'yakh persika. [^{14}C-kinetic studies of glycolate pathway in peach leaves.] - In : VAKLINOVA, S.G., VANKOVA-RADEVA, R., VASILEVA, V.S. (ed.) : Fotosinteticheskaya Assimilyatsiya CO_2 i Fotodykhanie. Pp. 126 - 132. Izdat. bolg. Akad. Nauk, Sofiya 1979. [In R.]

38609 - **MANSFIELD, T.A.** : Can we control the water requirements of plants ? - Trends biochem. Sci. *4* (2) : N27 - N28, 1979. [Ps.]

✳38610 - **MANSFIELD, T.A., JONES, M.B.** : Photosynthesis : Leaf and whole plant aspects. - In : HALL, M.A. (ed.) : Plant Structure, Function and Adaptation. Pp. 294 - 325. Macmillan, London 1976.

B38611 - **MARCELLE, R., CLIJSTERS, H., VAN POUCKE, M.** (ed.) : Photosynthesis and Plant Development. - Dr. W. Junk bv. Publ., The Hague - Boston - London 1979.

38612 - **MARCO, G. di, GREGO, S., TRICOLI, D.** : RuBP carboxylase-oxygenase in field--grown wheat. - J. exp. Bot. *30* : 851 - 861, 1979.

38613 - **MARCUS, R.A.** : Electron transfer and tunneling in chemical and biological systems. - In : GERISCHER, H., KATZ, J.J.(ed.) : Light-Induced Charge Separation in Biology and Chemistry. Pp. 15 - 43. Verlag Chemie, Weinheim - New York 1979. [Chl.]

38614 - **MARES, D.J., COOTE, M.A., POSSINGHAM, J.V.** : Membrane-bound plastid inclusions and chloroplast thylakoid formation in sunflower (*Helianthus annuus* L.). - Ann. Bot. *43* : 191 - 196, 1979.

38615 - **MARINETTI, T.D., OKAMURA, M.Y., FEHER, G.** : Localization of the primary quinone binding site in reaction centers from *Rhodopseudomonas sphaeroides* R-26 by photoaffinity labeling. - Biochemistry *18* : 3126 - 3133, 1979.

38616 - **MARKOWSKI, A., DUBERT, F.** : Zależność efektów neutralizacji dwutlenku siarki amoniakiem od stężenie siarki i azotu w pożywce oraz od stopnie szkodliwego działania SO_2 na rośliny słonecznika i fasoli. [Dependence of the effects of sulphur dioxide neutralization with ammonia on sulphur and nitrogen concentration in the nutrient medium and on the extent of damage caused by SO_2 in sunflower and bean plants.] - Acta agr. silv., Ser. agr. *18* (2) : 47 - 65, 1979. [Dry-matter accumulation; in Pol., aB : E, R.]

38617 - **MARKWELL, J.P., MILES, C.D., BOGGS, R.T., THORNBER, J.P.** : Solubilization of chloroplast membranes by zwitterionic detergents. Effect on photosystem II activity. - FEBS Lett. *99* : 11 - 14, 1979.

38618 - **MARKWELL, J.P., THORNBER, J.P., BOGGS, R.T.** : Higher plant chloroplasts : Evidence that all the chlorophyll exists as chlorophyll-protein complexes. - Proc. nat. Acad. Sci. USA *76* : 1233 - 1235, 1979.

38619 - **MAROC, J., GARNIER, J.** : Characterization of new strains of nonphotosynthetic mutants of *Chlamydomonas reinhardtii* II. Quinones and cytochromes *b*-559, *b*-563 and *c*-553 in twelve mutants having impaired photosystem II function. - Plant Cell Physiol. *20* : 1029 - 1040, 1979.

38620 - **MAROC, J., GARNIER, J.** : Photooxidation of the cytochrome *b*-559 in the presence of various substituted 2-anilinothiophenes and of some other compounds, in *Chlamydomonas reinhardtii*. - Biochim. biophys. Acta *548* : 374 - 385, 1979.

38621 - **MARÓTI, P., LAVOREL, J.** : Intensity and time-dependence of the carotenoid triplet quenching under light flashes of rectangular shape in *Chlorella*. - Photochem. Photobiol. *29* : 1147 - 1151, 1979.

38622 - MARSH, J.E., NASH, T.H. III. : Lichens in relation to the Four Corners
Power Plant in New Mexico. - Bryologist *82* : 20 - 28, 1979. [Oxygen uptake,
production.]

38623 - MARSHO, T.V., BEHRENS, P.W., RADMER, R.J. : Photosynthetic oxygen reduction
in isolated intact chloroplasts and cells from spinach. - Plant Physiol.
64 : 656 - 659, 1979.

38624 - MARTIN, B., ÖQUIST, G. : Seasonal and experimentally induced changes in the
ultrastructure of chloroplasts of *Pinus silvestris*. - Physiol. Plant. *46* :
42 - 49, 1979.

38625 - MARTIN, B.A., GAUGER, J.A., TOLBERT, N.E. : Changes in activity of ribulose-
-1,5-bisphosphate carboxylase/oxygenase and three peroxisomal enzymes during
tomato fruit development and ripening. - Plant Physiol. *63* : 486 - 489, 1979.

38626 - MARTIN, P.G. : Amino acid sequence of the small subunit of ribulose-1,5-
-bisphosphate carboxylase from spinach. - Aust. J. Plant Physiol. *6* :
401 - 408, 1979.

38627 - MARTINEZ-CARRASCO, R., THORNE, G.N. : Physiological factors limiting grain
size in wheat. - J. exp. Bot. *30* : 669 - 679, 1979. [Photosynthates.]

38628 - MARTSENYUK, P.P., KARNAUKHOV, V.N. : Lyuminestsentnye spektral'nye kharakteristiki kletok sinezelenykh vodorosleĭ (tsianobakteriĭ). [Luminescent spectral characteristics of blue-green algae cells (cyanobacteria).] - Gidrobiol. Zh. *15* (1) : 69 - 73, 1979. [Chl ; in R, ab : E.]

38629 - MARZOLA, D.L., BARTHOLOMEW, D.P. : Photosynthetic pathway and biomass energy production. - Science *205* : 555 - 559, 1979.

38630 - MASAKI, R., WADA, K., MATSUBARA, H. : Isolation and characterization of two
ferredoxin-NADP$^+$ reductases from *Spirulina platensis*. - J. Biochem. (Tokyo)
86 : 951 - 962, 1979.

38631 - MASAROVIČOVÁ, E. : Relationships between the CO_2 compensation concentration,
the slope of CO_2 curves of net photosynthetic rate and the energy of irradiance. - Biol. Plant. *21* : 434 - 439, 1979.

38632 - MASAROVIČOVÁ, E., MINARČIC, P. : Qualitative and quantitative analysis of
the *Fagus silvatica* L. leaves. I. Anatomical characteristic and photosynthetic activity. - Biológia (Bratislava) *34* : 513 - 521, 1979.

38633 - MASLENNIKOVA, V.G., TEREKHOVA, I.V., DOMAN, N.G. : Assimilyatsiya okisi
ugleroda kletkami khlorelly v zavisimosti ot kontsentratsii CO_2 v gazovoĭ
smesi. [Assimilation of carbon monoxide by *Chlorella* cells as a function
of CO_2 concentration in a gaseous medium.] - In : KORDYUM, V.A. (ed.) :
Rol' Nizshikh Organizmov v Krugovorote Veshchestv v Zamknutykh Ėkologicheskikh Sistemakh. Pp. 304 - 310. Naukova Dumka, Kiev 1979. [In R.]

38634 - MASLOV, V.G. : Primenenie spektroskopii vyzhiganiya provalov dlya issledovaniya fotoperenosa ėlektrona v reaktsionnykh tsentrakh fotosistemy 1 khlorelly. [Application of the hole burning spectroscopy for the investigation
of electron phototransfer in the reaction centres of the photosystem 1 of
Chlorella.] - Dokl. Akad. Nauk SSSR *246* : 1511 - 1513, 1979. [In R.]

*38635 - MASLOVA, N.F., GLADUNOV, I.M. : Vliyanie oblucheniya na fotosintez raste-
niĭ. [Effect of irradiation on plant photosynthesis.] - In : Fiziologiya
i Biokhimiya Rasteniĭ. Pp. 61 - 68. Shtiintsa, Kishinev 1975. [In R.]

*38636 - MASON, L., ZUBER, M.S. : Diallel analysis of maize for leaf angle, leaf
area, yield, and yield components. - Crop Sci. *16* : 693 - 696, 1976.

38637 - MATAGA, N., MASUHARA, H., KOBAYASHI, T. : [Nano- and picosecond chemistry
in relation to biological phenomena.] - Kakagu Sosetsu *24* : 253 - 271,
1979. [Ps; in Jap.]

38638 - MATHIS, P., BUTLER, W.L., SATOH, K. : Carotenoid triplet state and chlorophyll fluorescence quenching in chloroplasts and subchloroplast particles.
- Photochem. Photobiol. *30* : 603 - 614, 1979.

38639 - **MATHIS, P., CONJEAUD, H.** : Rapid reduction of P-700 photooxidized by a flash at low temperature in spinach chloroplasts. - Photochem. Photobiol. *29* : 833 - 837, 1979.

*38640 - **MATHUR, D.D., HENDERSHOTT, C.H., VINES, H.M.** : Efficiency of water utilization in Crassulacean acid metabolism plants when IN CAM *versus* OUT OF CAM. - Ind. J. Plant Physiol. *21* : 7 - 11, 1978. [Ps.]

*38641 - **MATHUR, D.D., NATARELLA, N.J., VINES, H.M.** : Elemental analyses of Crassulacean acid metabolism plant tissue. - Commun. Soil Sci. Plant Anal. *9* : 127 - 139, 1978.

*38642 - **MATHUR, D.D., VINES, H.M.** : Environmental effects on $\delta^{13}C$ shift in the leaves of *Sedum rubrotinctum*. - Commun. Soil Sci. Plant Anal. *9* : 843 - 850, 1978.

38643 - **MATHUR, D.D., VINES, H.M.** : Environmental effects on the elemental leaf composition of Crassulacean acid metabolism plants. - J. Plant Nutr. *1* : 407 - 416, 1979.

*38644 - **MATINYAN, I.G.** : O fotokhimicheskoĭ aktivnosti list'ev v svyazi s ikh vozrastom i fazoĭ razvitiya rasteniĭ. [Photochemical activity of leaves in relation to their growth phase and phase of plant development.] - Tr. bot. Inst. Akad. Nauk arm. SSR *20* (Voprosy Individual'nogo Razvitiya Vysshikh Rasteniĭ) : 136 - 138, 1977. [In R.]

38645 - **MATSUI, H., YUDA, E., NAKAGAWA, S.** : Physiological studies on the ripening of Delaware grapes. I. Effects of the number of leaves and changes in polysacharides or organic acids on sugar accumulation in the berries. - J. Jap. Soc. hort. Sci. *48* : 9 - 18, 1979.

38646 - **MATSUOKA, Y.** : [Air pollution and photosynthesis inhibition, damages in rice from sulfite gas.] - Kagaku to Seibutsu *17* : 225 - 226, 1979. [In Jap.]

38647 - **MATSUURA, K., MASAMOTO, K., ITOH, S., NISHIMURA, M.** : Effect of surface potential of the intramembrane electrical field measured with carotenoid spectral shift in chromatophores from *Rhodopseudomonas sphaeroides*. - Biochim. biophys. Acta *547* : 91 - 102, 1979.

38648 - **MATTHEWS, B.W., FENNA, R.E., BOLOGNESI, M.C., SCHMID, M.F., OLSON, J.M.** : Structure of a bacteriochlorophyll *a*-protein from the green photosynthetic bacterium *Prostecochloris aestuarii*. - J. mol. Biol. *131* : 259 - 285, 1979.

*38649 - **MAUDINAS, B.** : Mise au point sur la photoreduction d'hydrogène par les bactéries photosynthétiques. - Compt. rend. 103e Congr. nat. Soc. sav., Sect. Sci. *1* : 319 - 330, 1978.

38650 - **MAUNEY, J.R., GUINN, G., FRY, K.E., HESKETH, J.D.** : Correlation of photosynthetic carbon dioxide uptake and carbohydrate accumulation in cotton, soybean, sunflower and sorghum. - Photosynthetica *13* : 260 - 266, 1979.

38651 - **MAURO, S., LANNOYE, R., BARBER, J.** : Cation composition of the diffuse layer and the ability of ionophores to uncouple photosynthetic flow as monitored by millisecond delayed light emission. - Photobiochem. Photobiophys. *1* : 11 - 15, 1979.

38652 - **MAUSER, H.** : Zur Berechnung des Stoffumsatzes bei Photoreaktionen. - Z. Naturforsch. *34 C* : 1295 - 1296, 1979.

38653 - **MAUZERALL, D.** : Photoinduced electron transfer at the water-lipid bilayer interface. - In : GERISCHER, H., KATZ, J.J. (ed.) : Light-Induced Charge Separation in Biology and Chemistry. Pp. 241 - 257. Verlag Chemie, Weinheim - New York 1979.

38654 - **MAUZERALL, D.** : Multiple excitations and reaction yields in photosynthetic systems. - Photochem. Photobiol. *29* : 169 - 170, 1979.

38655 - **MAWSON, B.T., CUMMINS, W.R., COLMAN, B.** : The effect of abscisic acid on photosynthesis of isolated mesophyll cells. - Plant Physiol. *63* (Suppl.) : 80, 1979.

38656 - MAZHOROVA, L.E., PSHENOVA, K.V., MUTUSKIN, A.A. : Fotozavisimye prevrashche-
niya tsitokhroma f v khloroplastakh i list'yakh bobov, obrabotannykh poli-
enovymi antibiotikami. [Photo-dependent conversions of cytochrome f in
bean chloroplasts and leaves treated with polyene antibiotics.] - Biokhimiya
44 : 2005 - 2012, 1979. [In R, ab : E.]

38657 - MAZUR, T., PANAK, H., WOJNOWSKA, T., CIEĆKO, Z. : Intensywność fotosyntezy
u różnych odmian ziemniaków w zależności od poziomu nawożenia mineralnego.
[Photosynthetic rate in various potato cultivars in dependence on the level
of mineral fertilization.] - Zesz. nauk. Akad. rolniczo-tech. Olsztyne,
Rolnictwo 26 : 55 - 63, 1979. [In Pol., ab : E, R.]

38658 - McCABE, J., SHELP, B., URSINO, D.J. : Photosynthesis and photophosphorylation in radiation-stressed soybean plants and the relation of these processes to photoassimilate export. - Environ. exp. Bot. 19 : 253 - 261, 1979.

38659 - McCARTHY, J.J., CARPENTER, E.J. : Oscillatoria (Trichodesmium) thiebautii
(Cyanophyta) in the Central North Atlantic Ocean. - J. Phycol. 15 : 75 - 82,
1979. [Ps.]

38660 - McCARTHY, S.A., REBEIZ, C.A. : Detection of an inhibitor of protochlorophyll
biosynthesis in cucumber cotyledons. - Plant Physiol. 63 (Suppl.) : 97,
1979.

38661 - McCARTY, R.E. : Interactions between nucleotides and coupling factor 1 in
chloroplasts. - Trends biochem. Sci. 4 (2) : 28 - 30, 1979.

38662 - McCARTY, R.E. : Roles of a coupling factor for photophosphorylation in chloroplasts. - Annu. Rev. Plant Physiol. 30 : 79 - 104, 1979.

38663 - McCASHIN, B.G., CANVIN, D.T. : Photosynthetic and photorespiratory characteristics of mutants of Hordeum vulgare L. - Plant Physiol. 64 : 354 - 360,
1979.

38664 - McCRACKEN, I.J. : Changes in the carbohydrate concentration of pine seedlings after cool storage. - N. Zeal. J. Forest Sci. 9 : 34 - 43, 1979.

38665 - McCRAY, J.A., KIHARA, T. : Rates of reduced cytochrome c-ferricyanide binding and electron transfer. - Biochim. biophys. Acta 548 : 417 - 426, 1979.

38666 - McDONDALD, R.C., STEPHEN, R.C. : Effect of sowing and harvesting dates on
dry matter production of autumn-sown Tama ryegrass, ryecorn, and oats. -
N. Zeal. J. exp. Agr. 7 : 271 - 275, 1979.

38667 - McINTOSH, A.R., BOLTON, J.R. : CIDEP in the photosystems of green plant
photosynthesis. - Rev. Chem. Intermed. 3 : 121 - 129, 1979.

38668 - McINTOSH, A.R., MANIKOWSKI, H., BOLTON, J.R. : Observations of chemically
induced dynamic electron polarization in photosystem I of green plants and
algae. - J. phys. Chem. 83 : 3309 - 3313, 1979.

38669 - McINTOSH, A.R., MANIKOWSKI, H., WONG, S.K., TAYLOR, C.P.S., BOLTON, J.R. :
CIDEP observations in Photosystem I of green plant and algal photosynthesis. - Biochem. biophys. Res. Commun. 87 : 605 - 612, 1979.

38670 - McINTOSH, L., LINK, G., BOGORAD, L. : Cloning of the 32K photogene from
the maize chloroplast genome. - Plant Physiol. 63 (Suppl.) : 96, 1979.

38671 - McKEE, J.W.A., HAWKE, J.C. : The incorporation of [^{14}C] acetate into the
constituent fatty acids of monogalactosyldiglyceride by isolated spinach
chloroplasts. - Arch. Biochem. Biophys. 197 : 322 - 332, 1979.

38672 - McKENZIE, G.H., CH'NG, A.L., GAYLER, K.R. : Glutamine synthetase/glutamine :
α-ketoglutarate aminotransferase in chloroplasts from the marine alga Caulerpa simpliciuscula. - Plant Physiol. 63 : 578 - 582, 1979.

38673 - McKINION, J.M., WEAVER, R.E.C. : Simulation of plant response to primary
stress vectors. - Trans. ASAE 22 : 586 - 591, 597, 1979.

38674 - McKINNEY, D.W., BUCHANAN, B.B., WOLOSIUK, R.A. : Association of a thioredo-
doxin-like protein with chloroplast coupling factor (CF$_1$). - Biochem. biophys. Res. Commun. 86 : 1178 - 1184, 1979.

38675 - McKONE, H.T. : The rapid isolation of carotenoids from foods. - J. chem. Educ. *56* : 676, 1979.

38676 - McLAUGHLIN, S.B., McCONATHY, R.K. : Temporal and spatial patterns of carbon allocation in the canopy of white oak. - Can. J. Bot. *57* : 1407 - 1413, 1979.

38677 - McLAUGHLIN, S.B., McCONATHY, R.K., BESTE, B. : Seasonal changes in within--canopy allocation of ^{14}C-photosynthate by white oak. - Forest Sci. *25* : 361 - 370, 1979.

38678 - McLAUGHLIN, S.B., SHRINER, D.S., McCONATHY, R.K., MANN, L.K. : Effects of SO_2 dosage kinetics and exposure frequency on photosynthesis and transpiration of kidney beans (*Phaseolus vulgaris* L.). - Environ. exp. Bot. *19* : 179 - 191, 1979.

38679 - McMILLEN, G.G., McCLENDON, J.H. : Leaf angle : An adaptive feature of sun and shade leaves. - Bot. Gaz. *140* : 437 - 442, 1979. [Ps.]

38680 - McWILLIAM, J.R., MANOKARAN, W., KIPNIS, T. : Adaptation to chilling stress in sorghum. - In : LYONS, J.M., GRAHAM, D., RAISON, J.K. (ed.) : Low Temperature Stress in Crop Plants : The Role of the Membrane. Pp. 491 - 505. Academic Press, New York 1979. [Chl.]

38681 - MÉALLIER, P., PERCHERANCIER, J.P., POUYET, B. : Influence des herbicides sur le comportement photochimique des chlorophylles. A - Influence des amides substitués. - Chemosphere *8* : 903 - 908, 1979.

*38682 - MECHLER, B., OELZE, J. : Differentiation of the photosynthetic apparatus of *Chromatium vinosum*, strain D. I. The influence of growth conditions. - Arch. Microbiol. *118* : 91 - 97, 1978.

*38683 - MECHLER, B., OELZE, J. : Differentiation of the photosynthetic apparatus of *Chromatium vinosum*, strain D. II. Structural and functional differences. - Arch. Microbiol. *118* : 99 - 108, 1978.

*38684 - MECHLER, B., OELZE, J. : Differentiation of the photosynthetic apparatus of *Chromatium vinosum*, strain D. III. Analyses of spectral alteration. - Arch. Microbiol. *118* : 109 - 114, 1978.

38685 - MEENAKSHI, R.M., GNANARETHINAM, J.L. : Phytochemical aspects of onion (*Allium cepa*, LINN.) under water stress. - Plant Physiol. *63* (Suppl.) : 89, 1979. [Chl.]

38686 - MEGARD, R.O., COMBS, W.S. Jr., SMITH, P.D., KNOLL, A.S. : Attenuation of light and daily integral rates of photosynthesis attained by planktonic algae. - Limnol. Oceanogr. *24* : 1038 - 1050, 1979.

*38687 - MEHRETEAB, A., STRAUSS, G. : Energy transfer and energy losses in bilayer membrane vesicles (liposomes). - Photochem. Photobiol. *28* : 369 - 375, 1978. [Chl, Car.]

38688 - MEHTA, R.S., HAWXBY, K.W. : Effects of simazine on the blue-green alga *Anacystis nidulans*. - Bull. environm. Contam. Toxicol. *23* : 319 - 326, 1979. [Chl.]

38689 - MEINESZ, A. : Contribution à l'étude de *Caulerpa prolifera* (FORSSKÅL) LAMOUROUX (Chlorophycée-Caulerpale). III - Biomasse et productivité primaire dans une station des côtes continentales françaises de la Méditerranée. - Bot. mar. *22* : 123 - 127, 1979.

*38690 - MEISTER, A. : Messung von Absorptionsspektren *in vivo*. - Kulturpflanze *25* : 141 - 154, 1977.

38691 - MEISTER, A., BRECHT, E. : Aggregation der Chlorophylle und Phäophytine in synthetischen Pigment-Protein-Komplexen. - Biochem. Physiol. Pflanz. *174* : 305 - 317, 1979.

38692 - MELACK, J.M. : Temporal variability of phytoplankton in tropical lakes. - Oecologia *44* : 1 - 7, 1979. [Ps, Chl.]

38693 - MELACK, J.M. : Photosynthesis and growth of *Spirulina platensis* (*Cyanophyta*) in an equatorial lake (Lake Simbi, Kenya). - Limnol. Oceanogr. *24* : 753 - 760, 1979.

38694 - MELACK, J.M. : Photosynthetič rates in four tropical African fresh waters. - Freshwater Biol. *9* : 555 - 571, 1979.

38695 - MELCAREK, P.K., BROWN, G.N. : Chlorophyll fluorescence monitoring of freezing point exotherms in leaves. - Cryobiology *26* : 69 - 73, 1979.

*38696 - MELCHERS, G., SACRISTÁN, M.D., HOLDER, A.A. : Somatic hybrid plants of potato and tomato regenerated from fused protoplasts. - Carlsberg Res. Commun. *43* : 203 - 218, 1978. [RuBPC.]

38697 - MELEKHOV, E.I., DOLGIKH,T.A., BELIKOV, P.S. : Vremennoĭ khod fotosinteza v usloviyakh bystrogo i medlennogo nagreva. [Time-course of photosynthesis under rapid and slow rise in temperature.] - Fiziol. Rast. *26* : 167 - 173, 1979. [In R, ab : E.]

38698 - MELIS, A., DUYSENS, L.N.M. : Biphasic energy conversion kinetics and absorbance difference spectra of photosystem II of chloroplasts. Evidence for two different photosystem II reaction centers. - Photochem. Photobiol. *29* : 373 - 382, 1979.

38699 - MELIS, A., SCHREIBER, U. : The kinetic relationship between the C-550 absorbance change, the reduction of $Q(\Delta A_{320})$ and the variable fluorescence yield change in chloroplasts at room temperature. - Biochim. biophys. Acta *547* : 47 - 57, 1979.

38700 - MELKONIAN, M., ROBENEK, H. : The eyespot of the flagellate *Tetraselmis cordiformis* STEIN (*Chlorophyceae*) : Structural specialization of the outer chloroplast membrane and its possible significance in phototaxis of green algae. - Protoplasma *100* : 183 - 197, 1979.

38701 - MELZER, R., SACKEWITZ, H. : Probleme der Mutationsauslösung und Nutzung von Mutationen bei der Zuckerrübe (*Beta vulgaris* L.). -Arch. Züchtungsforsch. *9* : 65 - 72, 1979. [Chl.]

38702 - MENAUT, J.C., CESAR, J. : Structure and primary productivity of Lamto savannas, Ivory coast. - Ecology *60* : 1197 - 1210, 1979.

38703 - MERRILL, J.E., WAALAND, J.R. : Photosynthesis and respiration in a fast growing strain of *Gigartina exasperata* (HARVEY and BAILEY). - J. exp. mar. Biol. Ecol. *39* : 281 - 290, 1979.

*38704 - MESHINEV, T. : P"rvichna produktivnost na asotsiyata *Nardus stricta - Festuca fallax* v mestnostta Beglika, Zapadni Rodopi. [Primary productivity of the *Nardus stricta - Festuca fallax* association in the locality Beglika, Western Rhodopes.] - Fitologiya (Sofia) *8* : 47 - 53, 1977. [In Bulg., ab : E.]

38705 - METIVIER, J.R. : The effect of time of nitrate application upon the growth of barley cultivars of differing endogenous nitrogen levels : longer term experiments. - Ann. Bot. *43* : 753 - 764, 1979. [Leaf area formation.]

*B38706 - Metody Issledovaniya ·Fotokhimicheskikh Reaktsiĭ Fotosinteza *in Vitro* i *in Vivo*. [Methods of Studying Photochemical Reactions of Photosynthesis *in Vitro* and *in Vivo*.] - Akad. Nauk SSSR, Institut Fotosinteza, Pushchino 1975. [In R.]

38707 - METZGER, U., OHMANN, E. : Konzentrationsabhängige Effekte von Dibromthymochinon auf den photosynthetischen Elektronentransport von *Spinacea oleracea* L. - In : SCHÜTTE, H.R. (ed.) : Wirkungsmechanismen von Herbiziden und synthetischen Wachstumsregulatoren. Pp. 222 - 229. G. Fischer, Jena 1979.

38708 - METZNER, H., FISCHER, K., BAZLEN, O. : Isotope ratios in photosynthetic oxygen. - Biochim. biophys. Acta *548* : 287 - 295, 1979.

38709 - MEYER, J., KELLEY, B.C., COLBEAU, A., JOUANNEAU, Y., VIGNAIS, P.M. : Nitrogenase and hydrogenase activities in the phototrophic bacterium *Rhodopseudomonas capsulata*. - Physiol. vég. *17* : 670 - 671, 1979.

38710 - **MICHAEL, D., DICKMANN, D., NELSON, N.** : Photosynthesis, CO_2 compensation and stomatal conductance of young poplar plants grown under intensive culture. - Plant Physiol. *63* (Suppl.) : 121, 1979.

*38711 - **MICHAELIS, G., PRATJE, E.** : V. Cytoplasmic inheritance. - Progr. Bot. *40* : 261 - 275, 1978. [Chloroplast.]

*38712 - **MICHALSKI, W.P.** : Struktura i funkcja błony purpurowej bakterii słonolubnych z rodzaju *Halobacterium*. [Structure and function of purple membrane from halophilic bacteria of *Halobacterium* strain.] - Postępy Biochem. *23* : 297 - 319, 1977. [In Pol., ab : E.]

38713 - **MICHEL, B.E.** : Correction of thermal gradient errors in stem thermocouple hygrometers. - Plant Physiol. *63* : 221 - 224, 1979.

38714 - **MICHEL-BEYERLE, M.E., SCHEER, H., SEIDLITZ, H., TEMPUS, D., HABERKORN, R.** : Time-resolved magnetic field effect on triplet formation in photosynthetic reaction centers of *Rhodopseudomonas sphaeroides* R-26. - FEBS Lett. *100* : 9 - 12, 1979.

38715 - **MICKLE, A.M., WETZEL, R.G.** : Effectiveness of submersed angiosperm-epiphyte complexes on exchange of nutrients and organic carbon in littoral systems. III. Refractory organic carbon. - Aquat. Bot. *6* : 339 - 355, 1979.

38716 - **MIGINIAC-MASLOW, M., HOARAU, A.** : The adenine nucleotide levels and the adenylate energy charge values of different *Triticum* and *Aegilops* species. - Z. Pflanzenphysiol. *93* : 387 - 394, 1979. [Ps.]

38717 - **MIGINIAC-MASLOW, M., HOARAU, A., MOYSE, A.** : Hill reaction studies with protoplasts from cultivated wheats and their wild relatives. - Z. Pflanzenphysiol. *95* : 95 - 104, 1979.

38718 - **MIGNUCCI, J.S., BOYER, J.S.** : Inhibition of photosynthesis and transpiration in soybean infected by *Microsphaera diffusa*. - Phytopathology *69* : 227 - 230, 1979.

38719 - **MIKHAIL, E.H., EL-ZEFTAWI, B.M.** : Effect of soil types and rootstocks on root distribution, chemical composition of leaves and yield of Valencia oranges. - Aust. J. Soil Res. *17* : 335 - 342, 1979. [Chl.]

38720 - **MILBORROW, B.V.** : Antitranspirants and the regulation of abscisic acid content. - Aust. J. Plant Physiol. *6* : 249 - 254, 1979.

38721 - **MILBURN, J.A.** : An ideal viscous flow porometer. - J. exp. Bot. *30* : 1021 - 1034, 1979.

B38722 - **MILBURN, J.A.** : Water Flow in Plants. - Longman, London - New York 1979. [Ps.]

38723 - **MILES, C.D., MARKWELL, J.P., THORNBER, J.P.** : Effect of nuclear mutation in maize on photosynthetic activity and content of chlorophyll-protein complexes. - Plant Physiol. *64* : 690 - 694, 1979.

38724 - **MILES, D.** : Chlorophyll-protein mutants of maize. - Plant Physiol. *63* (Suppl.) : 160, 1979.

38725 - **MILICĂ, C.I., POPOVICI, I., SÎRBU, M., AIRINEI, A.** : Particularități ale solurilor şi hibrizilor de sfeclă de zahăr în absorbția energiei luminoase de către pigmenți. [Pecularities of sugar beet cultivars and hybrids in the radiant energy absorption by the pigments.] - Lucrări ştiinţ. Inst. agron. "Ion Ionescu de la Brad", Ser. Agron. *23* : 77 - 82, 1979. [In Roum., ab : E.]

38726 - **MILIUS, A., KYVASK, V.** : O nekotorykh pokazatelyakh fitoplanktona ozera Myannik"yarv. [Some characteristics of phytoplankton in Lake Mannikjarv.] - Eesti NSV Tead. Akad. Toim., Biol. *28* (2) : 134 - 136, 1979. [Chl; in R, ab : E, Est.]

38727 - MILLER, C.H., McCOLLUM, R.E., CLAIMON, S. : Relationships between growth
of bell peppers (*Capsicum annuum* L.) and nutrient accumulation during
ontogeny in field environments. - J. amer. Soc. hort. Sci. *104* : 852 - 857,
1979.

38728 - MILLER, G.W., DENNEY, A., WOOD, J.K., WELKIE, G.W. : Light-induced delta-
-aminolevulinic acid in dark-grown barley seedlings. - Plant Cell Physiol.
20 : 131 - 143, 1979.

38729 - MILLER, K.R. : Structure of a bacterial photosynthetic membrane. - Proc.
nat. Acad. Sci. USA *76* : 6415 - 6419, 1979.

38730 - MILLER, K.R. : The photosynthetic membrane. - Sci. Amer. *241* (4) : 102 -
109, 112 - 113, 1979.

38731 - MILLER, K.R., CUSHMAN, R.A. : A chloroplast membrane lacking photosystem II.
Thylakoid stacking in the absence of the photosystem II particle. - Biochim. biophys. Acta *546* : 481 - 497, 1979.

38732 - MILLER, P.D., VAUGHN, K.C., WILSON, K.G. : Induction, ultrastructure, isolation, and tissue culture of plastid mutants in carrot. - Plant Physiol.
63 (Suppl.) : 117, 1979.

38733 - MILLS, J.D., CROWTHER, D., SLOVACEK, R.E., HIND, G., McCARTY, R.E. : Electron transport pathways in spinach chloroplasts. Reduction of the primary
acceptor of Photosystem II by reduced nicotinamide adenine dinucleotide
phosphate in the dark. - Biochim. biophys. Acta *547* : 127 - 137, 1979.

38734 - MILLS, J.D., HIND, G. : Light-induced Mg^{2+} ATPase activity of coupling factor in intact chloroplasts. - Biochim. biophys. Acta *547* : 455 - 462, 1979.

38735 - MILLS, J.D., MITCHELL, P.D., BARBER, J. : The cyclic electron transport
pathway in chloroplasts. Reduction of plastoquinone by reduced nicotinamide
adenine dinucleotide phosphate in the dark. - Photobiochem. Photobiophys.
1 : 3 - 9, 1979.

38736 - MILLS, W.R., JOY, K.W. : A rapid method for isolation of purified physiologically active chloroplasts, and its uses. - Plant Physiol. *63* (Suppl.) :
5, 1979.

38737 - MILNE, R. : Water loss and canopy resistance of a young Sitka spruce plan-
tation. - Boundary-Layer Meteorol. *16* : 67 - 81, 1979.

38738 - MILOIKOVA-PEYCHEVA, S., GEORGIEV, G. : ^{14}C-incorporation in photosynthetic
products in maize plants in the restoring processes after phosphorus starvation. - In : Mineral Nutrition of Plants. Vol. II. Pp. 79 - 84. Publ.
House Central Cooperative Union, Sofia 1979.

*B38739 - MILTHORPE, F.L., MOORBY, J. : An Introduction to Crop Physiology. - Cambrid-
ge Univ. Press, Cambridge 1974. [Ps.]

B38740 - MILTHORPE, F.L., MOORBY, J. : An Introduction to Crop Physiology. 2nd Ed. -
Cambridge University Press, Cambridge - London - New York - New Rochelle -
Melbourne - Sydney 1979. [Ps, Chl.]

38741 - MINARČIC, P., HERICH, R., PAULECH, C. : Changes of surface membrane and
chloroplast ultrastructure of barley, after infection with powdery mildew.
- Phytopathol. Z. *94* : 97 - 102, 1979.

38742 - MINKOV,I.N., KIMENOV, G.P. : Vliyanie polozhitel'nykh ékstremal'nykh temperatur na raspredelenie ^{14}C sredi produktov fotosinteza v list'yakh
Haberlea rhodopensis FRIV. i *Ramonda serbica* PANC. [Effect of extreme positive temperatures on ^{14}C-distribution among the products of photosynthesis in the leaves of *Haberlea rhodopensis* FRIV. and *Ramonda serbica* PANC.] -
In : VAKLINOVA, S.G., VANKOVA-RADEVA, R., VASILEVA, V.S. (ed.) : Fotosinteticheskaya Assimilyatsiya CO_2 i Fotodykhanie. Pp. 98 - 103. Izdat. bolg.
Akad. Nauk, Sofiya 1979. [In R.]

38743 - MIRHADI, M.J., KOBAYASHI, Y. : Studies on the productivity of grain sorghum.
II. Effects of wilting treatments at different stages of growth on the deve-
lopment, nitrogen uptake and yield of irrigated grain sorghum. - Jap. J.
Crop Sci. *48* : 531 - 542, 1979.

38744 - MIRHADI, M.J., YOSHIDA, S., KOBAYASHI, Y. : Studies on the productivity of grain sorghum. I. Nitrogen nutrition of grain sorghum. - Jap. J. Crop Sci. *48* : 483 - 489, 1979.

38745 - MISHKIND, M., MAUZERALL, D., BEALE, S.I. : Diurnal variation *in situ* of photosynthetic capacity in *Ulva* is caused by a dark reaction. - Plant Physiol. *64* : 896 - 899, 1979.

38746 - MISHUSTINA, N.E., TIKHAYA, N.I., CHAPLYGINA, N.S. : (Na^+ + K^+)-ATFaznaya aktivnost' izolirovannykh membran pobegov galofita *Halocnemum strobilaceum*. [(Na^+ + K^+)-ATP-ase activity of membranes isolated from the halophyte *Halocnemum strobilaceum* shoots.] - Fiziol. Rast. *26* : 541 - 547, 1979. [In R, ab : E.]

38747 - MIŠTINOVÁ, A., MIŠTINA, T. : Variabilita počtu chloroplastov v stomatických bunkách lucerny (*Medicago sativa* L.). [Variability in chloroplast number in stomata cells of alfalfa (*Medicago sativa* L.).] - Vedecké Práce výskum. Ústavu rastlin. Výroby Piešťanoch *16* : 29 - 38, 1979. [In Slovak, ab : E, R.]

38748 - MISZALSKI, Z., ZIEGLER, I. : Increase in chloroplastic thiol groups by SO_2 and its effect on light modulation of NADP-dependent glyceraldehyde 3-phosphate dehydrogenase. - Planta *145* : 383 - 387, 1979.

38749 - MITCHELL, D.T., RICE, K.A. : Translocation of ^{14}C-labelled assimilates in cabbage during club root development. - Ann. appl. Biol. *92* : 143 - 152, 1979.

*38750 - MITCHELL, P. : Vectorial chemistry and the molecular mechanics of chemiosmotic coupling : power transmission by proticity. - Biochem. Soc. Trans. *4* : 399 - 430, 1976.

*38751 - MITCHELL, P. : Future trends : Protonmotive chemiosmotic mechanisms in oxidative and photosynthetic phosphorylation. - Trends biochem. Sci. *3* : N58, N60, N61, 1978.

38752 - MITCHELL, S.F., BURNS, C.W. : Oxygen consumption in the epilimnia and hypolimnia of two eutrophic,warm-monomictic lakes. - New Zealand J. mar. Freshw. Res. *13* : 427 - 441, 1979. [Ps.]

*38753 - MITROFANOV, B.A., GOĬSA, N.I., FERENTS, A.F. : Vliyanie mineral'nogo pitaniya na fotosintez ozimoĭ pshenitsy. [Effect of mineral nutrition on photosynthesis of winter wheat.] - In : Mineral'noe Pitanie i Produktivnost' Rasteniĭ. Pp. 170 - 177, 324. Naukova Dumka, Kiev 1978. [In R.]

38754 - MITSCH, W.J., EWEL, K.C. : Comparative biomass and growth of cypress in Florida Wetlands. - Amer. Midland Natur. *101* : 417 - 426, 1979.

38755 - MITSUI, A. : Biosaline research : The use of photosynthetic marine organisms in food and feed production. - In : HOLLAENDER, A., ALLER, J.C., EPSTEIN, E., SAN PIETRO, A., ZABORSKY,O.R.(ed.) : The Biosaline Concept. An Approach to the Utilization of Underexploited Resources. Environ. Sci. Res. Vol. 14. Pp. 177 - 215. Plenum Press, New York 1979.

38756 - MITSULOV, N. : Sistemi za avtomatichno upravlenie na fitoklimatichni kameri FK-3-II. Elektronna sistema za avtomatichno regulirane na temperaturata na v"zdushnoto pole. [Systems for automatic control of FK-3-II phytoclimatic chambers. Electronic system for automatic regulation of air temperature.] - Rasteniev. Nauki *16* (6) : 3 - 15, 1979. [In Bulg., ab : E, R.]

38757 - MIYACHI, S. : Light-enhanced dark CO_2 fixation. - In : GIBBS, M., LATZKO, E. (ed.) : Photosynthesis II. (Encycl. Plant Physiol. N.S. Vol. 6.) Pp. 68 - 76. Springer-Verlag, Berlin - Heidelberg - New York 1979.

38758 - MIYACHI, S., HOGETSU, D., MIYACHI, S. : Effects of environmental factors on photosynthetic carbon metabolism. - In : VAKLINOVA, S.G., VANKOVA-RADEVA, R., VASILEVA, V.S. (ed.) : Fotosinteticheskaya Assimilyatsiya CO_2 i Fotodykhanie. Pp. 15 - 30. Izdat. bolg. Akad. Nauk, Sofiya 1979.

38759 - MIYACHI, S., SHIRAIWA, Y. : Form of inorganic carbon utilized for photosynthesis in *Chlorella vulgaris* 11h cells. - Plant Cell Physiol. *20* : 341 - 348, 1979.

38760 - MIYAJI, K.-I., TAGAWA, H. : Longevity and productivity of leaves of a cultured annual *Glycine max* MERRILL. I. Longevity of leaves in relation to density and sowing time. - New Phytol. *82* : 233 - 244, 1979.

38761 - MIYAMOTO, K., HALLENBECK, P.C., BENEMANN, J.R. : Hydrogen production by the
thermophilic alga *Mastigocladus laminosus* : Effects of nitrogen, temperature, and inhibition of photosynthesis. - Appl. environ. Microbiol. *38* :
440 - 446, 1979.

38762 - MIYAMOTO, K., HALLENBECK, P.C., BENEMANN, J.R. : Nitrogen fixation by thermophilic blue-green algae (*Cyanobacteria*) : Temperature characteristics and
potential use in biophotolysis. - Appl. environm. Microbiol. *37* : 454 - 458,
1979. [Ps.]

38763 - MIYAMOTO, K., HALLENBECK, P.C., BENEMANN, J.R. : Solar energy, conversion
by nitrogen-limited cultures of *Anabaena cylindrica*. - J. Ferment. Technol.
57 : 287 - 293, 1979.

38764 - MIYANISHI, K., HOY, A.R., CAVERS, P.B. : A generalized law of self-thinning
in plant populations (Self-thinning in plant populations). - J. theor. Biol.
78 : 439 - 442, 1979.

38765 - MIYASAKA, T., WATANABE, T., FUJISHIMA, A., HONDA, K. : Highly efficient
quantum conversion at chlorophyll *a*-lecithin mixed monolayer coated electrodes. - Nature *277* : 638 ⊣ 640, 1979.

38766 - MIYAZAKI, T., MORITA, S., HATANO, M., NOZAWA, T. : Bacteriochlorophyll-*a*
types in chromatophore and subchromatophore preparations from *Rhodopseudomonas sphaeroides*. - J. Biochem. (Tokyo) *86* : 1411 - 1417, 1979.

38767 - MIZIORKO, H.M. : Ribulose-1,5-bisphosphate carboxylase. Evidence in support
of the existence of distinct CO_2 activator and CO_2 substrate sites. - J.
biol. Chem. *254* : 270 - 272, 1979.

*38768 - MLADENOVA, Ĭ., DANKOV, T., YANEV, T., ĬORDANOV, I. : V"zmozhnost za ranno
predskazvane na kheterozisa pri tsarevitsata. I. Sravnitelen analiz na
samooprasheni linii i tekhni khibridi v F_1. [Possibility for early prediction of heterosis in maize. I. Comparative analysis of self-pollinated lines of maize and their F_1 hybrids.] - Genet. Sel. *11* (2/3) : 94 - 105,
1978. [Ps, Chl; in Bulg., ab : E, R.]

38769 - MOGILEVA, G.A., ZELENSKIĬ, M.I., SAKHAROVA, O.V. : Osobennosti fotosinteticheskogo apparata *Triticum monococcum*. [Peculiarities of the photosynthetic
apparatus of *Triticum monococcum*.] - Byul. vsesoyuz. nauch.-issl. Inst. Ras-
tenievod.Im.N.I.Vavilova *87*:59 - 64, 1979. [In R.]

*38770 - MOHAMED, A.H., GNANAM, A. : Regulation of photosynthetic carbon flow by am-
monium ions in isolated bean leaf cells. - Plant Biochem. J. *4* : 1 - 9,
1977.

38771 - MOHAMED, A.H., GNANAM, A. : A possible mechanism of ammonium ion regulation
of photosynthetic carbon flow in higher plants. - Plant Physiol. *64* : 263 -
268, 1979.

38772 - MOHAMED, G.E.S., MARSHALL, C. : The pattern of distribution of phosphorus
and dry matter with time in spring wheat. - Ann. Bot. *44* : 721 - 730, 1979.
[Dry matter production.]

38773 - MOHANTY, P., MAYNE, B.C., KE, B. : Further characterization of a photosystem II particle isolated from spinach chloroplasts by triton treatment. Delayed light emission. - Biochim. biophys. Acta *545* : 285 - 295, 1979.

38774 - MOHR, W.P. : Pigment bodies in fruits of crimson and high pigment lines of
tomatoes. - Ann. Bot. *44* : 427 - 434, 1979. [Chloroplast.]

38775 - MOLL, W.A.W., DE WIT, B. : Chlorophyllase activity in plastid membranes of
bean leaves grown in darkness and in (intermittent) light. - Photosynthetica
13 : 146 - 154, 1979.

38776 - MOLNÁR, E.N., NOSEK, J.N. : Spatial processes in a grassland community,
I. Number of species and individuals, cover and biomass at the community
level. - Acta bot. Acad. Sci. hung. *25* : 339 - 348, 1979.

38777 - **MOLNÁR, J., KORSŐS, I.** : A *Lithocolletis blancardella* (F.)-fertőzöttség hatása az almalevél klorofilltartalmára és a termés mennyiségére. [Effect of *Lithocolletis blancardella* (F.) infestation on the chlorophyll content and yield of apple trees.] - Növenyvedelem (Budapest) *15* : 373 - 375, 1979. [In Hung.]

38778 - **MOLNÁR, P., SZABOLCS, J.** : Alkaline permanganate oxidation of carotenoid epoxides and furanoids. - Acta chim. Acad. Sci. hung. *99* : 155 - 173, 1979.

38779 - **MOMEN, N.N., CARLSON, R.E., SHAW, R.H., ARJMAND, O.** : Moisture-stress effects on the yield components of two soybean cultivars. - Agron. J. *71* : 86 - 90, 1979. [Resistances.]

*38780 - **MONCHOR, D., VACEK, K.** : Optical and fluorescence measurements of chlorophyll aggregation in polymer matrix (PVA). I. Chlorophyll *a*. - Stud. biophys. *72* : 117 - 118, 1978.

*38781 - **MONCHOR, D., VACEK, K.** : Optical and fluorescence measurements of chlorophyll aggregation in polymer matrix (PVA). II. Chlorophyll *b*. - Stud. biophys. *72* : 119 - 120, 1978.

38782 - **MONHEIMER, R.H.** : Effect of cysteine and methionine on sulfate uptake and primary productivity by axenic algal cultures and lake microplankton communities. - J. Phycol. *15* : 284 - 288, 1979.

38783 - **MONMA, E., TSUNODA, S.** : Photosynthetic heterosis in maize. - Jap. J. Breed. *29* : 159 - 165, 1979.

*38784 - **MOONEY, H.A., TROUGHTON, J.H., BERRY, J.A.** : Carbon isotope ratio measurements of succulent plants in southern Africa. - Oecologia *30* : 295 - 305, 1977.

38785 - **MOORE, P.D.** : Ecology of photosynthesis. - Nature *280* : 193 - 194, 1979.

38786 - **MOORE, R., BLACK, C.C. Jr.** : Nitrogen assimilation pathways in leaf mesophyll and bundle sheath cells of C_4 photosynthesis plants formulated from comparative studies with *Digitaria sanguinalis* (L.) SCOP. - Plant Physiol. *64* : 309 - 313, 1979.

38787 - **MOR, Y., HALEVY, A.H.** : Translocation of ^{14}C-assimilates in roses. I. The effect of the age of the shoot and the location of the source leaf. - Physiol. Plant. *45* : 177 - 182, 1979.

38788 - **MOREL, C.** : Rôle coordinateur du CAM dans le métabolisme intermédiaire. I. Fonction anaplérotique de l'enzyme malique et cycle des acides tricarboxyliques. - Physiol. vég. *17* : 697 - 712, 1979.

38789 - **MOREL, C., VALON, C.** : Rôle coordinateur du CAM dans le métabolisme intermédiaire. II. Photopériode et voies de synthèse des acides aminés. - Physiol. vég. *17* : 713 - 730, 1979.

38790 - **MORESHET, S., COHEN, Y., FUCHS, M.** : Effect of increasing foliage reflectance on yield, growth, and physiological behavior of a dryland cotton crop. - Crop Sci. *19* : 863 - 868, 1979. [Ps.]

38791 - **MORESHET, S., GREEN, G.C.** : Stomatal development and gas exchange in citrus fruit in relation to soil moisture. - Plant Physiol. *63* (Suppl.) : 122, 1979. [Ps.]

*38792 - **MORGAN, D.L, PRUITT, W.O., LOURENCE, F.J.** : Estimation of atmospheric radiation. - J. appl. Meteorol. *10* : 463 - 468, 1971.

38793 - **MORGAN, J.A., BROWN, R.H.** : Photosynthesis in grass species differing in carbon dioxide fixation pathways. II. A search for species with intermediate gas exchange and anatomical characteristics. - Plant Physiol. *64* : 257 - 262, 1979.

38794 - **MORGAN, K.C., KALFF, J.** : Effect of light and temperature interactions on growth of *Cryptomones erosa* (*Cryptophyceae*). - J. Phycol. *15* : 127 - 134, 1979. [Ps, Chl.]

38795 - MORITA, S. : [Chemical conversion of photoenergy in biological system.] - Kagaku, Zokan (Kyoto) *82* : 23 - 37, 1979. [Ps; in Jap.]

38796 - MORITA, S. : [Photosynthesis by photosynthetic bacteria.] - In : SHIBATA, K., UEMONSA, S., HARA, T. (ed.) : Koseibutsugaku. Vol. 1. Pp. 169 - 179. Bus. Center Acad. Soc. Japan, Tokyo 1979. [In Jap.]

38797 - MORONEY, J.V., McCARTY, R.E. : Reversible uncoupling of photophosphorylation by a new bifunctional maleimide. - J. biol. Chem. *254* : 8951 - 8955, 1979.

38798 - MOROT-GAUDRY, J.F., FARINEAU, J., JOLIVET, E. : Effect of leaf position and plant age on photosynthetic carbon metabolism in leaves of 8 and 16 day-old maize seedlings (W 64 A) with and without the gene *opaque 2*. - Photosynthetica *13* : 365 - 375, 1979.

38799 - MOROZOV, V.L. : Énergeticheskaya otsenka zapasov fitomassy travostoev v severnykh otrogakh Ganal'skogo khrebta (Kamchatka). [Energy content of phytomass reserves of grasslands in northern parts of Ganal mountain ridge (Kamchatka).] - Probl. Bot. *14*[(1) Flora i Rastitel'nost' Vysokogoriĭ]: 140 - 144, 1979. [In R.]

38800 - MOROZOV, V.L. : Radiatsionnyĭ rezhim krupnotravnykh labaznikovykh soobshchestv na Kamchatke. [Radiation regime of tall herb *Filipendula* communities in Kamchatka.] - Ékologiya *1979* (4) : 25 - 33, 1979. [In R.]

38801 - MOROZOV, V.L. : Produktivnost' krupnotrav'ya na Dal'nem Vostoke. [Productivity of tall herbaceous vegetation in the far east.] - Izv. sib. Otd. Akad. Nauk SSSR, Ser. biol. Nauk *1979* (2) : 32 - 39, 1979. [In R, ab : E.]

38802 - MORRILL, L.C., LOEBLICH, A.R. III. : An investigation of heterotrophic and photoheterotrophic capabilities in marine *Pyrrophyta*. - Phycologia *18* : 394 - 404, 1979.

38803 - MORRIS, S.C., GRAHAM, D., LEE, T.H. : Phytochrome control of chlorophyll synthesis in potato tubers. - Plant Sci. Lett. *17* : 13 - 19, 1979.

38804 - MORRISON, S.L., HUFFAKER, R.C., LOOMIS, R.S. : Photosynthesis and nitrate reduction in leaf slices of barley. - Plant Physiol. *63* (Suppl.) : 46, 1979.

38805 - MORRISON BAIRD, L.A., LEOPOLD, A.C., BRAMLAGE, W.J., WEBSTER, B.D. : Ultrastructural modifications associated with imbibition of the soybean radicle. - Bot. Gaz. *140* : 371 - 377, 1979. [Protoplasts.]

38806 - MOSKALENKO, A.A., LADYGIN, V.G. : Belkovyĭ sostav membran khloroplastov mutantov *Chlamydomonas reinhardti* s neaktivnymi fotosistemami 1 ili 2. [Protein composition of chloroplast membranes of mutants of *Chlamydomonas reinhardti* with inactive photosystem 1 or 2.] - Dokl. Akad. Nauk SSSR *249* : 1017 - 1019, 1979. [In R.]

✲38807 - MOTODA, S., KAWAMURA, T., NISHIZAWA, S. : Further report on the biological structure of the sea at long. 142° E in the North Pacific with particular reference to the interrelation between living and non-living organic matter. - In : SUGAHARA, K. (ed.) : The Kuroshio II Proceedings of the Second Symposium on the Results of the Cooperative Study of the Kuroshio and Adjacent Regions. Pp. 185 - 192. Saikon Publ., Tokyo 1972. [Chl.]

✲38808 - MOTTLEY, J. : Studies on the modes of action of *n*-alkylguanidines and triorganotins on photosynthetic energy conservation in the pea and the unicellular alga *Chlamydomonas reinhardi* DANGEARD. - Pestic. Biochem. Physiol. *9* : 340 - 350, 1978.

38809 - MOTTO, M., SORESSI, G.P., SALAMINI, F. : Growth analysis in a reduced leaf mutant of common bean (*Phaseolus vulgaris* L.). - Euphytica *28* : 593 - 600, 1979.

38810 - MOUDRIANAKIS, E.N., TIEFERT, M.A. : Stability of bound ADP functioning as a phosphoryl donor in ATP synthesis by chloroplasts. - J. biol. Chem. *254* : 9509 - 9517, 1979.

38811 - MOURA, I., MOURA, J.J.G., SANTOS, M.H., XAVIER, A.V., LE GALL, J. : Redox
studies on rubredoxins from sulphate and sulphur reducing bacteria. - FEBS
Lett. *107* : 419 - 421, 1979.

38812 - MOURIOUX, G., DOUCE, R. : Transport du sulfate à travers la double membrane
limitante, ou enveloppe, des chloroplastes d'épinard. - Biochimie *61* :
1283 - 1292, 1979.

38813 - MOUSSEAU, M. : Phytochrome involvement in CO_2 exchange during growth and
development of a quantitative short day plant, *Chenopodium polyspermum* in
different photoperiod. - In : MARCELLE, R., CLIJSTERS, H., VAN POUCKE, M.
(ed.) : Photosynthesis and Plant Development. Pp. 83 - 94. Dr. W. Junk bv.
Publ., The Hague - Boston - London 1979.

38814 - MOUTONNET, P., COUCHAT, P. : Mesure journalière sur cycle végétatif complet
des échanges gazeux : photosynthèse, transpiration et respiration nocturne
d'une culture de Maïs conduite sur colonnes de sol. - Physiol. Plant. *47* :
39 - 43, 1979.

38815 - MOWERY, P.C., LOZIER, R.H., CHAE, Q., TSENG, Y.-W., TAYLOR, M., STOECKE-
NIUS, W. : Effect of acid pH on the absorption spectra and photoreactions
of bacteriorhodopsin. - Biochemistry *18* : 4100 - 4107, 1979.

38816 - MUALLEM, A., MALKIN, S. : Anomalous oxygen uptake from isolated chloroplasts
inhibited in Photosystem II and without external electron donors. - Biochim. biophys. Acta *546* : 175 - 182, 1979.

38817 - MUCCIO, D.D., CASSIM, J.Y. : Interpretation of the absorption and circular
dichroic spectra of oriented purple membrane films. - Biophys. J. *26* :
427 - 440, 1979. [Bacteriorhodopsin.]

38818 - MUKAI, H., ALOI, K., KOIKE, I., IIZUMI, H., OHTSU, M., HATTORI, A. :
Growth and organic production of eelgrass (*Zostera marina* L.) in temperate
waters of the Pacific coast of Japan. I. Growth analysis in spring-summer.
- Aquat. Bot. *7* : 47 - 56, 1979.

38819 - MUKHERJI, S., BISWAS, A.K. : Modulation of chlorophyll, carotene and xantho-
phyll formation by penicillin, benzyladenine and embryonic axis in mung
bean (*Phaseolus aureus* L.) cotyledons. - Ann. Bot. *43* : 225 - 229, 1979.

38820 - MUKOHATA, Y. : [Photophosphorylation and biomembrane.] - In : SHIBATA, K.,
UEMONSA, S., HARA, T. (ed.) : Koseibutsugaku. Vol. 1. Pp. 143 - 153. Bus.
Center Acad. Soc. Japan, Tokyo 1979. [In Jap.]

38821 - MUKOHATA, Y. : [Membrane-binding pigments of halophilic bacteria.] - In :
MASUI, M., ONISHI, H., UNEMOTO, T. (ed.) : Koen Biseibutsu. Pp. 176 - 185.
Ishiyaku, Tokyo 1979. [In Jap.]

38822 - MULDOON, D.K., PEARSON, C.J. : Primary growth and re-growth of the tropical
tallgrass hybrid *Pennisetum* at different temperatures. - Ann. Bot. *43* :
709 - 717, 1979. [Dry-matter and leaf area accumulation.]

38823 - MÜLLER, F., SCHARF, H. : Der Einfluss der Vorsaatjarowisation auf den Troc-
kenmassegehalt und auf den Gehalt an gebundenen Wasser während des Winters
bei Gerste (*Hordeum vulgare* L.). - Arch. Züchtungsforsch. *9* : 101 - 108,
1979. [Dry-matter accumulation.]

38824 - MÜLLER, H.W., BALTSCHEFFSKY, M. : On the oligomycin-sensitivity and subunit composition of the ATPase complex from *Rhodospirillum rubrum.*-Z. Naturforsch. *34 C* : 229 - 232, 1979.

38825 - MÜLLER, H.W., SCHMITT, M., SCHNEIDER, E., DOSE, K. : Immunological and reconstitution studies on the adenosine triphosphatase complex from *Rhodospirillum rubrum.* - Biochim. biophys. Acta *545* : 77 - 85, 1979.

38826 - MÜLLER, P.J., SUESS, E. : Productivity, sedimentation rate, and sedimentary
organic matter in the oceans - I. Organic carbon preservation. - Deep-Sea
Res. *26 A* : 1347 - 1362, 1979. [Ps.]

38827 - MULLER, R.N. : Biomass accumulation and reproduction in *Erythronium albi-
dum.* - Bull. Torrey bot. Club *106* : 276 - 283, 1979.

38828 - MULLER, R.N., MILLER, J.E., SPRUGEL, D.G. : Photosynthetic response of
         field-grown soybeans to fumigations with sulphur dioxide. - J. appl. Ecol.
         16 : 567 - 576, 1979.

38829 - MÜLLER, W. : Zum „Obstgarten-Klima" (Strahlung und Temperatur) in Öster-
         reich. - Pflanzenschutz-Berichte 45 : 97 - 127, 1979. [Leaf energy balance.]

38830 - MÜLLER, W. : Oberflächentemperaturmessungen von Vegetationsoberflächen
         (im Rahmen des Integrierten Pflanzenschutzes). - Pflanzenschutz-Berichte
         45 : 129 - 143, 1979.

B38831 - MULLER, W.H. : Botany : A Functional Approach. Fourth Edition. - Macmillan
         Publ. Co., Inc., New York, Collier Macmillan Publ., London 1979. [Ps, Chl,
         Car.]

38832 - MULLET, J.E., ARNTZEN, C.J. : Simulation of grana stacking in a model mem-
         brane system. - Plant Physiol. 63 (Suppl.) : 53, 1979.

38833 - MULLET, J.E., BURKE, J., DITTO, C.L., WATSON, J.L., ARNTZEN, C.J. : Integral
         protein complexes of chloroplast thylakoids. - Plant Physiol. 63 (Suppl.) :
         29, 1979.

38834 - MULLIGAN, D.R., PATRICK, J.W. : Gibberellic-acid-promoted transport of as-
         similates in stems of Phaseolus vulgaris L. Localized versus remote site(s)
         of action. - Planta 145 : 233 - 238, 1979.

38835 - MULLIGAN, M., TOLBERT, N.E. : Localization and some properties of 3-phospho-
         glycerate phosphatase in spinach chloroplasts. - Plant Physiol. 63 (Suppl.)
         : 64, 1979.

*38836 - MUNZ, A., WITSCH, H. von : Artspezifische Wachstumsreaktionen einiger
         Grünalgen auf einen kombinierten DDT-Starklicht-Salzstress. - Ber. deut.
         bot. Ges. 91 : 665 - 674, 1978. [Chl.]

38837 - MURAKAMI, S. : [Chemical composition and molecular organization of chloro-
         plast thylakoids.] - Tanpakushitsu Kakusan Koso, Bessatsu 21 : 13 - 27, 1979.
         [In Jap.]

38838 - MURATA, N., KUME, N., OKADA, Y., HORI, T. : Preparation of girdle lamella-
         -containing chloroplasts from the diatom Phaeodactylum tricornutum. -
         Plant Cell Physiol. 20 : 1047 - 1053, 1979.

38839 - MUREĬ, I.A., SHUL'GIN, I.A. : Ob adaptatsiyakh i arkhitektonike fiziolo-
         gicheskikh protsessov v tselom rastenii. [Adaptations and architectonics
         of physiological processes in an entire plant.] - Biol. Nauki 1979 (12) :
         5 - 21, 1979. [In R.]

38840 - MURPHY, D.J., STUMPF, P.K. : Light-dependent induction of polyunsaturated
         fatty acid biosynthesis in greening cucumber cotyledons. - Plant Physiol.
         63 : 328 - 335, 1979.

38841 - MUSTÁRDY, L.A, JÁNOSSY, A.G.S. : Evidence of helical thylakoid arrangement
         by scanning electron microscopy. - Plant Sci. Lett. 16 : 281 - 284, 1979.

38842 - MUZTAR, A.J., SLINGER, S.J., BURTON, J.H. : Chemical composition of aquatic
         macrophytes. IV. Carotenoids, soluble sugars and starch in relation to their
         pigmenting, and ensiling potential. - Can. J. Plant Sci. 59 : 1093 - 1098,
         1979.

38843 - MVÉ AKAMBA, L., SIEGENTHALER, P.-A. : Effect of linolenate on photosynthe-
         sis by intact spinach chloroplasts. Conditions for the inhibition of ortho-
         phosphate uptake. - FEBS Lett. 99 : 6 - 10, 1979.

38844 - MVÉ AKAMBA, L., SIEGENTHALER, P.-A. : Effect of linolenate on photosynthe-
         sis by intact spinach chloroplasts II. Influence of preillumination of lea-
         ves on the inhibition of photosynthesis by linolenate. - Plant Cell Physiol.
         20 : 405 - 411, 1979.

38845 - MVÉ AKAMBA, L., SIEGENTHALER, P.A. : Effet de l'acide linolénique sur la
         photosynthèse de chloroplastes intacts de feuilles d'Épinard. IV. Modifica-
         tion de la distribution de l'incorporation de ^{14}C dans les esters phospha-
         tes du cycle de réduction du CO_2. - Physiol. vég. 17 : 587 - 595, 1979.

*38846 - NADAKAVUKAREN, M.J., McCRACKEN, D.A., BERTAGNOLLI, B.L. : Scanning electron
          microscopy of isolated chloroplasts. - J. submicrosc. Cytol. *9* : 247 - 250,
          1977.

38847 - NAGARAJAH, S. : Differences in cuticular resistance in relation to transpi-
          ration in tea (*Camellia sinensis*). - Physiol. Plant. *46* : 89 - 92, 1979.
          [Stomatal resistance.]

38848 - NAGARAJAH, S. : The effect of potassium deficiency on stomatal and cuticu-
          lar resistance in tea (*Camellia sinensis*). - Physiol. Plant. *47* : 91 - 94,
          1979.

38849 - NAGLE, B.J., VILLALON, B., BURNS, E.E. : Computer malfunction causes exten-
          sive errata in *Capsicum* data. - J. Food Sci. *44* : 1792 - 1793, 1979. [Car.]

38850 - NAGY-TÓTH, F., SORAN, V. : Data concerning the absorption spectrum of
          Scenedesmus acutiformis intact cells. - Rev. roum. Biol., Sér. Biol. vég.
          24 : 33 - 37, 1979.

38851 - NAĬDENOVA, Ts. : Vliyanie na kratkotraĭnoto zasushavane v"rkhu intenzivnost-
          ta na fotosintezata na letniya d"b. [Effect of brief drought on photosynthe-
          tic rate of *Quercus pedunculata* var. *tardiflora*.] - Gorskostop. Nauka *16*
          (3) : 3 - 12, 1979. [In Bolg., ab : E, R.]

38852 - NAIK, G.R., JOSHI, G.V. : Photosynthetic carbon fixation in iron-chlorotic
          and recovered green sugarcane leaves. - Plant Soil *53* : 505 - 511, 1979.

38853 - NAITO, K., IIDA, A., SUZUKI, H., TSUJI, H. : The effect of benzyladenine on
          changes in nuclease and protease activities in intact bean leaves during
          ageing. - Physiol. Plant. *46* : 50 - 53, 1979. [Chl.]

38854 - NAITO, K., TSUJI, H., HATAKEYAMA, I., UEDA, K. : Benzyladenine-induced in-
          crease in DNA content per cell, chloroplast size, and chloroplast number
          per cell in intact bean leaves. - J. exp. Bot. *30* : 1145 - 1151, 1979.

38855 - NAKAHARA, H., FUKUDA, K., INOKUCHI, H. : Absorption spectra and photoemis-
          sion of chlorophyll-*b* multilayers. - Chem. Lett. *1979* : 453 - 456, 1979.

38856 - NAKAMURA, H. : [Investigation on injury to rice plants from photochemical
          oxidants.] - Bull. nat. Inst. agr. Sci., Ser. D [Nogyo Gijutsu Kenkyusho
          Hokoku] *30* : 1 - 58, 1979. [In Jap., ab : E.]

38857 - NAKAMURA, Y., YAMADA, M. : The light-dependent step of *de novo* synthesis
          of long chain fatty acids in spinach chloroplasts. - Plant Sci. Lett. *14* :
          291 - 295, 1979.

38858 - NAKANO, Y., KAMIGATAGUCHI, Y. : [A simulation model to estimate evapotrans-
          piration from a soybean canopy.] - Sci. Bull. Fac. Agr. Kyushu Univ. *33* :
          197 - 207, 1979. [Resistances; in Jap., ab : E.]

38859 - NAKASEKO, K., GOTOH, K., ASANUMA, K. : [Comparative studies on dry matter
          production, plant type and productivity in soybean, Azuki bean and Kidney
          bean. I. Differences in dry matter accumulation patterns under the low po-
          pulation density.]- Jap. J. Crop Sci. *48* : 82 - 91, 1979. [In Jap., ab : E.]

38860 - NAKASEKO, K., GOTOH, K., ASANUMA, K. :[Comparative studies on dry matter
          production, plant type and productivity in soybean, Azuki bean and Kidney
          bean. II. Relationships between vertical distribution of leaf area and some
          morphological characteristics.]- Jap. J. Crop Sci. *48* : 92 - 98, 1979.
          [In Jap., ab : E.]

38861 - NAKATANI, H.Y., BARBER, J., MINSKI, M.J. : The influence of the thylakoid
          membrane surface properties on the distribution of ions in chloroplasts. -
          Biochim. biophys. Acta *545* : 24 - 35, 1979.

38862 - NAKAYAMA, K., YAMAOKA, T., KATOH, S. : Chromatographic separation of photo-
          systems I and II from the thylakoid membrane isolated from a thermophilic
          blue-green alga. - Plant Cell Physiol. *20* : 1565 - 1576, 1979.

38863 - NAKOS, G. : Lime-induced chlorosis in *Pinus radiata*. - Plant Soil *52* :
          527 - 536, 1979.

38864 - NALIN, C.M., CROSS, R.L., LUCAS, J.J., KOHLBRENNER, W.E. : Lack of evidence
for covalently-bound carbohydrates in energy-transducing ATPases from mitochondria, bacteria, and chloroplasts. - FEBS Lett. *104* : 209 - 214, 1979.

38865 - NANDA, H.P., KAR, R.K., KABI, T. : Symptomological changes through seasonal
cycle in mosaic virus infected papaya. - Geobios *6* : 235 - 237, 1979.
[Ps, Chl.]

38866 - NASH, T.H. III., SIGAL, L.L. : Gross photosynthetic response of lichens to
short-term ozone fumigations. - Bryologist *82* : 280 - 285, 1979.

*38867 - NASKIDASHVILI, I.K., NASKIDASHVILI, A.V. : Izuchenie vnekornevoĭ podkormki
mochevinoĭ sovmestno s yadokhimikatami. [Study of extra-root supply of urea
together with toxic substances.] - Nauch. Tr. voronezh. sel'skokhoz. Inst.
93 (Issledovaniya po Biologii, Agrotekhnike i Selektsii Plodovykh Kul'tur)
: 39 - 46, 1978. [Chl; in R.]

38868 - NASYROV, Yu.S. : Fiziologo-geneticheskie osnovy povysheniya urozhaĭnosti
sel'skokhozyaĭstvennykh kul'tur. [Physiological and genetical bases for
increasing crop productivity.] - Sel'skokhoz. Biol. *14* : 762 - 766, 1979.
[In R, ab : E.]

38869 - NAUMANN, W.-D., PLANCHER, B. : Untersuchungen zur klimatisierenden Bereg-
nung von Obstgehölzen. II. Nettoassimilation bei Apfelsämlingen als Modell-
pflanzen und bei Jungbäumen von "Golden Delicious". - Gartenbauwissenschaft *44* : 22 - 26, 1979.

38870 - NAVARA, J. : Die Veränderungen des Wassergehaltes in den Blättern der Aprikose während der Vegetationsperiode. - Biológia (Bratislava) *34* : 273 -
282, 1979. [Dry-matter accumulation.]

38871 - NAVARRO, S., GARCIA, A.L., GALINDO, L. : Clorofilasa en hojas de citrus.
I. Extracción, medida y factores que influyen en la actividad del enzima.
[Chlorophyllase in *Citrus* tree leaves. I. Extraction measurement and factors that influence its activity.] - An. Univ. Murcia, Ciencias *33* : 199 -
211, 1979. [In Span., ab : E.]

38872 - NAYAK, S.K., MURTY, K.S. : Effect of low light intensity on chlorophyll
content and RuDP carboxylase activity in rice. - Plant Biochem. J. *6* :
102 - 106, 1979.

38873 - NAYAK, S.K., MURTY, P.S.S., MURTY, K.S. : Photosynthesis and translocation
in rice during ripening as influenced by light intensities. - J. nucl. agr.
Biol. *8* : 23 - 25, 1979.

38874 - NAYLOR, A.W., GILES, L.J., MULBRY, W.W. III. : Long term effects of intermittent and flashing light on growth characteristics, chloroplast biochemistry and ultrastructure of plants. - Plant Physiol. *63* (Suppl.) : 98,
1979.

38875 - NAZAROV, S.K. : Fotosintez i novoobrazovanie listovogo apparata u rasteniĭ
Arktiki. [Photosynthesis and new formation of foliage in arctic plants.] -
Ėkologiya *1978* (6) : 76 - 79, 1978. [In R, ab : E.]

38876 - NAZAROVA,I.G., EVSTIGNEEV, V.B. : Obratimoe fotookislenie vodorastvorimykh
analogov khlorofilla : obrazovanie ėlektrodno-aktivnoĭ okislennoĭ formy.
[The reversible photooxidation of soluble chlorophyll analogues : formation
of an electrode-active oxidized compound.] - Biofizika *24* : 771, 1979.
[In R.]

*38877 - NECHAEVA, E.P.: Vliyanie sinego i krasnogo sveta na formirovanie fotosin-
teticheskogo apparata v prorostkakh yachmenya. [Effect of blue and red
radiation on the formation of photosynthetic apparatus in barley seedlings.]
- In : Vliyanie Fiziko-khimicheskikh Faktorov Sredy na Rasteniya. Pp. 94 -
102. Perm. gos. Univ., Perm' 1978. [In R.]

*38878 - NEDRANKO, L.V. : Vliyanie raznogo urovnya azota i fosfora v pitatel'noĭ
srede na sostoyanie khlorofill-belkovo-lipoidnogo kompleksa khloroplastov
tomatnykh rasteniĭ. [Effect of various levels of nitrogen and phosphorus
in the nutrient medium on the state of the chlorophyll-protein-lipoid complex in chloroplasts of tomato plants.] - Tr. kishinev. s.-kh. Inst. *114* :
23 - 26, 1974. [In R.]

*38879 - **NEDRANKO, L.V.** : O pigmentakh zelenogo lista tomatnykh rasteniĭ v svyazi s usloviyami mineral'nogo pitaniya. [Pigments of green leaves of tomato plants in relation to conditions of mineral nutrition.] - Tr. kishinev. s.-kh. Inst. *114* : 17 - 23, 1974. [In R.]

38880 - **NEGISI, K.** : Bark respiration rate in stem segments detached from young *Pinus densiflora* trees in relation to velocity of artificial sap flow. - J. jap. Forest. Soc. *61* : 88 - 93, 1979. [Photosynthates.]

*38881 - **NEIL, G., NICHOLAS, D.J.D., BOCKRIS, J.O'M., McCANN, J.F.** : The photosynthetic production of hydrogen. - Int. J. Hydrogen Energy *1* : 45 - 58, 1976.

38882 - **NELSON, N., EYTAN, E.** : Approach to the membrane sector of the chloroplast coupling device. - In : **MUKOHATA, Y., PACKER, L.** (ed.) : Cation Flux across Biomembranes. Pp. 409 - 415. Academic Press, New York - San Francisco - London 1979.

38883 - **NES, A.** : The effect of plant spacing on yield components of black currants. - Acta Agr. scand. *29* : 263 - 272, 1979.

38884 - **NESSLER, C.L., MAHLBERG, P.G.** : Plastids in laticifers of *Papaver*. I. Development and cytochemistry of laticifer plastids in *P. somniferum* L. (*Papaveraceae*). - Amer. J. Bot. *66* : 266 - 273, 1979.

38885 - **NESSLER, C.L., MAHLBERG, P.G.** : Plastids in laticifers of *Papaver* II. Enzyme cytochemistry of membrane-bound inclusions of laticifer plastids in *P. bracteatum* LINDL. (*Papaveraceae*). - Amer. J. Bot. *66* : 274 - 279, 1979.

38886 - **NESTSYAROVICH, M.D., NOVIKAVA, A.A., SHAVEL', S.Kh.** : Uplyŭ roznaĭ intensiŭnastsi asvyatlennya na rost i fotasintez seyantsaŭ nekatorykh drevavykh raslin. [Effect of light intensity on growth and photosynthesis of some woody plants seedlings.] - Vestsi Akad. Navuk belarus. SSR, Ser. biyal. Navuk *5* : 5 - 9, 1979. [In Belorus., ab : E.]

38887 - **NETA, P., SCHERZ, A., LEVANON, H.** : Electron transfer reactions involving porphyrins and chlorophyll *a*. - J. amer. chem. Soc. *101* : 3624 - 3629, 1979.

38888 - **NEUMANN, J., DRECHSLER, Z., SEARLE, G.F.W., BARBER, J.** : 2-p-toluidinonaphthalene-6-sulphonate (TNS). An energy-transfer inhibitor in chloroplasts. - FEBS Lett. *102* : 121 - 125, 1979.

38889 - **NG, B.H., ANDERSON, J.W.** : Light-dependent incorporation of selenite and sulphite into selenocysteine and cysteine by isolated pea chloroplasts. - Phytochemistry *18* : 573 - 580, 1979.

*B38890 - **NICHIPOROVICH, A.A.** (ed.) : Photosynthesis of Productive Systems. - Israel Program sci. Translations, Jerusalem 1967.

38891 - **NICHIPOROVICH, A.A.** : Potentsial'naya produktivnost' rasteniĭ i printsipy optimal'nogo ee ispol'zovaniya. [Potential plant productivity and the principles of its optimal use.] - Sel'skokhoz. Biol. *6* : 683 - 694, 1979. [In R.]

B38892 - **NICHIPOROVICH, A.A.** : Énergeticheskaya Éffektivnost' Fotosinteza i Produktivnost' Rasteniĭ. [Energetic Efficiency of Photosynthesis and Plant Productivity.] - Akad. Nauk SSSR, Nauch. Tsentr Biol. Issledovaniĭ, Pushchino 1979. [In R.]

38893 - **NICHOLS, R., HO, L.C.** : Respiration, carbon balance and translocation of dry matter in the corolla of rose flowers. - Ann. Bot. *44* : 19 - 25, 1979. [Ps.]

38894 - **NIEHAUS, F.** : Carbon dioxide as a constraint for global energy scenarios. - In : **BACH, W., PANKRATH, J., KELLOG, W.** (ed.) : Man's Impact on Climate. Pp. 285 - 297. Elsevier Sci. Publ. Comp., Amsterdam - Oxford - New York 1979.

38895 - **NIELSEN, A.M., SOJKA, G.A.** : Photoheterotrophic utilization of acetate by the wild type and an acetate-adapted mutant of *Rhodopseudomonas capsulata*. - Arch. Microbiol. *120* : 39 - 42, 1979. [Chl.]

38896 - NIELSEN , N.C., SMILLIE, R.M., HENNINGSEN, K.W., WETTSTEIN, D.von, FRENCH, C.S. : Composition and function of thylakoid membranes from grana-rich and grana-deficient chloroplast mutants of barley. - Plant Physiol. 63 : 174 - 182, 1979.

38897 - NIIMI, Y., TORIKATA, H. : Changes in photosynthesis and respiration during berry development in relation to the ripening of Delaware grapes. - J. jap. Soc. hort. Sci. 47 : 448 - 453, 1979.

38898 - NIKALAEVA, G.M., KUPERMAN, N.I. : Razmerkavanne svezhaŭtvoranykh malekul galaktalipidaŭ u fraktsyyakh, vyluchanykh z membran. [Distribution of fresh-ly formed galactolipid molecules in fractions isolated from membranes.] - Vestsi Akad. Navuk belarus. SSR, Ser. biyal. Navuk 1979 (2) : 28 - 31, 138, 1979. [In Belorus., ab : E, R.]

38899 - NIKITINA, K.A., YUDINA, T.G., GUSEV, M.V. : Morfologicheskaya geterogennost' pri razlichnykh usloviyakh zhizni i destruktsii tsianobakterii Anabaena variabilis. [Morphological heterogeneity of the cyanobacterium Anabaena variabilis under different conditions of its growth and destruction.] - Mikrobiologiya 48 : 873 - 879, 1979. [Chl, Car, Bil; in R, ab : E.]

38900 - NIKOLAEVA, L.F., FLOROVA, N.B., PORSHNEVA, E.B. : Spektral'nye formy khlo-rofilla, sinteziruemogo v otsutstvie sveta v prorostkakh reliktovykh khvoĭ-nykh. [Spectral forms of chlorophyll synthesized in the absence of light by seedlings of relict coniferous plants.] - Zh. obshch. Biol. 40 : 128 - 137, 1979. [In R, ab : E.]

38901 - NIKOLAEVA, M.K., OSIPOVA, O.P. : Funktsional'naya aktivnost' khloroplastov bobov, vyrashchennykh pri raznykh intensivnostyakh sveta. [Functional acti-vity of chloroplasts of Vicia faba plants grown under various light intensi-sities.] - Fiziol. Rast. 26 : 799 - 807, 1979. [In R, ab : E.]

38902 - NILSEN, K.N. : Enhanced stomatal conductance, water consumption, and growth of tomato (Lycopersicon esculentum MILL.) in response to far-red irradiance supplementation within controlled environments. - Plant Physiol. 63 (Suppl.) : 126, 1979.

38903 - NILSEN, S., HAUGSTAD, M., HOLMEN, A.T. : The effect of photosynthetic enhan-cement on photorespiration in Sinapis alba. - Physiol. Plant. 47 : 19 - 24, 1979.

38904 - NINET, L., RENAUT, J. : Carotenoids. - In : PEPPLER, H.J., PERLMAN, D. (ed.) : Microbial Technology. (2nd Ed.) Vol. 1. Pp. 529 - 544. Academic Press, New York - San Francisco - London 1979. [Industrial fermentation production.]

38905 - NISHI, N., KATAOKA, M., SOE, G., KAKUNO, T., UEKI, T., YAMASHITA, J., HORIO, T. : Disintegration of Rhodospirillum rubrum chromatophore membrane into photoreaction units, reaction centers, and ubiquinone-10 protein with mixture of cholate and deoxycholate. - J. Biochem. (Tokyo) 86 : 1211 - 1224, 1979.

38906 - NISHI, N., YAMASHITA, J. : [Bacterial photosynthesis.] - Tanpakushitsu Ka-kusan Koso, Bessatsu 21 : 225 - 242, 1979. [In Jap.]

38907 - NISHIDA, K. : Diurnal fluctuation of light and dark CO_2 fixation in peeled and unpeeled leaves of Bryophyllum daigremontiana. - Plant Cell Physiol. 20 : 259 - 261, 1979.

38908 - NISHIDA, K., HAYASHI, Y. : Deacidification of the leaves of Bryophyllum Calycinum under anaerobiosis : abnormal efflux of malate into the cyto-plasm. - Plant Cell Physiol. 20 : 1209 - 1215, 1979.

*38909 - NISHIMURA, M., AKAZAWA, T. : Reconstitution of spinach ribulose-1,5-diphos-phate carboxylase from separated subunits. - Biochem. biophys. Res. Commun. 59 : 584 - 590, 1974.

38910 - NISHIMURA , M., HIRAOKA, T. : Ultrastructural cytochemistry of chloroplast lamellae - DAB oxidation reaction. - Acta histochem. cytochem. 12 : 635, 1979.

38911 - NISHIZAWA, A.N., WOLOSIUK, R.A., BUCHANAN, B.B. : Chloroplast phenylalanine ammonia-lyase from spinach leaves. Evidence for light-mediated regulation *via* the ferredoxin/thioredoxin system. - Planta *145* : 7 - 12, 1979.

38912 - NITTA, T. : Requirement of alkaline hydrolysis for estimating chloroplast RNA using the orcinol reaction. - Bot. Mag. (Tokyo) *92* : 145 - 149, 1979.

38913 - NOACK, K., THOMSON, A.J. : Conformation and optical activity of all-*trans*, mono-*cis*, and di-*cis* carotenoids : temperature dependent circular dichroism. - Helv. chim. Acta *62* : 1902 - 1921, 1979.

*B38914 - NOBEL, P.S. : Introduction to Biophysical Plant Physiology. - W.H. Freeman and Comp., San Francisco 1974. [Ps.]

38915 - NOBEL, P.S., HARTSOCK, T.L. : Environmental influences on open stomates of a crassulacean acid metabolism plant, *Agave deserti*.- Plant Physiol. *63* : 63 - 66, 1979. [Resistances.]

38916 - NOBLE, R.D., JENSEN, K.F. : Effects of SO_2 and $SO_2 + O_3$ on photosynthesis. - Plant Physiol. *63* (Suppl.) : 151, 1979.

38917 - NODA, H., AMANO, H., OHTA, F., HORIGUCHI, H. : [Changes in bacterial flora and in chemical constituents of red rot infected-laver, *Porphyra yezoensis*, during treatment with histidine.] - Bull. jap. Soc. sci. Fish. *45* : 1163 - 1168, 1979. [Chl, Bil; in Jap., ab : E.]

38918 - NOKS, P.P., KONONENKO, A.A., RUBIN, A.B. : Funktsional'naya aktivnost' fotosinteticheskikh reaktsionnykh tsentrov iz *Rhodopseudomonas sphaeroides* pri fiksirovannoǐ gidratatsii preparatov.[Function activity in photosynthetic reaction centres from *Rhodopseudomonas sphaeroides* at fixed hydration levels of the preparation.] - Bioorg. Khim. *5* : 879 - 885, 1979. [In R, ab : E.]

38919 - NOLAN, W.G., LAZENBY, J.M. : Temperature-induced changes in membranes of isolated chloroplasts. - Plant Physiol. *63* (Suppl.) : 54, 1979.

38920 - NOLAND, T.L., KOZLOWSKI, T.T. : Influence of potassium nutrition on susceptibility of silver maple to ozone. - Can. J. Forest Res. *9* : 501 - 503, 1979. [Resistances.]

38921 - Nomenclature Committee of the International Union of Biochemistry : Nomenclature of iron-sulfur proteins. Recommendations, 1978. - Biochim. biophys. Acta *549* : 101 - 105, 1979.

38922 - NONOMURA, A.M. : Ultrastructural development of the parasitic red alga *Janczewskia morimotoi*. - J. Phycol. *15* (Suppl.) : 14, 1979. [Chloroplast.]

38923 - NORDLUND, S., ERIKSSON, U. : Nitrogenase from *Rhodospirillum rubrum*. Relation between "switch-off" effect and the membrane component. Hydrogen production and acetylene reduction with different nitrogenase component ratios. - Biochim. biophys. Acta *547* : 429 - 437, 1979.

*38924 - NOVAK, V.A., IVANKINA, N.G. : Svetoindutsirovannoe pogloshchenie ionov kletkami presnovodnykh rasteniǐ. [Light-induced influx of ions by the cells of freshwater plants.] - Fiziol. Rast. *25* : 315 - 322, 1975. [Ps; In R, ab : E.]

*38925 - NOVAK, V.A., IVANKINA, N.G. : Svetoindutsirovannaya bioélektricheskaya reaktsiya rasteniǐ. [Light-induced bioelectrical reaction of plants.] - Voprosy Biol. (Tomsk) *1977* : 153 - 163, 1977. [Ps; in R.]

*38926 - NOVAK, V.A., IVANKINA, N.G. : Zavisimost' svetoindutsirovannogo vnutrikletochnogo élektricheskogo potentsiala élodei ot protsessov fotosinteza. [Relation between light-induced intracellular electric potential in *Elodea* and photosynthetic processes.] - Tsitologiya *19* : 508 - 513, 1977. [In R, ab : E.]

38927 - NUGENT, J.H.A., EVANS, M.C.W. : Light-induced EPR signals at cryogenic temperatures in subchloroplast particles enriched in photosystem II. - FEBS Lett. *101* : 101 - 104, 1979.

38928 - NULTSCH, W., SCHUCHART, H., DILLENBURGER, M. : Photomovement of the red
        alga *Porphyridium cruentum* (AG.) *naegeli* I. Photokinesis. - Arch. Microbiol.
        122 : 207 - 212, 1979. [Ps.]

38929 - NULTSCH, W., SCHUCHART, H., HÖHL, M. : Investigations on the phototactic
        orientation of *Anabaena variabilis.* - Arch. Microbiol. *122* : 85 - 91, 1979.
        [Ps, Bil.]

38930 - NUTALL, W.F., ZANDSTRA, H.G., BOWREN, K.E. : Yield and N percentage of spring
        wheat as affected by phosphate fertilizer, moisture use, and available soil P
        and N. - Agron. J. *71* : 385 - 391, 1979.

38931 - OBIEFUNA, J.C., NDUBIZU, T.O.C. : Estimating leaf area of plantain. - Sci.
        Hort. *11* : 31 - 36, 1979.

38932 - OCHIAI, H., SHIBATA, H., FUJISHIMA, A., HONDA, K. : Photocurrent by immobi-
        lized chloroplast film electrode. - Agr. biol. Chem. (Tokyo) *43* : 881 - 883,
        1979.

*38933 - ODUMANOVA-DUNAEVA, G.A. : Vliyanie blizhneĭ infrakrasnoĭ radiatsii na foto-
        sintez rasteniĭ v usloviyakh svetokul'tury. [Effect of near infrared radia-
        tion on plant photosynthesis under light culture conditions.] - Bot. Zh.
        62 : 811 - 819, 1977. [In R, ab : E.]

38934 - OECHEL, W.C., LAWRENCE, W.T. : Energy utilization and carbon metabolism in
        mediterranean scrub vegetation of Chile and California. I. Methods: A trans-
        portable cuvette field photosynthesis and data acquisition system and repre-
        sentative results for *Ceanothus greggii.* - Oecologia *39* : 321 - 335, 1979.

38935 - OECHEL, W.C., MUSTAFA, J. : Energy utilization and carbon metabolism in me-
        diterranean scrub vegetation of Chile and California. II. The relationship
        between photosynthesis and cover in chaparral evergreen shrubs. - Oecologia
        41 : 305 - 315, 1979.

*38936 - OELZE-KAROW, H., MOHR, H. : Die Bedeutung des Phytochroms für die Entwick-
        lung der Kapazität für Photophosphorylierung. - Ber. deut. bot. Ges. *91* :
        605 - 610, 1978.

38937 - OESTERHELT, D. : Proton translocation by bacteriorhodopsin. - In : GERISCHER,
        H., KATZ, J.J. (ed.) : Light-Induced Charge Separation in Biology and Che-
        mistry. Pp. 493 - 501. Verlag Chemie, Weinheim - New York 1979.

38938 - OESTERHELT, D., HARTMANN, R. : The function of the purple membrane in *Halo-
        bacterium halobium.* - In : CARAFOLI, E., SEMENZA, G. (ed.) : Membrane Bio-
        chemistry. A Laboratory Manual on Transport and Bioenergetics. Pp. 154 -
        163. Springer-Verlag, Berlin - Heidelberg - New York 1979.

38939 - OETTMEIER, W. : A radioactive 1,4-benzoquinone as inhibitor of the DBMIB-
        -type in photosynthetic electron transport. - Z. Naturforsch. *34 C*: 242 -
        249, 1979.

38940 - OETTMEIER, W. : Quantitative structure activity relationship of diphenyl-
        amines as inhibitors of photosynthetic electron transport and photophospho-
        rylation. - Z. Naturförsch. *34 C*: 1024 - 1027, 1979.

38941 - OGAWA, M., KONISHI, M. : Kinetics of photoconversion of protochlorophyllide
        649 to chlorophyllide 676 at low temperature in etiolated cotyledons of
        Pharbitis nil. - Biochim. biophys. Acta *548* : 119 - 127, 1979.

38942 - OGAWA, T. : Stomatal responses to light and CO_2 in greening wheat leaves. -
        Plant Cell Physiol. *20* : 445 - 452, 1979. [Ps, Chl.]

38943 - OGAWA, Y., KING, R.W. : Indirect action of benzyladenine and other chemicals
        on flowering of *Pharbitis nil* CHOIS. Action by interference with assimilate
        translocation from induced cotyledons. - Plant Physiol. *63* : 643 - 649,
        1979.

*38944 - OGURA, N. : Dissolved organic matter in the sea, its production, utilization
        and decomposition. - In : SUGAWARA, K. (ed.) : The Kuroshio II. Proceedings

of the Second Symposium on the Results of the Cooperative Study of the Kuroshio and Adjacent Regions. Pp. 201 - 205. Saikon Publ., Tokyo 1972. [Chl.]

38945 - OHAD, I., CLAYTON, R.K., BOGORAD, L. : Photoreversible absorbance changes
in solutions of allophycocyanin purified from *Fremyella diplosiphon:* temperature dependence and quantum efficiency. - Proc. nat. Acad. Sci. USA *76* :
5655 - 5659, 1979.

38946 - OH-HAMA, T., HASE, E. : Photoinhibition of 5-aminolevulinic acid formation
under CO_2-free condition. - Plant Cell Physiol. *20* : 1407 - 1415, 1979.

38947 - OHKI, K., FUJITA, Y. : Photoreversible absorption changes of guanidine-HCl-
-treated phycocyanin and allophycocyanin isolated from the blue-green alga
Tolypothrix tenuis. - Plant Cell Physiol. *20* : 483 - 490, 1979.

38948 - OHKI, K., FUJITA, Y. : *In vivo* transformation of phycobiliproteins during
photobleaching of *Tolypothrix tenuis* to forms active in photoreversible
absorption changes. - Plant Cell Physiol. *20* : 1341 - 1347, 1979.

38949 - OHKI, K., YOUNG, C.T. : Photosynthesis, carbonic anhydrase and amino acids
related to zinc status in cotton. - In : MARCELLE, R., CLIJSTERS, H., VAN
POUCKE, M. (ed.) : Photosynthesis and Plant Development. Pp. 161 - 173.
Dr. W. Junk bv. Publ., The Hague - Boston - London 1979.

38950 - OHKI, Y., MUSASHI, A., TSUBO, Y. : A new type of mutant of *Euglena* which
produces permanently bleached progeny by darkness. - Experientia *35* : 489 -
490, 1979, [Ps, chloroplast.]

38951 - OHLROGGE, J.B., KUHN, D.N., STUMPF, P.K. : Subcellular localization of acyl
carrier protein in leaf protoplasts of *Spinacia oleracea.* - Proc. nat. Acad.
Sci. USA *76* : 1194 - 1198, 1979. [RuBPC.]

38952 - OHMANN, E. : Autotrophic carbon dioxide assimilation in prokaryotic microorganisms. - In : GIBBS, M., LATZKO, E. (ed.) : Photosynthesis II. (Encycl.
Plant Physiol. N.S. Vol. 6.) Pp. 54 - 67. Springer-Verlag, Berlin - Heidelberg - New York 1979.

38953 - OHMANN, E., METZGER, U. : Die Wirkung von Herbiziden auf den photosynthetischen Elektronentransport. - Wiss. Z. Martin-Luther Univ. Halle - Witten-
berg, math.-naturwiss. Reihe *28* (3) : 61 - 74, 1979.

38954 - OHNISHI, O. : Frequency of chlorophyll-deficient and other detrimental genes
in Japanese populations of buckwheat, *Fagopyrum esculentum* MOENCH. - Jap.
J. Genet. *54* : 259 - 270, 1979.

38955 - OKABE, K.-I., CODD, G.A., STEWART, W.D.P. : Hydroxylamine stimulates carboxylase activity and inhibits oxygenase activity of cyanobacterial RuBP
carboxylase/oxygenase. - Nature *279* : 525 - 527, 1979.

38956 - OKADA, M., HORIE, H. : Diurnal rhythm in the Hill reaction in cell-free
extracts of the green alga *Bryopsis maxima.* - Plant Cell Physiol. *20* :
1403 - 1406, 1979.

38957 - OKAMURA, M.Y., ISAACSON, R.A., FEHER, G. : Spectroscopic and kinetic properties of the transient intermediate acceptor in reaction centers of *Rhodopseudomonas sphaeroides.* - Biochim. biophys. Acta *546* : 394 - 417, 1979.

38958 - O'KELLEY, J.C., HARDMAN, J.K. : Flavin compounds as agents for the oxidation of plastocyanin in blue light. - Photochem. Photobiol. *29* : 829 - 832,
1979.

38959 - OKITA, T.W., PREISS, J. : Degradation of spinach leaf starch : Isolation
and characterization of amylases from chloroplasts and whole leaf homogena-
te. - Plant Physiol. *63* (Suppl.) : 66, 1979.

38960 - OKU, T., SUGAHARA, K., TOMITA, G. : Photosynthetic CO_2-fixing activity in
dark-grown spruce seedlings. - Plant Cell Physiol. *20* : 857 - 859, 1979.

38961 - O'LEARY, M.H., OSMOND, C.B. : Diffusional fractionation of carbon isotopes
during photosynthetic carboxylation. - Plant Physiol. *63* (Suppl.) : 38,
1979.

38962 - OLECH, K. : Niektóre aspekty wzrostu liści. [Some aspects of leaf growth.] - Wiadom. bot. *23* : 265 - 270, 1979. [Ps; in Pol.]

38963 - OLIVE, J., WOLLMAN, F.A., BENNOUN, P., RECOUVREUER, M. : Ultrastructure--function relationship in *Chlamydomonas reinhartii* thylakoids, by means of a comparison between the wild type and the F_{34} mutant which lacks the photosystem II reaction center. - Mol. Biol. Rep. *5* : 139 - 143, 1979.

38964 - OLIVER, D.J. : Photorespiratory glycolate and glycine metabolism by isolated soybean cells in the dark. - Plant Physiol. *63* (Suppl.) : 154, 1979.

38965 - OLIVER, D.J. : Mechanism of decarboxylation of glycine and glycolate by isolated soybean cells. - Plant Physiol. *64* : 1048 - 1052, 1979.

38966 - OLIVER, D.J. : The interaction between O_2 and CO_2 concentrations on the regulation of glycolate synthesis in tobacco leaf discs. - Plant Sci. Lett. *15* : 35 - 40, 1979.

38967 - OLIVER, D.J., THORNE, J.H., POINCELOT, R.P. : Rapid isolation of mesophyll cells from soybean leaves. - Plant Sci. Lett. *16* : 149 - 155, 1979. [Ps.]

38968 - OLLERENSHAW, J.H., INCOLL, L.D. : Leaf photosynthesis in pure swards of two grasses (*Lolium perenne* and *Lolium multiflorum*) subjected to contrasting intensities of defoliation. - Ann. appl. Biol. *92* : 133 - 142, 1979.

38969 - OLSEN, L.F., COX, R.P. : Light-induced proton transport by chloroplasts suspended in fluid media at sub-zero temperatures : kinetics and stoichiometry. - Europe. J. Biochem. *95* : 427 - 432, 1979.

38970 - OLSEN, L.F., COX, R.P. : The effect of some inhibitors of photosynthetic electron transport on the kinetics of redox changes of the reaction centre chlorophyll of Photosystem I (P-700) and of cytochrome *f* at sub-zero temperatures. - Europe. J. Biochem. *102* : 139 - 145, 1979.

38971 - OLSEN, L.F., COX, R.P. : The effect of intrathylakoid pH[*] on the rate of chloroplast electron transport reactions at subzero temperatures. - FEBS Lett. *103* : 250 - 252, 1979.

38972 - OLSON, J.M., THORNBER, J.P. : Photosynthetic reaction centers. - In : CAPALDI, R.A. (ed.) : Membrane Proteins in Energy Transduction. Pp. 279 - 340. Marcel Dekker Inc., New York - Basel 1979.

*38973 - OLSSON, I., ÖLUNDH, E. : On plankton production in Kungsbacka Fjord, an estuary on the Swedish west coast. - Mar. Biol. *24* : 17 - 28, 1974.

38974 - O'NEAL, S.W. : An investigation into environmental factors limiting growth in *Caulerpa paspaloides* (*Caulerpaceae*). - J. Phycol. *15* (Suppl.) : 14, 1979. [Ps.]

*38975 - ONG, H.T. : Roles of hormones in the responses of excised tomato cotyledons to mannitol induced water stress. - Biol. Plant. *20* : 318 - 323, 1978. [Chl.]

*38976 - ONG, H.T. : Gel electrophoresis patterns of proteins and peroxidases of excised tomato cotyledons subjected to mannitol induced water stress. - Biol. Plant. *20* : 330 - 334, 1978. [Chl.]

38977 - ONISHCHENKO, L.I. : Vliyanie mikroèlementov na razvitie khlorofillonosnoĭ sistemy list'ev èspartseta. [Effect of trace elements on the development of a chlorophyll-containing system of sainfoil leaves.] - Aktual'. Vopr. sovrem. Bot. (Kiev) *1979* : 117 - 121, 1979. [In R.]

38978 - ONO, T.-A., MURATA, N. : Temperature dependence of the photosynthetic activities in the thylakoid membranes from the blue-green alga *Anacystis nidulans*. - Biochim. biophys. Acta *545* : 69 - 76, 1979.

*38979 - OOHARA, H., YOSHIDA, N., MURAKAWA, E., CHANG, N.K. : The promoting effect and utilization of alcohol on legume and grass forage plants. I. The growth and production of alfalfa and orchard grass. - Res. Bull. Obihiro Univ. *7* : 295 - 310, 1971. [Root/shoot ratio.]

☆38980 - OOHARA, H., YOSHIDA, N., OOHARA, Y., CHANG, N.K. : The promoting effect
and utilization of alcohol on legume and grass forage plants. II. The chemical compositions of alfalfa and orchardgrass. - Res. Bul. Obihiro Univ.
7 : 472 - 487, 1972. [Chl, Car.]

☆38981 - OPANASENKO, V.K., MAL'YAN, A.N., MAKAROV, A.D. : Opredelenie konstant dis-
sotsiatsii i kompleksoobrazovaniya ADF i ATF s ionami nekotorykh metallov.
[Determination of constants of dissociation and complex formation of ADP
and ATP with ions of some metals.] - In : Itogi Issledovaniya Mekhanizma
Fotosinteza. Pp. 104 - 115. Institut Fotosinteza Akad. Nauk SSSR, Pushchino
1974. [In R.]

38982 - OREN, A., PADAN, E., MALKIN, S. : Sulfide inhibition of photosystem II in
cyanobacteria (blue-green algae) and tobacco chloroplasts. - Biochim. biophys. Acta 546 : 270 - 279, 1979.

38983 - OREN, R., GROMET-ELHANAN, Z. : Coupling factor ATPase complex of *Rhodospirillum rubrum*. Purification and characterization of an oligomycin and N,N'-
-dicyclohexylcarbodiimide-sensitive (Ca^{2+} + Mg^{2+})-ATPase. - Biochim. biophys. Acta 548 : 106 - 118, 1979.

38984 - ORITANI, T., ENBUTSU, T., YOSHIDA, R. : [Studies on nitrogen metabolism in
crop plants. XVI. Changes in photosynthesis and nitrogen metabolism in relation to leaf area growth of several rice varieties.] - Jap. J. Crop Sci.
48 : 10 - 16, 1979. [In Jap., ab E.]

38985 - ORON, G., SHELEF, G., LEVI, A. : Growth of *Spirulina maxima* on cow-manure
wastes. - Biotechnol. Bioeng. 21 : 2169 - 2173, 1979. [Chl.]

38986 - ORT, D.R., PARSON, W.W. : The quantum yield of flash-induced proton release
by bacteriorhodopsin containing membrane fragments. - Biophys. J. 25 :
341 - 353, 1979.

38987 - ORT, D.R., PARSON, W.W. : Enthalpy changes during the photochemical cycle
of bacteriorhodopsin. - Biophys. J. 25 : 355 - 364, 1979.

38988 - ORTNER, K.M., BINDER, J. : Modellierung und Prognose der Ertragsentwicklung
in Abhängigkeit vom Witterungsverlauf. - Monatsber. österr. Landwirtsch.
1979 (2) : 104 - 113, 1979.

38989 - ORTOIDZE, T.V., BORISEVITCH, G.P., VENEDIKTOV, P.S., KONONENKO, A.A.,
MATORIN, D.N., RUBIN, A.B. : Electric field stimulation of delayed fluorescence in dry films of chloroplasts and subchloroplast particles enriched in PSII. - Biochem. Physiol. Pflanzen 174 : 85 - 91, 1979.

38990 - ORWICK, P.L., SCHREIBER, M.M. : Interference of redroot pigweed (*Amaranthus
retroflexus*) and robust foxtail (*Setaria viridis* var. *robusta-alba* or var.
robusta-purpurea) in soybeans (*Glycine max*). - Weed Sci. 27 : 665 - 674,
1979. [Leaf-area distribution.]

38991 - OSAFUNE, T., SCHIFF, J.A. : Light-induced changes in a proplastid remnant
in dark grown resting *Euglena gracilis* var. *bacillaris* W_3BUL. - Plant Physiol. 63 (Suppl.) : 27, 1979.

38992 - OSAFUNE, T., SCHIFF, J.A. : Stigma and flagellar swelling in relation to
light and carotenoids in *Euglena gracilis* var. *bacillaris*. - Plant Physiol.
63 (Suppl.) : 28, 1979.

☆38993 - OSAKI, M., TANAKA, A. : [The ^{14}C-retention percentage in the rice plant.] -
Nippon Dojo Hiryogaku Zasshi 49 : 217 - 220, 1978. [In Jap.]

38994 - OSAKOVSKIĬ, V.L., ALEKSEEV, V.G. : Funktsional'naya aktivnost' khloroplastov pshenitsy v usloviyakh Severa. [Functional activity of wheat chloroplasts under conditions of the North.] - Izv. sib. Otd. Akad. Nauk SSSR,
Ser. biol. Nauk 1979 (2) : 112 - 117, 1979. [In R, ab : E.]

38995 - OSHIMA, R.J., BRAEGELMANN, P.K., FLAGLER, R.B., TESO, R.R. : The effects of
ozone on the growth, yield, and partitioning of dry matter in cotton. -
J. Environ. Qual. 8 : 474 - 479, 1979. [Growth analysis.]

38996 - OSMOND, C.B., LUDLOW, M.M., DAWIS,R., COWAN, I.R., POWLES,S.B.,WINTER, K. :
Stomatal responses to humidity in *Opuntia inermis* in relation to control
of CO_2 and H_2O exchange patterns. - Oecologia *41* : 65 - 76, 1979.

38997 - OSMOND, C.B., NOTT, D.L., FIRTH, P.M. : Carbon assimilation patterns and
growth of the introduced CAM plant *Opuntia inermis* in Eastern Australia. -
Oecologia *40* : 331 - 350, 1979.

*38998 - ÖSTRÖM, B. : Solubility of CO_2, total CO_2 content and primary production in
sea water. Relationships and formulae for calculations. - Bot. mar. *20* :
69 - 74, 1977.

38999 - OSTROVSKAYA, L.K. : Ul'trastuktura khloroplastov i kationnyĭ balans sredy.
[Chloroplast ultrastructure and cation balance of the medium.]- Fiziol.
Biokhim. kul't. Rast. *11* : 407 - 417, 1979. [In R, ab : E.]

39000 - OSTROVSKAYA, L.K. : Morfologicheskaya i funktsional'naya neodnorodnost'
membran khloroplastov. [Morphological and functional heterogeneity of chloroplast membranes.] - Usp. sovrem. Biol. *87* (1) : 93 - 107, 1979. [In R.]

39001 - OSTROVSKAYA, L.K., GAMAYUNOVA, M.S., SILAEVA, A.M., GRIGORA, M.Yu., MANUIL'-
SKAYA, S.V. : Structure and composition differences between grana and in-
tergrana thylakoids. - Photosynthetica *13* : 130 - 135, 1979.

39002 - OSZLÁNYI, J. : Energetická hodnota biomasy stromov s rozličným biosociolo-
gickým postavením. [Energetic value of biomass of trees with different bio-
sociological position.] - Lesnícky Časopis (Bratislava) *25* : 177 - 188,
1979. [In Slovak, ab : E, G, R.]

39003 - OSZLÁNYI, J. : Energetický ekvivalent nadzemnej biomasy drevín v dubovo-
-hrabovom lesnom ekosystéme. [Energetic equivalent of above-ground tree
biomass in an oak-hornbeam forest ecosystem.] - In : Bilancia Energie a
Vody v Pol'ných a Lesných Ekosystémoch. Pp. 42 - 48. Vysoká Škola pol'nohos-
podárska, Nitra 1979. [In Slovak.]

39004 - OSZLÁNYI, J. : Wood, bark and leaves energy values of *Carpinus betulus* L.,
Acer campestre L., *Quercus cerris* L. and *Q. petraea* LIEBL. - Biológia
(Bratislava) *34* : 775 - 784, 1979.

39005 - O'TOOLE, J.C., CRUZ, R.T., SEIBER, J.N. : Epicuticular wax and cuticular
resistance in rice. - Physiol. Plant.*47* : 239 - 244, 1979. [Resistances.]

39006 - OULTON, K., WILLIAMS, G.J. III, MAY, D.S. : Ribulose-1,5-bisphosphate carboxylase from altitudinal populations of *Taraxacum officinale*. - Photosynthetica *13* : 15 - 20, 1979.

39007 - OURISSON, G., ROHMER, M., ANTON, R. : From terpenes to sterols : macroevo-
lution and microevolution. - In : SWAIN, T., WALLER, G.R. (ed.) : Topics
in the Biochemistry of Natural Products. Pp. 131 - 162. Plenum Press,
New York - London 1979. [Car.]

39008 - OUTLAW, W.H. Jr., MANCHESTER, J. : Guard cell starch concentration quantita-
tively related to stomatal aperture. - Plant Physiol. *84* : 79 - 82, 1979.

39009 - OUTLAW, W.H. Jr., MANCHESTER, J., DiCAMELLI, C.A. : Histochemical approach
to properties of *Vicia faba* guard cell phosphoenolpyruvate carboxylase. -
Plant Physiol. *64* : 269 - 272, 1979.

39010 - OUTLAW, W.H. Jr., MANCHESTER, J., DiCAMELLI, C.A., RANDALL, D.D., RAPP, B.,
VEITH, G.M. : Photosynthetic carbon reduction pathway is absent in chloroplasts of *Vicia faba* guard cells. - Proc.nat.Acad.Sci.USA *76*: 6371 - 6375,
1979.

39011 - OVCHINNIKOV, Yu.A., ABDULAEV, N.G., FEIGINA, M.Yu., KISELEV, A.V., LOBA-
NOV, N.A. : The structural basis of the functioning of bacteriorhodopsin :
An overview. - FEBS Lett. *100* : 219 - 224, 1979.

39012 - OVERFIELD, R.E., WRAIGHT, C.A., DEVAULT, D. : Microsecond photooxidation
kinetics of cytochrome c_2 from *Rhodopseudomonas sphaeroides* : *in vivo*
and solution studies. - FEBS Lett. *105* : 137 - 142, 1979.

*39013 - **OVSYANNIKOV, A.S.** : Fiziologicheskie i biologicheskie osnovy plodonosheniya kryzhovnika. [Physiological and biological bases of fructification of *Ribes*.] - Sb. nauch. Rabot vsesoyuz. nauch.-issled. Inst. Sadovodstva (Michurinsk) *21* : 187 - 192, 1975. [Ps; in R.]

*39014 - **OVSYANNIKOV, A.S.** : Fotosinteticheskaya deyatel'nost' razlichnykh po pro-iskhozhdeniyu sortov grushi. [Photosynthetic activity of pear cultivars differing in their origin.] - Sb. nauch. Rabot vsesoyuz. nauch.-issled. Inst. Sadovodstva (Michurinsk) *25* : 1 - 5, 1977. [In R.]

39015 - **OVSYANNIKOV, A.S.** : Izuchenie fotosinteticheskoĭ deyatel'nosti u novykh i raĭonirovannykh sortov grushi v svyazi s urozhaem. [Photosynthetic activity in new and local pear cultivars in relation to yield.] - In : Biologicheskie Osnovy Produktivnosti Plodovykh Semechkovykh Kul'tur. Pp. 42 - 43. Nauka, Moskva 1979. [In R.]

39016 - **OWEN, T., CESS, R.D., RAMANATHAN, V.** : Enhanced CO_2 greenhouse to compensate for reduced solar luminosity on early Earth. - Nature *277* : 640 - 642, 1979. [Ps.]

39017 - **OWERS-NARHI, L., ROBINSON, S.J., DeROO, C.S., YOCUM, C.F.** : Reconstitution of cyanobacterial photophosphorylation by a latent Ca^{+2}-ATPase. - Biochem. biophys. Res. Commun. *90* : 1025 - 1031, 1979.

39018 - **PAAU, A.S., COWLES, J.R., ORO, J., BARTEL, A., HUNGERFORD, E.** : Separation of algal mixtures and bacterial mixtures with flow-microfluorometr using chlorophyll and ethidium bromide fluorescence. - Arch. Microbiol. *120* : 271 - 273, 1979.

39019 - **PACE, G.M., VOLK, R.J., JACKSON, W.A.** : Nitrate assimilation in corn seedlings as affected by CO_2-limited photosynthesis. - Plant Physiol. *63* (Suppl.) : 25, 1979.

39020 - **PACE, G.W., ARCHER, M.C., TANNENBAUM, S.R.** : Factors controlling ATP levels during photophosphorylation catalyzed by *Rhodospirillum rubrum* chromatophores. - Europe. J. appl. Microbiol. Biotechnol. *6* : 271 - 278, 1979.

39021 - **PACKER, L., KONISHI, T., SCHERRER, P., MEHLHORN, R.J., QUINTANILHA, A.T., SHIEH, P.K., PROBST, I., CARMELI, C., LANYI, J.K.** : Proton translocation by bacteriorhodopsin. - In : MUKOHATA, Y., PACKER, L. (ed.) : Cation Flux Across Biomembranes. Pp. 417 - 429. Academic Press, New York - San Francisco - London 1979.

39022 - **PACKER, L., TRISTRAM, S., HERZ, J.M., RUSSELL, C., BORDERS, C.L.** : Chemical modification of purple membranes : Role of arginine and carboxylic acid residues in bacteriorhodopsin. - FEBS Lett. *108* : 243 - 248, 1979.

39023 - **PACKHAM, N.K., JACKSON, J.B.** : Transport of local anaesthetica across chromatophore membranes. - Biochim. biophys. Acta *546* : 142 - 156, 1979.

39024 - **PADAN, E.** : Facultative anoxygenic photosynthesis in cyanobacteria. - Annu. Rev. Plant Physiol. *30* : 27 - 40, 1979.

39025 - **PAERL, H.W.** : Optimization of carbon dioxide and nitrogen fixation by the blue-green alga *Anabaena* in freshwater blooms. - Oecologia *38* : 275 - 290, 1979.

39026 - **PAILLOTIN, G., SWENBERG, C.E.** : Dynamics of excitons created by a single picosecond pulse. - In : Chlorophyll Organization and Energy Transfer in Photosynthesis. Pp. 201 - 215. Excerpta Medica, Amsterdam - Oxford - New York 1979.

39027 - **PAILLOTIN, G., SWENBERG, C.E., BRETON, J., GEACINTOV, N.E.** : Analysis of picosecond laser-induced fluorescence phenomena in photosynthetic membranes utilizing a master equation approach. - Biophys. J. *25* : 513 - 533, 1979.

39028 - PAILLOTIN, G., VERMEGLIO, A., BRETON, J. : Orientation of reaction center and antenna chromophores in the photosynthetic membrane of *Rhodopseudomonas viridis*. - Biochim. biophys. Acta *545* : 249 - 264, 1979.

39029 - PAIS, I. : Titan-ein neues Spurenelement in der Pflanzenernährung. - Tagungsber. Akad. Landwirtschaftswissensch. DDR *173* (Albert-Daniel-Thaer--Tagung) : 75 - 79, 1979. [Chl.]

39030 - PAIS, I., FEHÉR, M., FARKAS, E. : Role of titanium in the life of plants I. - Acta agron. Acad. Sci. hung. *28* : 378 - 383, 1979. [Chl.]

*39031 - PAKHOMOVA, L.M., BALAKHONTSEV, E.N., GIRFANOV, V.K. : Vliyanie mineral'-nykh élementov i regulyatorov rosta na ottok assimilyatov i produktivnost' sakharnoi svekly. [Effect of mineral elements and growth regulators on assimilates outflow and productivity of sugar beet.] - Fiziol. Biokhim. kul't. Rast. *10* : 151 - 155, 1978. [In R, ab : E.]

*39032 - PAL, U.R., SAXENA, M.C. : Leaf area and yield components - their relationships with yield in soybean [*Glycine max* (L.) MERR.] . - Acta agron. Acad. Sci. hung. *26* : 438 - 445, 1977.

39033 - PALIT, P., KUNDU, A., MANDAL, R.K., SIRCAR, S.M. : Productivity of rice plant in relation to photosynthesis, photorespiration and translocation. - Indian J. Plant Physiol. *22* : 66 - 73, 1979.

39034 - PALIT, P., KUNDU, A., MANDAL, R.K., SIRCAR, S.M. : Source-sink control of dry matter production and photosynthesis in rice plant after flowering. - Indian J. Plant Physiol. *22* : 87 - 91, 1979.

39035 - PALLARDY, S.G., KOZLOWSKI, T.T. : Early root and shoot growth of *Populus* clones. - Silvae Genet. *28* : 153 - 156, 1979. [Net assimilation rate.]

39036 - PALLARDY, S.G., KOZLOWSKI, T.T. : Stomatal response of *Populus* clones to light intensity and vapor pressure deficit. - Plant Physiol. *64* : 112 - 114, 1979. [Resistances.]

39037 - PALLAS, J.E. Jr. : Carbon dioxide. - In : TIBBITTS, T.W., KOZLOWSKI, T.T. (ed.) : Controlled Environment Guidelines for Plant Research. Pp. 207 - 228. Academic Press, New York 1979.

39038 - PALLETT, K.E., DODGE, A.D. : Sites of action of photosynthetic inhibitor herbicides ; experiments with trypsinated chloroplasts. - Pestic. Sci. *10* : 216 - 220, 1979.

39039 - PALLETT, K.E., DODGE, A.D. : The role of light and oxygen in the action of the photosynthetic inhibitor herbicide monuron. - Z. Naturforsch. *34 C* : 1058 - 1061, 1979.

39040 - PALMER, R.G., SHERIDAN, M.A., TABATABAI, M.A. : Effects of genotype, temperature, and illuminance on chloroplast ultrastructure of a chlorophyll mutant in soybeans. - Cytologia *44* : 881 - 891, 1979.

39041 - PAN, R.L., IZAWA, S. : Photosystem II energy coupling in chloroplasts with H_2O_2 as electron donor. - Biochim. biophys. Acta *547* : 311 - 319, 1979.

39042 - PANDA, B.B., SHARMA, C.B.S.R. : Organophosphate induced chlorophyll mutations in *Hordeum vulgare*. - Theor. appl. Genet. *55* : 253 - 255, 1979.

39043 - PANETTA, F.D. : Shade tolerance as reflected in population structure of the woody weed, groundsel bush (*Baccharis halimifolia* L.). - Aust. J. Bot. *27* : 609 - 615, 1979.

39044 - PANIGRAHI, P.K., BISWAL, U.C. : Ageing of chloroplasts *in vitro* I. Quantitative analysis of the degradation of pigments, proteins and nucleic acids. - Plant Cell Physiol. *20* : 775 - 779, 1979.

39045 - PANIGRAHI, P.K., BISWAL, U.C. : Ageing of chloroplasts *in vitro* II. Changes in absorption spectra and the DCPIP Hill reaction. - Plant Cell Physiol. *20* : 781 - 787, 1979.

39046 - PAONE, D.A.M., STEVENS, S.E. Jr. : Physiological parameters of *Agmenellum quadruplicatum* strain PR-6 during growth on limiting nitrogen. - Plant Physiol. *63* (Suppl.) : 25, 1979. [Chl, Bil.]

39047 - PAPAGEORGIOU, G.C. : Applications of fluorescence quenching to biological problems : "High-energy state" quenching of chloroplast fluorescence. - Zagadnienia Biofiz. współczes. *4* : 21 - 26, 1979.

39048 - PAPAGEORGIOU, G.C. : Molecular and functional aspects of immobilized chloroplast membranes. - In : BARBER, J. (ed.) : Photosynthesis in Relation to Model Systems. Pp. 211 - 242. Elsevier, Amsterdam - New York - Oxford 1979.

39049 - PAQUES, M., BONOTTO, S., SIRONVAL, C. : Presence of filamentous structures connecting chloroplasts in the hairs of the sterile whorls of *Acetabularia mediterranea*. - In : BONOTTO, S., KEFELI, V., PUISEUX-DAO, S. (ed.) : Developmental Biology of *Acetabularia*. Pp. 179 - 182. Elsevier/North-Holland Biomedical Press, Amsterdam - New York - Oxford 1979.

39050 - PAQUES, M., SIRONVAL, C., BONOTTO, S. : On chloroplast movement in the stalk of *Acetabularia mediterranea*. - In : BONOTTO, S., KEFELI, V., PUISEUX-DAO, S. (ed.) : Developmental Biology of *Acetabularia*. Pp. 155 - 163. Elsevier/North-Holland Biomedical Press, Amsterdam - New York - Oxford 1979.

39051 - PARADIES, H.H. : Crystalization of coupling factor 1 (CF_1) from spinach chloroplast. - Biochem. biophys. Res. Commun. *91* : 685 - 692, 1979.

39052 - PÂRJOL, L., PICU, I., HURDUC, N. : Cercetări privind sistemul de fertiliza-re la soia în condiții de bacterizare și irigare. [Fertilizing system on soybean in the conditions of bacterizing and irrigation.] - An. Inst. Cercet. Cereale Plante tehnice Fundulea *44* : 171 - 188, 1979. [In Roum., ab:E, R.]

39053 - PÂRJOL, L., POPA, F.G., HURDUC, N. : Cercetări privind rezistența la sece-tă a unor soiuri și linii de fasole. [Drought resistance of some bean cultivars and lines.] - An. Inst. Cercet. Cereale Plante tehnice Fundulea *44* : 415 - 426, 1979. [Dry-matter accumulation; in Roum., ab : E, R.]

39054 - PARK, I.-K., TSUNODA, S. : Effect of low temperature on chloroplast structure in cultivars of rice. - Plant Cell Physiol. *20* : 1449 - 1453, 1979.

*39055 - PARKER, R.R., SIBERT, J. : Studies on a production system using a large volume floating pond. - In : 10th European Symposium Marine Biology. Vol. 2. Pp. 457 - 466. Ostend, Belgium 1975. [Chl.]

39056 - PARKINSON, K.J., DAY, W. : The use of orifices to control the flow rate of gases. - J. appl. Ecol. *16* : 623 - 632, 1979. [CO_2.]

39057 - PÄRNIK, T., VIIL, J. : Intrafoliaceous CO_2 fluxes at low and high light intensities. - In : VAKLINOVA, S.G., VANKOVA-RADEVA, R., VASILEVA, V.S. (ed.) : Fotosinteticheskaya Assimilyatsiya CO_2 i Fotodykhanie. Pp. 80 - 85. Izdat. bolg. Akad. Nauk, Sofiya 1979.

39058 - PARSONS, J.E., PHENE, C.J., BAKER, D.N., LAMBERT, J.R., McKINION, J.M. : Soil water stress and photosynthesis in cotton. - Physiol. Plant. *47* : 185 - 189, 1979.

39059 - PARSONS, L.R. : Breeding for drought resistance : What plant characteristics impart resistance ? - HortScience *14* : 590 - 593, 1979. [Canopy structure.]

39060 - PARSONS, T.R. : The Strait of Georgia Programme. - In : DUNBAR, M.J. (ed.) : Marine Production Mechanisms. Pp. 133 - 149. Cambridge Univ. Press, London 1979. [Chl.]

39061 - PARSONS, T.R. : Some ecological, experimental and evolutionary aspects of the upwelling ecosystem. - South African J. Sci. *75* : 536 - 540, 1979. [Primary productivity.]

39062 - PARTHIER, B. : The equivocal role of phytohormones (cytokinins) in chloroplast development. - Biochem. Physiol. Pflanzen *174* : 173 - 214, 1979.

39063 - PASS, R.G., HARTLEY, D.E. : Net photosynthesis of three foliage plants under low irradiation levels. - J. amer. Soc. hort. Sci. *104* : 745 - 748, 1979.

39064 - PASSERA, C. : Effetto del glicidato di sodio sulla fissazione fotosintetica di $^{14}CO_2$ in *Chlorella vulgaris*. [Effect of sodium glycidate on the photosynthetic fixation of $^{14}CO_2$ in *Chlorella vulgaris*.] - Agr. ital. (Pisa) *1979* : 323 - 340, 1979. [In Ital.]

39065 - PASSERA, C., GHISI, R. : Il cammino del ^{14}C e il ruolo della fotorespirazione in due varietà di orzo a differente contenuto di proteina fogliare. [Distribution of ^{14}C and role of photorespiration in two varieties of barley with different leaf protein contents.] - Agrochimica *23* : 195 - 207, 1979. [In Ital., ab : E, F, G, Span.]

39066 - PATE, J.S., LAYZELL, D.B., ATKINS, C.A. : Economy of carbon and nitrogen in a nodulated and nonnodulated (NO_3-grown) legume. - Plant Physiol. *64* : 1083 - 1088, 1979. [Photosynthates.]

39067 - PATE, J.S., LAYZELL, D.B., McNEIL, D.L. : Modeling the transport and utilization of carbon and nitrogen in a nodulated legume. - Plant Physiol. *63* : 730 - 737, 1979.

39068 - PATEL, C.L., GHILDYAL, B.P., TOMAR, V.S. : Lysimetry studies on the water use and growth of rice under different soil-water regimes. - Indian J. agr. Sci. *49* : 90 - 95, 1979. [Dry-matter accumulation.]

39069 - PATRA, H.K., MISHRA, D. : Pyrophosphatase, peroxidase and polyphenoloxidase activities during leaf development and senescence. - Plant Physiol. *63* : 318 - 323, 1979. [Chl.]

39070 - PATRICK, J.W. : Auxin-promoted transport of metabolites in stems of *Phaseolus vulgaris* L. Further studies on effects remote from the site of hormone application. - J. exp. Bot. *30* : 1 - 13, 1979.

39071 - PATTERSON, D.T. : Methodology and terminology for the measurement of light in weed studies - a review. - Weed Sci. *27* : 437 - 443, 1979.

39072 - PATTERSON, D.T. : The effects of shading on the growth and photosynthetic capacity of itchgrass (*Rottboellia exaltata*). - Weed Sci. *27* : 549 - 553, 1979.

39073 - PATTERSON, D.T., DUKE, S.O. : Effect of growth irradiance on the maximum photosynthetic capacity of water hyacinth [*Eichhornia crassipes* (MART.) SOLMS]. - Plant Cell Physiol. *20* : 177 - 184, 1979.

39074 - PATTERSON, D.T., FLINT, E.P. : Effects of chilling on cotton (*Gossypium hirsutum*), velvetleaf (*Abutilon theophrasti*), and spurred anoda (*Anoda cristata*). - Weed Sci. *27* : 473 - 479, 1979. [Ps.]

39075 - PATTERSON, G.M.L., HARRIS, D.O., COHEN, M.S. : Inhibition of photosynthetic and mitochondrial electron transport by a toxic substance isolated from the alga *Pandorina morum*. - Plant Sci. Lett. *15* : 293 - 300, 1979.

39076 - PATTERSON, T.G., MOSS, D.N. : Senescence in field-grown wheat. - Crop Sci. *19* : 635 - 640, 1979. [Ps, Chl.]

39077 - PAUCĂ-COMĂNESCU, M., TĂCINĂ, A. : Variaţia conţinutului dé clorofilă în diferite ecosisteme forestiere. [Variation of chlorophyll content in various forest ecosystems.] - Stud. Cercet. Biol., Ser. Biol. veg. *31* : 129 - 141, 1979. [In Roum., ab : E.]

39078 - PAUL, F., COLBEAU, A., VIGNAIS, P.M. : Phosphorylation coupled to H_2 oxidation by chromatophores from *Rhodopseudomonas capsulata*. - FEBS Lett. *106* : 29 - 33, 1979. [Ps.]

39079 - PAUL, J.S., KROHNE, S.D., BASSHAM, J.A. : Stimulation of CO_2 incorporation and glutamine synthesis by 2,4-D in photosynthesizing leaf-free mesophyll cells. - Plant Sci. Lett. *15* · 17 - 24, 1979.

39080 - PAUL, M.H., PLANCHON, C., ÉCOCHARD, R. : Étude des relations entre le développement foliaire, le cycle de développement et la productivité chez le Soja. - Ann. Amélior. Plant. *29* : 479 - 492, 1979.

*39081 - PAULECH, C. : Vplyv múčnatky trávnej na fotosyntézu izogenických línií jač-
         meňa. [Influence of powdery mildew on the photosynthesis of isogenic barley
         lines.] - Acta bot. slov. Acad. Sci.slov. Ser A 4 : 303 - 312, 1978. [In
         Slovak, ab : E, R.]

*39082 - PAULECH, C., MINARČIC, P. : Príspevok k poznaniu vplyvu huby Erysiphe gra-
         minis DC. na štruktúru a funkciu chloroplastov jačmeňa. [Effect of Ery-
         siphe graminis DC. on chloroplast structure and function in barley.] -
         Sb. vys. Školy zem. Praze, Fak. agron. 1978 (Ochrana Rostlin v Zemědělské
         Velkovýrobě) : 150 - 168, 1978. [In Slovak, ab : E, R.]

*39083 - PAUTOVA, V.N., KOZHOVA, O.M., IZMEST'EVA,L.R., KRASHCHUK, L.S., ZAUSAEVA,
         N.A., DAVYDOVA, I.K., LOPATINA, N.I., ROMANENKO, T.I. : Osobennosti gidro-
         khimicheskogo i gidrobiologicheskogo rezhimov Bratskogo vodokhranilishcha
         v period stabilizatsii. [Peculiarities of hydrochemical and hydrobiological
         regimes of Bratsk water reservoir in stabilization period.] - In : Gidro-
         biologicheskie i Ikhtiologicheskie Issledovaniya v Vostochnoī Sib.iri. Pp.
         20 - 36, 210 - 211. Irkutsk. gos. Univ., Irkutsk 1978. [ChI; in R.]

*39084 - PAVLINOVA, O.A., TURKINA, M.V. : Biosintez i fiziologicheskaya rol' sakha-
         rozy v rastenii. [Biosynthesis and physiological role of saccharose in the
         plant.] - Fiziol. Rast. 25 : 1025 - 1041, 1978. [Ps; in R, ab : E.]

*39085 - PEARMAN,G.I.,FRANCEY, R.J., FRASER, P.J.B. : Climatic implications of stable
         carbon isotopes in tree rings. - Nature 260 : 771 - 773, 1976. [$\delta^{13}C$.]

*39086 - PEARMAN, G.I., GARRATT, J.R. : Global aspects of carbon dioxide. - Search
         3 (3) : 67 - 73, 1972.

 39087 - PEARMAN, I., THOMAS, S.M., THORNE, G.N. : Effect of nitrogen fertilizer on
         photosynthesis of several varieties of winter wheat. - Ann. Bot. 43 :
         613 - 621, 1979.

 39088 - PEARSON, C.J. : Daily cycles of photosynthesis, respiration and transloca-
         tion. - In : MARCELLE, R., CLIJSTERS, H., VAN POUCKE, M. (ed.) : Photo-
         synthesis and Plant Development. Pp. 125 - 136. Dr. W. Junk bv. Publ.,
         The Hague - Boston - London 1979.

 39089 - PEARSON, L.C. : Effects of temperature and moisture on phenology and pro-
         ductivity of Indian ricegrass. - J. Range Manage. 32 : 127 - 134, 1979.
         [Growth analysis.]

 39090 - PECK, R.A., KIRKHAM, M.B. : Water relations and yield of winter wheat grown
         under three water regimes in the High Plains. - Proc. Oklahoma Acad. Sci.
         59 : 53 - 59, 1979. [Stomatal resistance.]

 39091 - PEDERSEN, J.B. : Determination of the primary reactions of photosynthesis
         from transient ESR signals. - FEBS Lett.97 : 305 - 310, 1979.

 39092 - PEISER, G.D., YANG, S.F. : Sulfite-mediated destruction of β-carotene. -
         J. agr. Food Chem. 27 : 446 - 449, 1979.

 39093 - PEISKER, M. : Conditions for low, and oxygen-independent, CO_2 compensation
         concentrations in C_4 plants as derived from a simple model. - Photosyn-
         thetica 13 : 198 - 207, 1979.

 39094 - PEISKER, M., TICHÁ, I., APEL, P. : Variations in the effect of temperature
         on oxygen dependence of CO_2 gas exchange in wheat leaves. - Biochem. Phy-
         siol. Pflanzen 174 : 391 - 397, 1979.

 39095 - PELLIN, M.J., KAUFMANN, K.J., WASIELEWSKI, M.R. : In vitro duplication of
         the primary light-induced charge separation in purple photosynthetic bac-
         teria. - Nature 278 : 54 - 55, 1979.

 39096 - PENNAK, R.W., LAVELLE, J.W. : In situ measurements of net primary produc-
         tion in a Colorado mountain stream. - Hydrobiologia 66 : 227 - 235, 1979.

 39097 - PENNING DE VRIES, F.W.T., WITLAGE, J.M., KREMER, D. : Rates of respira-
         tion and of increase in structural dry matter in young wheat, ryegrass
         and maize plants in relation to temperature, to water stress and to their
         sugar content. - Ann. Bot. 44 : 595 - 609, 1979. [Dry-matter accumulation.]

39098 - PEOPLES, M.B., FRITH, G.J.T., DALLING, M.J. : Proteolytic enzymes in green
wheat leaves. IV. Degradation of ribulose 1,5-bisphosphate carboxylase by
acid proteinases isolated on DEAE-cellulose. - Plant Cell Physiol. *20* :
252 - 258, 1979.

39099 - PEOPLES, T.R., KOCH, D.W. : Role of potassium in carbon dioxide assimilation in *Medicago sativa* L. - Plant Physiol. *63* : 878 - 881, 1979.

39100 - PERIASAMY, N., LINSCHITZ, H. : Photodisaggregation of chlorophyll *a* and *b*
dimers. - J. amer. chem. Soc. *101* : 1056 - 1057, 1979.

39101 - PERRY, W.B., BOSWELL, J.T., STANFORD, J.A. : Critical problems with extraction of ATP for bioluminescence assay of plankton biomass. - Hydrobiologia
65 : 155 - 163, 1979.

39102 - PERSANOV, V.M., GOGOTOV, I.N. : Vydelenie molekulyarnogo vodoroda kletkami
zelenoĭ vodorosli *Chlorella vulgaris*. [Hydrogen evolution by green alga
Chlorella vulgaris cells.] - Fiziol. Rast. *26* : 560 - 567, 1979. [In R,
ab : E.]

39103 - PESCHEK, G.A. : Anaerobic hydrogenase activity in *Anacystis nidulans*. H_2-
-dependent photoreduction and related reactions. - Biochim. biophys. Acta
548 : 187 - 202, 1979.

39104 - PESCHEK, G.A. : Aerobic hydrogenase activity in *Anacystis nidulans*. The
oxyhydrogen reaction. - Biochim. biophys. Acta *548* : 203 - 215, 1979.

39105 - PESCHEK, G.A. : The role of the Calvin cycle for anoxygenic CO_2 photoassimilation in *Anacystis nidulans*. - FEBS Lett. *106* : 34 - 38, 1979.

39106 - PESTEMER, W. : Biological determination of photosynthetic inhibitors in
soils and water and application of bioassays to herbicide investigations. -
Z. Naturforsch. *34 C* : 964 - 965, 1979.

39107 - PETELLE, M., HAINES, B., HAINES, E. : Insect food preferences analyzed
using $^{13}C/^{12}C$ ratios. - Oecologia *38* : 159 - 166, 1979.

*39108 - PETERSON, B.J. : Phytoplankton production and phosphorus supply in Cayuga
lake (1968 - 1973). - Hydrobiologia *54* : 113 - 127, 1977. [Ps, Chl.]

39109 - PETERSON, C.A., RAUSER, W.E. : Callose deposition and photoassimilate export in *Phaseolus vulgaris* exposed to excess cobalt, nickel, and zinc. -
Plant Physiol. *63* : 1170 - 1174, 1979.

39110 - PETERSON, R.B., KE, B. : Presence of phycobilins in heterocysts isolated
from *Anabaena variabilis*. - Plant Physiol. *63* (Suppl.) : 28, 1979.

39111 - PETKE, J.D., MAGGIORA, G.M., SHIPMAN, L., CHRISTOFFERSEN, R.E. : Stereo-
electronic properties of photosynthetic and related systems - V. *Ab initio*
configuration interaction calculations on the ground and lower excited singlet and triplet states of ethyl chlorophyllide *a* and ethyl pheophorbide
a. - Photochem. Photobiol. *30* : 203 - 223, 1979.

39112 - PETR, J., HNILICA, P., SCHMIDT, J. : Tvorba výnosu jarního ječmene - odno-
žování, tvorba klasů a zrn v klasu. [Spring barley yield formation - tillering, formation of ears and of grains per ear.] - Rost. Výroba (Praha)
25 : 433 - 444, 1979. [In Czech, ab : E, G, R.]

39113 - PETREA, V. : L'action du Na-dodécylsulfuricum des eaux polluées sur quelques processus physiologiques chez l'algue *Chlorella vulgaris*. - Rev. roum.
Biol.,Sér. Biol. vég. *24* : 39 - 41, 1979. [Ps.]

39114 - PETROSYAN, G.P., SAAKYAN, R.G., SAKUNTS, L.E. : Soderzhanie pigmentov v
list'yakh i yagodakh vinograda v zavisimosti ot kolichestva pogloshchen-
nogo natriya v meliorirovannom solontse-solonchake. [Content of pigments
in the leaves and berries of grape vine plants in relation to the amount
of absorbed sodium in reclaimed solonetz-solonchik soil.] - Biol. Zh.
Armenii *32* (1) : 25 - 30, 1979. [In R, ab : Armen., E.]

39115 - PETROV, V.E., SEĬFULLINA, N.Kh., LOSEVA, N.L., KLEMENT'EVA, G.S. : Spektral'naya zavisimost' reaktivatsii fotosinteza assimiliruyushcheĭ kletki
posle teplovogo povrezhdeniya. [The wavelength dependence of photosynthesis

reactivation in the assimilating heat-injured cell.] - Fiziol. Rast. *26* : 1219 - 1225, 1979. [In R, ab : E.]

39116 - PETROVIĆ, S.M., KOLAROV, L.A., PERIŠIĆ-JANJIĆ,N.U.:Separation of chloroplast leaf pigments by chromatography on starch and cellulose thin layers. - J. Chromatogr. *171* : 522 - 526, 1979.

39117 - PETRUKHIN, Yu.A., STEPANOVA, S.P. : Fotosintez list'ev khlorofituma v za-visimosti ot stepeni razvitiya v nikh beloĭ tkani. [Photosynthesis of *Chlo-rophytum* leaves depending upon the degree of the white tissue development.] - Biol. Nauki *1979* (5) : 89 - 91, 1979. [In R.]

39118 - PETTY, K., JACKSON, J.B., DUTTON, P.L. : Factors controlling the binding of two protons per electron transferred through the ubiquinone qnd cyto-chrome b/c_2 segment of *Rhodopseudomonas sphaeroides* chromatophores. - Bio-chim. biophys. Acta *546* : 17 - 42, 1979.

39119 - PETTY, K.M., JACKSON, J.B. : Correlation between ATP synthesis and the de-cay of the carotenoid band shift after single flash activation of chroma-tophores from *Rhodopseudomonas capsulata*. - Biochim. biophys. Acta *547* : 463 - 473, 1979.

39120 - PETTY, K.M., JACKSON, J.B. : Kinetic factors limiting the synthesis of ATP by chromatophores exposed to short flash excitation. - Biochim. biophys. Acta *547* : 474 - 483, 1979.

39121 - PETTY, K.M., JACKSON, J.B. : Two protons transferred per ATP synthesised after flash activation of chromatophores from photosynthetic bacteria. - FEBS Lett. *97* : 367 - 372, 1979.

39122 - PEVERLY, J.H., JOHNSON, R.L. : Nutrient chemistry in herbicide-treated ponds of different fertility. - J. environ. Qual. *8* : 294 - 300, 1979.

39123 - PFISTER, K., ARNTZEN, C.J. : The mode of action of photosystem II-specific inhibitors in herbicide-resistant weed biotypes. - Z. Naturforsch. *34 C* : 996 - 1009, 1979.

39124 - PFISTER, K., DITTO, C., ARNTZEN, C.J. : Alterations in the photosynthesis II complex of herbicide-tolerant weed biotypes. - Plant Physiol. *63* (Suppl.) : 41, 1979.

39125 - PFISTER, K., RADOSEVICH, S.R., ARNTZEN, C.J. : Modification of herbicide binding to photosystem II in two biotypes of *Senecio vulgaris* L. - Plant Physiol. *64* : 995 - 999, 1979.

39126 - PHAN, C.T., BRACH, E.J., JASMIN, J.J. : Studies on the detection of lettuce maturity : anatomical observations arrd reflectance measurements in the vi-sible range (350-650 nm). - Can. J. Plant Sci. *59* : 1067 - 1075, 1979.

39127 - PHELOUNG, P., BRADY, C.J. : Soluble and fraction 1 protein in leaves of C3 and C4 grasses. - J. Sci. Food Agr. *30* : 246 - 250, 1979.

*39128 - PHILIP, J.R. : The use of point quadrats, with special reference to stem--like organs. - Aust. J. Bot. *14* : 105 - 125, 1966. [Growth analysis.]

39129 - PHIPPS, R.H., FULFORD, R.J. : Relationship between the production of forage maize grown at different plant densities and accumulated temperature and Ontario heat units. - Maydica *24* : 235 - 246, 1979. [Dry-matter accumula-tion.]

39130 - PICKARD, W.F., MINCHIN, P.E.H., TROUGHTON, J.H. : Real time studies of carbon-11 translocation in moonflower. III. Further experiments on the ef-fects of a nitrogen atmosphere, water stress, and chilling; and a qualita-tive theory of stem translocation. - J. exp. Bot. *30* : 307 - 318, 1979.

*39131 - PIECHA, W. : Przebieg fotosyntezy i transpiracji u niektórych roślin mo-tylkowych. [Course of photosynthesis and transpiration of some legumes.] - Zesz. nauk. Akad. rol.-tech. Olsztynie, Rolnictwo *22* : 11 - 18, 1977. [In Pol.. ab E, R.]

39132 - PIERRE, J.N., QUEIROZ, O. : Regulation of glycolysis and level of the Crassulacean acid metabolism. - Planta *144* : 143 - 151, 1979.

39133 - PIHAKASKI, K., PIHAKASKI, S. : Effects of chilling on the ultrastructure and net photosynthesis of *Pellia epiphylla*. - Ann. Bot. *43* : 773 - 781, 1979.

39134 - PILL, W.G., LAMBETH, V.N., HINCKLEY, T.M. : Effects of Cycocel and nitrogen form on tomato water relations, ion composition, and yield. - Can. J. Plant Sci. *59* : 391 - 397, 1979. [Resistances.]

39135 - PINEAU, B., LEDOIGT, G., MAILLEFER, C., LEFORT-TRAN, M. : Présence de sous-unités de la RubPcase dan les enveloppes des chloroplastes d'épinard. - Plant Sci. Lett. *16* : 337 - 343, 1979.

39136 - PINEVICH, A.V., DESNITSKIĬ, A.G. : K probleme èvolyutsionnogo proiskhozhdeniya nekotorykh kletochnykh organell. [Evolutionary origin of some cellular organelles.] - Tsitologiya *21* : 755 - 767, 1979. [Chloroplast; in R, ab : E.]

*39137 - PINKHASOV, Yu.I. : Sravnitel'noe izuchenie intensivnosti fotosinteza razlichnykh chasteĭ i organov u tselykh rasteniĭ khlopchatnika v ontogeneze. [Comparative study of photosynthetic rate in various parts and organs of intact cotton plants in ontogeny.] - Fiziol. Rast. *25* : 1151 - 1157, 1978. [In R, ab : E.]

39138 - PINKHASOV, Yu.I., DZHAFAROV, M.I., DZHUMANKULOV, Kh.D.·: Fotosintez i produktivnost' khlopchatnika pod deĭstviem khlorkholinkhlorida. [Photosynthesis and productivity of cotton plants as affected by chlorocholine chloride.] - Fiziol. Rast. *26* : 1265 - 1272, 1979. [In R, ab : E.]

39139 - PINTÉR, L. : A kukorica (*Zea mays* L.) hibridek levélfelületének gyors meghatározási lehetőségei hazai biológiai viszonyok között. [Possibilities of a rapid determination of leaf area in maize (*Zea mays* L.) hybrids under the biological conditions·of Hungary.] - Növénytermelés *28* : 397 - 401, 1979. [In Hung.

39140 - PINTER, L., KALMAN, L. : Effects of defoliation on lodging and yield of maize hybrids. - Exp. Agr. *15* : 241 - 245, 1979. [Production.]

39141 - PINTHUS, M.J., MEIRI, J. : Effects of the reversal of day and night temperatures on tillering and on the elongation of stems and leaf blades of wheat. - J. exp. Bot. *30* : 319 - 326, 1979. [Dry-matter production.]

39142 - PIRIE, N.W. : The efficiency of protein production by different farming systems. - In : HEWITT, E.J., CUTTING, C.V. (ed.) : Nitrogen Assimilation of Plants. Pp. 613 - 624. Academic Press. London - New York - San Francisco 1979. [Photosynthates.]

39143 - PISICĂ-DONOSE, A., DORNESCU, D., ROŞU, A., SIMINICEANU, E. : Cercetări privind corelaţia proceselor de creştere cu producţia la grîul de toamnă Bezostaia 1 în condiţii de fertilizare deferită.'[Correlation between growth and production of winter wheat Bezostaya 1 under different fertilization.] - Stud. Cerc. Biol., Ser. Biol. veg. *31* : 115 - 121, 1979. [In Roum., ab : E.]

39144 - PITOMBEIRA, J.B., HOUSLEY, T.L., OHLROGGE, A.J., COUNCE, P.A. : The influence of DNBP on ¹⁴C-sucrose accumulation and ¹⁴C-assimilate transport. - Plant Physiol. *63* (Suppl.) : 45, 1979.

39145 - PIZZOLATO, T.D., FRICK, H. : Pyrimidine metabolism in *Lemna minor*. II. Specific inhibition of plastid replication in a higher plant by cytidine deoxyriboside. - Plant Physiol. *63* : 979 - 983, 1979.

39146 - PLACE, N.A., MORGAN, M.S., RUTKOSKI, A., NEWMAN, D.W., JAWORSKI, J.G. : Fatty-acid metabolism in senescing and regreening soybean cotyledons. - Planta *147* : 246 - 250, 1979. [Chl.]

39147 - PLANCHON, C. : Photosynthesis, transpiration, resistance to CO_2 transfer, and water efficiency of flag leaf of bread wheat, durum wheat and triticale. - Euphytica *28* : 403 - 408, 1979.

39148 - PLANTE-CUNY, M.-R. : La chlorophylle *a* fonctionnelle dans les substrats meubles marins, indice de la biomasse du microphytobenthos. - Rapp. Comm. int. Mer Médit. *25/26* : 285 - 290, 1979.

39149 - PLATT, S.G., HENRIQUES, F., RAND, L. : Effects of virus infection on the chlorophyll content, photosynthetic rate and carbon metabolism of *Tolmiea menziesii*. - Physiol. Plant Pathol. *15* : 351 - 365, 1979.

39150 - PLATT, S.G., RAND, L. : Thin-layer chromatographic separation of ^{14}C-labeled metabolites from photosynthate. - J. Liquid Chromatogr. *2* : 239 - 253, 1979.

39151 - PLATT-ALOIA, K.A., THOMSON, W.W. : Membrane bound inclusions in epidermal plastids of developing sesame leaves and cotyledons. - New Phytol. *83* : 793 - 799, 1979.

39152 - PLESNIČAR, M., STANKOVIČ, Ž. : Wavelength-dependent photophosphorylation catalysed by Photosystem 1 or Photosystem 2 in isolated pea chloroplasts. - Photosynthetica *13* : 359 - 364, 1979.

39153 - PLUMB-DHINDSA, P.L., DHINDSA, R.S., THORPE, T.A. : Non-autotrophic CO_2 fixation during shoot formation in tobacco callus. - J. exp. Bot. *30* : 759 - 767, 1979.

39154 - POINCELOT, R.P. : Carbonic anhydrase. - In : GIBBS, M., LATZKO, E. (ed.) : Photosynthesis II. (Encycl. Plant Physiol. N.S. Vol. 6.) Pp. 230 - 238. Springer-Verlag, Berlin - Heidelberg - New York 1979.

39155 - POKROVSKAYA, T.N. : Ustoĭchivost' "makrofitnykh" ozer k antropogennym evtrofiruyushchim vozdeĭstviyam. [Resistance of "macrophytic" lakes to antropogenic eutrofication activities.] - Izv. Akad. Nauk SSSR, Ser. geograf. *1979* (4) : 37 - 46, 1979. [Ps production; in R.]

39156 - POLLARD, A., WYN JONES, R.G. : Enzyme activities in concentrated solutions of glycinebetaine and other solutes. - Planta *144* : 291 - 298, 1979. [Ps.]

39157 - POLLHAMER, E. : A martonvásári tavaszi árpák termőképessége és terméselemzése. [Productivity and yield analysis of summer barley varieties of Martonvásár.] - Növénytermelés *28* : 193 - 204, 1979. [In Hung., ab : E.]

39158 - POLLOCK, R.B., KANEMASU, E.T. : Estimating leaf-area index of wheat with LANDSAT data. - Remote Sensing Environ. *8* : 307 - 312, 1979.

39159 - PONOMAREVA, R.P. : Deĭstvie ostrogo γ-oblucheniya na fotosinteticheskiĭ apparat berezy i sosny. [Effect of acute γ-irradiation on the photosynthetic apparatus of birch and pine trees.] - Tr. Inst. prikl. Geofiz. *38* : 68 - 75, 1979. [In R.]

*39160 - PONZI, R., PIZZOLONGO, P. : The ultrastructure of suspensor cells of *Ipomoea purpurea* ROTH. - J. submicrosc. Cytol. *4* : 199 - 204, 1972. [Plastids.]

*39161 - PONZI, R., PIZZOLONGO, P. : Ultrastructure of plastids in the suspensor cells of *Ipomoea purpurea* ROTH. - J. submicrosc. Cytol. *5* : 257 - 263, 1973.

39162 - POOLE, R.T., CONOVER, C.A. : Influence of shade and nutrition during production and dark storage simulating shipment on subsequent quality and chlorophyll content of foliage plants. - HortScience *14* : 617 - 619, 1979.

39163 - POOVAIAH, B.W., NUKAYA, A. : Polygalacturonase and cellulase enzymes in the normal Rutgers and mutant *rin* tomato fruits and their relationship to the respiratory climacteric. - Plant Physiol. *64* : 534 - 537, 1979. [Chl.]

*B39164 - POPOV,K.I.,ĬORDANOV,I.T.,DILOVA,S.A.,USHEVA,N.Ya.,STANEV,V.P.,PETKOVA,R.A., CHICHEV.P.N.,ZEĬNALOV,Yu.A. (ed.): Bibliografiya po Teme "Vliyanie Sostoyaniya Pigment-Belkovogo Kompleksa na Intensivnost' i Produktivnost' Fotosinteza". [Bibliography on the Topics "Influence of the State of the Pigment-Protein Complex on the Rate and Productivity of Photosynthesis".] - Inst. Fiziol. Rast. M.Popova, bolg. Akad. Nauk, Sofiya 1970. [In R.]

39165 - POPOVA, I.A. : O variabel'nosti velichiny fotosinteticheskoĭ edinitsy. [Variability of photosynthetic unit size.] - Bot. Zh. *64* : 1474 - 1478, 1979. [In R.]

39166 - POPOVA, L.P., DIMITROVA, O.D. : Influence of ferredoxin on the activity of ribulosediphosphate and phosphoenylpyruvate carboxylases in isolated proto-plasts of pea and maize. - Dokl. bolg. Akad. Nauk *32* : 1559 - 1562, 1979.

39167 - POPOVA-STAEVSKA, L.P., DIMITROVA, O.D. : Activation of phosphoenolpyruvate carboxylase in C-3 and C-4 plants by glucollate. - Dokl. bolg. Akad. Nauk *32* : 955 - 958, 1979.

39168 - POPOVIC, R.B., KYLE, D.J., COHEN, A.S., ZALIK, S. : Stabilization of thyla-koid membranes by spermine during stress-induced senescence of barley leaf discs. - Plant Physiol. *64* : 721 - 726, 1979.

39169 - PORATH, D. : Pathways of plastid differentiation in *Spirodela oligorrhiza*. - New Phytol. *82* : 733 - 737, 1979.

39170 - PORTER, J.W., SPURGEON, S.L. : Enzymatic synthesis of carotenes. - Pure appl. Chem. *51* : 609 - 622, 1979.

39171 - POSTIUS, S., KIRSCHSTEIN, M., SENGER, H. : Die Populationsverteilung von Chlorophyllfluoreszenz und Zellvolumen während des Entwicklungszyklus von *Scenedesmus obliquus*. - Ber. deut. bot. Ges. *92* : 731 - 740, 1979.

39172 - POTTER, J.F., ROSS, G.J.S. : Maximum likelihood estimation of breakpoints and the comparison of the goodness of fit with that of conventional curves. - In : LYONS, J.M., GRAHAM, D., RAISON, J.K. (ed.) : Low Temperature Stress in Crop Plants : The Role of the Membrane. Pp. 535 - 542. Academic Press, New York - San Francisco - London 1979. [Chl.]

39173 - POTTER, J.R., BREEN, P.J. : Partitioning of photosynthate to leaf starch and sucrose. - Plant Physiol. *63* (Suppl.) : 44, 1979.

39174 - POULSEN, C. : The cyanogen bromide fragments of the large subunit of ribu-losebisphosphate carboxylase from barley. - Carlsberg Res. Commun. *44* : 163 - 189, 1979.

39175 - POULSEN, C., MARTIN, B., SVENDSEN, I. : Partial amino acid sequence of the large subunit of ribulosebisphosphate carboxylase from barley. - Carlsberg Res. Commun. *44* : 191 - 199, 1979.

39176 - POW, T., KRASNA, A.I. : Photoproduction of hydrogen from water in hydroge-nase-containing algae. - Arch. Biochem. Biophys. *194* : 413 - 421, 1979.

39177 - POWERS, C.D., WURSTER, C.F., ROWLAND, R.G. : DDE inhibition of marine algal cell division and photosynthesis per cell. - Pestic. Biochem. Physiol. *10* : 306 - 312, 1979.

39178 - POWLES, S.B., CRITCHLEY, C., OSMOND, C.B. : Photoinhibition in the absence of photorespiration. - Plant Physiol. *63* (Suppl.) : 152, 1979.

39179 - POWLES, S.B., OSMOND, C.B., THORNE, S.W. : Photoinhibition of intact atta-ched leaves of C$_3$ plants illuminated in the absence of both carbon dioxide and of photorespiration. - Plant Physiol. *64* : 982 - 988, 1979.

39180 - PRADEL, J., CLEMENT-METRAL, J. : A 4-vinylprotochlorophyllide complex accu-mulated by "Phofil" mutant of *Rhodopseudomonas spheroides*. Spectral proper-ties and macromolecular conformation. - Photosynthetica *13* : 29 - 36, 1979.

39181 - PRAHL, C. : Photosynthesis and respiration of some littoral marine algae from Greenland. - Phycologia *18* : 166 - 168, 1979.

39182 - PRASAD, B.J., RAO, D.N. : Influence of nitrogen dioxide (NO$_2$) on photosyn-thetic apparatus and net primary productivity of wheat plants. - Acta bot. indica *7* : 16 - 21, 1979.

39183 - PREISS, J., LEVI, C. : Metabolism of starch in leaves. - In : GIBBS, M., LATZKO, E. (ed.) : Photosynthesis II. (Encycl. Plant Physiol. N.S. Vol. 6.) Pp. 282 - 312. Springer-Verlag, Berlin - Heidelberg - New York 1979.

39184 - PRENZEL, U., LICHTENTHALER, H.K. : Separation of prenyllipids by high per-formance liquid chromatography. - In : APPELQVIST, L.-Å., LILJENBERG, C. (ed.) : Advances in the Biochemistry and Physiology of Plant Lipids. Pp. 319 - 325. Elsevier/North-Holland Biomedical Press, Amsterdam 1979. [Chl, Car.]

39185 - PREZELIN, B.B., MATLICK, A. : Time-course for photoadaptation of photosynthesis in a dinoflagellate. - J. Phycol. *15* (Suppl.) : 26, 1979.

39186 - PREZELIN, B.B., SWEENEY, B.M. : Photoadaptation of photosynthesis in two
bloom-forming dinoflagellates. - In : TAYLOR, D.L., SELIGER, H.H. (ed.) :
Toxic Dinoflagellate Blooms. Vol. 1. Pp. 101 - 106. Elsevier/North Holland,
New York 1979.

39187 - PRIMAK, A.P., SHELEPOVA, V.M., SHMANAEVA, T.N. : Vliyanie uslovii ponizhen-
noi osveshchennosti na rost i soderzhanie khlorofilla v prorostkakh razlichnykh po svetotrebovatel'nosti sortov tomata. [Effect of conditions of
reduced light on the growth and level of chlorophyll in varieties of tomato
seedlings with different light requirements.] - Tr. vsesoyuz. nauch.-issled.
Inst. Selektsii Semenovod. ovoshch. Kul'tur *1979* (9) : 86 - 95, 1979.
[In R.]

39188 - PRINS, H.B.A., HELDER, R.J. : Photosynthetic use of HCO_3^- by *Potamogeton* and
Elodea. - Plant Physiol. *63* (Suppl.) : 61, 1979.

39189 - PRINS, H.B.A., SNEL, J.F.H., HELDER, R.J., ZANSTRA, P.E. : Photosynthetic
bicarbonate utilization in the aquatic angiosperms *Potamogeton* and *Elodea*.-
Hydrobiol. Bull. *13* : 106 - 111, 1979.

39190 - PRISTUPA, N.A. : Lokalizatsiya ketosakharov v assimiliruyushchikh i provo-
dyashchikh tkanyakh lista sakharnoi svekly. [Localization of ketosugars in
assimilatory and conducting tissues of the sugar-beet leaf.] - Fiziol. Rast.
26 : 584 - 592, 1979. [In R, ab : E.]

*B39191 - Problemy Biosinteza Khlorofilla. [Problems of Chlorophyll Biosynthesis.]-
Nauka i Tekhnika, Minsk 1971. [In R.]

39192 - PROHÁSZKA, K., HORVÁTH, R. : A szerves és műtrágyák hatása a fiatal rozsnö-
vények festékanyag-tartalmára. [Effect of organic and mineral fertilizers
affecting the content of pigments in young rye plants.] - Bot. Közlem. *66* :
109 - 113, 1979. [In Hung., ab : G.]

39193 - PRONINA, N.B. : Vliyanie sveta na aktivnost' adenilatkinazy i pterinbelko-
vykh kompleksov iz khloroplastov gorokha. [Effect of light on the activity
of adenylate kinase and pterin-protein complexes from pea chloroplasts.] -
Izv. Akad. Nauk SSSR, Ser. biol. *1979* (2) : 228 - 237, 1979. [In R, ab : E.]

39194 - PRONINA, N.B., LADONIN, V.F. : Deistvie 2,4-D na sopryazhenie transporta
ělektronov s sintezom ATF v khloroplastakh yachmenya i gorokha i snizhenie
toksicheskogo deistviya gerbitsida pod vliyaniem fosfornykh udobrenii.
[Effect of 2,4-D on the connection of electron transport with the synthesis
of ATP in barley and pea chloroplasts.] - Sel'skokhoz. Biol. *14* : 41 - 44,
1979. [In R, ab : E.]

39195 - PROSUNKO, V.M., KOZEL, A.I. : Opredelenie ploshchadi listovoi poverkhnosti
risa raschetnym metodom. [The determination of rice leaf surface using the
calculation method.] - Sel'skokhoz. Biol. *14* : 232 - 234, 1979. [In R,
ab : E.]

39196 - PRUDER, G.D., BOLTON, E.T. : The role of CO_2 enrichment of aerating gas
in the growth of an estuarine diatom. - Aquaculture *17* : 1 - 15, 1979.
[Ps.]

39197 - PUISEUX-DAO, S., HOURSIANGOU-NEUBRUN, D., DUBACQ, J.P. : Cytoplasmic strea-
ming and chloroplast differentiation in *Acetabularia*. - In : BONOTTO, S.,
KEFELI, V., PUISEUX-DAO, S. (ed.) : Developmental Biology of *Acetabularia*.
Pp. 141 - 154. Elsevier/North-Holland Biomedical Press, Amsterdam - New
York - Oxford 1979.

39198 - PULLIN, C.A., BROWN, R.G., EVANS, E.H. : Detection of allophycocyanin in
photosystem I preparations from the blue-green alga, *Chlorogloea fritschii*.
- FEBS Lett. *101* : 110 - 112, 1979.

39199 - PUNZ, W. : Der Einfluss isolierter und kombinierter Schadstoffe auf die
Flechtenphotosynthese. - Photosynthetica *13* : 428 - 433, 1979.

39200 - PUNZ, W. : The effect of single and combined pollutants on lichen water content. - Biol. Plant. *21* : 472 - 474, 1979. [Ps.]

39201 - PUPILLO, P., BOSSI, P. : Two forms of NADP-dependent malic enzyme in expanding maize leaves. - Planta *144* : 283 - 289, 1979.

39202 - PUPILLO, P., FAGGIANI, R. : Subunit structure of three glyceraldehyde 3-phosphate dehydrogenases of some flowering plants. - Arch. Biochem. Biophys. *194* : 581 - 592, 1979.

39203 - PUROHIT, A.N., PODOLNY, V.Z., CHETVERIKOV, A.G. : Association between onset of generative physe and amplitude of ESR signal I in wheat leaves. - Naturwissenschaften *66* : 473 - 475, 1979.

39204 - PUROHIT, K., McFADDEN, B.A. : Ribulose 1,5-bisphosphate carboxylase and oxygenase from *Thiocapsa roseopersicina* : Activation and catalysis. - Arch. Biochem. Biophys. *194* : 101 - 106, 1979.

39205 - PUROHIT, K., McFADDEN, B.A., LAWLIS, V.B. : Ribulose bisphosphate carboxylase/oxygenase from *Thiocapsa roseopersicina*. - Arch. Microbiol. *121* : 75 - 82, 1979.

39206 - PUROHIT, S.S. : Studies with a new growth regulator : dikegulac. 1. Effects on seed germination, seedling growth and chlorophyll biosynthesis. - Comp. Physiol. Ecol. *4* : 264 - 266, 1979.

*39207 - PUSHNYAK, L.F. : Vliyanie mineral'nogo pitaniya na pigmenty fotosinteticheskogo apparata v svyazi s produktivnost'yu rastenii podsolnechnika. [Effect of mineral nutrition on pigments of the photosynthetic apparatus in relation to sunflower plants productivity.] - In : Mineral'noe Pitanie i Produktivnost' Rastenii. Pp. 178 - 185, 324. Naukova Dumka, Kiev 1978. [In R.]

39208 - PUTT, D.A., HOUGH, R.A. : Measurement of CO_2 compensation point of the macroalgae *Chara vulgaris* (*Characeae*). - J. Phycol. *15* (Suppl.) : 26, 1979.

39209 - PYRINA, I.L. : Primary production of phytoplankton in the Volga. - In : MORDUKHAI-BOLTOVSKOI, P.D. (ed.) : The·River Volga and Its Life. Pp. 180 - 194. Dr. W.Junk bv. Publ., The Hague - Boston - London 1979.

39210 - PYRINA, I.L., TRIFONOVA, I.S. : Issledovaniya produktivnosti fitoplanktona Ladozhskogo ozera. [Phytoplankton productivity of the Ladoga lake.] - Gidrobiol. Zh. *15* (4) : 26 - 31, 1979. [Chl; in R, ab : E.]

39211 - QUADIR, A., HARRISON, P.J., DeWREEDE, R.E. : The effects of emergence and subemergence of the photosynthesis and respiration of marine macrophytes. - Phycologia *18* : 83 - 88, 1979.

39212 - QUARRIE, S.A., JONES, H.G. : Genotypic variation in leaf water potential, stomatal conductance and· abscisic acid concentration in spring wheat subjected to artificial drought stress. - Ann. Bot. *44* : 323 - 332, 1979.

39213 - QUEBEDEAUX, B. : Symbiotic N_2 fixation and its relationship to photosynthetic carbon fixation in higher plants. - In : GIBBS, M., LATZKO, E. (ed.) : Photosynthesis II. (Encycl. Plant Physiol. N.S. Vol. 6.) Pp. 472 - 480. Springer-Verlag, Berlin - Heidelberg - New York 1979.

39214 - QUEBEDEAUX, B. : Oxygen concentration effects on assimilate partitioning and energy production in developing soybean seeds. - Plant Physiol. *63* (Suppl.) : 39, 1979.

*39215 - QUEBEDEAUX, B., GIAQUINTA, R.T. : Oxygen effects on metabolite distribution of $^{14}CO_2$-derived assimilates in developing soybean seeds. - Plant Physiol. *61* (Suppl.) : 8, 1978.

39216 - QUEIROZ, O. : Les horloges biologiques. Organisation du métabolisme chez les végetaux supérieurs et adaptation au climat. - Bull. Soc. bot. Fr. *126* (Actual.bot.1) : 5 - 19, 1979. [CAM.]

39217 - **QUEIROZ, O.** : CAM : Rhythms of enzyme capacity and activity as adaptive mechanisms. - In : GIBBS, M., LATZKO, E. (ed.) : Photosynthesis II. (Encycl.
Plant Physiol. N.S. Vol. 6.) Pp. 126 - 139. Springer-Verlag, Berlin - Heidelberg - New York 1979.

39218 - **RABE, R., KREEB, K.H.** : Enzyme activities and chlorophyll and protein content in plants as indicators of air pollution. - Environ. Pollut. *19* :
119 - 137, 1979.

39219 - **RABIE, R.K., ARIMA, Y., KUMAZAWA, K.** : Effect of combined nitrogen on the
translocation pattern of photosynthetic assimilates in nodulated soybean
plant as revealed by ^{14}C. - In : Mineral Nutrition of Plants. Vol. I. Pp.
297 - 311. Publ. House Central Cooperative Union, Sofia 1979.

*B39220 - **RACKER, E.** : A New Look at Mechanisms in Bioenergetics. - Academic Press,
New York - San Francisco - London 1976.

39221 - **RACKER, E., VIOLAND, B., O'NEAL, S., ALFONZO, M., TELFORD, J.** : Reconstitution, a way of biochemical research; some new approaches to membrane-bound
enzymes. - Arch. Biochem. Biophys. *198* : 470 - 477, 1979.

39222 - **RADEMAKER, H., HOFF, A.J., DUYSENS, L.N.M.** : Magnetic field-induced increase of the yield of (bacterio)chlorophyll emission of some photosynthetic
bacteria and of *Chlorella vulgaris*. - Biochim. biophys. Acta *546* : 248 -
255, 1979.

39223 - **RADEVA, V., TOPCHIEVA, A.** : Vliyanie na periodichnoto pochveno zasushavane
v"rkhu natrupvaneto na sukho veshchestvo, dobiva na semena i nyakoi fizio-
logichni pokazateli pri lyutsernata. [Periodical soil drought as affecting
dry matter accumulation, seed yield and some physiological characteristics
in alfalfa.] - Rasteniev"dni Nauki *16* (1) : 13 - 25, 1979. [Ps; in Bulg.,
ab : E, R.]

39224 - **RADIN, J.W., PARKER, L.L.** : Water relations of cotton plants under nitrogen
deficiency. II. Environmental interactions on stomata. - Plant Physiol.
64 : 499 - 501, 1979. [Stomatal resistance.]

39225 - **RADMER, R.** : Mass spectrometric determination of hydroxylamine photooxidation by illuminated chloroplasts. - Biochim. biophys. Acta *546* : 418 - 425,
1979.

39226 - **RADOSEVICH, S.R., STEINBACK, K.E., ARNTZEN, C.J.** : Effect of photosystem II
inhibitors of thylakoid membranes of two common groundsel (*Senecio vulgaris*)
biotypes. - Weed Sci. *27* : 216 - 218, 1979.

39227 - **RADUNZ, A.** : Binding of antibodies onto the thylakoid membrane. V. Distribution of proteins and lipids in the thylakoid membrane. - Z. Naturforsch.
34 C : 1199 - 1204, 1979.

39228 - **RAFFERTY, C.N.** : Light-induced perturbation of aromatic residues in bovine
rhodopsin and bacteriorhodopsin. - Photochem. Photobiol. *29* : 109 - 120,
1979.

39229 - **RAFFERTY, C.N., BOLT, J., SAUER, K., CLAYTON, R.K.** : Photooxidation of
antenna bacteriochlorophyll in chromatophores from carotenoidless mutant
Rhodopseudomonas sphaeroides and the attendant loss of dimeric exciton interactions. - Proc. nat. Acad. Sci. USA *76* : 4429 - 4432, 1979.

39230 - **RAFFERTY, C.N., CLAYTON, R.K.** : Linear dichroism and the orientation of
reaction centers of *Rhodopseudomonas sphaeroides* in dried gelatin films. -
Biochim. biophys. Acta *545* : 106 - 121, 1979.

39231 - **RAFFERTY, C.N., CLAYTON, R.K.** : The orientations of reaction center transition moments in the chromatophore membrane of *Rhodopseudomonas sphaeroides*,
based on new linear dichroism and photoselection measurements. - Biochim.
biophys. Acta *546* : 189 - 206, 1979.

39232 - RAFII, Z.E., ASHTON, F.M., GLENN, R.K. : Metabolic sites of action of fluridone in isolated mesophyll cells. - Weed Sci. 27 : 422 - 426, 1979. [Ps.]

39233 - RAGAN, M.A., JENSEN, A. : Quantitative studies on brown algal phenols. III. Light-mediated exudation of polyphenols from Ascophyllum nodosum (L.) LE JOL. - J. exp. mar. Biol. Ecol. 36 : 91 - 101, 1979. [Ps.]

*39234 - RAGGI, V. : CO_2 release in light, dark and upon light-dark transition (CO_2 outburst) in rust-infected Bean leaves. - Phytopathol. mediterranea 17 : 135 - 138, 1978.

*39235 - RAGHAVENDRA, A.S. : Kranz leaf anatomy and C_4 dicarboxylic acid pathway of photosynthesis in Spinifex squarrosus L. - Indian J. exp. Biol. 15 : 645 - 648, 1977.

39236 - RAGHAVENDRA, A.S. : Variation with age in the carboxylation pattern by leaves of Amaranthus paniculatus and Oryza sativa. - Plant Physiol. 63 (Suppl.) : 159, 1979.

*39237 - RAGHAVENDRA, A.S., DAS, V.S.R. : Diversity in the biochemical and biophysical characteristics of C_4 dicotyledonous plants. - Indian J. Plant Physiol. 19 : 101 - 112, 1976.

*39238 - RAGHAVENDRA, A.S., RAJENDRUDU, G., DAS, V.S.R. : Simultaneous occurrence of C_3 and C_4 photosynthesis in relation to leaf position in Mollugo nudicaulis. - Nature 273 : 143 - 144, 1978.

39239 - RAGHAVENDRA, A.S., VALLEJOS, R.H. : Regulation of phosphoenolpyruvate carboxylase in C_4 plants: Inhibition by pyrophosphate. - Plant Physiol. 63 (Suppl.) : 4, 1979.

39240 - RAINS, D.W. : Salt tolerance of plants: Strategies of biological systems. - In : HOLLAENDER, A., ALLER, J.C., EPSTEIN, E., SAN PIETRO, A., ZABORSKY, O.R. (ed.):The Biosaline Concept.An Approach to the Utilization of Underexploited Resources. Pp. 47 - 67. Plenum Press, New York - London 1979. [Ps.]

39241 - RAISON; J., BERRY, J., BJÖRKMAN, O. : Changes in chloroplast lipid fluidity during acclimation of plants to high and low growth temperature. - Plant Physiol. 63 (Suppl.) : 141, 1979.

39242 - RAISON, J.K., BERRY, J.A. : Viscotropic denaturation of chloroplast membranes and acclimation to temperature by adjustment of lipid viscosity. - Carnegie Inst. Year Book 78 : 149 - 152, 1979.

*39243 - RAKHIMOV, G.T. : Aktivnost' izolirovannykh khloroplastov v zavisimosti ot vozrasta list'ev. [Activity of isolated chloroplasts depending on leaf age.] - Uzb. biol. Zh. 17 (4) : 30 - 32, 1973. [In R.]

*39244 - RAKHTSEENKA, I.N., BUDKEVICH, T.A. : Uzaemaŭplyŭ kanyushyny chyrvonaĭ, lyutsěrny i tsimafeeŭki na ikh rost i praduktsyĭnasts' u zmeshanykh pasevakh. [Growth and productivity of mixed populations as affected by interactions among red clover, alfalfa and timothy grass.] - Vestsi Akad. Navuk belarus. SSR, Ser. biyal. Navuk 1978 (4) : 10 - 15, 138, 1978. [In Belorus., ab : E, R.]

39245 - RAMAKRISHNAN, T.V., FRANCIS, F.J. : Stability of carotenoids in model aqueous systems. - J. Food Qual. 2 : 177 - 189, 1979.

39246 - RAMAKRISHNAN, T.V., FRANCIS, F.J.' : Coupled oxidation of carotenoids in fatty acid esters of varying unsaturation. - J. Food Qual. 2 : 277 - 287, 1979.

39247 - RAMASWAMY, N.K., NAIR, P.M. : Inhibition of O_2 evolution by NADP in isolated potato tuber chloroplast and its identity with 3-(3,4-dichlorophenyl)-1,1-dimethyl urea inhibition site. - Arch. Biochem. Biophys. 193 : 56 - 62, 1979.

39248 - RAMATI, A., LIPHSCHITZ, N., WAISEL, Y. : Osmotic adaptation in Panicum repens. Differences between organ, cellular and subcellular levels. - Physiol. Plant. 45 : 325 - 331, 1979. [Chloroplast.]

39249 - RAMINA, A., HACKETT, W.P., SACHS, R.M. : Flowering in *Bougainvillea*. A function of assimilate supply and nutrient diversion. - Plant Physiol. *64* : 810 - 813, 1979.

39250 - RAMÍREZ, R., BATZÁN, T., VICENTE, C. : Regulation of ferredoxin-NADP⁺-reductase from *Evernia prunastri* by several nucleotides. - Phyton *37* : 81 - - 84, 1979.

39251 - RANGNEKAR, P.: Effects of calcium deprivation on the intracellular distribution of the early products of photosynthesis in tomato leaf cells. - In: MARCELLE, R., CLIJSTERS, H., VAN POUCKE, M. (ed.) : Photosynthesis and Plant Development. Pp. 185 - 192. Dr.W.Junk bv.- Publ., The Hague - Boston - - London 1979.

39252 - RAO, K.K., HALL, D.O. : Hydrogen production from isolated chloroplasts. - In : BARBER, J. (ed.) : Photosynthesis in Relation to Model Systems. Pp. 299 - 329. Elsevier, Amsterdam - New York - Oxford 1979.

39253 - RAO, K.R., KUMAR, N.R., REDDY, A.N. : Studies of photosynthesis in some liverworts. - Bryologist *82* : 286 - 289, 1979.

*39254 - RAO, P.G. : Influence of riboflavin on growth, respiration, and chlorophyll and protein contents in green gram (*Phaseolus radiatus* LINN.). - Curr. Sci. *42* : 580 - 581, 1973.

39255 - RAO, S.S., KWIATKOWSKI, R.E., JURKOVIC, A.A. : Distribution of bacteria and chlorophyll *a* at a nearshore station in Lake Ontario. - Hydrobiologia *66* : 33 - 39, 1979.

39256 - RAO, S.V.R., RAO, M.S., SINGH, V.P. : Effect of DDT and sumithion on the photosynthesis of phytoplankton. - Nat. Acad. Sci. Lett. *2* : 325 - 326, 1979.

39257 - RAO, S.V.R., SINGH, V.P., MALL, L.P. : The effect of sewage and industrial waste discharges on the primary production of a shalow turbulent river. - Water Res. *13* : 1017 - 1021, 1979.

39258 - RAPS, S. : Thylakoid polypeptide profiles of PSI and PSII reaction center mutants of *Scenedesmus obliquus*. - Plant Physiol. *63* (Suppl.) : 54, 1979.

39259 - RASCHKE, K. : Movements of stomata. - In : HAUPT, W., FEINLEIB, M.E. (ed.) : Physiology of Movements. (Encycl. Plant Physiol. N.S. Vol.7.) Pp. 383 - - 441. Springer-Verlag, Berlin - Heidelberg - New York 1979.

39260 - RASCIO, N., CASADORO, G. : Sunflower etioplast membranes. - J. Ultrastr. Res. *68* : 325 - 327, 1979.

39261 - RASKIN, V.I. : Izmenenie kvantovogo vykhoda fluorestsentsii khlorofillida v protsesse vosstanovleniya protokhlorofillida v ětiolirovannykh list'yakh. [Change in quantum yield of chlorophyllide fluorescence in the process of protochlorophyllide reduction in etiolated leaves.] - Dokl. Akad. Nauk SSSR *245* : 1487 - 1489, 1979. [In R.]

39262 - RASPOPOV, I.M. : Vegetation der großen seichten Seen im Nordwesten der UdSSR und ihre Produktion. - Arch. Hydrobiol. *86* : 242 - 253, 1979.

39263 - RATHNAM, C.K.M. : Metabolic regulation of carbon flux during C₄ photosynthesis. II. *In situ* evidence for refixation of photorespiratory CO₂ by C₄ phosphoenolpyruvate carboxylase. - Planta *145* : 13 - 23, 1979.

39264 - RATHNAM, C.K.M., CHOLLET, R. : Phosphoenolpyruvate carboxylase reduces photorespiration in *Panicum milioides*, a C₃-C₄ intermediate species. - Arch. Biochem. Biophys. *193* : 346 - 354, 1979.

39265 - RATHNAM, C.K.M., CHOLLET, R. : Photosynthetic carbon metabolism in *Panicum milioides*, a C₃-C₄ intermediate species: Evidence for a limited C₄ dicarboxylic acid pathway of photosynthesis. - Biochim. biophys. Acta *548* : 500 - 519, 1979.

39266 - RATHNAM, C.K.M., CHOLLET, R. : Evidence for C₄ photosynthesis in *Panicum milioides*, a C₃-C₄ intermediate species with reduced photorespiration. - Plant Physiol. *63* (Suppl.) : 38, 1979.

B39267 - RATINEN, H. : Fluorescence Spectra from Dark- and Light-Grown Pea Leaves
         at 77 K. - Res. Rep. Dep. Phys. Univ. Oulu *1* : 1 - 11 + 7 pp., 1979.

39268 - RATYNI, A.I., RIZNICHENKO, G.Yu., CHAMOROVSKIĬ, S.K., VOROB'EVA, T.N., PYT'-
        EVA, N.F., RUBIN, A.B. : Funktsional'naya organizatsiya èlektrontransport-
        noĭ tsepi khromatoforov *Rhodospirillum rubrum* v otsutstvie èkzogennogo do-
        nora èlektronov. [Functional organization of the electron transport chain
        in chromatophores of *Rhodospirillum rubrum* in the absence of exogenous
        electron donor.] - Biofizika *24* : 671 - 675, 1979. [In R, ab : E.]

39269 - RAU, W., SCHROTT, E.L. : Light-mediated biosynthesis in plants. - Photochem.
        Photobiol. *30* : 755 - 765, 1979. [Chl.]

39270 - RAWSON, H.M. : Vertical wilting and photosynthesis, transpiration, and wa-
        ter use efficiency of sunflower leaves. - Aust. J. Plant Physiol. *6* :
        109 - 120, 1979.

39271 - RAWSON, H.M., BAGGA, A.K. : Influence of temperature between floral ini-
        tiation and flag leaf emergence on grain number in wheat. - Aust. J. Plant
        Physiol. *6* : 391 - 400, 1979. [Dry-matter accumulation.]

39272 - RAWSON, H.M., CRAVEN, C.L. : Variation between short-duration mungbeans
        cultivars (*Vigna radiata* (L.) WILCZEK) in response to temperature and pho-
        toperiod. - Indian J. Plant Physiol. *22* : 127 - 136, 1979. [Dry-matter ac-
        cumulation.]

39273 - RAWSON, J.R.Y., BOERMA, C.L. : Hybridization of EcoRI chloroplast DNA frag-
        ments of *Euglena* to pulse labeled RNA from different stages of chloroplast
        development. - Biochem. biophys. Res. Commun. *89* : 743 - 749, 1979.

39274 - RAY, R.C., MISHRA, D. : Bleaching effect of ethylenediaminetetraacetic
        acid on detached rice leaves. - Nat. Acad. Sci. Lett. *2* : 247 - 248, 1979.
        [Chl.]

39275 - RAY, T.B., BLACK, C.C. : The C_4 pathway and its regulation. - In : GIBBS,
        M., LATZKO, E. (ed.) : Photosynthesis II. (Encycl. Plant Physiol. N.S. Vol.
        6.) Pp. 77 - 101. Springer-Verlag, Berlin - Heidelberg - New York 1979.

39276 - RAY, T.B., MAYNE, B.C., PETERS, G.A., TOIA, R. : Action spectra for photo-
        synthesis and acetylene reduction in the *Azolla-Anabaena* symbiosis. - Plant
        Physiol. *63* (Suppl.) : 112, 1979.

39277 - RAY, T.B., MAYNE, B.C., TOIA, R.E. Jr., PETERS, G.A. : *Azolla-Anabaena* re-
        lationship. VIII. Photosynthetic characterization of the association and
        individual partners. - Plant Physiol. *64* : 791 - 795, 1979.

39278 - REBEIZ, C.A., BAZZAZ, M.B. : Cell-free agriculture: The concept and its
        initial implementation. - Biotechnol. Bioeng. Symp. *8* (Biotechnol. Energy
        Prod. Conserv.) : 453 - 471, 1979. [Ps.]

39279 - REBEIZ, C.A., BELANGER, F., COHEN, C.E., McCARTHY, S.A. : Greening of
        higher plants. - Illinois Res. *21* (1) : 3 - 4, 1979.

39280 - REDDY, M.R., PRASAD, R. : Effect of nitrogen doses and raw direction on
        LAI, light transmission, plant height and dry matter production of wheat
        cultivars grown in pure and mixed stand. - Biol. Plant. *21* : 85 - 91,
        1979.

*39281 - REDDY, T.P., VAIDYANATH, K. : Synergistic interaction of gamma rays and
        some metallic salts in the induction of chlorophyll mutations in rice. -
        Mutat. Res. *52* : 361 - 365, 1978.

39282 - REDLINGER, T., APEL, K. : Isolation and characterization of four proto-
        chlorophyllide-binding polypeptides in barley. - Plant Physiol. *63* (Suppl.)
        : 97, 1979.

39283 - REED, M.L. : Intracellular location of carbonate dehydratase (carbonic
        anhydrase) in leaf tissue. - Plant Physiol. *63* : 216 - 217, 1979.

39284 - REGER, B.J., YATES, I.E. : Distribution of photosynthetic enzymes between
        mesophyll, specialized parenchyma and bundle sheath cells of *Arundinella
        hirta*. - Plant Physiol. *63* : 209 - 212, 1979.

39285 - REICH, R. : Intrinsic probes of charge separation. - In : GERISCHER, H.,
         KATZ, J.J. (ed.) : Light-Induced Charge Separation in Biology and Chemistry.
         Pp. 361 - 387. Verlag Chemie, Weinheim - New York 1979. [Chl, Car.]

39286 - REICHE, H., BARD, A.J. : Heterogeneous photosynthetic production of amino
         acids from methane-ammonia-water at Pt/TiO$_2$. Implications in chemical evo-
         lution. - J. amer. chem. Soc. 101 : 3127 - 3128, 1979.

*39287 - REILLY, A., REILLY, C. : Copper-induced chlorosis in Becium Homblei (De
         WILD.) DUVIGN. & PLANCKE. - Plant Soil 38 : 671 - 674, 1973.

39288 - REIMER, S., LINK, K., TREBST, A. : Comparison of the inhibition of photo-
         synthetic reactions in chloroplasts by dibromothymoquinone, bromonitrothy-
         mol and ioxynil. - Z. Naturforsch. 34 C : 419 - 426, 1979.

39289 - REIMER, S., SELMAN, B.R. : Tentoxin-induced binding of adenine nucleotides
         to soluble spinach chloroplast coupling factor 1. - Biochim. biophys. Acta
         545 : 415 - 423, 1979.

39290 - REIMER, T.O., BARGHOORN, E.S., MARGULIS, L. : Primary productivity in an
         early Archean microbial ecosystem. - Precambrian Res. 9 : 93 - 104, 1979.

39291 - REINKE, D.C., DeNOYELLES, F. : The effect of cadmium on primary production
         of natural phytoplankton communities with varying light intensities and
         exposure times. - J. Phycol. 15 (Suppl.) : 22, 1979.

39292 - REINMAN, S., THORNBER, J.P. : The electrophoretic isolation and partial
         characterization of three chlorophyll-protein complexes from blue-green
         algae. - Biochim. biophys. Acta 547 : 188 - 197, 1979.

39293 - REMENNIKOV, V.G., SAMUILOV, V.D. : Netsiklicheskiĭ perenos èlektronov i
         generatsiya membrannogo potentsiala v khromatoforakh Rhodospirillum rubrum.
         [Non-cyclic electron transport and generation of membrane potential in chro-
         matophores of Rhodospirillum rubrum.] - Biol. Nauki 1979 (5) : 45 - 52,
         1979. [In R.]

39294 - REMENNIKOV, V.G., SAMUILOV, V.D. : Generatsiya membrannogo potentsiala pri
         funktsionirovanii polnoĭ i sokrashchennoĭ sistem tsiklicheskogo perenosa
         èlektronov v khromatoforakh Rhodospirillum rubrum. [Membrane potential ge-
         neration at the functioning of the complete and shortened systems of cyclic
         electron transfer in Rhodospirillum rubrum chromatophores.] - Biol. Nauki
         1979 (10) : 24 - 29, 1979. [In R.]

39295 - REMENNIKOV, V.G., SAMUILOV, V.D. : Photooxidase activity of Rhodospirillum
         rubrum chromatophores and reaction center complexes. The role of non-cyc-
         lic electron transfer in generation of the membrane potential. - Biochim.
         biophys. Acta 546 : 220 - 235, 1979.

39296 - REMENNIKOV, V.G., SAMUILOV, V.D. : Two regimens of electrogenic cyclic re-
         dox chain operation in chromatophores of non-sulfur purple bacteria. A
         study using antimycin A. - Biochim. biophys. Acta 548 : 216 - 223, 1979.

39297 - REMENNIKOV, V.G., SAMUILOV, V.D. : Photooxidase activity of isolated chro-
         matophores and intact cells of phototrophic bacteria. - Arch. Mikrobiol.
         123 : 65 - 71, 1979. [Ps.]

39298 - RENGER, G. : A rapid vectorial back reaction at the reaction centers of
         Photosystem II in Tris-washed chloroplasts induced by repetitive flash
         excitation. - Biochim. biophys. Acta 547 : 103 - 116, 1979.

39299 - RENGER, G. : Studies about the mechanism of herbicidal interaction with
         photosystem II in isolated chloroplasts. - Z. Naturforsch. 34 C : 1010 -
         - 1014, 1979.

39300 - RENGER, G., DIFIORE, D., LUURING, B., GRÄBER, P. : The variation of the
         electrochromic difference spectrum at various stages of the chloroplast de-
         velopment. - Z. Naturforsch. 34 C : 120 - 124, 1979.

39301 - RENGER, G., TIEMANN, R. : Studies on the proton transport at system II in
         trypsin-treated spinach chloroplasts. - Biochim. biophys. Acta 545 :
         316 - 324, 1979.

39302 - RENTHAL, R., HARRIS, G.J., PARRISH, R. : Reaction of the purple membrane with a carbodiimide. - Biochim. biophys. Acta 547 : 258 - 269, 1979.

39303 - RENTZEPIS, P.M. : Energy transfer mechanisms studied by picosecond spectroscopy. - In : GERISCHER, H., KATZ, J.J. (ed.) : Light-Induced Charge Separation in Biology and Chemistry. Pp. 471 - 492. Verlag Chemie, Weinheim - New York 1979. [Ps.]

39304 - RESTAINO, F., SCARAMUCCI, S., INTERLANDI, G., MARCHESINI, A. : Dosaggio enzimatico dell'ossigeno disciolto nei liquidi. Nota I: Velocita fotosintetica in cultivar di *Cichorium endivia* L. [Enzymatic determination of oxygen in aqueous solutions. Note I: Photosynthetic rate in cultivar of *Cichorium endivia* L.] - Atti Soc. ital. Sci. nat. Mus. Civ. Stor. Nat. Milano 120 : 132 - 140, 1979. [In Ital., ab : E.]

39305 - RESTALL, C., CHAPMAN, D., QUINN, P.J. : Hydrogenation of the polyunsaturated lipids of chloroplast membranes and the effects on photosynthetic functions. - Biochem. Soc. Trans. 7 : 366 - 369, 1979.

39306 - RESTALL, C.J., WILLIAMS, P., PERCIVAL, M.P., QUINN, P.J., CHAPMAN, D. : The modulation of membrane fluidity by hydrogenation processes. III. The hydrogenation of biomembranes of spinach chloroplasts and a study of the effect of this on photosynthetic electron transport. - Biochim. biophys. Acta 555 : 119 - 130, 1979.

39307 - RETZLAFF, G., HILTON, J.L., JOHN, J.B.S. : Inhibition of photosynthesis by bentazon in intact plants and isolated cells in relation to the pH. - Z. Naturforsch. 34 C : 944 - 947, 1979.

39308 - REUTER, J.E. : Seasonal distribution of phytoplankton biomass in a nearshore area of the central basin of Lake Erie, 1975-1976. - Ohio J. Sci. 79 : 218 - 226, 1979.

39309 - REYNOLDS, J.F., CUNNINGHAM, G.L., SYVERTSEN, J.P. : A net CO_2 exchange model for *Larrea tridentata*. - Photosynthetica 13 : 279 - 286, 1979.

39310 - RHODES, P., HYMAN, F., KUNG, S.-D., SCHAEFFER, G. : Biochemical characterization of a yellow mutation of tobacco. - Plant Physiol. 63 (Suppl.) : 160, 1979. [Chl.]

39311 - RICHARD, E.P.,Jr., GOSS, J.R., ARNTZEN, C.J. : Glyphosate does not inhibit photosynthetic electron transport and phosphorylation in pea (*Pisum sativum*) chloroplasts. - Weed Sci. 27 : 684 - 688, 1979.

39312 - RICHARDS, R.A., THURLING, N. : Genetic analysis of drought stress response in rapeseed (*Brassica campestris* and B. *napus*). III. Physiological characters. - Euphytica 28 : 755 - 759, 1979. [Chl.]

39313 - RICHARDSON, D.H.S., NIEBOER, E., LAVOIE, P., PADOVAN, D. : The role of metal-ion binding in modifying the toxic effects of sulphur dioxide on the lichen *Umbilicaria muhlenbergii*. II. ^{14}C-fixation studies. - New Phytol. 82 : 633 - 643, 1979.

*39314 - RICHARDSON, J.L., JIN, L.T. : Algal productivity of natural and artificially enriched fresh waters in Malaya. - Verh. int. Ver. theor. angew. Limnol. 19 : 1383 - 1389, 1975. [Ps.]

39315 - RICKLE, G.K., CUSANOVICH, M.A. : The kinetics of photooxidation of c-type cytochromes by *Rhodospirillum rubrum* reaction centers. - Arch. Biochem. Biophys. 197 : 589 - 598, 1979.

39316 - RIDLEY, S.M., RIDLEY, J. : Interaction of chloroplasts with inhibitors. Location of carotenoid synthesis and inhibition during chloroplast development. - Plant Physiol. 63 : 392 - 398, 1979.

*39317 - RIEMANN, B., MATHIESEN, H. : Danish research into phytoplankton primary production. - Folia limnol. scand. 17 (Danish Limnology. Reviews and Perspectives) : 49 - 54, 1977.

39318 - RIJGERSBERG, C.P., AMESZ, J., THIELEN, A.P.G.M., SWAGER, J.A. : Fluorescence emission spectra of chloroplasts and subchloroplast preparations at low temperature. - Biochim. biophys. Acta *545* : 473 - 482, 1979 .

39319 - RIJGERSBERG, C.P., MELIS, A., AMESZ, J., SWAGER, J.A. : Quenching of chlorophyll fluorescence and photochemical activity of chloroplasts at low temperature. - In : Chlorophyll Organization and Energy Transfer in Photosynthesis. Pp. 305 - 322. Excerpta Medica, Amsterdam - Oxford - New York 1979.

39320 - RIKHIREVA, G.T., PULATOVA, M.K., NAZAROVA, N.M., CHEKULAEVA, L.N., PLAKUNOVA, V.G., BALASHOV, S.P. : Izuchenie metodom ÉPR struktury Mn-soderzhashchikh tsentrov v purpurnykh membranakh i ikh roli v fotoindutsirovannykh prevrashcheniyakh bakteriorodopsinovykh kompleksov. [ESR study of the structure of Mn-containing centres in purple membranes and its role in photoinduced transformations of bacteriorhodopsin complex.] - Biofizika ·*24* : 1003 - 1009, 1979. [In R, ab : E.]

39321 - RIPER, D.M., OWENS, T.G., FALKOWSKI, P.G. : Chlorophyll turnover in *Skeletonema costatum*, a marine plankton diatom. - Plant Physiol. *64* : 49 - 54, 1979.

39322 - RISCH, N., BROCKMANN, H. Jr., GLOE, A. : Strukturaufklärung von neuartigen Bacteriochlorophyllen aus *Chloroflexus aurantiacus*. - Liebigs Ann. Chem. *1979* : 408 - 418, 1979.

39323 - RISCH, N., REICH, H. : Partialsynthese eines stereochemisch einheitlichen Bacteriochlorophylls d. - Tetrahedron Lett. *1979* : 4257 - 4260, 1979.

39324 - RIVERA, E.R., SMITH, B.N. : Crystal morphology and ^{13}carbon/^{12}carbon composition of solid oxalate in cacti. - Plant Physiol. *64* : 966 - 970, 1979.

39325 - RIVKIN, R.B. : Effects of lead on growth of the marine diatom *Skeletonema costatum*. - Mar. Biol. *50* : 239 - 247, 1979. [Ps, Chl.]

*39326 - RIVOAL, J.C., BRIAT, B., CAMMACK, R., HALL, D.O., RAO, K.K., DOUGLAS, I.N., THOMSON, A.J. : The low temperature magnetic circular dichroism spectra of iron-sulphur proteins. 1.Oxidised rubredoxin. - Biochim. biophys. Acta *493* : 122 - 131, 1977.

39327 - ROBARTS, R.D. : Underwater light penetration, chlorophyll *a* and primary production in a tropical African lake (Lake McIlwaine, Rhodesia). - Arch. Hydrobiol. *86* : 423 - 444, 1979.

39328 - ROBERTS, S.W., KNOERR, K.R., STRAIN, B.R. : Comparative field water relations of four co-occurring forest tree species. - Can. J. Bot. *57* : 1876 - 1882, 1979. [Stomatal resistance.]

39329 - ROBINSON, H.H., YOCUM, C.F. : Photochemical generation of reduced quinone catalysts of Photosystem I cyclic photophosphorylation activity. - Photochem. Photobiol. *29* : 135 - 140, 1979.

39330 - ROBINSON, H.H., YOCUM, C.F. : Reversal of the high affinity DBMIB inhibition site of chloroplast electron transport by bovine serum albumin. - Plant Physiol. *63* (Suppl.) : 55, 1979.

39331 - ROBINSON, J.M., GIBBS, M. : Light dependent H_2O_2 production in spinach chloroplast lamellar membranes measured in the presence or absence of NADP reduction. - Plant Physiol. *63* (Suppl.) : 40, 1979.

39332 - ROBINSON, S.J., YOCUM, C.F. : Photosynthetic properties of spheroplast preparations of the cyanobacterium *Spirulina platensis*. - Plant Physiol. *63* (Suppl.) : 29, 1979.

39333 - ROBINSON, S.P., EDWARDS, G.E., WALKER, D.A. : Established methods for the isolation of intact chloroplasts. - In : REID, E. (ed.) : Plant Organelles. Pp. 13 - 24. Ellis Horwood, Chichester 1979.

39334 - ROBINSON, S.P., McNEIL, P.H., WALKER, D.A. : Ribulose bisphosphate carboxylase - lack of dark inactivation of the enzyme in experiments with protoplasts. - FEBS Lett. *97* : 296 - 300, 1979.

39335 - ROBINSON, S.P., WALKER, D.A. : Rapid separation of the chloroplast and cy-
typlasmic fractions from intact leaf protoplasts. - Arch. Biochem. Biophys.
196 : 319 - 323, 1979.

39336 - ROBINSON, S.P., WALKER, D.A. : The control of 3-phosphoglycerate reduction
in isolated chloroplasts by the concentrations of ATP, ADP and 3-phospho-
glycerate. - Biochim. biophys. Acta *545* : 528 - 536, 1979.

39337 - ROBINSON, S.P., WALKER, D.A. : The site of sucrose synthesis in isolated
leaf protoplasts. - FEBS Lett. *107* : 295 - 299, 1979. [Ps.]

39338 - ROBISON, P.D., MARTIN, M.N., TABITA, F.R. : Differential effects of metal
ions on *Rhodospirillum rubrum* ribulosebisphosphate carboxylase/oxygenase
and stoichiometric incorporation of HCO_3^- into a cobalt(III)-enzyme complex.
- Biochemistry *18* : 4453 - 4458, 1979.

39339 - ROBISON, P.D., TABITA, F.R. : Modification of ribulose bisphosphate carboxylase from *Rhodospirillum rubrum* with tetranitromethane. - Biochem. biophys. Res. Commun. *88* : 85 - 91, 1979.

39340 - RODIONOV, A.V., SHKROB, A.M. : Gidroliz al'dimina retinalya v bakterio-
rodopsine, indutsirovannyǐ ionami serebra. [Ag^+-induced hydrolysis of the
retinal aldimine in bacteriorhodopsin.] - Bioorg. Khim. *5* : 376 - 394,
1979. [In R, ab : E.]

39341 - ROEMER, S.C., HOAGLAND, K.D. : Seasonal attenuation of quantum irradiance
(400 - 700 nm) in three Nebraska reservoirs. - Hydrobiologia *63* : 81 - 92,
1979.

39342 - ROGERS, C.S. : The effect of shading on coral reef structure and function.
- J. exp. mar. Biol. Ecol. *41* : 269 - 288, 1979. [Ps.]

39343 - ROGERS, C.S. : The productivity of San Cristobal Reef, Puerto Rico. -
Limnol. Oceanogr. *24* : 342 - 349, 1979.

39344 - ROHATGI, A., SINGH, S.P. : Isolation and characterization of pigment mutants of the blue-green alga *Aphanothece stagnina*. - Mol. gen. Genet.
169 : 59 - 62, 1979.

39345 - ROHWER, F., FLÜCKIGER, W. : Effect of atrazine on growth, nitrogen fixation and photosynthetic rate of *Anabaena cylindrica*. - Angew. Bot. *53* :
59 - 64, 1979.

39346 - ROMAGOUX, J.-C. : Caracteristiques du microphytobenthos d'un lac volcani-
que meromictique (Lac Pavin, France). I. Biomasse chlorophyllienne et
déterminisme du cycle annuel. - Int. Rev. ges. Hydrobiol. *64* : 303 - 343,
1979.

39347 - ROMANENKO, V.I., EĬRIS, M.P., KUDRYAVTSEV, V.M., PUBIENES, M.A. : Intensivnost' fotosinteza fitoplanktona v vodokhranilishchakh Kuby. [Photosynthe-
tic rate of phytoplankton in water reservoirs of Cuba.] - Tr. Inst. Biol.
vnutr. Vod *37* (40) (Mikrobiologicheskie i Khimicheskie Protsessy Destruktsii Organicheskogo Veshchestva v Vodoemakh) : 21 - 59, 1979. [In R.]

39348 - ROMANENKO, V.I., EĬRIS, M.P., KUDRYAVTSEV, V.M., PUBIENES, M.A. : Mikrobiologicheskie protsessy krugovorota organicheskogo veshchestva v vodokhranilishche Anabanil'ya (Kuba). [Microbiological processes of the cycle
of organic matter in the water reservoir Anabanilja (Cuba).] - Gidrobiol.
Zh. *15* (5) : 12 - 18, 1979. [Ps; in R, ab : E.]

39349 - ROMANYUK, V.A., ZVALINSKIĬ, V.I. : Poslesvechenie morskoǐ zelenoǐ vodorosli *Ulva fenestrata* v millisekundnom i detsisekundnom intervalakh zatukhaniya. [Postluminescence of marine green alga *Ulva fenestrata* in millisecond and decisecond intervals of attenuation.]- Biofizika *24* : 1022 -
1025, 1979. [In R, ab : E.]

39350 - RÖMHELD, V., MARSCHNER, H. : Fine regulation of iron uptake by the Fe-effi-
cient plant *Helianthus annuus*. - In : HARLEY, J.L., SCOTT-RUSSELL, R. (ed.)
: The Soil-Root Interface. Pp. 405 - 417. Academic Press, London - New
York - San Francisco, 1979. [Chl.]

39351 - ROSA, L. : Interaction between exogenous ribose 5-phosphate and the Benson-
-Calvin cycle in intact spinach chloroplasts. - Plant Sci. Lett. *16* : 211 -
218, 1979.

39352 - ROSA, L. : The ATP/2e ratio during photosynthesis in intact spinach chloroplasts. - Biochem. biophys. Res. Commun. *88* : 154 - 162, 1979.

39353 - ROSE, R.J. : The association of chloroplast DNA with photosynthetic membrane vesicles from spinach chloroplasts. - J. Cell Sci. *36* : 169 - 183,
1979.

39354 - ROSEFF, S.J., BERNARD, J.M. : Seasonal changes in carbohydrate levels in
tissues of *Carex lacustris*. - Can. J. Bot. *57* : 2140 - 2144, 1979.

*39355 - ROSEN, J.A., PIKE, C.S., GOLDEN, M.L. : Zinc, iron and chlorophyll meta-
bolism in zinc-toxic corn. - Plant Physiol. *59* : 1085 - 1087, 1977.

39356 - ROSHCHINA, V.D., ROSHCHINA, V.V., KOTOVA, I.N. : Deĭstvie èkstraktov tsi-
kuty na dvizhenie khloroplastov i nekotorye fotosinteticheskie reaktsii.
[The effect of extracts from *Cicuta virosa* on chloroplast movement and some photosynthetic reactions.] - Fiziol. Rast. *26* : 147 - 152, 1979. [In R,
ab : E.]

39357 - ROSHCHINA, V.V. : O deĭstvii dibrotimokhinona na fotosinteticheskiĭ èlek-
tronnyĭ transport. [Effect of dibromothymoquinone on photosynthetic electron transport.] - Biokhimiya *44* : 477 - 481, 1979. [In R, ab : E.]

39358 - ROSHCHINA, V.V., LADYGIN, V.G. : Opredelenie tsitokhromov v khloroplastakh
mutantov *Chlamydomonas reinhardii*. [Determination of cytochromes in chloroplasts of *Chlamydomonas reinhardii* mutants.] - Prikl. Biokhim. Mikrobiol. *15* : 883 - 888, 1979. [In R, ab : E.]

39359 - ROSINSKI, J., RIGBI, M., SIEGELMAN, H.W. : Purification and properties of
phycobilisomes from *Cyanobacteria*. - Plant Physiol. *63* (Suppl.) : 145,
1979.

39360 - ROSS, S.M., TYREE, M.T. : Mason and Maskell's diffusion analogue reconciled
with a translocation theory. - Ann. Bot. *44* : 637 - 640, 1979. [Photosyn-
thates.]

39361 - ROTHER, J.A., FAY, P. : Some physiological-biochemical characteristics of
planktonic blue-green algae during bloom formation in three Salopian meres.
- Freshw. Biol. *9* : 369 - 379, 1979. [Ps.]

39362 - ROTHSCHILD, K.J., CLARK, N.A. : Polarized infrared spectroscopy of oriented purple membrane. - Biophys. J. *25* : 473 - 487, 1979. [Fourier transformations.]

39363 - ROUGHAN, P.G., HOLLAND, R., SLACK, C.R. : On the control of long-chain-
-fatty acid synthesis in isolated intact spinach (*Spinacia oleracea*) chloroplasts. - Biochem. J. *184* : 193 - 202, 1979. [Ps, Chl.]

39364 - ROUGHAN, P.G., HOLLAND, R., SLACK, C.R., MUDD, J.B. : Acetate is the pre-
ferred substrate for long-chain fatty acid synthesis in isolated spinach
chloroplasts. - Biochem. J. *184* : 565 - 569, 1979. [Chl.]

39365 - ROUGHAN, P.G., MUDD, J.B., McMANUS, T.T., SLACK, C.R. : Linoleate and α-li-
nolenate synthesis by isolated spinach (*Spinacia oleracea*) chloroplasts. -
Biochem. J. *184* : 571 - 574, 1979.

39366 - ROWE, J., REID, J. : Some aspects of carbon relations in the barley-*Helminthosporium teres* complex. I. The effects of infection upon carboxylation *in vivo* and *in vitro*. - Can. J. Bot. *57* : 195 - 207, 1979.

39367 - ROWE, J., REID, J. : Some aspects of carbon relations in the barley-*Helminthosporium teres* complex. The effects of infection upon net accumulation of carbon in the tissues. - Can. J. Bot. *57* : 208 - 214, 1979.

39368 - ROWE, M.P., SCHWARZ, O.J. : Effect of paraquat on the distribution of starch in the stem of *Pinus taeda* seedlings. - Plant Physiol. *63* (Suppl.) : 65, 1979.

39369 - ROY, H., ADARI, H., COSTA, K.A. : Characterization of free subunits of ribulose-1,5-bisphosphate carboxylase. - Plant Sci. Lett. *16* : 305 - 318, 1979.

39370 - ROY, S., LEGENDRE, L. : DCMU-enhanced fluorescence as an index of photosynthetic activity in phytoplankton. - Mar. Biol. *55* : 93 - 101, 1979.

*39371 - RUBIN, P.M., ZETOONEY, E., McGOWAN, R.E. : Uptake and utilization of sugar phosphates by *Anabaena flos-aquae*. - Plant Physiol. *60* : 407 - 411, 1977. [Ps.]

39372 - RUBINSHTEĬN, A.I., CHERNAVSKIĬ, D.S. : Ob èntropii izlucheniya i ee roli v protsesse fotosinteza. [Radiation entropy and its role in photosynthesis.] - Biofizika *24* : 1010 - 1015, 1979. [In R, ab : E.]

*39373 - RUBY, R.H. : Delayed fluorescence from *Chlorella* : II. Effects of electron transport inhibitors DCMU and NH_2OH. - Photochem. Photobiol. *26* : 293 - 298, 1977.

39374 - RUDENKO, T.I., SHMELEVA, V.L., MAKAROV, A.D. : Vliyanie intensivnosti sveta na konformatsionnye kolebaniya khloroplastov. [Effect of illuminance on conformational oscillations of chloroplasts.] - Fiziol. Rast. *26* : 184 - 187, 1979. [In R.]

39375 - RUDENKO, T.I., SHMELEVA, V.L., MAKAROV, A.D. : Izuchenie konformatsionnykh kolebaniĭ khloroplastov iz rasteniĭ gorokha, vyrashchennykh pri razlichnoĭ intensivnosti osveshcheniya. [A study of conformational oscillations of chloroplasts from pea plants grown under various light intensities.] - Fiziol. Rast. *26* : 1150 - 1155, 1979. [In R, ab : E.]

39376 - RÜDIGER, W., BENZ, J. : Influence of aminotriazol on the biosynthesis of chlorophyll and phytol. - Z. Naturforsch. *34 C* : 1055 - 1057, 1979.

39377 - RUDOLPH, K., RASCHE, E. : A leaf bioassay for semiquantitative determination of the chlorosis inducing toxin from *Pseudomonas phaseolicola*. - Phytopathol. Z. *96* : 215 - 221, 1979.

39378 - RUGGIU, D., SARACENI, C., DE BORTOLI, T., NAKANISHI, M. : Primary production in Lago di Mergozzo (N.Italy) and implications of phytoplankton cell size. - Mem. Ist. ital. Idrobiol. "Dott. Marco De Marchi" *37* : 223 - 246, 1979.

39379 - RÜHLE, W., WILD, A. : Measurements of cytochrome *f* and P-700 in intact leaves of *Sinapis alba* grown under high-light and low-light conditions. - Planta *146* : 377 - 385, 1979.

39380 - RÜHLE, W., WILD, A. : The intensification of absorbance changes in leaves by light-dispersion. Differences between high-light and low-light leaves. - Planta *146* : 551 - 557, 1979.

39381 - RUNDEL, P.W., BRATT, G.C., LANGE, O.L. : Habitat ecology and physiological response of *Sticta filix* and *Pseudocyphellaria delisei* from Tasmania. - Bryologist *82* : 171 - 180, 1979. [Ps.]

39382 - RUNDEL, P.W., RUNDEL, J.A., ZIEGLER, H., STICHLER, W. : Carbon isotope ratios of central Mexican Crassulaceae in natural and greenhouse environments. - Oecologia *38* : 45 - 50, 1979.

39383 - RUNDEL, P.W., STICHLER, W., ZANDER, R.H., ZIEGLER, H. : Carbon and hydrogen isotope ratios of bryophytes from acid and humid regions. - Oecologia *44* : 91 - 94, 1979.

39384 - RUPP, H., DE LA TORRE, A., HALL, D.O. : The electron spin relaxation of the electron acceptors of photosystem I reaction centre studied by microwave power saturation. - Biochim. biophys. Acta *548* : 552 - 564, 1979.

*39385 - RUPP, H., RAO, K.K., HALL, D.O., CAMMACK, R. : Electron spin relaxation of
         iron-sulphur proteins studied by microwave power saturation. - Biochim.
         biophys. Acta *537* : 255 - 269, 1978. [Ferredoxins.]

39386 - RUTHERFORD, A.W., EVANS, M.C.W. : The high potential semiquinone-iron
         signal in *Rhodopseudomonas viridis* is the specific quinone secondary elec-
         tron acceptor in the photosynthetic reaction centre. - FEBS Lett. *104* :
         227 - 230, 1979.

39387 - RUTHERFORD, A.W., EVANS, M.C.W. : A high potential semiquinone-iron type
         EPR signal in *Rhodopseudomonas viridis*. - FEBS Lett. *100* : 305 - 308,
         1979.

39388 - RUTHERFORD, A.W., HEATHCOTE, P., EVANS, M.C.W. : Electron-paramagnetic-
         -resonance measurements of the electron-transfer components of the reaction
         centre of *Rhodopseudomonas viridis*. Oxidation-reduction potentials and
         interactions of the electron acceptors. - Biochem. J. *182* : 515 - 523,
         1979.

39389 - RYABUSHKINA, I.A. : Raspredelenie ^{14}C v produktakh fotosinteza u proto-
         plastov i u diskov iz list'ev tabaka. [Distribution of ^{14}C-photosynthates
         in protoplasts and leaf discs of tobacco plants.] - Fiziol. Rast. *26* :
         1135 - 1142, 1979. [In R, ab : E.]

39390 - RYAN, F.J., COVERT, N.L. : Integrity of isolated spinach and *Petunia* cells.
         - Plant Physiol. *63* (Suppl.) : 65, 1979. [Ps.]

39391 - RYBIN, I.A., SHAVNIN, S.A., MIKHEEVA, S.A. : O svetoindutsiruemoǐ ėlektro-
         reaktsii list'ev bobov. [On light-induced electric response of bean leaves.]
         - Fiziol. Biokhim. kul't. Rast. *11* : 345 - 350, 1979. [Ps; in R, ab : E.]

39392 - RYBKINA, G.V., BIGLOVA, S.G., PAL'M, G.G. : O kharaktere ob"emnykh izme-
         neniǐ khloroplastov pri izmenenii osmoticheskogo potentsiala sredy. [Cha-
         racter of volume changes of chloroplasts during changing of osmotic poten-
         tial of the medium.] - Uch. Zapiski kazan. gos. pedag. Inst. *195* (Faktory
         Sredy i Rastenie) : 40 - 50, 1979. [In R.]

39393 - RYBKINA, G.V., LOSEVA, N.L., BIGLOVA, S.G., PAL'M, G.G. : K izucheniyu
         korrelyatsiǐ vodnogo balansa i fotokhimicheskoǐ aktivnosti khloroplastov.
         [Correlations of water balance and photochemical activity of chloroplasts.]
         - Uch. Zapiski kazan. gos. pedag. Inst. *195* (Faktory Sredy i Rastenie) :
         30 - 39, 1979. [In R.]

B39394 - RYCHNOVSKÁ, M. (ed.) : Function of Grasslands in Spring Region - Kameničky
         Project. Progress Report on MAB Project No 91. - Botanical Institute, Cze-
         choslovak Academy of Sciences, Brno 1979. [Ps.]

39395 - RYLE, G.J.A., POWELL, C.E., GORDON, A.J. : The respiratory costs of nitro-
         gen fixation in soyabean, cowpea, and white clover. II. Comparisons of the
         cost of nitrogen fixation and the utilization of combined nitrogen. - J.
         exp. Bot. *30* : 145 - 153, 1979. [Ps.]

39396 - RYRIE, I.J., CRITCHLEY, C., TILLBERG, J.-E. : Structure and energy-linked
         activities in reconstituted bacteriorhodopsin-yeast ATPase proteoliposomes.
         - Arch. Biochem. Biophys. *198* : 182 - 194, 1979.

39397 - SABATER, B., MARTÍN, M. : Improvement of ATP yield in the acid-base tran-
         sition of barley chloroplasts by treatment with butylated hydroxytoluene. -
         Plant Cell Physiol. *20* : 683 - 687, 1979.

39398 - SACHS, R.M., HACKETT, W.P., RAMINA, A., MALOOF, C. : Photosynthetic assi-
         milation and nutrient diversion as controlling factors in flower initia-
         tion in *Bougainvillea* (San Diego red) and *Nicotiana tabacum* cv. Wis. 38. -
         In : MARCELLE, R., CLIJSTERS, H., VAN POUCKE, M. (ed.) : Photosynthesis
         and Plant Development. Pp. 95 - 101. Dr. W.Junk bv. Publ., The Hague -
         Boston - London 1979.

39399 - **SAHAI, R., AGRAWAL, N., KHOSLA, N.** : Effect of fertilizer factory effluent
on seed germination, seedling growth and chlorophyll content of *Phaseolus
radiatus* LINN. - Trop. Ecol. *20* : 155 - 162, 1979.

39400 - **SAHU, J.K., ADHIKARY, S.P., PATTNAIK, H.** : Effect of organic substrates
on chlorophyll and carotenoid production of *Anabaena* sp under light and
dark conditions. - J. indian bot. Soc. *58* : 358 - 362, 1979.

39401 - **SAINIS, J.K., SANE, P.V., SINGH, P., NAIK, M.S.** : Evolution of $^{14}CO_2$ from
[1,4-^{14}C]succinate in wheat leaves during nitrate assimilation. - Indian
J. Biochem. Biophys. *16* : 440 - 442, 1979. [Photorespiration.]

*39402 - **SAKAI, S.** : [Effect of carbon dioxide concentration on photosynthesis of
single leaf in tea plant.] - Chagyo Gijutsu Kenkyu [Study Tea] *53* : 17 - 22,
1977. [In Jap., ab : E.]

*39403 - **SAKALO, N.D.** : Struktura i aktivnost' fotosinteticheskogo apparata kukuruzy
v raznykh usloviyakh fosfornogo pitaniya. [Structure and activity of photosynthetic apparatus of maize under different nitrogen nutrition.] - In :
Mineral'noe Pitanie i Produktivnost' Rasteniĭ. Pp. 43 - 47, 317. Naukova
Dumka, Kiev 1978. [In R.]

39404 - **SAKHAROVA, O.V.** : Fotofosforiliruyushchaya aktivnost' izolirovannykh khloroplastov v ontogeneze u raznykh vidov pshenits. [Activity of photophosphorylation in isolated chloroplasts during ontogeny of different wheat species.]
- Byull. vsesoyuz. nauch.-issled. Inst. Rastenievod. im. N.I. Vavilova *87* :
70 - 76, 1979. [In R.]

39405 - **SĂLĂGEANU, V.** : Expériences de culture en masse de l'algue *Dunaliella viridis* dans le laboratoire. - Rev. roum. Biol., Sér. Biol. vég. *24* : 125 -
- 126, 1979.

39406 - **SALAMON, Z., MARTYŃSKI, T.** : Influence of chlorophyll *a* on intermolecular
interactions in liquid crystal. - Biophys. Chem. *9* : 369 - 374, 1979.

*39407 - **SALAZAR, C.R.** : Determinación de la fotosíntesis de dos variedades comer-
ciales de papaya (*Carica papaya* L.) y su posible relacion con la produc-
cion y calidad de los frutos. [Determination of photosynthesis in commer-
cial varieties of papaya (*Carica papaya* L.) and its possible relationship
to production and quality of the fruits.] - Rev. Inst. Colomb. Agropecu.
13 : 291 - 295, 1978. [In Span., ab : E.]

39408 - **SALCHEVA, G., GEORGIEVA, D., VANKOVA-RADEVA, R., GRAMATIKOVA,Kh., MECHEVA,
R.** : Vliyanie molibdena na soderzhanie plastidnykh pigmentov, fotokhimi-
cheskuyu aktivnost' i aktivnost' nitratreduktazy ozimoĭ pshenitsy v zavisimosti ot temperatury vyrashchivaniya rasteniĭ. [Influence of molybdenum
on the content of plastid pigments, photochemical activity of chloroplasts
and the activity of nitrate reductase of the winter wheat depending on
growth temperature.]-In : VAKLINOVA, S.G., VANKOVA-RADEVA, R., VASILEVA,
V.S. (ed.) : Fotosinteticheskaya Assimilyatsiya CO_2 i Fotodykhanie. Pp.
114 - 120. Izdat. bolg. Akad. Nauk, Sofiya 1979. [In R.]

39409 - **SALCHEVA, G., GHEORGIEVA, D., FEDINA, I., GRAMATIKOVA, H., PETKOVA, M.,
VAKLINOVA, S.** : The influence of molibdenum on the photosynthesis and nitrogen metabolism in winter wheat, grown at different temperature conditions. I. Content of the plastid pigments, photochemical activity of the
chloroplasts and intensity of the photosynthesis. - In : Mineral Nutrition
of Plants. Vol.II. Pp. 52 - 58. Publ. House Central Cooperative Union, Sofia
1979.

39410 - **SALCHEVA, G., GHEORGIEVA, D., VUNKOVA-RADEVA, R., POPOVA, L., FEDINA, I.,
PETKOVA, M.** : Influence of molybdenum on the functional state of chloroplasts, the activity of nitrate reductase and the yield of winter wheat
grown on water logged soil. - In : Mineral Nutrition of Plants. Vol.II.
Pp. 66 - 73. Publ. House Central Cooperative Union, Sofia 1979.

*39411 - **SAL'NIKOV, A.I., ALEKSANDROVA, G.A., BYNOV, F.A., AGAPOVA, M.V., TETYUEV,
V.A.** : Vliyanie ėlektricheskogo polya na nekotorye fiziologo-biokhimicheskie protsessy i urozhaĭ ogurtsov v zakrytom grunte. [Effect of electric

field on some physiological and biochemical processes and the yield of cucumbers in a greenhouse.] - In : Vliyanie Fiziko-khimicheskikh Faktorov Sredy na Rasteniya. Pp. 19 - 25. Perm. gos. Univ., Perm' 1978. [In R.]

39412 - SALONEN, K. : Hiilenmääritys hydrobiologisessa tutkimuksessa: Menetelmät ja niiden sovellutuksia. [Determination of carbon in hydrobiological studies: Methods and applications.] - Univ. Helsinki 1979. [In Fin., ab : E.]

39413 - SALONEN, K. : A versatile method for the rapid and accurate determination of carbon by high temperature combustion. - Limnol. Oceanogr. *24* : 177 - - 183, 1979.

39414 - SALONEN, K., HOLOPAINEN, A.-L. : A comparison of methods for the estimation of phytoplankton primary production. - Int. Rev. ges. Hydrobiol. *64* : 147 - - 155, 1979.

39415 - SALVADOR, G.F., BENEY, G., NIGON, V. : The influence of lincomycin and 3,(3,4 dichlorophenyl),1,1-dimethyl urea (DCMU) on chlorophyll and Δ-amino-levulinic acid synthesis during greening of *Euglena gracilis*. - Biol. cell. *35* : 289 - 294, 1979.

39416 - SALVUCCI, M., HOLADAY, S., BOWES, G. : Enzymes associated with C_4 acid metabolism in submerged aquatic macrophyte (SAM) species. - Plant Physiol. *63* (Suppl.) : 2, 1979.

39417 - SAMARAKOON, A.B., RAUSER, W.E. : Carbohydrate levels and photoassimilate export from leaves of *Phaseolus vulgaris* exposed to excess cobalt, nickel, and zinc. - Plant Physiol. *63* : 1165 - 1169, 1979.

*39418 - SAMEOTO, D.D. : Distribution of sound scattering layers caused by euphausiids and their relationship to chlorophyll *a* concentrations in the Gulf of St. Lawrence estuary. - J. Fish. Res. Board Can. *33* : 681 - 687, 1976.

39419 - SAMMONS, D.J., PETERS, D.B., HYMOWITZ, T. : Screening soybeans for drought resistance. II. Drought box procedure. - Crop Sci. *19* : 719 - 722, 1979. [Plant growth rate.]

39420 - SAMSUDDIN, Z., IMPENS, I. : Photosynthesis and diffusion resistance to carbon dioxide in *Hevea brasiliensis* MUEL. AGR. clones. - Oecologia *37* : 361 - - 363, 1979.

39421 - SAMSUDDIN, Z., IMPENS, I. : Relationship between leaf age and some carbon dioxide exchange characteristics of four *Hevea brasiliensis* MUELL. ARG. clones. - Photosynthetica *13* : 208 - 210, 1979.

39422 - SAMSUDDIN, Z., IMPENS, I. : The development of photosynthetic rate with leaf age in *Hevea brasiliensis* MUELL. ARG. clonal seedlings. - Photosynthetica *13* : 267 - 270, 1979.

39423 - SAMSUDDIN, Z., IMPENS, I. : Photosynthetic rates and diffusion resistances of seven *Hevea brasiliensis* MUELL. ARG. clones. - Biol. Plant. *21* : 154 - - 156, 1979.

39424 - SÁNCHEZ-DÍAZ, M.F., MOONEY, H.A. : Resistance to water transfer in desert shrubs native to Death Valley, California. - Physiol. Plant. *46* : 139 - - 146, 1979. [Stomatal resistance.]

39425 - SANDERS, J.G. : Effects of arsenic speciation and phosphate concentration on arsenic inhibition of *Skeletonema costatum (Bacillariophyceae.)* - J. Phycol. *15* : 424 - 428, 1979. [Ps.]

39426 - SANDMANN, G., KUNERT, K.-J., BÖGER, P. : Biological systems to assay herbicidal bleaching. - Z. Naturforsch. *34 C* : 1044 - 1046, 1979.

39427 - SANE, P.V., JOHANNINGMEIER, U., TREBST, A. : The inhibition of photosynthetic electron flow by DCCD. An indication for proton channels. - FEBS Lett. *108* : 136 - 140, 1979.

39428 - SANIEWSKI, M., ANTOSZEWSKI, R., RUDNICKI, R.M. : Translocation of ^{14}C-photosynthates into the bulb as related to flowering of hyacinths. - Bull. Acad. pol. Sci. Sér. Sci. biol. Cl. V *27* : 781 - 786, 1979.

39429 - SANKHLA, N., HUBER, W., EDER, A. : Effect of fusicoccin on chlorophyll bio-synthesis. - Biochem. Physiol. Pflanzen *174* : 296 - 304, 1979.

39430 - SANO, H., SPAETH, E., BURTON, W.G. : Messenger RNA of the large subunit of ribulose-1,5-bisphosphate carboxylase from *Chlamydomonas reinhardi*. Iso-lation and properties. - Europe. J. Biochem. *93* : 173 - 180, 1979.

39431 - SANTAKUMARI, M., REDDY, C.S., REDDY, A.R.C., DAS, V.S.R. : CAM behavior in a grain legume, chickpea. - Naturwissenschaften *66* : 54 - 55, 1979.

39432 - SANTARIUS, K.A., HEBER, U., KRAUSE, G.H. : Untersuchungen über die physio-logisch-biochemischen Ursachen von Empfindlichkeit und Resistenz von Bio-membranen gegenüber extremen Temperaturen und hohen Salzkonzentrationen. - Ber. deut. bot. Ges. *92* : 209 - 223, 1979. [Chloroplast.]

39433 - SANTARIUS, K.A., MÜLLER, M. : Investigations on heat resistance of spinach leaves. - Planta *146* : 529 - 538, 1979. [Ps, Chl.]

39434 - SANTILLAN, R.A., OCUMPAUGH, W.R., MOTT, G.O. : Estimating forage yield with a disk meter. - Agron. J. *71* : 71 - 74, 1979. [Growth analysis.]

39435 - SANZ-MUÑOZ, M., POZO-HERNANDEZ, C. : Chlorophylls in *Pinus pinea* germina-ting seeds. - An. real Acad. Farm. *45* : 581 - 588, 1979.

39436 - SAPOZHNIKOV, D.I., POPOVA, O.F., POPOVA, I.A., MASLOVA, T.G. : Vliyanie predvaritel'nogo progreva list'ev na reaktsii violaksantinovogo tsikla, indutsiruemye krasnym svetom. [Effect of leaf pre-heating on the reactions of violaxanthin cycle induced by red light.] - Fiziol. Rast. *26* : 239 - - 243, 1979. [In R, ab : E.]

39437 - SARIĆ, M., KRSTIĆ, B. : Contents of chlorophyll, carotenoid and intensity of CO_2 absorption in the leaves of different colour of *Hedera helix* L. - Bull. Acad. serbe Sci. Arts, Cl.Sci. nat. math., Sci. nat. *68* (19) : 31 - - 36, 1979.

39438 - SARKAR, S.K., MALHOTRA, S.S. : Effects of SO_2 on organic acid content and malate dehydrogenase activity in jack pine needles. - Biochem. Physiol. Pflanzen *174* : 438 - 445, 1979.

39439 - SARMA, N.P., PATNAIK, A., JACHUCK, P.J. : Azide mutagenesis in rice - effect of concentration and soaking time on induced chlorophyll mutation frequency. - Environ. exp. Bot. *19* : 117 - 121, 1979.

39440 - SASAKI, K., NAGAI, S. : Growth, vitamin B_{12}, and photopigment formations of *Rhodopseudomonas sphaeroides* S growing on propionate media under dark and light conditions. - Europe. J. appl. Microbiol. Biotechnol. *7* : 201 - 210, 1979.

39441 - SATO, F., ASADA, K., YAMADA, Y. : Photoautotrophy and the photosynthetic potential of chlorophyllous cells in mixotrophic cultures. - Plant Cell Physiol. *20* : 193 - 200, 1979.

39442 - SATO, K. : [The growth responses of soybean plant to photoperiod and tem-perature. III. The effects of photoperiod and temperature on the develop-ment and anatomy of photosynthetic organ.] - Jap. J. Crop Sci. *48* : 66 - 74, 1979. [In Jap., ab : E.]

39443 - SATO, K., SHIMIZU, S. : The conditions for bacteriochlorophyll formation and the ultrastructure of a methanol-utilizing bacterium, *Protaminobacter ruber*, classified as non-photosynthetic bacteria. - Agr. biol. Chem. *43* : 1669 - 1675, 1979.

39444 - SATOH, K. : [Chlorophyll-protein complexes.] - Tampakushitsu Kakusan Koso, Bessatsu *21* : 40 - 51, 1979. [In Jap.]

39445 - SATOH, K. : Properties of light-harvesting chlorophyll *a/b*-protein, and photosystem I chlorophyll *a*-protein, purified from digitonin extracts of spinach chloroplasts by isoelectrofocusing. - Plant Cell Physiol. *20* : 499 - - 512, 1979.

39446 - SATOH, K. : Polypeptide composition of the purified Photosystem II pigment--protein complex from spinach. - Biochim. biophys. Acta *546* : 84 - 92, 1979.

*39447 - SATOH, K., KATOH, S. : Parallel time courses of electrochromic shifts of carotenoids and cytochrome f photooxidation in intact *Bryopsis* chloroplasts. - Plant Cell Physiol. *18* : 1077 - 1087, 1977.

39448 - SATOH, K., KATOH, S. : Two electrogenic mechanisms contributing to the 560 nm absorption changes in intact *Bryopsis* chloroplasts. - Biochim. biophys. Acta *545* : 454 - 465, 1979.

39449 - SATTER, R.L. : Leaf movements and tendril curling. - In : HAUPT, W., FEIN-LEIB, M.E. (ed.) : Physiology of Movements. Pp. 442 - 484. Springer-Verlag, Berlin - Heidelberg - New York 1979. [Chloroplast.]

39450 - SAUER, A., HEISE, K.-P. : A possible mechanism for a light-driven regulation of the fatty acid composition in galactolipids of chloroplasts. - Z. Naturforsch. *34 C* : 815 - 819, 1979.

39451 - SAUER, K. : Photosynthesis - the light reactions. - Annu. Rev. phys. Chem. *30* : 155 - 178, 1979.

*39452 - SAUER, K., AUSTIN, L.A. : Bacteriochlorophyll-protein complexes from the light-harvesting antenna of photosynthetic bacteria. - Biochemistry *17* : 2011 - 2019, 1978.

39453 - SAUER, K., MATHIS, P., ACKER, S., VAN BEST, J.A. : Absorption changes of P-700 reversible in milliseconds at low temperature in Triton-solubilized Photosystem I particles. - Biochim. biophys. Acta *545* : 466 - 472, 1979.

39454 - SAVAGE, M.J. : Terminology pertaining to photosynthesis. - Crop Sci. *19* : 424, 1979.

39455 - SAVIDGE, G. : Photosynthetic characteristics of marine phytoplankton from contrasting physical environments. - Mar. Biol. *53* : 1 - 12, 1979.

*39456 - SAWARD, D., STIRLING, A., TOPPING, G. : Experimental studies on the effects of copper on a marine food chain. - Mar. Biol. *29* : 351 - 361, 1975. [Ps, Chl.]

*39457 - SAXENA, P.N., TEWARI, A., KHAN, M.A. : Effect of *Anacystis nidulans* on the physico-chemical and biological characteristics of raw sewage. - Proc. indian Acad. Sci. *B 79* (3) : 139 - 146, 1974.

39458 - SAYRE, R.T., HOMANN, P.H. : A light-dependent oxygen consumption induced by Photosystem II of isolated chloroplasts. - Arch. Biochem. Biophys. *196* : 525 - 533, 1979.

39459 - SAYRE, R.T., HOMANN, P.H. : Correlation between photorespiratory activity and hydrogenase content in *Chlorella* strains. - Plant Physiol. *63* (Suppl.) : 153, 1979.

39460 - SAYRE, R.T., KENNEDY, R.A., PRINGNITZ, D.J. : Photosynthetic enzyme activities and localization in *Mollugo verticillata* populations differing in the levels of C_3 and C_4 cycle operation. - Plant Physiol. *64* : 293 - 299, 1979.

39461 - SCHÄFER, W., DANNOWSKI, M., WURL, B. : Bestimmung des Temperaturoptimums der CO_2-Aufnahmerate bei Zuckerrüben. - Arch. Acker- Pflanz. *23* : 289 - 296, 1979.

39462 - SCHAPENDONK, A.H.C.M., VREDENBERG, W.J. : Activation of the reaction II component of P515 in chloroplasts by pigment system 1. - FEBS Lett. *106* : 257 - 261, 1979.

39463 - SCHAPENDONK, A.H.C.M., VREDENBERG, W.J., TONK, W.J.M. : Studies on the kinetics of the 515 nm absorbance changes in chloroplasts. Evidence for the induction of a fast and a slow P515 response upon saturating light flashes. - FEBS Lett. *100* : 325 - 330, 1979.

*39464 - SCHAREN, A.L., SCHAEFFER, G.W., KRUPINSKY, J.M., SHARPE, F.T. Jr. : Effects of flag leaf axial lesions caused by *Septoria nodorum* on ^{14}C translocation and yield of wheat. - Physiol. Plant Pathol. *6* : 193 - 198, 1975.

39465 - SCHEER, H., FORMANEK, H., RÜDIGER, W. : The conformation of bilin chromophores in biliproteins : Ramachandran-type calculations. - Z. Naturforsch. *34 C* : 1085 - 1093, 1979.

39466 - SCHEIBE, R., BECK, E. : On the mechanism of light activation of the NADP-dependent malate dehydrogenase in spinach chloroplasts. - Plant Physiol. *63* (Suppl.) : 2, 1979.

39467 - SCHEIBE, R., BECK, E. : On the mechanism of activation by light of the NADP-dependent malate dehydrogenase in spinach chloroplasts. - Plant Physiol. *64* : 744 - 748, 1979.

39468 - SCHIDLOWSKI, M. : Antiquity and evolutionary status of bacterial sulfate reduction : Sulfur isotope evidence. - Origins Life *9* : 299-311,1979. [Ps.]

39469 - SCHIDLOWSKI, M., APPEL, P.W.U., EICHMANN, R., JUNGE, C.E. : Carbon isotope geochemistry of the 3.7×10^9-yr-old Isua sediments, West Greenland : Implications for the archaean carbon and oxygen cycles. - Geochim. cosmochim. Acta *43* : 189 - 199, 1979.

39470 - SCHIMZ, A., HILDEBRAND, E. : Chemosensory responses of *Halobacterium halobium*. - J. Bacteriol. *140* : 749 - 753, 1979. [Bacteriorhodopsin.]

39471 - SCHLIMME, E., DE GROOT, E.J., SCHOTT, E., STROTMANN, H., EDELMANN, K. : Photophosphorylation of base-modified nucleotide analogs by spinach chloroplasts. - FEBS Lett. *106* : 251 - 256, 1979.

39472 - SCHLOSS, J.V., PHARES, E.F., LONG, M.V., NORTON, I.L., STRINGER, C.D., HARTMAN, F.C. : Isolation, characterization, and crystallization of ribulosebisphosphate carboxylase from autotrophically grown *Rhodospirillum rubrum*. - J. Bacteriol. *137* : 490 - 501, 1979.

39473 - SCHMID, G.H., THIBAULT, P. : Evidence for a rapid oxygen-uptake in tobacco chloroplasts. - Z. Naturforsch. *34 C* : 414 - 418, 1979.

39474 - SCHMID, G.H., THIBAULT, P. : Characterization of a light-induced oxygen-uptake in tobacco protoplasts. - Z. Naturforsch. *34 C* : 570 - 575, 1979. [Ps.]

39475 - SCHMIDT, A. : Photosynthetic assimilation of sulfur compounds. - In : GIBBS, M., LATZKO, E. (ed.) : Photosynthesis II. (Encycl. Plant Physiol. N.S. Vol. 6.) Pp. 481 - 496. Springer-Verlag, Berlin - Heidelberg - New York 1979.

39476 - SCHMIDT, A., CHRISTEN, U. : Distribution of thioredoxins in cyanobacteria. - Z. Naturforsch. *34 C* : 1272 - 1274, 1979. [Carbohydrate metabolism.]

39477 - SCHMIDT, B., KIES, L., WEBER, A. : Die Pigmente von *Cyanophora paradoxa*, *Gloeochaete wittrockiana* and *Glaucocystis nostochinearum* . - Arch. Protistenkunde *122* : 164 - 170, 1979.

39478 - SCHMIDT, G.W., DEVILLERS-THIERY, A., DESRUISSEAUX, H., BLOBEL, G., CHUA, N.-H. : NH_2-terminal amino acid sequences of precursor and mature forms of the ribulose-1,5-bisphosphate carboxylase small subunit from *Chlamydomonas reinhardtii*. - J. Cell Biol. *83* : 615 - 622, 1979.

39479 - SCHMIDT, H.-L., WINKLER, F.J. : Einige Ursachen der Variationsbreite von δ^{13}C-Werten bei C_3- und C_4-Pflanzen. - Ber. deut. bot. Ges. *92* : 185 - 191, 1979.

39480 - SCHMITZ, K., SRIVASTAVA, L.M. : Long distance transport in *Macrocystis integrifolia* I. Translocation of ^{14}C-labeled assimilates. - Plant Physiol. *63* : 995 - 1002, 1979.

39481 - SCHMITZ, K., SRIVASTAVA, L.M. : Long distance transport in *Macrocystis integrifolia* II. Tracer experiments with ^{14}C and ^{32}P. - Plant Physiol. *63* : 1003 - 1009, 1979.

39482 - SCHNARRENBERGER, C. : Pflanzen arbeiten mit Verlust. Photosynthese und Lichtatmung - zwei untrennbare Prozesse. - Umschau Wiss. Techn. *79* : 210 - 217, 1979. [Ps.]

39483 - SCHNEIDER, E., MÜLLER, H.-W., RITTINGHAUS, K., THIELE, V., SCHWULÉRA, U., DOSE, K. : Properties of the F_0F_1 ATPase complex from *Rhodospirillum rubrum* chromatophores solubilized by Triton X-100. - Europe. J. Biochem. *97* : 511 - 517, 1979.

39484 - SCHNEIDER, H.A.W., BOGORAD, L. : Spectral response curves for the formation of phycobiliproteins, chlorophyll and δ-aminolevulinic acid in *Cyanidium caldarium*. - Z. Pflanzenphysiol. *94* : 449 - 459, 1979.

39485 - SCHOBERT, B., ELSTNER, E.F. : *In vivo* photobleaching of chlorophyll and carotenoids as a consequence of fatty acid oxidation in *Phaeodactylum tricornutum*. - Plant Physiol. *63* (Suppl.) : 62, 1979.

39486 - SCHÖNBOHM, E. : Durch Phytochrom aktivierbare kontraktile Plasmaelemente : Lokalisation in der *Mougeotia*-Zelle. 5.Mitteilung zur Mechanik der Chloroplastenbewegung. - Z. Pflanzenphysiol. *93* : 185 - 188, 1979.

39487 - SCHÖNBOHM, E., BRÜHL, K.L. : Ein neuer Inversionstyp der Schwachlichtbewegung des *Mougeotia*-Chloroplasten im längsschwingenden polarisierten Licht. - Ber. deut. bot. Ges. *92* : 305 - 311, 1979.

39488 - SCHÖNBOHM, E., HAUPT, W. : Light-oriented chloroplast movement in *Mougeotia* sp. - Acta protozool. *18* : 195, 1979.

39489 - SCHÖNBOHM, E., HELLWIG, H. : Zum Photorezeptorproblem der Schwachlichtbewegung des *Mougeotia*-Chloroplasten im Blau bzw. Hellrot bei niederen Temperaturen. - Ber. deut. bot. Ges. *92* : 749 - 762, 1979.

39490 - SCHÖNBOHM, E., SCHÖNBOHM, E., LÜCKE, G. : Die Entstehung symmetrischer und asymmetrischer Phytochrom-Gradienten bei *Mougeotia* und deren Bedeutung für die Chloroplastenorientierung im Stark- und Schwachlicht. - Ber. deut. bot. Ges. *92* : 297 - 304, 1979.

39491 - SCHÖNFELD, M., MONTAL, M., FEHER, G. : Functional reconstitution of photosynthetic reaction centers in planar lipid bilayers. - Proc. nat. Acad. Sci. USA *76* : 6351 - 6355, 1979.

39492 - SCHÖNHERR, J., SCHMIDT, H.W. : Water permeability of plant cuticles. Dependence of permeability coefficients of cuticular transpiration on vapor pressure saturation deficit. - Planta *144* : 391 - 400, 1979. [Resistances.]

39493 - SCHOPF, R. : Zur Nahrungsqualität von Fichtennadeln für forstliche Schadinsekten 11. Aufnahme und Auscheidung ^{14}C-markierter Verbindungen durch Larven von *Gilpinia hercyniae* (*Hym., Diprionidae*) nach Verfütterung radioaktiver Fichtennadeln. - Z. angew. Entomol. *87* : 262 - 276, 1979. [Photosynthates.]

39494 - SCHRAUDOLF, H., ŠONKA, J. : Effects of 5-bromo-deoxyuridine on chloroplast structure in gametophytes of *Anemia phyllitidis* L. SW. - Europe. J. Cell Biol. *19* : 135 - 138, 1979.

39495 - SCHRECKENBACH, T. : The properties of bacteriorhodopsin and its incorporation into artificial systems. - In : BARBER, J. (ed.) : Photosynthesis in Relation to Model Systems. Pp. 189 - 209. Elsevier, Amsterdam - New York - Oxford 1979.

*39496 - SCHRECKENBACH, T., OESTERHELT, D. : Photochemical and chemical studies on the chromophore of bacteriorhodopsin. - Fed. Proc. *36* : 1810 - 1814, 1977.

39397 - SCHREIBER, U. : Cold-induced uncoupling of energy transfer between phycobilins and chlorophyll in *Anacystis nidulans*. Antagonistic effects of monovalent and divalent cations, and of high and low pH. - FEBS Lett. *107* : 4 - 9, 1979.

39498 - SCHREIBER, U., AVRON, M. : Properties of ATP-driven reverse electron flow in chloroplasts. - Biochim. biophys. Acta *546* : 436 - 447, 1979.

39499 - SCHREIBER, U., RIJGERSBERG, C.P., AMESZ, J. : Temperature-dependent rever-
        sible changes in phycobilisome-thylakoid membrane attachment in *Anacystis
        nidulans*. - FEBS Lett. *104* : 327 - 331, 1979.

39500 - SCHUBER, M., KLUGE, M. : Crassulaceen-Säurestoffwechsel (CAM) bei mittel-
        europäischen Sukkulenten : Ökologische Untersuchungen an *Sempervivum*-Arten.
        - Flora *168* : 205 - 216, 1979.

39501 - SCHULZE, E.-D., KÜPPERS, M. : Short-term and long-term effects of plant
        water deficits on stomatal response to humidity in *Corylus avellana* L. -
        Planta *146* : 319 - 326, 1979. [Ps.]

39502 - SCHUMACHER, A., DREWS, G. : Effects of light intensity on membrane diffe-
        rentiation in *Rhodopseudomonas capsulata*. - Biochim. biophys. Acta *547* :
        417 - 428, 1979.

39503 - SCHUMM, F., KREEB, K.H. : Die Nettophotosynthese von Flechtentransplantaten
        als Maß für die Immissionsbelastung der Luft. - Angew. Bot. *53* : 31 - 39,
        1979.

39504 - SCHWARTZBACH,S.D., FREYSSINET, G., SCHIFF, J.A., HECKER, L.I., BARNETT,
        W.E. : Isolation of plastid ribosomes from *Euglena*. - In : COLOWICK, S.P.,
        KAPLAN, N.O. (ed.) : Methods in Enzymology. Vol. 59. Pp. 434 - 437. Academic
        Press, New York - San Francisco - London 1979.

39505 - SCHWARTZBACH, S.D., LEE, K., HORRUM, M.A. : Nutritional repression of light
        induced chloroplast development in *Euglena*. - Plant Physiol. *63* (Suppl.) :
        160, 1979.

39506 - SCHWARTZBACH, S.D., SCHIFF, J.A. : Events surrounding the early development
        of *Euglena* chloroplasts 13. Photocontrol of protein synthesis. - Plant
        Cell Physiol. *20* : 827 - 838, 1979.

39507 - SCHWARZ, O.J., RYAN, F.J., RAYBIN, R.A., BANNER, W.A. : Carbon fixation
        and enzyme activities in *Pinus virginiana* after paraquat treatment. - Plant
        Physiol. *63* (Suppl.) : 65, 1979.

*39508 - SCHWERTNER, H.A., BIALE, J.B. : Lipid composition of plant mitochondria
        and of chloroplasts. - J. Lipid Res. *14*˙ : 235 - 242, 1973.

39509 - SCOTT, B.D. : Seasonal variations of phytoplankton production in an estuary
        in relation to coastal water movements. - Aust. J. mar. Freshwater Res.
        30 : 449 - 461, 1979.

39510 - SCOTT, H.D., BATCHELOR, J.T. : Dry weight and leaf area production rates
        of irrigated determinate soybeans. - Agron. J. *71* : 776 - 782, 1979.

39511 - SCOTT, H.D., GEDDES, R.D. : Plant water stress of soybean (*Glycine max*)
        and common cocklebur (*Xanthium pensylvanicum*) : A comparison under field
        conditions.- Weed Sci. *27* : 285 - 289, 1979. [Resistances.]

39512 - SCOTT, J.A., FRENCH, N.R., LEETHAM, J.W. : Patterns of consumption in grass-
        lands. - In : FRENCH, N.R. (ed.) : Perspectives in Grassland Ecology.
        Pp. 89 - 106. Springer-Verlag, New York - Heidelberg - Berlin 1979. [Ener-
        gy utilization.]

B39513 - SCOTT, T.K. (ed.) : Plant Regulation and World Agriculture. - Plenum Press,
        New York - London 1979. [Ps.]

39514 - SCULLEY, M.J., DUNIEC, J.T., THORNE, S.W. : Reconciliation of theory and
        experiment on 90° selective scattering spectra as a measure of intact or
        broken granal or agranal chloroplasts. - FEBS Lett. *98* : 377 - 380, 1979.

39515 - SEARLE, G.F.W., BARBER, J. : The interaction of an amphipathic fluorescen-
        ce probe, 2-p-toluidinonaphthalene-6-sulphonate, with isolated chloroplasts.
        - Biochim. biophys. Acta *545* : 508 - 518, 1979.

39516 - SEARLE, G.F.W., TREDWELL, C.J. : Picosecond fluorescence from photosynthe-
        tic systems *in vivo*. - In : Chlorophyll Organization and Energy Transfer
        in Photosynthesis. Pp. 257 - 281. Excerpta Medica, Amsterdam - Oxford -
        New York 1979.

39517 - SEARLE, G.F.W., TREDWELL, C.J., BARBER, J., PORTER, G. : Picosecond time-
-resolved fluorescence study of chlorophyll organisation and excitation
energy distribution in chloroplasts from wild -type barley and a mutant
lacking chlorophyll *b*. - Biochim. biophys. Acta *545* : 496 - 507, 1979.

39518 - SEBALD, W., HOPPE, J., WACHTER, E. : Amino acid sequence of the ATPase
proteolipid from mitochondria, chloroplasts and bacteria (wild type and
mutants). - In : QUAGLIARIELLO, E., PALMIERI, F., PAPA, S., KLINGENBERG, M.
(ed.) : Function and Molecular Aspects of Biomembrane Transport. Pp. 63 -
74. Elsevier/North-Holland Biomedical Press, Amsterdam - New York - Shannon 1979.

*39519 - SEEFELD, K.-P., MÖBIUS, D., KUHN, H. : Electron transfer in monolayer
assemblies with incorporated ruthenium (II) complexes. - Helv. chim. Acta
60 : 2608 - 2632, 1977. [Ps model.]

39520 - SEELY, G.R. : Association of chlorophyll with surfactants on a particle
surface. - In : Proc. Int. Sol. Energy Soc. Silver Jubilee Congr. Vol. 1.
Pp. 112 - 116. Pergamon, Elmsford 1979.

39521 - SEELY, G.R. : Chlorophyll on plasticized particles - a new model system. -
Biotechnol. Bioeng. Symp. *8* (Biotechnol. Energy Prod. Conserv.) : 473 -
481, 1979.

39522 - SEELY, G.R. : Properties of chlorophyll on plasticized polyethylene particles. - In : Chlorophyll Organization and Energy Transfer in Photosynthesis. Pp. 41 - 59. Excerpta Medica, Amsterdam - Oxford - New York 1979.

B39523- SEEMANN, J., CHIRKOV, Y.I., LOMAS, J., PRIMAULT, B. : Agrometeorology. -
Springer-Verlag, Berlin - Heidelberg - New York 1979.

39524 - SEEMANN, J.R., DOWNTON, W.J.S., BERRY, J.A. : Field studies of acclimation
to high temperature : winter ephemerals in Death Valley. - Carnegie Inst.
Year Book *78* : 157 - 162, 1979. [Ps, Chl.]

39525 - SEIP, K.L., LUNDE, G., MELSOM, S., MEHLUM, E., MELHUUS, A., SEIP, H.M. :
A mathematical model for the distribution and abundance of benthic algae
in a Norwegian fjord. - Ecol. Modelling *6* : 133 - 166, 1979.

39526 - SEITOVA, T.A., RUSINOVA, N.G., DOMAN, N.G. : Svoĭstva i chetvertichnaya
struktura ribulozodifosfatkarboksilazy iz list'ev volosnetsa sitnikovogo.
[Properties and quaternary structure of ribulosebisphosphate carboxylase
from the leaves of *Elymus* (*Psathyrostachys*) *junceus*.]-Biokhimiya *44* :
1891 - 1898, 1979. [In R, ab : E.]

39527 - SEITZ, K. : Light induced changes in the centrifugability of chloroplasts :
Different action spectra and different influence of inhibitors in the low
and high intensity range. - Z. Pflanzenphysiol. *95* : 1 - 12, 1979.

39528 - SEITZ, K. : Light dependent control of cytoplasmic streaming and chloroplast movement. - Acta protozool. *18* : 197 - 198, 1979.

39529 - SEITZ, K. : Cytoplasmic streaming and cyclosis of chloroplasts. - In :
HAUPT, W., FEINLEIB, M.E. (ed.) : Physiology of Movements. (Encycl. Plant
Physiol. N.S. Vol. 7.) Pp. 150 - 169. Springer-Verlag, Berlin - Heidelberg
- New York 1979.

*39530 - SELIGER, H.H., LOFTUS, M.E.,SUBBA RAO,D.V.:Dinoflagellate accumulations
in Chesapeake Bay. - In : LoCICERO, V.R. (ed.) : Proceedings of The First
International Conference on Toxic Dinoflagellate Blooms. Pp. 181 - 205.
Massachusetts Science and Technology Foundation, Wakefield 1975. [Ps, Chl.]

39531 - SELMAN, B.R., SELMAN-REIMER, S. : Tentoxin-induced adenine nucleotide exchange with soluble and thylakoid membrane-bound chloroplast coupling
factor 1. - FEBS Lett. *97* : 301 - 304, 1979.

39532 - SELVARAJ, K.V. : Effect of moisture regimes and topping on the chlorophyll
content of cotton. - Food Farming Agr. *11* (1) : 6 - 7, 1979.

39533 - SEMENENKO, V.E., AVRAMOVA, S., GEORGIEV, D., PRONINA, N.A. : O svetovoĭ
         zavisimosti karboangidraznoĭ aktivnosti kletok *Chlorella* i *Scenedesmus*.
         [Light dependence of carbonic anhydrase activity in *Chlorella* and *Scene-
         desmus* cells.] - Fiziol. Rast. *26* : 1069 - 1075, 1979. [In R, ab : E.]

*39534 - SEMENENKO, V.E., ZVEREVA, M.G., KASATKINA, T.I., TSOGLIN, L.N. : Tempera-
         turnaya induktsiya svetozavisimoĭ beloksinteziruyushcheĭ sistemy *Chlorella*,
         soprovozhdayushchayasya uvelicheniem funktsional'noĭ aktivnosti fotosin-
         teticheskogo apparata. [Temperature induction of light dependent protein
         synthesizing system of *Chlorella* accompanied by enhancement of functional
         activity of the photosynthetic apparatus.] - In : Itogi Issledovaniya
         Mekhanizma Fotosinteza. Pp. 179 - 196. Institut Fotosinteza Akad. Nauk SSSR,
         Pushchino 1974. [Ps; in R.]

39535 - SEMENOVA, G.A., LADYGIN, V.G., TAGEEVA, S.V. : Issledovanie DNK-soderzha-
         shchikh oblasteĭ khloroplastov v zigotakh khlamidomonady metodom elektron-
         noĭ radioavtografii. [Study of the DNA-containing regions of chloroplasts
         in *Chlamydomonas* zygotes using electron autoradiography.] - Dokl. Akad.
         Nauk SSSR *245* : 719 - 721, 1979. [In R.]

39536 - SEMIKHATOVA, O.A., ZALENSKIĬ, O.V. : Ob izuchenii gazoobmena v issledova-
         niyakh produktsionnogo protsessa rasteniĭ. [Gas exchange measurements in
         the study of plant productivity.] - Bot. Zh. *64* : 3 - 9, 1979. [In R,
         ab : E.]

B39537 - SEN, D.N., CHAWAN, D.D., BANSAL, R.P. (ed.) : Structure, Function and
         Ecology of Stomata. - Bishen Singh Mahendra Pal Singh, Dehra Dun 1979.
         [Chloroplast, resistances.]

39538 - SENGER, H., BISHOP, N.I. : Observations on the photohydrogen producing
         activity during the synchronous cell cycle of *Scenedesmus obliquus*. -
         Planta *146* : 53 - 62, 1979.

39539 - SENSER, M., BECK, E. : Kälteresistenz der Fichte II. Einfluß von Photo-
         periode und Temperatur auf die Struktur und photochemischen Reaktionen
         von Chloroplasten. - Ber. deut. bot. Ges. *92* : 243 - 259, 1979.

*39540 - SENTER, S.D. : Carotenoids of pecan nutmeats. - HortScience *10* : 592,
         1975.

39541 - SEO, S.W., CHAMURA, S. : [Studies on the characters of the improved semi-
         -dwarf, high-yielding indica rice varieties. I. Characters of the ripening
         from the viewpoint of sink, source, and reserve carbohydrates.] - Jap. J.
         Crop Sci. *48* : 365 - 370, 1979. [In Jap., ab : E.]

39542 - SERRA, J.L., LLAMA, M.J., RAO, K.K., HALL, D.O. : Properties of the
         monomeric and dimeric forms of *Chromatium* hydrogenase. - Biochem. Soc.
         Trans. *7* : 225 - 226, 1979.

39543 - ŠESTÁK, Z. : Changes in activities of photosystems during leaf ontogenesis.
         - In : MARCELLE, R., CLIJSTERS, H., VAN POUCKE, M. (ed.) : Photosynthesis
         and Plant Development. Pp. 21 - 29. Dr. W.Junk bv. Publ., The Hague -
         Boston - London 1979.

39544 - ŠESTÁK, Z., ČATSKÝ, J. : Bibliography of reviews and methods of photo-
         synthesis - 44, 45, 46. - Photosynthetica *13* : 211 - 229, 337 - 351, 474 -
         478, 1979.

B39545 - ŠESTÁK, Z., ČATSKÝ, J. (ed.) : Photosynthesis Bibliography. Vol. 5/1 -
         1974, Vol. 5/2 - 1974. - Dr. W.Junk b.v. Publ., The Hague 1979.

39546 - SETTER, T.L., BRUN, W.A., BRENNER, M.L. : Source/sink interactions in soy-
         beans. I. A possible role of ABA. - Plant Physiol. *63* (Suppl.) : 43, 1979.

39547 - SEXTON, J.C., ALLEN, M.F., MOORE, T.S. Jr. : Enhanced organic and total
         phosphorus and chlorophyll concentrations of *Bouteloua gracilis* with my-
         corrhizal infection. - Plant Physiol. *63* (Suppl.) : 130, 1979.

39548 - SHABEL'SKAYA, É.F. : K metodike matseratsii rastitel'noĭ tkani. [Method of
         maceration of plant tissue.]- Wiss. Z. pädag. Hochschule "Karl Liebknecht"
         Potsdam *23* : 165 - 175, 1979. [Chloroplast.]

39549 - SHACKEL, K.A., HALL, A.E. : Reversible leaflet movements in relation to
drought adaptation of cowpeas, *Vigna unguiculata* (L.) WALP. - Aust. J.
Plant Physiol. *6* : 265 - 276, 1979. [Canopy architecture.]

39550 - SHAFIROVICH, V.Ya., KHANNANOV, N.K., MORAVSKIĬ, A.P. : Fotosensibilizirovannoe okislenie ionov dvukhvalentnogo margantsa pod deĭstviem vidimogo
sveta. [Photosensitized oxidation of bivalent manganese ions under the
action of visible light.] - Dokl. Akad. Nauk SSSR *244* : 650 - 654, 1979.
[Ps; in R.]

*B39551 - SHAKHOV, A.A. (ed.) : Khloroplasty i Mitokhondrii. [Chloroplasts and Mi-
tochondria.] - Nauka, Moskva 1969. [In R.]

39552 - SHANER, D.L., LYON, J.L. : Somatal cycling in *Phaseolus vulgaris* L. in
response to glyphosphate. - Plant Sci. Lett. *15* : 83 - 87, 1979. [Ps.]

39553 - SHANMUGASUNDARAM, S., LAKSHMANAN, M. : Heterotrophic growth in the blue-
-green alga *Anacystis nidulans*. - Experientia *35* : 609 -610, 1979. [Chl,
Car, Bil.]

39554 - SHAPOSHNIKOVA, M.G., PAKSHINA, E.V., SHUBIN, V.V., KRASNOVSKIĬ, A.A. :
Aktivatsiya i ingibirovanie fotoindutsirovannogo pogloshcheniya protonov
v khromatoforakh *Rhodospirillum rubrum* detergentami i rastvoritelyami.
[Stimulation and inhibition of light-induced proton uptake in *Rhodospirillum rubrum* chromatophores by detergents and solvents.] - Biofizika *24* :
554 - 555, 1979. [In R, ab : E.]

39555 - SHARENKOVA, Kh. : Vliyanie temperatury na fotosinteticheskuyu produktivnost' *Spirulina platensis* (GOM) GEITL. [Effect of temperature on the photosynthetic productivity of *Spirulina platensis* (GOM) GEITL.] - In :
VAKLINOVA, S.G., VANKOVA-RADEVA, R., VASILEVA, V.S. (ed.) : Fotosinteticheskaya Assimilyatsiya CO_2 i Fotodykhanie. Pp. 171 - 175. Izdat. bolg.
Akad. Nauk, Sofiya 1979. [In R.]

39556 - SHARKEY, T.D., RASCHKE, K. : Separation and measurement of direct and indirect effects of light on stomata. - Plant Physiol. *63* (Suppl.) : 60,
1979.

39557 - SHARMA, R., SOPORY, S.K., GUHA-MUKHERJEE, S. : Phytochrome regulation of
peroxidase activity in maize. IV. Photosynthetic independence of peroxidase enhancement. - Plant Cell Physiol. *20* : 1003 - 1012, 1979.

39558 - SHARMA, R.K., GRIFFING, B., SCHOLL, R.L. : Variations among races of *Arabidopsis thaliana* (L.) HEYNH for survival in limited carbon dioxide. -
Theor. appl. Genet. *54* : 11 - 15, 1979. [Ps.]

39559 - SHARP, J.H., UNDERHILL, P.A., HUGHES, D.J. : Interaction (allelopathy)
between marine diatoms : *Thalassiosira pseudonana* and *Phaeodactylum tricornutum*. - J. Phycol. *15* : 353 - 362, 1979. [Chl.]

39560 - SHARP, R.E., DAVIES, W.J. : Solute regulation and growth by roots and
shoots of water-stressed maize plants. - Planta *147* : 43 - 49, 1979. [Sto-
matal resistance.]

39561 - SHARP, R.E., OSONUBI, O., WOOD, W.A., DAVIES, W.J. : A simple instrument
for measuring leaf extension in grasses, and its application in the study
of the effects of water stress on maize and sorghum. - Ann. Bot. *44* :
35 - 45, 1979. [Stomatal resistance.]

39562 - SHATILOV, I.S., CHAPOVSKAYA, G.V., ZAMARAEV, A.G. : Fotosinteticheskiĭ
potentsial i urozhaĭ zernovykh kul'tur. [Photosynthetic potential and
grain crop yield.] - Izv. Timiryazev. sel'sko-khoz. Akad. *1979* (4) : 18 - 30,
1979. [In R, ab : E.]

*39563 - SHATKOVSKIĬ, T.A., SHISHKANU, G.V., DOROKHOV, B.L. : Svetovoĭ rezhim pal'-
metnoĭ krony yabloni. [Light regime of a palmette crown of apple tree.] -
Sadovod. Vinograd. Vinodel. Mold. *1971* (6) : 6 - 8, 1971. [In R.]

39564 - SHEATH,R.G.,HELLEBUST,J.A.,SAWA,T.:Effects of low light and darkness on structural transformations in plastids of the *Rhodophyta*. - Phycologia *18* : 1 - 12, 1979.

39565 - SHEEHY, J.E., COBBY, J.M., RYLE, G.J.A. : The growth of perennial ryegrass : A model. - Ann. Bot. *43* : 335 - 354, 1979. [Ps.]

39566 - SHEEHY, J.E., FISHBECK, K., PHILLIPS, D.A. : Screening alfalfa for photosynthesis and nitrogen fixation. - Plant Physiol. *63* (Suppl.) : 85, 1979.

39567 - SHEEHY, J.E., WOODWARD, F.I., JONES, M.B., WINDRAM, A. : Microclimate, photosynthesis and growth of lucerne (*Medicago sativa* L.) I. Microclimate and photosynthesis. - Ann. Bot. *44* : 693 - 707, 1979.

39568 - SHEEN, S.J., DeJONG, D.W., CHAPLIN, J.F. : Polyphenol accumulation in chlorophyll mutants of tobacco under two cultural practices. - Beitr. Tabakforsch. *10* : 57 - 64, 1979.

39569 - SHELDRAKE, A.R., NARAYANAN, A. : Growth, development and nutrient uptake in pigeonpeas (*Cajanus cajan*).- J.agr.Sci.*92*:513-526,1979.[Growth analysis.]

39570 - SHELP, B., McCABE, J., URSINO, D.J. : Radiation-induced changes in the export and distribution of photoassimilated carbon in soybean plants. - Environ. exp. Bot. *19* : 245 - 252, 1979.

39571 - SHEPHERD, H.S., BOYNTON, J.E., GILLHAM, N.W. : Mutations in nine chloroplast loci of *Chlamydomonas* affecting different photosynthetic functions. - Proc. nat. Acad. Sci. USA *76* : 1353 - 1357, 1979 .

*39572 - SHEREVERYA, N.I. : Rol' geterotrofnogo pitaniya i fotosinteza v geterozise gibridov kukuruzy. [Role of heterotrophic nutrition and photosynthesis in heterotic maize hybrids.] - In : Geterosis Sel'skokhozyaĭstvennykh Rasteniĭ, ego Fiziologo-Biokhimicheskie i Biofizicheskie Osnovy. Pp. 68 - 74. Kolos, Moskva 1975. [In R.]

39573 - SHERIDAN, R.P. : Seasonal variation in sun-shade ecotypes of *Plectonema notatum* (*Cyanophyta*). - J. Phycol. *15* : 223 - 226, 1979. [Chl.]

39574 - SHERMAN, L.A., CUNNINGHAM, J. : Photosynthetic characteristics of temperature-sensitive-high fluorescence mutants of the blue-green alga, *Synechococcus cedrorum*. - Plant Sci. Lett. *14* : 121 - 131, 1979.

39575 - SHERMAN, W.V., EICKE, R.R., STAFFORD, S.R., WASACZ, F.M. : Branching in the bacteriorhodopsin photochemical cycle. - Photochem. Photobiol. *30* : 727 - 729, 1979.

*39576 - SHETH, M., RAMKRISHNA, D., FREDRICKSON, A.G. : Stochastic models of algal photosynthesis in turbulent channel flow. - AICHE J. *23* : 794 - 804, 1977.

39577 - SHEVCHENKO, A.I. : Pigment-protein complexes of Photosystem I from different chloroplast membranes. - In : Mineral Nutrition of Plants. Vol. II. Pp. 346 - 355. Publ. House Central Cooperative Union, Sofia 1979.

39578 - SHEYTANOV, H.E., MANOLOVA, N.I. : New method for tracing the decrease of O_2 concentration in a leaf "*in vivo*" in dark. - Dokl. bolg. Akad. Nauk *32* : 469 - 472, 1979.

39579 - SHI, Jiao-nai, WU, Min-xian, CHA, Jing-juan : [Studies on plant phosphoenolpyruvate carboxylase. I. Separation and properties of PEP carboxylase isoenzymes.] - Acta phytophysiol. sin. *5* : 255 - 235, 1979. [In Chin., ab : E.]

39580 - SHIBA, T., SIMIDU, U., TAGA, N. : Another aerobic bacterium which contains bacteriochlorophyll *a*. - Bull. jap. Soc. sci. Fisheries *45* : 801, 1979.

39581 - SHIBA, T., SIMIDU, U., TAGA, N. : Distribution of aerobic bacteria which contain bacteriochlorophyll *a*. - Appl. environm. Microbiol. *38* : 43 - 45, 1979.

39582 - SHIBATA, H., FUJIHARA, T., UCHIDA, I., OCHIAI, H. : [Studies on chloroplast development in radish cotyledons (5). Metabolism of 4-thiouridine in etiolated radish seedlings.] - Bull. Fac. Agr. Shimane Univ. *13* : 90 - 94, 1979. [In Jap., ab : E.]

39583 - SHIIO, I., UJIGAWA-TAKEDA, K. : Regulation of phosphoenolpyruvate carboxylase by synergistic action of aspartate and 2-oxoglutarate. - Agr. biol.
Chem. *43* : 2479 - 2485, 1979.

39584 - SHIMAZAKI, K., SUGAHARA, K. : Specific inhibition of photosystem II activity in chloroplasts by fumigation of spinach leaves with SO_2. - Plant
Cell Physiol. *20* : 947 - 955, 1979.

39585 - SHIMIZU, K., KIKUCHI, T., SUGANO, N., NISHI, A. : Carotenoid and steroid
syntheses by carrot cells in suspension culture. - Physiol. Plant. *46* :
127 - 132, 1979.

39586 - SHIMOKAWA, K. : [Biochemistry and physiology of ethylene (17).] - Nogyo
Oyobi Engei *54* : 1183 - 1187, 1979. [Chl; in Jap.]

39587 - SHIMOKAWA, K. : Preferential degradation of chlorophyll *b* in ethylene-trea-
ted fruits of "Satsuma" mandarin. - Sci. Hort. *11* : 253 - 256, 1979.

39588 - SHIN, M. : [Photosystem I and the related problems.] - Tanpakushitsu Kakusan Koso, Bessatsu *21* : 104 - 113, 1979. [In Jap.]

39589 - SHIN, M., YOKOYAMA, Z., ABE, A., FUKASAWA, H. : Properties of common wheat
ferredoxin, and a comparison with ferredoxins from related species of
Triticum and *Aegilops*. - J. Biochem. (Tokyo) *85* : 1075 - 1081, 1979.

39590 - SHIOI, Y., SASA, T. : Immobilization of photochemically-active chloroplasts onto diethylaminoethyl-cellulose. - FEBS Lett. *101* : 311 - 315,
1979.

39591 - SHIPILOVA, S.V., ZAPROMETOV, M.N. : Fenilalaninammiak-liaza v khloroplastakh chaĭnogo rasteniya. [Phenylalanine-ammonialyase in chloroplasts of
a tea plant.] - Dokl. Akad. Nauk SSSR *246* : 239 - 242, 1979. [In R.]

39592 - SHIPMAN, L.L., HOUSMAN, D.L. : Förster transfer rates for chlorophyll *a*. -
Photochem. Photobiol. *29* : 1163 - 1167, 1979.

*39593 - SHIRAHASHI, K., HAYAKAWA, S., SUGIYAMA, T. : Cold lability of pyruvate,
orthophosphate dikinase in the maize leaf. - Plant Physiol. *62* : 826 - 830,
1978. [Ps.]

39594 - SHIRAIWA, Y., MIYACHI, S. : Enhancement of ribulose 1,5-bisphosphate carboxylation reaction by carbonic anhydrase. - FEBS Lett. *106* : 243 - 246,
1979.

39595 - SHIROYA, M., KURA, M. : Translocation of organic compounds in sunflower
III. Effect of oligomycin, ouabain and phlorizin on translocation of sugars. - Plant Cell Physiol. *20* : 369 - 374, 1979.

39596 - SHISHIDO, Y., HORI, Y. : Studies on translocation and distribution of
photosynthetic assimilates in tomato plants. III. Distribution pattern as
affected by air and root temperatures in the night. - Tohoku J. agr. Res.
30 : 87 - 94, 1979.

*B39597 - SHLYK, A.A. (ed.) : Biosintez i Sostoyanie Khlorofillov v Rastenii. [Bio-
synthesis and State of Chlorophylls in a Plant.] - Nauka i Tekhnika, Minsk
1975. [In R, ab : E.]

39598 - SHLYK, A.A., MANANKINA, E.E. : Vliyanie ingibitorov belkovogo sinteza na
sostoyanie pigmentnogo fonda v tsikle razvitiya sinkhronnoĭ kul'tury khlorelly. [Effect of protein synthesis inhibitors on the pigment pool state
in the developmental cycle of synchronous *Chlorella* culture.] - Dokl. Akad.
Nauk belorus. SSR *23* : 937 - 940, 959, 1979. [In R, ab : E.]

39599 - SHMAT'KO, I.G., GULYAEV, B.I., SHVEDOVA, O.E., GOLIK, K.N., LATASHENKO,
O.P. : Parametry vodnogo rezhima i gazoobmena sortov ozimoĭ pshenitsy pri
ukhudshenii vodoobespechennosti. [Parameters of water regime and gas exchange in winter wheat varieties under decreasing water supply.] - Fiziol. Biokhim. kul't. Rast. *11* : 312 - 317, 332, 1979. [Ps; in R, ab : E.]

39600 - SHMAT'KO, I.G., SHVEDOVA, O.E. : Regulyatornye faktory vodoobmena kul'tur-
        nykh rasteniĭ. [Regulatory factors of cultivated plant water exchange.] -
        Fiziol. Biokhim. kul't. Rast. *11* : 460 - 470, 1979 . [Resistances; in R,
        ab : E.]

*39601 - SHMAT'KO, I.G., SHVEDOVA, O.E., LATASHENKO, O.P., KIRNOS, P.S. : Posled-
        stvie vliyaniya defitsita vlagi na postuplenie i raspredelenie azota v ras-
        teniyakh ozimoĭ pshenitsy. [After-effect of water deficit on the uptake
        and distribution of nitrogen in winter wheat plants.] - In : Mineral'noe
        Pitanie i Produktivnost' Rasteniĭ. Pp. 211 - 214, 326. Naukova Dumka, Kiev
        1978. [Chl; in R.]

39602 - SHMELEVA, V.L., IVANOV, B.N., BITYUKOVA, L.V. : Svetozavisimoe pogloshche-
        nie kisloroda izolirovannymi khloroplastami gorokha pri fotovosstanovlenii
        NADP+. [Light-dependent oxygen consumption by isolated pea chloroplasts
        under NADP+ photoreduction.] - Biokhimiya *44* : 911 - 916, 1979. [In R,
        ab : E.]

39603 - SHOMER-ILAN, A., BEER, S., WAISEL, Y. : Biochemical basis of ecological
        adaptation. - In : GIBBS, M., LATZKO, E. (ed.) : Photosynthesis II. (En-
        cycl. Plant Physiol. N.S. Vol. 6.) Pp. 190 - 201. Springer-Verlag, Berlin
        - Heidelberg - New York 1979.

39604 - SHOMER-ILAN, A., NEUMANN-GANMORE, R., WAISEL, Y. : Biochemical speciali-
        zation of photosynthetic cell layers and carbon flow paths in *Suaeda monoi-
        ca*. - Plant Physiol. *64* : 963 - 965, 1979.

39605 - SHOMER-ILAN, A., NISSENBAUM, A., GALUN, M., WAISEL, Y. : Effect of water
        regime on carbon isotope composition of lichens. - Plant Physiol. *63* :
        201 - 205, 1979. [Ps.]

39606 - SHORTESS, D.K., AMBY, R.P. : Pigment, free amino acid and chloroplast
        protein analyses of the pale green-13 mutant in maize. - Maydica *24* :
        215 - 221, 1979.

39607 - SHOSHAN, V., SELMAN, B.R. : The interaction of energy transfer inhibitors
        with the adenine nucleotide binding sites on soluble chloroplast coupling
        factor 1. - FEBS Lett. *107* : 413 - 418, 1979.

39608 - SHOSHAN, V., SHAVIT, N. : ATP synthesis and hydrolysis in chloroplast mem-
        branes. Differential inhibition by antibodies to chloroplast coupling fac-
        tor 1. - Europe. J. Biochem. *94* : 87 - 92, 1979.

39609 - SHROTRI, C.K., TEWARI, M.N., RATHORE, V.S. : Effect of zinc on chlorophyll,
        sugars and starch contents in maize *Zea mays* L. - Indian J. exp. Biol. *17* :
        58 - 60, 1979.

39610 - SHUBIN, L.M., BEKASOVA, O.D., EVSTIGNEEV, V.B. : Razreshenie struktury
        spektra pogloshcheniya sine-zelenykh vodorosleĭ metodom izmereniya 2-oĭ
        proizvodnoĭ pri -196 °C. [Resolution of the structure of absorption spec-
        trum of blue-green algae by the method of measuring the 2nd derivative at
        -196 °C.] - Biofizika *24* : 472 - 475, 1979. [In R, ab : E.]

*B39611 - SHUL'GIN, I.A. : Rastenie i Solntse. [Plant and Sun.] - Gidrometeoizdat,
        Leningrad 1973. [Ps; in R, ab : E.]

39612 - SHUTILOVA, N.I., KADOSHNIKOVA, I.G., KOZLOVSKAYA, N.G., KLEVANIK, A.V.,
        ZAKRZHEVSKIĬ, D.A. : Optimizatsiya usloviĭ vydeleniya trekh tipov pigment-
        -belkovolipidnykh kompleksov khloroplastov gorokha pri solyubilizatsii
        s pomoshchyu tritona X-100. [Optimization of isolation conditions for
        three types of pigment-protein-lipid complexes from chloroplasts during
        solubilization with *Triton X-100*.] - Biokhimiya *44* : 1160 - 1171, 1979.
        [In R, ab : E.]

39613 - SHUVALOV, V.A., ASADOV, A.A. : Arrangement and interaction of pigment
        molecules in reaction centers of *Rhodopseudomonas viridis*. Photodichroism
        and circular dichroism of reaction centers at 100 K. - Biochim. biophys.
        Acta *545* : 296 - 308, 1979.

39614 - SHUVALOV, V.A., DOLAN, E., KE, B. : Spectral and kinetic evidence for two
early electron acceptors in Photosystem I. - Proc. nat. Acad. Sci. USA 76 :
770 - 773, 1979.

39615 - SHUVALOV, V.A., KE, B., DOLAN, E. : Kinetic and spectral properties of the
intermediary electron acceptor A_1 in photosystem I. Subnanosecond spectroscopy. - FEBS Lett. 100 : 5 - 8, 1979.

39616 - SHUVALOV, V.A., KLEVANIK, A.V., SHARKOV, A.V., KRYUKOV, P.G. : Pikosekund-
naya spektroskopiya reaktsionnykh tsentrov fotosistemy I zelenykh rastenii.
[Picosecond spectroscopy of the reaction centres of photosystem I of green
plants.] - Dokl. Akad. Nauk SSSR 248 : 756 - 759, 1979. [In R.]

39617 - SHUVALOV, V.A., KLEVANIK, A.V., SHARKOV, A.V., KRYUKOV, P.G., BACON, K.E. :
Picosecond spectroscopy of photosystem 1 reaction centers. - FEBS Lett.
107 : 313 - 316, 1979.

39618 - SICHER, R.C., BAHR, J.T., JENSEN, R.G. : Measurement of ribulose 1,5-bis-
phosphate from spinach chloroplasts. - Plant Physiol. 64 : 876 - 879, 1979.

39619 - SICHER, R.C., JENSEN, R.G. : Regulation of dark CO_2 fixation in intact spinach chloroplasts. - Plant Physiol. 63 (Suppl.) : 64, 1979.

39620 - SICHER, R.C., JENSEN, R.G. : Photosynthesis and ribulose 1,5-bisphosphate
levels in intact chloroplasts. - Plant Physiol. 64 : 880 - 883, 1979.

39621 - SICHKAR, V.I. : Eksperimental'nye mutatsii u soi i ikh selektsionnoe znachenie. Soobshchenie I. Khlorofil'nye mutatsii. [Experimental mutations in
soybean and their selection value. I. Chlorophyll mutations.] - Genetika
15 : 96 - 102, 1979. [In R, ab : E.]

39622 - SIDERER, Y., MALKIN, S. : Flash-induced Mn^{2+} oxidation observed by ESR
spectrometry in lettuce chloroplasts. - FEBS Lett. 104 : 335 - 338, 1979.

39623 - SIEBERTZ, H.P., HEINZ, E., LINSCHEID, M., JOYARD, J., DOUCE, R. : Characterization of lipids from chloroplast envelopes. - Europe. J. Biochem. 101 :
429 - 438, 1979.

39624 - SIEFERMANN-HARMS, D., NINNEMANN, H. : Microscale separation of photosynthetically active pigment-protein complexes by isoelectric focusing. - Plant
Physiol. 63 (Suppl.) : 42, 1979.

39625 - SIEFERMANN-HARMS, D., NINNEMANN, H. : The separation of photosynthetically
active PS-I and PS-II containing chlorophyll-protein complexes by isoelectric focusing of bean thylakoids on polyacrylamide gel plates. - FEBS Lett.
104 : 71 - 77, 1979.

39626 - SIEFERT, E., PFENNIG, N. : Chemoautotrophic growth of Rhodopseudomonas species with hydrogen and chemotrophic utilization of methanol and formate. -
Arch. Microbiol. 122 : 177 - 182, 1979. [Chl.]

39627 - SIEGENTHALER, P.-A., MVÉ AKAMBA, L. : Effect of linolenate on photosynthesis
by intact spinach chloroplasts I. Inhibition of orthophosphate uptake and
of 3-P-glyceraldehyde efflux across the chloroplast envelope. - Plant Cell
Physiol. 20 : 395 - 404, 1979.

39628 - SIGAL, L.L., TAYLOR, O.C. : Preliminary studies of the gross photosynthetic
response of lichens to peroxyacetylnitrate fumigations. - Bryologist 82 :
564 - 575, 1979.

39629 - SIGALAT, C., KOUCHKOVSKY, Y. de : Short-time ATP synthesis in isolated
chloroplasts measured by the luminescence of firefly extracts. - In :
SCHRAM, E., STANLEY, P. (ed.) : Proceedings. International Symposium on
Analytical Applications of Bioluminescence and Chemiluminescence. Pp. 367 -
- 391. State Printing & Publishing, Inc., Westlake Village 1979.

39630 - SIGRIST-NELSON, K., AZZI, A. : The proteolipid subunit of chloroplast adenosine triphosphatase complex. Mobility, accessibility, and interactions
studied by a spin label technique. - J. biol. Chem. 254 : 4470 - 4474,
1979.

39631 - **SIGRIST-NELSON, K., AZZI, A.** : Purification of the chloroplast-membrane di-
cyclohexylcarbodi-imide-binding proteolipid by ion-exchange chromatography.
- Biochem. J.*177* : 687 - 692, 1979. [Chl.]

39632 - **SILAEVA, A.M., SILAEV, A.V.** : Metody kolichestvennogo analiza élektronno-
-mikroskopicheskikh izobrazheniĭ khloroplastov. [Methods for quantitative
analysis of electron-microscopic images of chloroplasts.] - Fiziol. Biokhim.
kul't. Rast. *11* : 547 - 562, 1979. [in R, ab : E.]

*39633 - **SILKIN, V.A., BELYANIN, V.N.** : Èksperimental'noe izuchenie vzaimodeĭstviya
dvukh faktorov v fotobiosinteze vodoroslevykh kletok. [Experimental study
of the interaction of two factors in photobiosynthesis by algal cells.] -
In : Limitirovanie i Ingibirovanie Protsessov Rosta i Mikrobiologicheskogo
Sinteza. Pp. 166 - 173. Nauch. Tsentr biol. Issled., Pushchino 1976. [Chl;
in R.]

*39634 - **SILKIN, V.A., BELYANIN, V.N., TRENKENSHU, R.P.**: Kolichestvennoe opisanie
protsessa rosta mikrovodrosleĭ v svetolimitiruemoĭ kul'ture. [Quantitative
description of the growth process of microalgae in culture with limited
light supply.] - In : Intensivnaya Svetokul'tura Rasteniĭ. Pp. 172 - 181.
Inst. Fiz. sib. Otd. Akad. Nauk SSSR, Krasnoyarsk 1977. [Chl; in R.]

39635 - **SILSBURY, J.H.** : Growth, maintenance and nitrogen fixation of nodulated
plants of subterranean clover (*Trifolium subterraneum* L.). - Aust. J. Plant
Physiol. *6* : 165 - 176, 1979.

39636 - **SILSBURY, J.H., ADEM, L., BAGHURST, P., CARTER, E.D.** : A quantitative exa-
mination of the growth of swards of *Medicago truncatula* cv. Jemalong. -
Aust. J. agr. Res. *30* : 53 - 63, 1979. [Dry-matter production.]

39637 - **SILVIUS, J.E., CHATTERTON, N.J., KREMER, D.F.** : Photosynthate partitioning
in soybean leaves at two irradiance levels. Comparative responses of accli-
mated and unacclimated leaves. - Plant Physiol. *64* : 872 - 875, 1979.

39638 - **SILVIUS, J.E., SNYDER, F.W.** : Photosynthate partitioning and enzymes of sucrose metabolism in sugarbeet roots. - Physiol. Plant. *46* : 169 - 173, 1979.

39639 - **SILVOLA, J., HANSKI, I.** : Carbon accumulation in. a raised bog. Simulation
on the basis of laboratory measurements of CO_2 exchange. - Oecologia *37* :
285 - 295, 1979.

39640 - **SILVOLA, J., HEIKKINEN, S.** : CO_2 exchange in the *Empetrum nigrum-Sphagnum
fuscum* community. - Oecologia *37* : 273 - 283, 1979.

39641 - **SIMON, J.-P.** : Adaptation and acclimation of higher plants at the enzyme
level : Latitudinal variations of thermal properties of NAD malate dehydrogenase in *Lathyrus japonicus* WILLD. (*Leguminosae*). - Oecologia *39* : 273 -
- 287, 1979.

39642 - **SIMON, J.-P.** : Differences in thermal properties of NAD malate dehydrogenase
in genotypes of *Lathyrus japonicus* WILLD. (*Leguminosae*) from maritime and
continental sites. - Plant Cell Environm. *2* : 23 - 33, 1979.

39643 - **SIMON, J.-P.** : Adaptation and acclimation of higher plants at the enzyme
level : Speed of acclimation for apparent energy of activation of NAD malate
dehydrogenase in *Lathyrus japonicus* WILLD. (*Leguminosae*) . - Plant Cell
Environm. *2* : 35 - 38, 1979.

39644 - **SIMON, J.-P.** : Adaptation and acclimation of higher plants at the enzyme
level : Temperature-dependent substrate binding ability of NAD malate dehydrogenase in four populations of *Lathyrus japonicus* WILLD. (*Leguminosae*).
- Plant Sci. Lett. *14* : 113 - 120, 1979.

39645 - **SIMONIS, W., LEE-KADEN, J.** : Selektive Wirkungen von Lindan (γ-1,2,3,4,5,6-
Hexachlorcyclohexan) an Photosynthese, Aminosäure-Membrantransport und Pro-
teinsynthese bei *Anacystis nidulans*. - Z. Naturforsch. *34 C* : 1062 - 1065,
1979.

39646 - **SIMPSON, D.J.** : Freeze-fracture studies on barley plastid membranes. III.
Location of the light-harvesting chlorophyll-protein. - Carlsberg. Res.
Commun. *44* : 305 - 336, 1979.

39647 - **SINCLAIR, J.** : The interaction of oxygen, carbon dioxide and nitrite with the photosynthetic electron transport chain of *Chlorella*. - Plant Physiol. *63* (Suppl.) : 40, 1979.

39648 - **SINCLAIR, J., SARAI, A., GARLAND, S.** : A backflow of electrons around Photosystem II in *Chlorella* cells. - Biochim. biophys. Acta *546* : 256 - 269, 1979.

39649 - **SINCLAIR, T.R., JOHNSON, M.N., DRAKE, G.M., VAN HOUTTE, R.C.** : Mobile laboratory for continuous, long-term gas exchange measurements of 39 leaves. - Photosynthetica *13* : 446 - 453, 1979.

39650 - **SINCLAIR, T.R., RAND, R.H.** : Mathematical analysis of cell CO_2 exchange under high CO_2 concentrations. - Photosynthetica *13* : 239 - 244, 1979.

*39651 - **SINDEN, S.L.** : Control of potato greening with household detergents. - Amer. Potato J. *48* : 53 - 56, 1971. [Chl.]

39652 - **SINGH, H., RAI, B., ASNANI, V.L.** : Influence of brachytic-2 dwarfing gene on the exprassion of leaf area index and light transmission in maize. - Indian J. Genet. Plant Breed. *39* : 419 - 424, 1979.

39653 - **SINGH, R., GANGULEE, R., ROYCHOWDHURY, J.** : Photosynthetic production and Hill reaction in healthy and virus infected papaya leaves. - Nat. Acad. Sci. Lett. (India) *2* : 3 - 4, 1979.

*39654 - **SINGH, R., SINGH, R.B., SRIVASTAVA, R.P.** : Changes in chlorophyll, carbohydrates and primary productivity of cucumber leaves as influenced by cucumber mosaic virus. - Indian J. exp. Biol. *15* : 82 - 83, 1977.

39655 - **SINGHAL, N.C., SINGH, M.P.** : Monosomic analysis of photosynthetic area above the flag leaf node in *T. aestivum* L. - Genet. agr. *33* : 237 - 244, 1979.

*39656 - **SINHA, S.K., AGGARWAL, P.K., CHATURVEDI, G.S.** : Physiological and biochemical analysis of adaptability in wheat. - In : Proceedings of the Fifth International Wheat Genetics Symposium. Vol.2. Pp. 946 - 953. Indian Soc. Genet. Plant Breeding, New Delhi 1978. [Stomatal resistance.]

39657 - **SIREVÅG, R., CASTENHOLZ, R.** : Aspects of carbon metabolism in *Chloroflexus*. - Arch. Microbiol. *120* : 151 - 153, 1979.

*39658 - **SIRONVAL, C.** : Biomass from algae: production and utilization. - In : Symposium Biological Solar Energy Conversion. Pp. 39 - 46. Limburgs Universitair Centrum, Diepenbeek 1978.

39659 - **SIVAKUMAR, M.V.K., SEETHARAMA, N., SINGH, S., BIDINGER, F.R.** : Water relations, growth, and dry matter accumulation of sorghum under post-rainy season conditions. - Agron. J. *71* : 843 - 847, 1979. [Stomatal resistance.]

39660 - **SIVAKUMAR, M.V.K., SHAW, R.H.** : Attenuation of radiation in moisture-stressed and unstressed soybeans. - Iowa State J. Res. *53* : 251 - 257, 1979. [Growth analysis.]

39661 - **SIVAKUMAR, M.V.K., SHAW, R.H.** : Stomatal conductance and leaf-water potential of soybeans under moisture stress. - Iowa State J. Res. *54* : 17 - 27, 1979.

39662 - **SIVAKUMARAN, S., HALL, M.A.** : Hormones in relation to stress recovery in *Populus robusta* cuttings. - J. exp. Bot. *30* : 53 - 63, 1979. [Stomatal resistance.]

39663 - **SKICKO, J., REGAN, D.L.** : Possibility for nitrogen removal from waste water by algae with carbon dioxide as a supplementary carbon source. - Progr. Water Tech. *11* : 405 - 411, 1979.

39664 - **SKRABKA, H., STACHURSKA, A., SZUWALSKA, Z.** : Dynamika przyrostu masy i produktywność stokłosy bezostnej i stokłosy uniolowatej przy zróżnicowanym nawożeniu azotem w doświadczeniu polowym.Cz.I. Wskaźniki produktywności i plony. [Dynamics of mass increase and productivity of smooth bromegrass and rescue grass with different nitrogen fertilization in field experiments. Part I. Indices of productivity and yields.] - Acta agrobot. *32* : 53 - 68, 1979. [In Pol., ab : E.]

39665 - SKRE, O., OECHEL, W.C. : Moss production in a black spruce *Picea mariana* forest with permafrost near Fairbanks, Alaska, as compared with two permafrost-free stands. - Holarctic Ecol. *2* : 249 - 254, 1979. [Ps.]

39666 - SKULACHEV, V.P. : Transduction of light into electric energy in bacterial photosynthesis: a study by means of orthodox electrometer techniques. - In : BARBER, J. (ed.) : Photosynthesis in Relation to Model Systems. Pp. 175 - 188. Elsevier, Amsterdam - New York - Oxford 1979.

39667 - SLABBERS, P.J., SORBELLO HERRENDORF, V., STAPPER, M. : Evaluation of simplified water-crop yield models. - Agr. Water Manage. *2* : 95 - 129, 1979. [Ps,resistances.]

39668 - SLAWYK, G. : ^{13}C and ^{15}N uptake by phytoplankton in the Antarctic upwelling area: results from the Antiprod I cruise in the Indian Ocean sector. - Aust. J. mar. Freshwater Res. *30* : 431 - 448, 1979.

39669 - SLAWYK, G., COLLOS, Y., AUCLAIR, J.C. : Reply to comment by Fisher *et al.* - Limnol. Oceanogr. *24* : 595 - 597, 1979. [^{13}C uptake determination.]

39670 - SLIFKIN, M.A., GARTY, H., SHERMAN, W.V., CAPLAN, S.R. : Modulation-excitation spectrophotometry of bacteriorhodopsin. - Int. J. biol. Macromol. *1* : 61 - 64, 1979.

39671 - SLIFKIN, M.A., GARTY, H., SHERMAN, W.V., VINCENT, M.F.P., CAPLAN, S.R. : Light-induced conductivity changes in purple membrane suspensions. - Biophys. Struct. Mech. *5* : 313 - 320, 1979.

*39672 - SLOBODYAN, S.N. : K voprosu o primenenii gidrazida maleinovoĭ kisloty v sveklovodstve. [Use of maleic acid hydrazide in sugar beet production.] - In : Pitanie Rasteniĭ i Primenenie Udobreniĭ. Pp. 33 - 35, 100. Kishinev. sel'skokhoz. Inst. Im. M.V. Frunze, Kishinev 1977. [Ps; in R.]

39673 - SLOOTEN, L., BRANDERS, C. : The influence of energy-transfer inhibitors of proton permeability and photophosphorylation in normal and preilluminated *Rhodospirillum rubrum* chromatophores. - Biochim. biophys. Acta *547* : 79 - 90, 1979.

39674 - SLOVACEK, R.E., CROWTHER, D., HIND, G. : Cytochrome function in the cyclic electron transport pathway of chloroplasts. - Biochim. biophys. Acta *547* : 138 - 148, 1979.

39675 - SLOVACEK, R.E., HIND, G. : Factors affecting CO_2-fixation in intact chloroplasts. - Plant Physiol. *63* (Suppl.) : 40, 1979.

39676 - SMILLIE, R.M. : Coloured components of chloroplast membranes as intrinsic membrane probes for monitoring the development of heat injury in intact tissues. - Aust. J. Plant Physiol. *6* : 121 - 133, 1979.

39677 - SMILLIE, R.M., MELCHERS, G., WETTSTEIN, D. von : Chilling resistance of somatic hybrids of tomato and potato. - Carlsberg Res. Commun. *44* : 127 - - 132, 1979. [Ps.]

39678 - SMILLIE, R.M., NOTT, R. : Heat injury in leaves of alpine, temperate and tropical plants. - Aust. J. Plant Physiol. *6* : 135 - 141, 1979. [Chl.]

39679 - SMILLIE, R.M., NOTT, R. : Assay of chilling injury in wild and domestic tomatoes based on photosystem activity of the chilled leaves. - Plant Physiol. *63* : 796 - 801, 1979.

39680 - SMIRNOV, A.S., GRODZINSKIĬ, D.M. : Kompartmentatsiya sakharozy i geksoz v svyazi s ikh ottokom iz list'ev sakharnoĭ svekly v temnote. [Compartmentation of sucrose and hexoses caused by their outflow from sugar beet leaves in the dark.] - Fiziol. Biokhim. kul't. Rast. *12* : 573 - 576, 1979. [In R, ab : E.]

39681 - SMIRNOVA, I.A., TIKHONOV, A.N., KONSTANTINOV, A.A., RUGGE, Ė.K. : O svobodnoradikal'nykh tsentrakh v khromatoforakh i preparatakh reaktsionnykh tsentrov *Rhodospirillum rubrum*. [Nature of free radical centres in chromatophores and reaction centre preparations from *Rhodospirillum rubrum*.] - Biofizika *24* : 761 - 762, 1979. [In R, ab : E.]

*39682 - SMITH, A. : Sward growth in relation to pattern of defoliation. - J. brit. Grassl. Soc. *23* : 294 - 298, 1968. [Canopy structure.]

*39683 - SMITH, A., ARNOTT, R.A., MACAULEY, J.R. : Sward productivity under alternate or repeated cutting of adjacent small areas. - J. brit. Grassl. Soc. *30* : 201 - 208, 1975.

*39684 - SMITH, A., MACAULEY, J.R. : Sward productivity within micro-patterns of height and frequency of defoliation. - J. brit. Grassl. Soc. *30* : 279 - 288, 1975.

39685 - SMITH, A.J. : Sun shines on photosynthetic prokaryotes. - Nature *281* : 254 - 255, 1979.

39686 - SMITH, B.B., REBEIZ, C.A. : Chloroplast biogenesis. XXIV. Intrachloroplastic localization of the biosynthesis and accumulation of protoporphyrin-IX, magnesium-protoporphyrin monoester, and longer wavelength metalloporphyrins during greening. - Plant Physiol. *63* : 227 - 231, 1979.

39687 - SMITH, B.N., MARTIN, G.E. II, BOUTTON, T.W. : Carbon isotopic evidence for the evolution of C_4 photosynthesis. - In : KLEIN, E.R., KLEIN, P.D. (ed.) : Stable Isotopes : Proceedings of the Third International Conference. Pp. 231 - 237. Academic Press, New York - San Francisco - London 1979.

39688 - SMITH, B.N., OTTO, C.B., MARTIN, G.E. II, BOUTTON, T.W. : Photosynthetic strategies of desert plants. - In : GOODIN, J.R., NORTHINGTON, D.K. (ed.) : Arid Land Plant Resources. Pp. 474 - 481. Int. Cen. Arid and Semi-Arid Land Studies, Lubbock 1979.

39689 - SMITH, C., DOO, A., BOWN, A.W. : The influence of pH on kinetic parameters of coleoptile phosphoenolpyruvate carboxylase. Relationship to auxin-stimulated dark fixation. - Can. J. Bot. *57* : 543 - 547, 1979.

39690 - SMITH, D.L., ROGAN, P.G. : Growth of the shoot apex of *Agropyron repens* (L.) BEAUV. during successive plastochrons. - Ann. Bot. *44* : 27 - 34, 1979.

39691 - SMITH, J.A., BERRY, J.K. : Optical diffraction analysis for estimating foliage angle distribution in grassland canopies. - Aust. J. Bot. *27* : 123 - 133, 1979.

39692 - SMITH, K.K., GOOD, R.E., GOOD, N.F. : Production dynamics for above and belowground components of a New Jersey *Spartina alterniflora* tidal marsh. - Estuarine coastal mar. Sci. *9* : 189 - 201, 1979.

*B39693 - SMITH, K.M. (ed.) : Porphyrins and Metalloporphyrins. - Elsevier Scientific Publishing Company, Amsterdam - Oxford - New York 1975. [Chl.]

39694 - SMITH, M.G., ROBINSON, J.M., GIBBS, M. : Effects of ascorbate, catalase, ribose-5-P and H_2O_2 on $^{14}CO_2$ assimilation by the spinach chloroplast at different O_2 partial pressures. - Plant Physiol. *63* (Suppl.) : 3, 1979.

39695 - SMITH, S.M., ELLIS, R.J. : Processing of small subunit precursor of ribulose bisphosphate carboxylase and its assembly into whole enzyme are stromal events. - Nature *278* : 662 - 664, 1979.

39696 - SMITH, V.H. : Nutrient dependence of primary productivity in Lakes. - Limnol. Oceanogr. *24* : 1051 - 1064, 1979.

39697 - SMITH, W.O.,Jr., BARBER, R.T. : A carbon budget for the autotrophic ciliate *Mesodinium rubrum*. - J. Phycol. *15* : 27 - 33, 1979. [Ps.]

39698 - SMODLAKA, N., KVEDER, S. : A contribution to the analytical determination of photosynthetic pigments of marine phytoplankton. - Rapp. Comm. int. Mer méditer. *25/26* (9) : 93 - 95, 1979.

39699 - SMOLOV, A.P., POLEVAYA, V.S. : Dykhanie i stanovlenie fotosinteza kul'tury tkani ruty na razlichnykh stadiyakh rostovogo tsikla. [Respiration and evolution of the photosynthesis of a rue tissue culture at different stages of the growth cycle.] - Dokl. Akad. Nauk SSSR *244* : 781 - 783, 1979. [In R.]

39700 - SNOZZI, M., BACHOFEN, R. : Characterisation of reaction centers and their phospholipids from *Rhodospirillum rubrum*. - Biochim. biophys. Acta *546* : 236 - 247, 1979.

39701 – SNYDERS, R., NADAKAVUKAREN, M.J. : Effect of X-irradiation on chlorophyll content and photosynthesis. – Plant Physiol. *63* (Suppl.) : 161, 1979.

39702 – SOCHANOWICZ, B., KANIUGA, Z. : Photosynthetic apparatus in chilling-sensitive plants. IV. Changes in ATP and protein levels in cold and dark stored and illuminated tomato leaves in relation to Hill reaction activity . – Planta *144* : 153 – 159, 1979.

39703 – SOCHANOWICZ, B., KANIUGA, Z. : Photosynthetic apparatus in chilling-sensitive plants. V. Changes in protein fractions of leaves and isolated chloroplasts following cold and dark storage and illumination of tomato leaves. – Planta *145* : 137 – 143, 1979.

39704 – SOE, G., NISHI, N., KAKUNO, T., YAMASHITA, J., HORIO, T. : Conversion of Ca^{2+}-ATPase activity into Mg^{2+}- and Mn^{2+}-ATPase activities with coupling factor purified from *Rhodospirillum rubrum* chromatophores. – In : MUKOHATA, Y., PACKER, L. (ed.) : Cation Flux across Biomembranes. Pp. 243 – 248. Academic Press, New York – San Francisco – London 1979 .

*39705 – SOKOLOVA, A.B. : Zavisimost' intensivnosti fotosinteza *Ascophyllum nodosum* ZE JOLIS ot temperatury. [Dependence of photosynthetic rate of *Ascophyllum nodosum* on temperature.] – In : Ėkologiya i Fiziologiya Rasteniĭ. Vyp.1. Pp. 48 – 52. Kalinin 1974. [In R.]

39706 – SOKOLOVE, P.M. : Conditions limiting the use of ionophore A23187 as a probe of divalent cation involvement in biological reactions. Evidence from the slow fluorescence quenching of Type A spinach chloroplasts. – Biochim. biophys. Acta *545* : 155 – 164, 1979.

39707 – SOKOLOVE, P.M., MARSHO, T.V. : Effect of stromal K^+ concentration on electron transport in intact chloroplasts. – Plant Physiol. *63* (Suppl.) : 28, 1979.

39708 – SOKOLOVE, P.M., MARSHO, T.V. : The effect of valinomycin on electron transport in intact spinach chloroplasts. – FEBS Lett. *100* : 179 – 184, 1979.

39709 – SOLÁROVÁ, J., POSPÍŠILOVÁ, J. : Diffusive conductances of adaxial and abaxial epidermes: Response to photon flux density during development of water stress in primary bean leaves. – Biol. Plant. *21* : 446 – 451, 1979.

39710 – SOLDATINI, G.F. : Changes of glycolate oxidase activity with leaf age in *Zea mays* L. – Z. Pflanzenphysiol. *94* : 267 – 271, 1979. [Chl.]

*B39711 – Solnechnaya Radiatsiya i Produktivnost' Rastitel'nogo Pokrova. [Solar Radiation and Productivity of a Plant Stand.] – Inst. Phys. Astron. Acad. Sci. eston. SSR, Tartu 1972. [In R, ab : E.]

39712 – SOMERVILLE, C.R., OGREN, W.L. : Gas exchange analysis of a photosynthesis/ /photorespiration mutant of *Arabidopsis thaliana*. – Plant Physiol. *63* (Suppl.) : 152, 1979.

39713 – SOMERVILLE, C.R., OGREN, W.L. : A phosphoglycolate phosphatase-deficient mutant of *Arabidopsis*. – Nature *280* : 833 – 836, 1979.

*39714 – SOMMER, U. : Untersuchung über den Zusammenhang zwischen Periphytonproduktion und Schilfdichte. – Sitzungsber. Österr. Akad. Wissen., math.-naturw. Kl., Abt. I *185* : 249 – 258, 1976.

*39715 – SOMMER, U. : Produktionsanalysen am Periphyton im Schilfgürtel des Neusiedler Sees. – Sitzungsber. Österr. Akad. Wissen., math.-naturwiss. Kl., Abt. I *186* : 219 – 246, 1977.

39716 – SØNDERGAARD, M. : Light and dark respiration and the effect of the lacunal system on refixation of CO_2 in submerged aquatic plants. – Aquat. Bot. *6* : 269 – 283, 1979.

39717 – SØNDERGAARD, M., SAND-JENSEN, K. : Carbon uptake by leaves and roots of *Littorella uniflora* (L.) ASCHERS. – Aquat. Bot. *6* : 1 – 12, 1979.

39718 – SØNDERGAARD, M., SAND-JENSEN, K. : The delay in ^{14}C fixation rates by three submerged mycrophytes. A source of error in the ^{14}C technique. – Aquat. Bot. *6* : 111 – 119, 1979.

39719 - SONG, Hong-yu, CHEN, Han-cai, WU, Meng-gan, CHEN , Bing-jian, Yu, Bao-lin :
[Functional association between hydrogenase and nitrogenase in photosynthetic bacterium *Rhodopseudomonas capsulata*.] - Acta phytophysiol. sin. *5* :
237 - 243, 1979. [In Chin., ab : E.]

39720 - SONG, Hong-yu, WU, Meng-gan, Yu, Bao-lin, CHEN, Han-cai, CHEN, Bing-jian :
[Physiological control of nitrogen-fixing activity in photosynthetic bacterium *Rhodopseudomonas capsulata*.] - Acta phytophysiol. sin. *5* : 141 -
150, 1979. [Ps; in Chin., ab : E.]

39721 - SONNEVELD, A., RADEMAKER, H., DUYSENS, L.N.M. : Chlorophyll *a* fluorescence
as a monitor of nanosecond reduction of the photooxidized primary donor
$P-680^+$ of Photosystem II. - Biochim. biophys. Acta *548* : 536 - 551, 1979.

39722 - SORSA, K. : Primary production of epipelic algae in Lake Suomunjärvi, Fin-
nish North Karelia. - Ann. bot. fenn. *16* : 351 - 366, 1979.

39723 - SOSIŃSKA, A., KACPERSKA-PALACZ, A. : Ribulosediphosphate and phosphoenolpyruvate carboxylase activities in winter rape as related to cold acclimation. - Z. Pflanzenphysiol. *92* : 455 - 458, 1979.

39724 - SPALDING, M.H., SCHMITT, M.R., KU, S.B., EDWARDS, G.E. : Intracellular localization of some key enzymes of Crassulacean acid metabolism in *Sedum
praealtum*. - Plant Physiol. *63* : 738 - 743, 1979.

39725 - SPALDING, M.H., STUMPF, D.K., KU, M.S.B., BURRIS, R.H., EDWARDS, G.E. :
Crassulacean acid metabolism and diurnal variations of internal CO_2 and O_2
concentrations in *Sedum praealtum* DC. - Aust. J. Plant Physiol. *6* : 557 -
567, 1979.

39726 - SPALDING, M.H., STUMPF, D.K., KU, S.B., EDWARDS , G.E. : Diurnal variations
in internal CO_2 and O_2 concentrations in *Sedum praealtum* in relation to
Crassulacean acid metabolism. - Plant Physiol. *63* (Suppl.) : 63, 1979.

39727 - ŠPÁNIK, F., OČKAY, S., ŠÍPOŠOVÁ, M. : Využívanie slnečného žiarenia porastom
jarného jačmeňa. [Utilization of solàr radiation by spring barley stands.] -
Rostl. Výroba (Praha) *25* : 15 - 22, 1979. [In Slovak, ab : E, G, R.]

39728 - SPEARING, A.M., KARLANDER, E.P. : Effects of light and low temperatures
on chlorophyll content and metabolism of *Chlorella sorokiniana* SHIHIRA and
KRAUSS. - Environm. exp. Bot. *19* : 237 - 243, 1979 .

39729 - SPERLING, W., RAFFERTY, C.N., KOHL, K.-D., DENCHER, N.A. : Isometric composition of bacteriorhodopsin under different environmental light conditions. - FEBS Lett. *97* : 129 - 132, 1979.

39730 - SPILLER, H., BÖGER, P. : Stabilization of high coupling ratios in spheroplast preparations of the blue-green alga *Nostoc muscorum*. - Plant Physiol.
63 (Suppl.) : 30, 1979.

*39731 - SPIRO, T.G. : Raman spectra of biological materials. - In : MOORE, C.B.
(ed.) : Chemical and Biological Applications of Lasers. Vol. 1. Pp. 29 -
70. Academic Press, New York - London 1974. [Chl, Car.]

39732 - SPOTTS, R.A., FERREE, D.C. : Photosynthesis, transpiration, and water potential of apple leaves infected by *Venturia inaequalis*. - Phytopathology
69 : 717 - 719, 1979.

39733 - SPREITZER, R.J., METS, L. : The recovery of light sensitive acetate requi-
ring mutants of *Chlamydomonas reinhardii*. - Plant Physiol. *63* (Suppl.) :
64, 1979. [RuBPC.]

39734 - SPRUIT, C.J.P., BOUTEN, L., TRIENEKENS, T. : Far red reversibility of the
induction of rapid chlorophyll accumulation in dark grown seedlings. -
Acta bot. neer. *28* : 213 - 220, 1979.

39735 - SPUDICH, J.L., STOECKENIUS, W. : Photosensory and chemosensory behavior
of *Halobacterium halobium* (phototaxis, chemotaxis, bacteriorhodopsin,
halobacteria). - Photobiochem. Photobiophys. *1* : 43 - 53, 1979.

39736 - SPYROPOULOS, C.G., LAMBIRIS, M.P. : Influence of temperature on the effects of water stress on *Quercus* species. - Ann. Bot. *44* : 215 - 220, 1979. [Chl.]

39737 - SQUIRE, G.R. : The response of stomata of pearl millet (*Pennisetum typhoides* S. and H.) to atmospheric humidity. - J. exp. Bot. *30* : 925 - 933, 1979. [Resistances.]

*39738 - SREE RAMULU, K. : Induced chlorophyll chimeras and mutations in *Sorghum*. - Madras agr. J. *57* : 727 - 732, 1970.

*39739 - SREE RAMULU, K. : Mutagenicity of radiations and chemical mutagens in *Sorghum*. - Theor. appl. Genet. *40* : 257 - 260, 1970. [Chl.]

*39740 - SRINIVAS, K., PATIL, S.V. : Effect of spacing and fertility levels on growth and yield of sunflower (*Helianthus annuus* L). - Mysore J. agr. Sci. *11* : 41 - 45, 1977. [Growth analysis.]

39741 - SRIVASTAVA, D.K., SINGH, M. : Photosynthesis, ribulose diphosphate carboxylase activity and chlorophyll content in relation to yield in rice (*Oryza sativa* L.). - J. nucl. agr. Biol. *8* : 41 - 43, 1979.

*39742 - SRIVASTAVA, L.M., VESK, M., SINGH, A.P. : Effect of chloramphenicol on membrane transformations in plastids. - Can. J. Bot. *49* : 587 - 593, 1971. [Chloroplast.]

39743 - SSYMANK, V., BÜHRMANN, H., ROBINSON, D.G. : Effect of inhibitors of RNA and protein synthesis on chloroplast development in a pigment mutant of *Scenedesmus obliquus* (algae). - Biol. cell. *36* : 59 - 64, 1979.

39744 - STADNICHUK, I.N., GUSEV, M.V. : Fikobiliproteidy sinezelenykh, krasnykh i kriptofitovykh vodorosleǐ. [Phycobiliproteins of blue-green, red and cryptophycean algae.] - Biokhimiya *44* : 579 - 593, 1979. [In R, ab : E.]

39745 - STAEHELIN, L.A., ARNTZEN, C.J. : Effects of ions and gravity forces on the supramolecular organization and excitation energy distribution in chloroplast membranes. - In : Chlorophyll Organization and Energy Transfer in Photosynthesis. Pp. 147 - 175. Excerpta Medica, Amsterdam - Oxford - New York 1979.

39746 - STAEHELIN, L.A., McDONNEL, A., CARTER, D.P. : Supramolecular organization of chloroplast membranes as revealed by freeze-fracture and freeze-etch electron microscopy. - In : BAILEY, G.W. (ed.) : 37th Ann. Proc. Electron Microscopy Soc. Amer. Pp. 180 - 183. San Antonio 1979.

39747 - STAMP, P. : Pigmentgehalte und Aktivitäten photosynthetisch wirksamer Enzyme in Blättern junger Maispflanzen in Abhängigkeit von der Temperatur zur Zeit der Kornausreife. - Z. Acker- Pflanzenbau *148* : 230 - 238, 1979.

39748 - STAMPER, J.H., ALLEN, J.C. : A model of the daily photosynthetic rate in a tree. - Agr. Meteorol. *20* : 459 - 481, 1979.

39749 - STĂNCIULESCU, G., PODOLEANU, M., IANOŞI, S. : Efectul combinat al regimului de irigare şi fertilizare asupra unor însuşiri de producţie morfo-fiziologice la hibrizii de tomate timpurii. [The combined effect of irrigation and fertilization conditions on some yield and morpho-physiological features of the early tomato hybrids.] - An. Inst. Cercetări Pentru Leguminocult. Floricult. *5* : 227 - 239, 1979. [Dry-matter production; in Roum., ab : E, F.]

39750 - STANEV, V., PANDEV, S., KUDREV, T. : Influence of the different supply with N, P, K, Ca, Mg and S on the transport of the assimilates and on the distribution of ^{14}C in the basic fractions of the organic compounds in sunflower. - In : Mineral Nutrition of Plants. Vol.I. Pp. 65 - 70. Publ. House Central Cooperative Union, Sofia 1979.

39751 - STANEV, V., TSONEV, Ts.: Vliyanie na azotniya i fosforniya nedostig v"rkhu difuzionnite s"protivleniya na CO_2, kompensatsionniya punkt i temperaturnata zavisimost na fotosintezata pri sl"nchogleda. [Influence of nitrogen and phosphorus deficit on CO_2 diffusion resistance, compensation point and photosynthetic temperature dependence in sunflower.] - Fiziol. Rast. (Sofia) *5* (3) : 12 - 18, 1979. [In Bulg., ab : E, R.]

39752 - STANEV, V.P. : On the calorific value of the dry matter and on the efficiency coefficient of PAR in sunflower and maize plants. - Dokl. bolg. Akad. Nauk *32* : 1547 - 1550, 1979.

39753 - **STANKYAVICHENE, N.A., NORVAĬSHENE, Yu.T., ZHVINENE, N.A., OGORODNIKENE, V.N.:** Izuchenie metodom gazo-adsorbtsionnoĭ khromatografii vliyaniya atsetona, khloroforma i diètilovogo èfira na gazoobmen u dekorativnykh rasteniĭ. [A gas-adsorption chromatography study of the effect of acetone, chloroform and diethyl ether on gas exchange in ornamental plants.]- Liet. TSR Moksly Akad. Darb. Ser.C Biol. Mokslai *1979* (3) : 31 - 38, 1979. [Ps, in R, ab : Latvian.]

*39754 - **STANLEY, J.L., PATTERSON, G.W.** : Sterols and fatty acids of some non-photosynthetic angiosperms. - Phytochemistry *16* : 1611 - 1612, 1977. [Ps.]

39755 - **STARCK, Z., KOZIŃSKA, M., SZANIAWSKI, R.** : Photosynthesis in tomato plants with modified source-sink relationship. - In : MARCELLE, R., CLIJSTERS, H., VAN POUCKE, M. (ed.) : Photosynthesis and Plant Development. Pp. 233 - 241. Dr. W. Junk bv. Publ., The Hague - Boston - London 1979.

39756 - **STARZECKI, W.** : The influence of CO_2 concentration on net photosynthesis of green-house tomato seedling (Pagham-cross variety). - Bull. Acad. pol. Sci., Sér. Sci. biol. II *27* : 289 - 294, 1979.

39757 - **STAUFFER, R.E., LEE, G.F., ARMSTRONG, D.E.** : Estimating chlorophyll extraction biases. - J. Fish. Res. Board Can. *36* : 152 - 157, 1979.

*39758 - **STAWICKI, S., SZEMPLIŃSKA, E.** : Wpływ aflatoksyny B_1 na zdolność kiełkowania ziarna pszenicy oraz zawartość w liścieniach białka, beta-karotenu i chlorofilu. [Effect of aflatoxin B_1 on the germinating capability of wheat grain as well as on protein, beta-carotene and chlorophyll contents in cotyledons.] - Prace Kom. Nauk roln. Kom. Nauk leśn. poznań. Towarz. Przyjac. Nauk *35* : 317 - 322, 1973. [In Pol., ab : E.]

*B39759 - **STEEMANN-NIELSEN, E.** : Marine Photosynthesis. - Elsevier Scientific Publishing Company, Amsterdam - Oxford - New York 1975.

39760 - **STEER, B.T.** : Integration of photosynthetic carbon metabolism and nitrogen metabolism on a daily basis. - In : MARCELLE, R., CLIJSTERS, H., VAN POUCKE, M. (ed.) : Photosynthesis and Plant Development. Pp. 309 - 320. Dr.W.Junk bv.-Publ., The Hague - Boston - London 1979.

39761 - **STEER, B.T., DARBYSHIRE, B.** : Some aspects of carbon metabolism and translocation in onions. - New Phytol. *82* : 59 - 68, 1979.

39762 - **STEINBACK, K., MULLET, J.E., ARNTZEN, C.J.** : Phosphorylation of chloroplast membrane proteins. - Plant Physiol.*63*(Suppl.) : 27, 1979.

39763 - **STEINBACK, K.E., BURKE, J.J., ARNTZEN, C.J.** : Evidence for the role of surface-exposed segments of the light-harvesting complex in cation-mediated control of chloroplast structure and function. - Arch. Biochem. Biophys. *195* : 546 - 557, 1979.

39764 - **STEINECK, O., HAEDER, H.E.** : The effect of potassium on growth and yield components of plants. - In : Potassium Research - Review and Trends. Pp. 165 - 187. International Potash Research Institute, Bern 1979. [Ps, Chl, chloroplast.]

*39765 - **STEINHÜBEL, G.** : Rozmery buniek asimilačného parenchymu ako vnútorný činiteľ rýchlosti fotosyntézy juvenilných jedincov smreka a sosny. [Cell dimensions in assimilation parenchyma as an internal factor of photosynthetic rate of juvenile pine and spruce plants.] - Čas. slezského Muz. Ser. C *27* (1) : 45 - 58, 1978. [In Slov., ab : G.]

39766 - **STEINITZ, Y., MAZOR, Z., SHILO, M.** : A mutant of the cyanobacterium *Plectonema boryanum* resistant to photooxidation. - Plant Sci. Lett. *16* : 327 - - 335, 1979. [Chl, Car.]

39767 - **STEMLER, A.** : A dynamic interaction between the bicarbonate ligand and Photosystem II reaction center complexes in chloroplasts. - Biochim. biophys. Acta *545* : 36 - 45, 1979.

39768 - **STENBERG, R.W.** : Enhanced production of a marine macrophyte (*Gracilaria tikvahiae*) by carbon dioxide (CO_2). - J. Phycol. *15* (Suppl.) : 13, 1979.

*39769 - STEPHENS, P.J., THOMSON, A.J., DUNN, J.B.R., KEIDERLING, T.A., RAWLINGS,
         RAO, K.K., HALL, D.O. : Circular dichroism and magnetic circular dichroi
         of iron-sulfur proteins. - Biochemistry 17 : 4770 - 4778, 1978. [Ferredo
         xin.]

*39770 - STEPHENS, P.J., THOMSON, A.J., KEIDERLING, T.A., RAWLINGS, J., RAO, K.K.
         HALL, D.O. : Cluster characterization in iron-sulfur proteins by magneti
         circular dichroism. - Proc. nat. Acad. Sci. USA 75 : 5273 - 5275, 1978.
         [Ferredoxin.]

 39771 - STEPONKUS, P.L. : Effects of freezing and cold acclimation on membrane
         structure and function. - In : MUSSELL, H., STAPLES, R. (ed.) : Stress P
         siology in Crop Plants. Pp. 144 - 158. John Willey & Sons, New York 1979
         [Chloroplast.]

 39772 - STEUP, M., LATZKO, E. : Intracellular localization of phosphorylases in
         spinach and pea leaves. - Planta 145 : 69 - 75, 1979. [Chloroplast.]

 39773 - STEVENS, R.A., MARTIN, E.S. : The structure of guard cells and substomat
         ion-adsorbent bodies. - In : SEN, D.N., CHAWAN, D.D., BANSAL, R.P. (ed.)
         Structure, Function and Ecology of Stomata. Pp. 7 - 21. Bishen Singh Ma-
         hendra Pal Singh, Dehra Dun 1979. [Chloroplast.]

 39774 - STEWART, A.C., BENDALL, D.S. : Preparation of an active, oxygen-evolving
         photosystem 2 particle from a blue-green alga. - FEBS Lett. 107 : 308 -
         - 312, 1979.

 39775 - STEWART, R.N., DERMEN, H. : Ontogeny in monocotyledons as revealed by st
         dies of the developmental anatomy of periclinal chloroplast chimeras. -
         Amer. J. Bot. 66 : 47 - 58, 1979.

 39776 - STEWART, W.D.P. : N2 fixation and photosynthesis in microorganisms. - In
         GIBBS, M., LATZKO, E. (ed.) : Photosynthesis II. (Encycl. Plant Physiol.
         N.S. Vol.6.) Pp. 457 - 471. Springer-Verlag, Berlin - Heidelberg - New
         York 1979.

 39777 - STIDHAM, M.A., HEATH, R.L. : Quantum yield for proton uptake in spinach
         chloroplasts. - Plant Physiol. 63 (Suppl.) : 30, 1979.

*39778 - STILLWELL, W.L., TIEN, H.T. : Oxygen production from chlorophyll-liposom
         Nature 273 : 406, 1978.

 39779 - ŞTIRBAN, M., ALBU, N. : Biosinteza şi acumularea pigmenţilor asimilatori
         la plantele de griu prevenite din seminţe tratate cu ultrasunete şi micr
         elemente. [Biosynthesis and storage of assimilatory pigments in wheat pl
         resulting from seeds treated with ultrasounds and trace elements.] - Cor
         trib. bot. Univ. Babeş-Bolyai Cluj-Napoca, Grad. Bot. 1979 : 271 - 276,
         1979. [In Roum., ab : E.]

 39780 - STOCKBURGER, M., KLUSMANN, W., GATTERMANN, H., MASSIG, G., PETERS, R. :
         Photochemical cycle of bacteriorhodopsin studied by resonance Raman spec
         troscopy. - Biochemistry 18 : 4886 - 4900, 1979.

 39781 - STOCKNER, J.G., CLIFF, D.D., SHORTREED, K.R.S. : Phytoplankton ecology c
         the Strait of Georgia, British Columbia. - J. Fish. Res. Board Can. 36 :
         657 - 666, 1979. [Chl.]

 39782 - STOECKENIUS, W. : A model for the function of bacteriorhodopsin. - In :
         CONE, R.A., DOWLING, J.E. (ed.) : Membrane Transduction Mechanisms. Pp.
         - 47. Raven Press, New York 1979.

 39783 - STOECKENIUS, W., LOZIER, R.H., BOGOMOLNI, R.A. : Bacteriorhodopsin and
         purple membrane of halobacteria. - Biochim. biophys. Acta 505 : 215 - 2
         1979.

 39784 - STOEV, K.D., SLAVCHEVA, T. : O vliyanii vazhneĭshikh ėkologicheskikh fal
         torov na fotosintez list'ev vinograda. [Study of complex effect of esser
         al ecological factors on vine photosynthesis.] - Fiziol. Rast. 26 : 441
         - 445, 1979. [In R, ab : E.]

39785 - STOICOVICI, L., GALLÓ, Ş. : Growth, dry matter production and mineral metabolism of *Festuca pratensis* L and *Festuca rubra* L grown in mixture on a li-
med podzolic soil. - Rev. roum. Biol. Sér. Biol. vég. *24* : 43 - 49, 1979.
[Growth analysis.]

*B39786 - STONE, J.F. (ed.) : Plant Modification for More Efficient Water Use. - Else-
vier Scientific Publishing Company, Amsterdam - Oxford - New York 1975.
[Ps.]

39787 - STORCH, T.A., DIETRICH, G.A. : Seasonal cycling of algal nutrient limitation in Chautauqua Lake, New York. - J. Phycol. *15* : 399 - 405, 1979.
[Ps.]

*39788 - STOYANOV, I., KUDREV, T. : Effect of magnesium deficiency in separate phases
of development on the growth and productivity of the Wisconsin 641 AA maize.
- Dokl. bolg. Akad. Nauk *29* : 1815 - 1817, 1976. [Dry-matter accumulation.]

39789 - STRAŠKRABA, M., DESORTOVÁ, B., FOTT, J. : Zur Methodik der Bestimmung und
Bewertung des Chlorophyll-*a* in Oberflächengewässern. - Acta hydrochim.
hydrobiol. *7* : 569 - 590, 1979,

39790 - STRATTON, G.W., CORKE, C.T. : The effect of cadmium ion on the growth, photosynthesis, and nitrogenase activity of *Anabaena inaequalis*. - Chemosphere
8 : 277 - 282, 1979.

39791 - STRATTON, G.W., CORKE, C.T. : The effect of nickel on the growth, photosynthesis, and nitrogenase activity of *Anabaena inaequalis*. - Can. J. Microbiol.
25 : 1094 - 1099, 1979.

39792 - STRATTON, G.W., HUBER, A.L., CORKE, C.T. : Effect of mercuric ion on the
growth, photosynthesis, and nitrogenase activity of *Anabaena inaequalis*. -
Appl. environm. Microbiol. *38* : 537 - 543, 1979.

39793 - STREETER, J.G., JEFFERS, D.L. : Distribution of total non-structural carbohydrates in soybean plants having increased reproductive load. - Crop
Sci. *19* : 729 - 734, 1979.

39794 - STROSS, R.G. : Density and boundary regulations of the *Nitella* meadow in
Lake George, New York. - Aquat. Bot. *6* : 285 - 300, 1979. [Ps.]

39795 - STROTMANN, H., BICKEL-SANDKÖTTER, S., EDELMANN, K., ECKSTEIN, F., SCHLIMME,
E., BOOS, K.S., LÜSTORFF, J. : Thiophosphate analogs of ADP and ATP as substrates in partial reactions of energy conversion in chloroplasts. - Biochim. biophys. Acta *545* : 122 - 130, 1979.

39796 - STROTMANN, H., BICKEL-SANDKÖTTER, S., SHOSHAN, V. : Kinetic analysis of
light-dependent exchange of adenine nucleotides on chloroplast coupling
factor CF_1. - FEBS Lett. *101* : 316 - 320, 1979.

39797 - STRZAŁKA, K. : Polypeptides from different fractions of chloroplast membranes of etiolated bean leaves exposed to light and chloramphenicol. -
Bull. Acad. pol. Sci., Sér. Sci. biol. Cl.II *27* : 155 - 159, 1979.

39798 - STRZAŁKA, K., KWIATKOWSKA, M. : Transport of proteins from cytoplasm into
plastids in chloramphenicol-treated bean leaf discs. Autoradiographic evidence. - Planta *146* : 393 - 398, 1979.

39799 - STUFF, R.G., HODGES, H.F., DALE, R.F., NYGUIST, W.E., NELSON, W.L., SCHEE-
RINGA, K.L. : Measurement of short-period corn growth. - Dep. Agron. agr.
Exp. Sta. Purdue Univ. Res. Bull. *961* : 1 - 20, 1979. [Crop growth rate.]

39800 - STULEN, I. : Influence of CO_2 on the functioning of nitrate reductase in
radish (a C_3 plant) and corn (a C_4 plant) seedlings. - In : HEWITT, E.J.,
CUTTING, C.V. (ed.) : Nitrogen Assimilation of Plants. Pp. 582 - 584.
Academic Press, London New York - San Francisco 1979.

39801 - STUMMANN, B.M. : Correlation of the 680 to 672 nm spectral shift and the
halving of the apparent molecular weight for chlorophyll(ide) holochrome
from barley. - Physiol. Plant. *45* : 122 - 126, 1979.

39802 - SUBBA RAO, B.L., GHOSH, A., JOHN, V.T. : Effect of rice tungro virus on
chlorophyll and anthocyanin pigments in two rice cultivars. - Phytopathol.
Z. *94* : 367 - 371, 1979.

39803 - SUBUDHI, B.P.R., SINGH, P.K. : Effect of phosphorus and nitrogen on growth, chlorophyll, amino nitrogen, soluble sugar contents and algal heterocysts of water fern *Azolla pinnata*. - Biol. Plant. *21* : 401 - 406, 1979.

39804 - SUD'INA, O.G., LOS', S.I. : Minlyvist' biliproteïdiv u predstavnykiv rodu *Nostoc* ADANSON v umovakh kul'tury. [Variability of biliproteids of the genus *Nostoc* ADANSON representatives in culture.] - Ukr. bot. Zh. *36* : 238 - - 242, 286, 1979. [In Ukr., ab : E, R.]

39805 - SUD'INA, O.G., LOZOVA, G.I., SYDORENKO, P.G., GOLOD, M.G., DOVBYSH, K.P., L'VIVS'KA, N.R., MARTYN, G.M., UVAROV, G.O., DONTSOVA, I.G. : Ul'trastruk-turna i biokhimichna kharakteristyka pigmentnogo aparatu klityn suspenziĭ-noï kul'tury *Nicotiana tabacum* L. [Ultrastructural and biochemical charac-teristics of pigment apparatus in cells of the *Nicotiana tabacum* L. suspen-sion culture.] - Ukr.bot. Zh. *36* : 105 - 110, 190, 1979. [Chl; in Ukr., ab : E, R.]

39806 - SUGAHARA, K., UCHIDA, S., TAKIMOTO, M. : [Effect of sulfite ions on water-
-soluble chlorophyll proteins.] - Res.Rep. nat. Inst. environm. Stud. *10* : 35 - 47, 1979. [In Jap., ab : E.]

39807 - SUGIMOTO, H. : [Infra-red thermometer and harvest predicting in agricultu-re and forestry.] - Keisoku Gidsutsu [Instrum. Automat.] *7*(9) : 27 - 31, 1979. [In Jap.]

39808 - SUGIMOTO, Y., NIKKI, I. : [Responses of pasture grasses to nitrogen ferti-lization. III. Effect of nitrogen fertilizer rate, leaf nitrogen concentra-tion and chlorophyll content on photosynthetic activities of some subtro-pical grass species.] - J. jap. Soc. Grassl. Sci. *25* : 121 - 127, 1979. [In Jap., ab : E.]

39809 - SUGIYAMA, T., SCHMITT, M.R., KU, S.B., EDWARDS, G.E. : Differences in cold lability of pyruvate,Pi dikinase among C_4 species. - Plant Cell Physiol. *20* : 965 - 971, 1979.

39810 - SUGIYAMA, Y., MUKOHATA, Y. : Changes in subunit construction of chloroplast coupling factor I with detachment from the membrane and addition of divalent cation. - In : MUKOHATA, Y., PACKER, L. (ed.) : Cation Flux across Biomem-branes. Pp. 261 - 273. Academic Press, New York - San Francisco - London 1979.

39811 - SUGIYAMA, Y., MUKOHATA, Y. : Modification of one lysine by pyridoxal phos-phate completely inactivates chloroplast coupling factor 1 ATPase. - FEBS Lett. *98* : 276 - 280, 1979.

39812 - SULLIVAN, C.Y., ROSS, W.M. : Selecting for drought and heat resistance in grain sorghum. - In : MUSSELL, H., STAPLES, R. (ed.) : Stress Physiology in Crop Plants. Pp. 264 - 281. John Willey & Sons, New York 1979. [Ps.]

39813 - SUMAYAO, C.R., KANEMASU, E.T. : Temperature and stomatal resistance of soybean leaves. - Can. J. Plant Sci. *59* : 153 - 162, 1979. [Growth analy-sis.]

39814 - SUNDQVIST, C., RYBERG, H. : Structure of protochlorophyll-containing plas-tids in the inner seed coat of pumpkin seeds (*Cucurbita pepo*). - Physiol. Plant. *47* : 124 - 128, 1979.

39815 - SUNG, F.J.M., KRIEG, D.R. : Relative sensitivity of photosynthetic assimi-lation and translocation of ^{14}carbon to water stress. - Plant Physiol. *64* : 852 - 856, 1979.

39816 - SUSKE, G., WAGNER, W., FOLLMANN, H. : NADPH-dependent thioredoxin reductase and a new thioredoxin from wheat. - Z. Naturforsch. *34* C : 214 - 221, 1979.

39817 - SÜSS, K.-H. : Isolation and partial characterization of membrane-bound ferredoxin-NADP$^+$-reductase from chloroplasts. - FEBS Lett. *101* : 305 - 310, 1979.

39818 - SUYAMA, K., HORI, K., ADACHI, S. : Interference by phytol derivatives in the gas chromatographic analysis of fatty acids in the lipids of plant shoots. - J. Chromatogr. *174* : 234 - 238, 1979. [Chl.]

*39819 - SUZUKI, T., HASEGAWA, K. : Separation of carotenoids, steroids and the related substances on lipophilic Sephadex. - Agr. biol. Chem. 38 : 871 - 872, 1974.

39820 - ŠVACHULA, V., ŠVACHULOVÁ, J. : Vliv extrémních vláhových podmínek na změny obsahu chlorofylů v listech cukrovky. [Effect of extreme moisture conditions on change in the content of chlorophylls in sugar beet leaves.] - Rost.Výroba (Praha) 25 : 1 - 8, 1979. [In Czech, ab : E, G, R.]

39821 - SVOBODA, J., TAYLOR, H.W. : Persistence of cesium-137 in Arctic lichens, Dryas integrifolia, and lake sediments. - Arctic alpine Res. 11 : 95 - 108, 1979. [Growth analysis.]

39822 - SWARTHOFF, T., AMESZ, J. : Photochemically active pigment-protein complexes from the green photosynthetic bacterium Prosthecochloris aestuarii. - Biochim. biophys. Acta 548 : 427 - 432, 1979.

39823 - SWARTZMAN, G.L. : A comparison of plankton simulation models emphasizing their applicability to impact assessment. - J. Environm. Management 9 : 145 - 163, 1979. [Ps.]

39824 - SWARTZMAN, G.L. : Simulation modeling of material and energy flow through an ecosystem: methods and documentation. - Ecol. Model. 7 : 55 - 81, 1979.

39825 - SWEENEY, B.M. : Endogenous rhythms in the movement of plants. - In : HAUPT, W., FEINLEIB, M.E. (ed.) : Physiology of Movements. (Encycl. Plant Physiol. N.S. Vol.7.) Pp. 71 - 93. Springer-Verlag, Berlin - Heidelberg - - New York 1979. [Chloroplast.]

39826 - SWEENEY, B.M., PRÉZELIN, B.B., WONG, D., GOVINDJEE : In vivo chlorophyll a fluorescence transients and the circadian rhythm of photosynthesis in Gonyaulax polyedra. - Photochem. Photobiol. 30 : 309 - 311, 1979.

39827 - SYMONS, M., NUYTEN, A., SYBESMA, C. : On the calibration of the carotenoid band shift with diffusion potentials. - FEBS Lett. 107 : 10 - 14, 1979.

39828 - SYVERTSEN, J.P., CUNNINGHAM, G.L. : The effects of irradiating adaxial or abaxial leaf surface on the rate of net photosynthesis of Perezia nana and Helianthus annuus. - Photosynthetica 13 : 287 - 293, 1979.

39829 - SZALONTAI, B., CSATORDAY, K. : Changes in phycocyanin-carotenoid association during nitrate starvation of Anacystis nidulans. - Biochem. biophys. Res. Commun. 88 : 1294 - 1300, 1979.

39830 - SZAREK, S.R. : The occurrence of Crassulacean Acid Metabolism: A supplementary list during 1976 to 1979. - Photosynthetica 13 : 467 - 473, 1979.

39831 - SZAREK, S.R. : Primary production in four North American deserts: indices of efficiency. - J. Arid Environm. 2 : 187 - 209, 1979. [Ps.]

39832 - SZARVAS, T., POZSÁR, B.I. : Formation of formaldehyde via photosynthesis in maize leaves. Detection of endogenous formaldehyde. - In : Proc. 19th hung. annu. Meet. Biochem. Pp. 35 - 38. Hung. Acad. Sci., Budapest 1979.

*39833 - TABACHNIK, N.F. : Effects of aldehydes on chlorophyll content of Euglena gracilis. - Chemistry 46 (7) : 25 - 26, 1973.

39834 - TABITA, F.R., COLLETTI, C. : Carbon dioxide assimilation in cyanobacteria: Regulation of ribulose 1,5-bisphosphate carboxylase. - J. Bacteriol. 140: 452 - 458, 1979.

39835 - TAGEEVA, S.V., LADYGIN, V.G., SEMENOVA, G.A. : Vzaimosvyaz' mezhdu spektral'nymi formami khlorofilla i strukturnoĭ organizatsieĭ membran khloroplastov Chlamydomonas reinhardii. [Interrelation between the spectral forms of chlorophyll and the structural organization of the membranes of Chlamydomonas reinhardii chloroplasts.] - Dokl. Akad. Nauk SSSR 246 : 1513 - - 1516, 1979. [In R.]

39836 - **TAGUCHI, S.** : Light utilization efficiencies of phytoplankton in the tropical North Pacific Ocean. - Bull. Plankton Soc. Jap. *26* : 1 - 10, 1979.
[Ps.]

*39837 - **TAKABE, T., AKAZAWA, T.** : Catalytic role of subunit A in ribulose-1,5-di-
phosphate carboxylase from *Chromatium* strain D. - Arch. Biochem. Biophys.
157 : 303 - 308, 1973.

39838 - **TAKABE, T., NISHIMURA, M., AKAZAWA, T.** : Isolation of intact chloroplasts
from spinach leaf by centrifugation in gradients of the modified silica
"Percoll". - Agr. biol. Chem. *43* : 2137 - 2142, 1979.

39839 - **TAKABE, T., OSMOND, C.B., SUMMONS, R.E., AKAZAWA, T.** : Effect of oxygen
on photosynthesis and biosynthesis of glycolate in photoheterotrophically
grown cells of *Rhodospirillum rubrum*. - Plant Cell Physiol. *20* : 233 - 241,
1979.

39840 - **TAKAGI, S.** : [Effect of β-carotene on spinach lipoxygenase activity.] -
Sci. Rep. Fac. Agr. Okayama Univ. *53* : 29 - 35, 1979. [In Jap., ab : E.]

39841 - **TAKAHAMA, U.** : Stimulation of lipid peroxidation and carotenoid bleaching
by deuterium oxide in illuminated chloroplast fragments : Participation
of singlet molecular oxygen in the reactions. - Plant Cell Physiol. *20* :
213 - 218, 1979.

39842 - **TAKAMIYA, K.-I., DUTTON, P.L.** : Ubiquinone in *Rhodopseudomonas sphaeroides*.
Some thermodynamic properties. - Biochim. biophys. Acta *546* : 1 - 16,
1979.

39843 - **TAKEDA, G.** : Ecological analysis of photosynthesis of barley and wheat. -
Jap. agr. Res. quart. *13* : 180 - 185, 1979.

39844 - **TAKEMATSU, N., KISHINO, M., OKAMI, N.** : Optical properties and modeling
of turbidity components in sea water (19). - Mer (Tokyo) *17* : 117 - 126,
1979. [Chl.]

39845 - **TAKEMOTO, J., BACHMANN, R.C.** : Orientation of chromatophores and sphero-
plast-derived membrane vesicles of *Rhodopseudomonas sphaeroides* : Analysis by localization of enzyme activities. - Arch. Biochem. Biophys. *195* :
526 - 534, 1979.

39846 - **TAKRURI, I., BOULTER, D.** : The amino acid sequence of ferredoxin from
Triticum aestivum (wheat). - Biochem. J. *179* : 373 - 378, 1979.

39847 - **TAKRURI, I.A.H., BOULTER, D.** : The amino acid sequence of ferredoxin from
Sambucus nigra. - Phytochemistry *18* : 1481 - 1484, 1979.

39848 - **TAMAI, H., SHIOI, Y., SASA, T.** : Purification and characterization of
δ-aminolevulinic acid dehydratase from *Chlorella regularis*. - Plant Cell
Physiol. *20* : 435 - 444, 1979.

39849 - **TAMAI, H., SHIOI, Y., SASA, T.** : Studies on chlorophyllase of *Chlorella
prototheCoides* IV. Some properties of the purified enzyme. - Plant Cell
Physiol. *20* : 1141 - 1145, 1979.

39850 - **TAMAS, I.A., SHERMAN, D.B., BECKER, J.D., OBERLANDER, R.M.** : Chlorophyll-
-dependent oxidation of indoleacetic acid and its analogs. Factors affecting decarboxylation and oxygen uptake. - In : MARCELLE, R., CLIJSTERS, H.,
VAN POUCKE, M. (ed.) : Photosynthesis and Plant Development. Pp. 205 - 217.
Dr. W.Junk b.v. Publ., The Hague - Boston - London 1979.

39851 - **TAMKIVI, R.P., AVARMAA, R.A.** : Lyuminestsentsiya, ispuskaemaya so vtorogo
vozbuzhdennogo èlektronnogo sostoyaniya (S$_2$) khlorofilla *a*. [Luminescence
emitted from the second excited electronic state (S$_2$) of chlorophyll *a*.] -
Biofizika *24* : 540 - 541, 1979. [In R, ab : E.]

39852 - **TANAKA, A., FUJITA, K.** : Growth, photosynthesis and yield components in
relation to grain yield of the field bean. - J. Fac. Agr. Hokkaido Univ.
59 : 145 - 238, 1979.

39853 - **TANAKA, K., MATSUO, K., NAKAIZUMI, Y., MORIOKA, Y., TAKASHITA, Y., TACHI-
BANA, Y., SAWAMURA, Y., KOHDA, S.** : Structure-activity relationships in

tetronic acids and their copper (II) complexes. - Chem. Pharm. Bull. *27* : 1901 - 1906, 1979. [Chl.]

*39854 - TANGEN, K., BRETTUM, P. : Phytoplankton and pelagic primary productivity in Øvre Heimdalsvatn. - Holarct. Ecol. *1* : 128 - 147, 1978.

39855 - TANIYAMA, T. : Studies on injurious effects of air pollutants on crop plants. XIV Effects of air pollutants on apparent photosynthesis of rice and corn plants. - Rep. environm. Sci., Mie Univ. *4* : 77 - 83, 1979.

*39856 - TANIYAMA, T., NOMURA, T. : [Effects of various synthetic detergents on photosynthesis, dry matter and grain production in rice plant.] - Rep. environm. Sci. Mie Univ. *3* : 93 - 104, 1978. [In Jap., ab : E.]

*39857 - TAYLOR, O.C. : Air pollutant injury to plant processes. - HortScience *10* : 501 - 504, 1975. [Ps.]

39858 - TAYO, T.O., MORGAN, D.G. : Factors influencing flower and pod development in oil-seed rape (*Brassica napus* L.). - J. agr. Sci. *92* : 363 - 373, 1979. [Photosynthates.]

39859 - TAZAWA, M., FUJII, S., KIKUYAMA, M. : Demonstration of light-induced potential change in *Chara* cells lacking tonoplast. - Plant Cell Physiol. *20* : 271 - 280, 1979. [Ps.]

39860 - TCHANG, F., MAZLIAK, P. : Séparation des mitochondries, microsomes, étioplastes et glyoxysomes à partir d'un homogénat de cotylédons de graines de Tournesol (*Helianthus annuus* L.). - Compt. rend.Acad.Sci. Paris, Sér. D *289* : 1325 - 1328, 1979.

39861 - TÉCHY, F., AGHION, J. : Réduction photosensibilisée de NAD par un sulfure à travers des membranes lipidiques contenant de la chlorophylle *a* et une quinone. - Compt. rend. Acad. Sci. Paris, Sér. D *289* : 903 - 906, 1979.

39862 - TECHY, F., DINANT, M., AGHION, J. : Interactions between photosynthetic pigments bound to lipid and protein particles. Spectroscopic properties. - Z. Naturforsch. *34 C* : 582 - 587, 1979.

39863 - TEDRO, S.M., MEYER, T.E., KAMEN, M.D. : Primary structure of a high potential, four-iron-sulfur ferredoxin from the photosynthetic bacterium *Rhodospirillum tenue*. - J. biol. Chem. *254* : 1495 - 1500, 1979.

39864 - TEH, K.H., SWANSON, C.A. : Comparative inhibition of photosynthesis and translocation by sulfur dioxide in bush bean. - Plant Physiol. *63* (Suppl.) : 34, 1979.

39865 - TELEWSKI, F.W., JAFFE, M.J. : A new inexpensive light monitoring instrumen for field and laboratory use. - Plant Physiol. *63* (Suppl.) : 141, 1979.

39866 - TENG, L.C., SHEN, T.C., GOH, S.H. : The flavoring compound of the leaves of *Pandanus amaryllifolius*. - Econ. Bot. *33* : 72 - 74, 1979. [Car.]

39867 - TENHUNEN, J.D., MEYER, A., LANGE, O.L., GATES, D.M. : Field test of a physiologically based steady-state photosynthesis model for whole leaves (WHOLEPHOT). - Plant Physiol. *63* (Suppl.) : 122, 1979.

39868 - TENHUNEN, J.D., WEBER, J.A., YOCUM, C.S., GATES, D.M. : Solubility of gases and the temperature dependency of whole leaf affinities for carbon dioxide and oxygen. An alternative perspective. - Plant Physiol. *63* : 916 - 923, 1979.

39869 - TENHUNEN, J.D., WESTRIN, S.S. : Development of a photosynthesis model with an emphasis on ecological applications. IV. Wholephot - whole leaf photosynthesis in response to four independent variables. - Oecologia *41* : 145 - 162, 1979.

39870 - TERADA, H. : [The mechanism of action of uncouplers.] - Seikagaku *51* : 211 - 232, 1979. [In Jap.]

39871 - TERADA, T., ICHIMURA, S. : Environmental properties and the distribution of phytoplankton biomass and photosynthesis in a small eutrophic estuary of Shimoda Bay. - Mer (Tokyo) *17* : 137 - 144, 1979. [Chl.]

39872 - TERADA, T., ICHIMURA, S. : Phytoplankton photosynthesis in a eutrophic estuary with special reference to salinity gradient. - Mer (Tokyo) *17* : 171 - 177, 1979.

39873 - TERAMURA, A.H. : Differences in photosynthetic capacity among three local populations of *Plantago lanceolata* L. - Plant Physiol. *63* (Suppl.) : 127, 1979.

39874 - TERAMURA, A.H., DAVIES, F.S., BUCHANAN, D.W. : Comparative photosynthesis and transpiration in excised shoots of rabbiteye blueberry. - HortScience *14* : 723 - 724, 1979.

39875 - TERAMURA, A.H., STRAIN, B.R. : Localized populational differences in the photosynthetic response to temperature and irradiance in *Plantago lanceolata*. - Can. J. Bot. *57* : 2559 - 2563, 1979.

39876 - TEREKHOVA, I.V., CHERNYAD'EV, I.I., GORONKOVA, O.I., DOMAN, N.G. : Spirulina, kharakteristika i vozmozhnaya regulyatsiya puteĭ fotosinteticheskoĭ assimilyatsii ugleroda. [*Spirulina*, characteristics and possible regulation of pathways of photosynthetic carbon assimilation.] - In : KORDYUM, V.A. (ed.) : Rol' Nizshikh Organizmov v Krugovorote Veshchestv v Zamknutykh Ėkologicheskikh Sistemakh. Pp. 327 - 332. Naukova Dumka,Kiev 1979.[In R.]

39877 - TERESHKOVA, G.M. : Pigmentnye kharakteristiki i zhiznesposobnost' khlorelly v usloviyakh temnovogo perioda. [Pigment characteristics and viability of *Chlorella* under dark period conditions.] - In : KORDYUM, V.A. (ed.) : Rol' Nizshikh Organizmov v Krugovorote Veshchestv v Zamknutykh Ėkologicheskikh Sistemakh. Pp. 230 - 232. Naukova Dumka, Kiev 1979. [Chl; in R.]

39878 - TERNER, J., HSIEH, C.-L., BURNS, A.R., EL-SAYED, M.A. : Time-resolved resonance Raman spectroscopy of intermediates of bacteriorhodopsin : The bK_{590} intermediate. - Proc. nat. Acad. Sci. USA *76* : 3046 - 3050, 1979.

39879 - TERNER, J., HSIEH, C.-L., BURNS, A.R., EL-SAYED, M.A. : Time-resolved resonance Raman characterization of the bO_{640} intermediate of bacteriorhodopsin. Reprotonation of the Schiff base. - Biochemistry *18* : 3629 - 3634, 1979.

39880 - TERPSTRA, W., GOEDHEER, J.C. : Chlorophyllase and photosystem. I. Methylviologen photoreduction by DCPIP-ascorbate in the presence of photosynthetic membrane fragments, chlorophyll-Triton X-100 or chlorophyll-chlorophyllase. - Physiol. Plant. *45* : 367 - 372, 1979.

39881 - TERRY, N. : The use of mineral nutrient stress in the study of limiting factors in photosynthesis. - In : MARCELLE, R., CLIJSTERS, H., VAN POUCKE, M. (ed.) : Photosynthesis and Plant Development. Pp. 151 - 160. Dr. W.Junk b.v. Publ., The Hague - Boston - London 1979.

*39882 - TERSKOV, I.A. : Parametricheskoe upravlenie biosintezom. [Parametric biosynthesis control.] - Vestnik Akad. Nauk SSSR *1976* (7) : 61 - 70, 1976. [Ps; in R.]

39883 - TERSKOV, I.A., FURYAEV, E.A., BELYANIN, V.N. : Mikrospektrofotometricheskie pokazateli kletok i vozrastnaya struktura populyatsii *Porphyridium cruentum* na razlichnykh urovnyakh obluchennosti periodicheskoĭ kul'tury. [Microspectrophotometrical characteristics of cells and age structure of a population of *Porphyridium cruentum* under different irradiances of the periodical culture.]-Izv. sib. Otd. Akad. Nauk SSSR, Ser. biol. Nauk *1979* : 97 - 104, 1979. [Chl, Bil; in R, ab : E.]

39884 - TEŞU, V., TOMA, L.D., MERLESCU, E. : Efectul salinizării asupra unor procese fiziologice la sfecla de zahăr. [Salinity effect on some physiological processes in sugar beet.] - Lucrări ştiinţ. Inst. agron. "Ion Ionescu de la Brad", Ser. Agron. *23* : 73 - 76, 1979. [Chl; in Roum., ab : E.]

39885 - TETLEY, R.M., BISHOP, N.I. : The differential action of metronidazole on nitrogen fixation, hydrogen metabolism, photosynthesis and respiration in *Anabaena* and *Scenedesmus*. - Biochim. biophys. Acta *546* : 43 - 53, 1979.

*39886 - THAYER, G.W., ADAMS, S.M., LaCROIX, M.W. : Structural and functional aspects of a recently established *Zostera marina* community. - In : CRONIN, L.E. (ed.) : Estuarine Research. Vol. 1. Pp. 518 - 540. Academic Press, New York - London 1975.

39887 - THEG, S.M., SAYRE, R.T. : Characterization of chloroplast manganese by electron paramagnetic resonance spectroscopy. - Plant Sci. Lett. *16* : 319 - 326, 1979.

39888 - THIAGARAJAH, M.R., HUNT, L.A., HUNTER, R.B. : Effect of short-term temperature fluctuations on leaf photosynthesis in corn (*Zea mays*). - Can. J. Bot. *57* : 2387 - 2393, 1979.

39889 - THIMANN, K.V. : Food plants and plant hormones in our future. - In : SCOTT, T.K. (ed.) : Plant Regulation and World Agriculture. Pp. 1 - 10. Plenum Press, New York - London 1979. [Stomatal resistance.]

39890 - THIMANN, K.V., MALIK, N., SATLER, S. : Stomatal aperture and the senescence of leaves. - In : SCOTT, T.K. (ed.) : Plant Regulation and World Agriculture. Pp. 319 - 326. Plenum Press, New York - London 1979. [Chl.]

39891 - THIMANN, K.V., SATLER, S. : Relation between senescence and stomatal opening: Senescence in darkness. - Proc. nat. Acad. Sci. USA *76* : 2770 - 2773, 1979. [Chl, stomatal resistance.]

39892 - THIMANN, K.V., SATLER, S.O. : Relation between leaf senescence and stomatal closure: Senescence in light. - Proc. nat. Acad. Sci. USA *76* : 2295 - - 2298, 1979. [Chl, stomatal resistance.]

39893 - THIMANN, K.V., SATLER, S.O. : Relation between leaf senescence and stomatal aperture. - Plant Physiol. *63* (Suppl.) : 72, 1979.

39894 - THOMAS, P., HUMMEL, R.L., SMITH, J.W. : Rotational Raman spectroscopy for the remote sensing of carbon dioxide. - J. Air Pollut. Control Assoc. *29* : 390 - 391, 1979.

39895 - THOMAS, R.J., STANTON, D.S., GRUSAK, M.A. : Radioctive tracer study of sporophyte nutrition in hepatics. - Amer. J. Bot. *66* : 398 - 403, 1979. [Ps, Chl.]

39896 - THOMAS, W.H. : Anomalous nutrient-chlorophyll interrelationships in the offshore eastern tropical Pacific Ocean. - J. mar. Res. *37* : 327 - 335, 1979.

39897 - THOMPSON, J.A., FENTON, I.G. : Influence of plant population on yield and yield components of irrigated sunflower in southern New South Wales. - Aust. J. exp. Agr. anim. Husb. *19* : 570 - 574, 1979.

39898 - THOMPSON, N. : Turbulence measurements above a pine forest. - Boundary Layer Meteorol. *16* : 293 - 310, 1979.

39899 - THOMPSON, R., TAYLOR, H. : Field plots for the practical estimation of potential yield. - Sci. Horticult. *10* : 309 - 316, 1979.

39900 - THOMPSON, R.G., FENSOM, D.S., ANDERSON, R.R., DROUIN, R., LEIPER, W. : Translocation of C-11 from leaves of *Helianthus, Heracleum, Nymphoides, Ipomoea, Tropaeolum, Zea, Fraxinus, Ulmus, Picea,* and *Pinus:* comparative shapes and some fine structure profiles. - Can. J. Bot. *57* : 845 - 863, 1979.

39901 - THORNBER, J.P., BARBER, J. : Photosynthetic pigments and models for their organization *in vivo*. - In : BARBER, J. (ed.) : Photosynthesis in Relation to Model Systems. Pp. 27 - 70. Elsevier, Amsterdam - New York - Oxford 1979.

39902 - THORNBER, J.P., MARKWELL, J.P., REINMAN, S. : Plant chlorophyll-protein complexes: recent advances. - Photochem. Photobiol. *29* : 1205 - 1216, 1979.

B39903 - THORNE, D.W., THORNE, M.D. (ed.) : Soil, Water and Crop Production. - AVI Publishing Company, Westport 1979. [Ps.]

39904 - THORNE, J.H. : Assimilate redistribution from soybean pod walls during seed development. - Agron. J. *71* : 812 - 816, 1979.

39905 - THORNLEY, J.H.M. : Wheat grain growth : Anthesis to maturity. - Aust. J. Plant Physiol. *6* : 187 - 194, 1979. [Ps.]

39906 - THORPE, N., WILLMER, C.M., MILTHORPE, F.L. : Stomatal metabolism : carbon dioxide fixation and labelling patterns during stomatal movement in *Commelina cyanea*. - Aust. J. Plant Physiol. *6* : 409 - 416, 1979.

39907 - THRASH, R.J., FANG, H.L.-B., LEROI, G.E. : On the role of forbidden low--lying excited states of light-harvesting carotenoids in energy transfer in photosynthesis. - Photochem. Photobiol. *29* : 1049 - 1050, 1979.

*39908 - THURLOW, D.L., DAVIS, R.B., SASSEVILLE, D.R. : Primary productivity, phytoplankton populations and nutrient bioassays in China Lake, Maine, U.S.A. - Verh. int. Ver. theoret. angew. Limnol. *19* (Part 2) : 1029 - 1036, 1975. [Ps, Chl.]

39909 - THURNAUER, M.C. : *Esr* study of the photoexcited triplet state in photosynthetic bacteria. - Rev. Chem. Intermed. *3* : 197 - 230, 1979.

39910 - THURNAUER, M.C., BOWMAN, M.K., MORRIS, J.R. : Time-resolved electron spin echo spectroscopy applied to the study of photosynthesis. - FEBS Lett. *100* : 309 - 312, 1979.

39911 - TIBBITTS, T.W. : Humidity and plants. - BioScience *29* : 358 - 363, 1979. [Ps.]

39912 - TIEFERT, M.A., MOUDRIANAKIS, E.N. : Role of AMP in photophosphorylation by spinach chloroplasts. - J. biol. Chem. *254* : 9500 - 9508, 1979.

39913 - TIEMANN, R., RENGER, G., GRÄBER, P., WITT, H.T. : The plastoquinone pool as possible hydrogen pump in photosynthesis. - Biochim. biophys. Acta *546* : 498 - 519, 1979.

39914 - TIEN, H.T. : An alternative hypothesis for the direction of hydrogen ion movement and energy transduction in *H. halobium*. - Biochem. biophys. Res. Commun. *89* : 226 - 232, 1979.

39915 - TIEN, H.T. : Photoeffects in pigmented bilayer lipid membranes. - In : BARBER, J. (ed.) : Photosynthesis in Relation to Model Systems. Pp. 115 - 173. Elsevier, Amsterdam - New York - Oxford 1979.

39916 - TIEN, H.T., STILLWELL, W. : *H. halobium* : II. *In vivo* studies. - Bioelectrochem. Bioenerg. *6* : 525 - 535, 1979. J. electroanal. Chem. *104* : 525 - - 535, 1979.

39917 - TIESZEN, L.L., HEIN, D., QVORTRUP, S.A., TROUGHTON, J.H., IMBAMBA, S.K. : Use of $\delta^{13}C$ values to determine vegetation selectivity in East African herbivores. - Oecologia *37* : 351 - 359, 1979.

39918 - TIESZEN, L.L., SENYIMBA, M.M., IMBAMBA, S.K., TROUGHTON, J.H. : The distribution of C_3 and C_4 grasses and carbon isotope discrimination along an altitudinal and moisture gradient in Kenya. - Oecologia *37* : 337 - 350, 1979.

39919 - TIJSSEN, S.B. : Diurnal oxygen rhythm and primary production in the mixed layer of the Antlantic Ocean at 20° N. - Neth. J. Sea Res. *13* : 79 - 84, 1979.

39920 - TIKHONOV, A.N., RUUGE, É.K. : Issledovanie élektronnogo transporta v fotosinteticheskikh sistemakh metodom élektronnogo paramagnitnogo rezonansa. VIII. Vzaimodeĭstvie dvukh fotosistem i kinetika okislitel'no-vosstanovitel'nykh prevrashcheniĭ P700 v razlichnykh rezhimakh impul'snogo osveshcheniya. [Electron paramagnetic resonance study of electron transport in photosynthetic systems. VIII. The interaction between two photosystems and kinetics of P700 redox transients under various conditions of flash excitation.] - Mol. Biol. (Moskva) *13* : 1085 - 1097, 1979. [In R, ab : E.]

39921 - TILLBERG, E., HOLMVALL, M., ERICSSON, T. : Growth cycles in *Lemna gibba* cultures and their effects on growth rate and ultrastructure. - Physiol. Plant. *46* : 5 - 12, 1979. [Chloroplast.]

*39922 - **TILNEY-BASSETT, R.A.E.** : The control of plastid inheritance in *Pelargonium*. IV. - Heredity *37* : 95 - 107, 1976. [Chloroplast.]

39923 - **TILNEY-BASSETT, R.A.E., ABDEL-WAHAB, O.A.L.** : Maternal effects and plastid inheritance. - In : NEWTH, D.R., BALLS, M. (ed.) : Maternal Effects in Development. Pp. 29 - 45. Cambridge University Press, Cambridge 1979.

39924 - **TILZER, M.M., HORNE, A.J.** : Diel patterns of phytoplankton productivity and extracellular release in ultra-oligotrophic Lake Tahoe. - Int. Rev. ges. Hydrobiol. *64* : 157 - 176, 1979.

39925 - **TIMMER, L.W., GARNSEY, S.M., GRIMM, G.R., EL-GHOLL, N.E., SCHOULTIES, C.L.** : Wilt and dieback of Mexican lime caused by *Fusarium oxysporum*. - Phytopathology *69* : 730 - 734, 1979. [Chl.]

39926 - **TING, I.P., LaPRE, L., LAZZARO, C.** : Effect of reduced oxygen on gross CO_2 uptake. - Plant Physiol. *63* (Suppl.) : 39, 1979.

39927 - **TINGEY, S.V., ANDERSEN, W.R., RINEHART, C.A.** : Variation for immuno-quantitated concentration of ribulose-1,5-bisphosphate carboxylase in barley cultivars. - Plant Physiol. *63* (Suppl.) : 8, 1979.

39928 - **TINKER, R.W., BRACH, E.J., LaCROIX, L.J., MACK, A.R., POUSHINSKY, G.** : Classification of land use and crop maturity, types, and disease status by remote reflectance measurements. - Agron. J. *71* : 992 - 1000, 1979. [Chl.]

39929 - **TISCHER, W., STROTMANN, H.** : Some properties of the DCMU-binding site in chloroplasts. - Z. Naturforsch. *34 C* : 992 - 995, 1979.

39930 - **TISCHNER, R., LORENZEN, H.** : Activation of the nitrate reducing system in synchronous *Chlorella* 211-8k (high temperature strain). - Biochem. Physiol. Pflanzen *174* : 99 - 105, 1979. [Ps.]

39931 - **TISON, D.L., LINGG, A.J.** : Dissolved organic matter utilization and oxygen uptake in algal-bacterial microcosms. - Can. J. Microbiol. *25* : 1315 - 1320, 1979. [Ps.]

39932 - **TITUS, J.E., ADAMS, M.S.** : Coexistence and the comparative light relations of the submersed macrophytes *Myriophyllum spicatum* L. and *Vallisneria americana* MICHX. - Oecologia *40* : 273 - 286, 1979. [Ps.]

39933 - **TITUS, J.E., ADAMS, M.S.** : Comparative carbohydrate storage and utilization patterns in the submersed macrophytes, *Myriophyllum spicatum* and *Vallisneria americana*. - Amer. Midland Naturalist *102* : 263 - 272, 1979.

39934 - **TITUS, J.E., ADAMS, M.S., GUSTAFSON, T.D., STONE, W.H., WESTLAKE, D.F.** : Evaluation of differential infrared gas analysis for measuring gas exchange by submersed aquatic plants. - Photosynthetica *13* : 294 - 301, 1979.

39935 - **TOBIESSEN, P.L., SLACK, N.G., MOTT, K.A.** : Carbon balance in relation to drying in four epiphytic mosses growing in different vertical ranges. - Can. J. Bot. *57* : 1994 - 1998, 1979.

39936 - **TODOROVA, A.D., CHICHEV, P., ÏORDANOV, I.** : Vliyanie khloramfenikola i tetratsiklina na vklyuchenie ugleroda v osnovnye fotoprodukty u *Scenedesmus acutus* MEYEN. [Effect of chloramphenicol and tetracycline on carbon incorporation into the main products of photosynthesis in *Scenedesmus acutus* MEYEN.] - In : VAKLINOVA, S.G., VANKOVA-RADEVA, R., VASILEVA, V.S. (ed.) : Fotosinteticheskaya Assimilyatsiya CO_2 i Fotodykhanie. Pp. 109 - 113. Izdat. bolg. Akad. Nauk, Sofiya 1979. [In R.]

39937 - **TOETZ, D.W.** : Biological and water quality effects of artificial mixing of Arbuckle Lake, Oklahoma, during 1977. - Hydrobiologia *63* : 255 - 262, 1979. [Chl.]

39938 - **TOIVONEN, P.M.A., HOFSTRA, G.** : The interaction of copper and sulphur dioxide in plant injury. - Can. J. Plant Sci. *59* : 475 - 479, 1979. [Stomatal resistance.]

39939 - TOLBERT, N.E. : Glycolate metabolism by higher plants and algae. - In :
GIBBS, M., LATZKO, E. (ed.) : Photosynthesis II. (Encycl. Plant Physiol. N.S.
Vol.6.) Pp. 338 - 352. Springer-Verlag, Berlin - Heidelberg - New York
1979.

39940 - TOLLENAAR, M., DAYNARD, T.B., HUNTER, R.B. : Effect of temperature on rate
of leaf appearance and flowering date in maize. - Crop Sci. 19 : 363 - 366,
1979.

39941 - TOLLIN, G., CASTELLI, F., CHEDDAR, G., RIZZUTO, F. : Laser photolysis studies of quinone reduction by chlorophyll a in alcohol solution. - Photochem. Photobiol. 29 : 147 - 152, 1979.

39942 - TOLSTOY, A. : Chlorophyll a in relation to phytoplankton volume in some Swedish lakes. - Arch. Hydrobiol. 85 : 133 - 151, 1979.

39943 - TOMA, L.-D., IFTENI, L., NĂDEJDE, M. : Cercetări asupra influenţei unor
erbicide cu volum redus faţă de conţinutul de pigmenţi asimilatori la păr.
[Effect of some herbicides having a reduced volume on the content of assimilatory pigments in pear tree.] - Lucrări ştiinţ. Inst. agron. "Ion Ionescu de la Brad", Ser. Horticult. 23 : 19 - 22, 1979. [In Roum., ab : E.]

39944 - TOMAS, C.R. : Olisthodiscus luteus (Chrysophyceae). III. Uptake and utilization of nitrogen and phosphorus. - J. Phycol. 15 : 5 - 12, 1979. [Chl.]

39945 - TOMATI, U., GALLI, E. : Water stress and - SH-dependent physiological activities in young maize plants. - J. exp. Bot. 30 : 557 - 563, 1979. [Ps.]

39946 - TOMBESI, L., ROSSI, C. de, INDIATI, R. : Effetti indotti dalle differenti
condizioni di fertilità chimica sui processi fotosintetico e respiratorio
e sulla composizione dei tessuti vegetali. [Effects of different chemical
fertilization on photosynthetic and respiratory processes and on plant tissues composition.] - Ann. Ist. sper. Nutr. Piante 9 (1) : 3 - 18, 1978/79.
[In Ital., ab : E, F, G.]

*39947 - TOMOAIA, M., IOANETTE, A., CHIFU, E. : Behaviour of β-apo-8'-carotenal at
the oil/water and air/water interfaces. - In : WOLFRAM, E. (ed.) : Proceedings of the International Conference on Colloid and Surface Science.
Pp. 559 - 566. Akad. Kiadó, Budapest 1975.

39948 - TOMOAIA-COTIŞEL, M., ALBU, I., CHIFU, E. : Adsorption of carotene and albumin at the oil/water interface. - Stud. Univ. Babeş-Bolyai, Chem. 24 (2) :
68 - 73, 1979.

39949 - TOMOAIA-COTISEL, M., CHIFU, E. : Mixed insoluble monolayers with β-apo-8'-
-carotenoic acid ethyl ester. - Gaz. chim. ital. 109 : 371 - 375, 1979.

39950 - TONK, W.J.M., SCHAPENDONK, A.H.C.M., VREDENBERG, W.J. : A double-compartment
mixing cuvette for measuring light- and chemically-induced absorbance changes in suspensions of energy-conserving particles. - J. biochem. biophys.
Methods 1 : 193 - 194, 1979. [Chloroplasts.]

B39951 - TRANQUILLINI, W. : Physiological Ecology of the Alpine Timberline. Tree
Existence at High Altitudes with Special Reference to the European Alps.
(Ecological Studies. Vol. 31.) - Springer-Verlag, Berlin - Heidelberg -
- New York 1979. [Ps.]

*B39952 - Transport Assimilyatov i Otlozhenie Veshchestv v Zapas u Rasteniĭ. [Photo-
synthate Transport and Accumulation of Organic Substances in Plants.] -
Akad. Nauk SSSR, Vladivostok 1973. [In R, ab : E.]

39953 - TRAVIS, A.J., MANSFIELD, T.A. : Stomatal responses to light and CO_2 are
dependent on KCl concentration. - Plant Cell Environm. 2 : 319 - 323, 1979.
[Malic enzyme.]

39954 - TRAVIS, A.J., MANSFIELD, T.A. : Reversal of the CO_2-responses of stomata
by fusicoccin. - New Phytol. 83 : 607 - 614, 1979.

39955 - TREBST, A. : Inhibition of photosynthetic electron flow by phenol and diphenylether herbicides in control and trypsin-treated chloroplasts. - Z.
Naturforsch. 34 C : 986 - 991, 1979.

39956 - TREBST, A., DRABER, W. : Structure activity correlations of recent herbicides in photosynthetic reactions. - Adv. Pestic. Sci. 2 : 223 - 234, 1979.

39957 - TREBST, A., REIMER, S., DRABER, W., KNOPS, H.J. : The effect of analogues
of dibromothymoquinone and of bromonitrothymol on photosynthetic electron
flow. - Z. Naturforsch. 34 C : 831 - 840, 1979.

39958 - TRÉMOLIÈRES, A., GUILLOT-SALOMON, T., DUBACQ, J.-P., JACQUES, R., MAZLIAK,
P., SIGNOL, M. : The effect of monochromatic light on α-linolenic and
trans-3-hexadecenoic acids biosynthesis, and its correlation to the development of the plastid lamellar system. - Physiol. Plant. 45 : 429 - 436,
1979.

*39959 - TRENKENSHU, R.P., BELYANIN, V.N., GRIBOVSKAYA, I.V. : Rost i produktivnost'
vodorosli Platymonas viridis v periodicheskoĭ kul'ture. [Growth and productivity of an alga Platymonas viridis in periodical culture.] - In : Mate-
rialy IX Vsesoyuznogo Rabochego Soveshchaniya po Voprosam Krugovorota Veshchestv v Zamknutoĭ Sisteme na Osnove Zhiznedeyatel'nosti Nizshikh Organizmov. Pp. 135 - 137. Naukova Dumka, Kiev 1976. [Dry-matter accumulation;
in R.]

*39960 - TRENKENSHU, R.P., TERSKOV, I.A., FURYAEV, E.A., YARUNTSOV, S.A. : Rostovye
i produktsionnye pokazateli vodorosli Porphyridium cruentum v plotnykh
kul'turakh. [Growth and production characteristics of the alga Porphyridium cruentum in dense cultures.] - In : Intensivnaya Svetokul'tura Rasteniĭ.
Pp. 191 - 200. Inst. Fiz. sib. Otd. Akad. Nauk SSSR, Krasnoyarsk 1977. [Ps;
in R.]

39961 - TRIBUTSCH, H. : Solar bacterial biomass bypasses efficiency limits of photosynthesis. - Nature 281 : 555 - 556, 1979.

39962 - TROTTER, D.M., HENDRICKS, A.C. : Attached, filamentous algal communities.
- Amer. Soc. Testing Materials spec. tech. Pub. 690 (Methods Meas. Periphyton Communities) : 58 - 69, 1979. [Ps, Chl.]

39963 - TROUGHTON, J.H. : $\delta^{13}C$ as an indicator of carboxylation reactions. - In :
GIBBS, M., LATZKO, E. (ed.) : Photosynthesis II. (Encycl. Plant Physiol.
N.S. Vol.6.) Pp. 140 - 149. Springer-Verlag, Berlin - Heidelberg - New
York 1979.

39964 - TROXLER, R.F., BROWN, A.S., BROWN, S.B. : Bile pigment synthesis in plants.
Mechanism of ^{18}O incorporation into phycocanobilin in the unicellular rho-
dophyte, Cyanidium caldarium. - J. biol. Chem. 254 : 3411 - 3418, 1979.

39965 - TROXLER, R.F., BROWN, A.S., ZILINSKAS, B.A. : Amino acid sequence at the
amino terminus of allophycocyanin I, II., and III from Nostoc sp. phycobilisomes. - Plant Physiol. 63 (Suppl.) : 98, 1979.

39966 - TROXLER, R.F., OFFNER, G.D. : δ-Aminolevulinic acid synthesis in a Cyanidium caldarium mutant unable to make chlorophyll a and phycobiliproteins.
- Arch. Biochem. Biophys. 195 : 53 - 65, 1979.

*39967 - TRUNOV, I.A. : Vzaimosvyaz' aktivnosti kornevoĭ sistemy s produktivnost'-
yu fotosinteza u plodonosyashchikh rasteniĭ yabloni. [Relationship of activity of root system and productivity of photosynthesis in fruit-bearing
apple plants.] - Sb. nauch. Tr. vsesoyuz. nauch.-issled. Inst. Sadovod.
Im. I.V.Michurina 27 (Sovershenstvovanie Sortimenta i Agrotekhnicheskikh
Priemov v Sadovodstve) : 62 - 66, 128, 1978. [In R.]

39968 - TSANKOV, B., KURTEV, P., BRAIKOV, D., PANDELIEV, S. : Fiziologicheska aktivnost na listata ot sorta Bolgar v zavisimost ot metamernoto im razpolo-
zhenie i nachina na otglezhdane na lozite. [Leaf physiological activity in
cultivar Bolgar as related to metameric position and mode of grapevine cultivation.] - Fiziol. Rast. (Sofiya) 55(4) : 93 - 100, 1979. [In Bulg., ab :
E, R.]

39969 - TSCHERMAK-WOESS, E. : Über Plastidenstapel bei Botrydiopsis alpina sowie
Anlage und Vermehrung der Stigmen bei dieser und Heterococcus (Xanthophy-
ceae). - Plant System. Evol. 131 : 179 - 192, 1979. [Chloroplast.]

39970 - TSCHISMADIA,I., MOORE, F.D. : Biosynthesis in isolated *Acetabularia* chloroplasts. IV. Plastoquinones. - In : BONOTTO, S., KEFELI, V., PUISEUX-DAO, S. (ed.) : Developmental Biology of *Acetabularia*. Pp. 183 - 194. Elsevier/North-Holland Biomedical Press, Amsterdam - New York - Oxford 1979.

39971 - TSEL'NIKER, Yu.L. : Resistances to CO_2 uptake at light saturation in forest tree seedlings of different adaptation to shade. - Photosynthetica *13* : 124 - 129, 1979.

39972 - TSEL'NIKER, Yu.L., MAĬ, V.V. : Protsessy rosta i fotosinteticheskaya aktivnost' lista osiny *Populus tremula* L. [Growth and photosynthetic activity of *Populus tremula* L. leaf.] - Fiziol. Rast. *26* : 1062 - 1068, 1979. [In R, ab : E.]

39973 - TSENOVA, E., FEDINA, I. : Vliyanie diurona na aktivnost' glitseral'degid-
-3-fosfatdegidrogenazy v zeleneyushchikh prorostkakh gorokha. [Effect of DCMU on glyceraldehyde-3-phosphate dehydrogenase activity in green pea seedlings.] - In : VAKLINOVA, S.G., VANKOVA-RADEVA, R., VASILEVA, V.S. (ed.) : Fotosinteticheskaya Assimilyatsiya CO_2 i Fotodykhanie. Pp. 62 - 67. Izdat. bolg. Akad. Nauk, Sofiya 1979. [In R.]

39974 - TSENOVA, M. : Vklyuchenie ^{14}C v glikolevuyu kislotu u fotosinteziruyushchikh izolirovannykh khloroplastov v zavisimosti ot kontsentratsii bikarbonata i ferritsianida. [^{14}C incorporation into glycolic acid in isolated chloroplasts depending on the concentration of bicarbonate and ferricyanide.] - In : VAKLINOVA, S.G., VANKOVA-RADEVA, R., VASILEVA, V.S. (ed.) : Fotosinteticheskaya Assimilyatsiya CO_2 i Fotodykhanie. Pp. 141 - 147. Izdat. bolg. Akad. Nauk, Sofiya 1979. [In R.]

39975 - TSOGLIN, L.N., EVSTRATOV, A.V., SEMENENKO, V.E. : Primenenie mikrovodoroslei dlya biosinteza mechenykh ^{13}C-soedinenii. [Use of microalgae for biosynthesis of labelled ^{13}C-compounds.] - Fiziol. Rast. *26* : 215 - 218, 1979. [In R, ab : E.]

39976 - TSUCHIYA, T., IWAKI, H. : Impact of nutrient enrichment in a waterchestnut ecosystem at Takahama-Iri bay of Lake Kasumigaura, Japan. II. Role of waterchestnut in primary productivity and nutrient uptake. - Water Air Soil Pollut. *12* : 503 - 510, 1979.

39977 - TSUJI, H., NAITO, K., HATAKEYAMA, I., UEDA, K. : Benzyladenine-induced increase in DNA content per cell, chloroplast size, and chloroplast number per cell in intact bean leaves. - J. exp. Bot. *30* : 1145 - 1151, 1979.

39978 - TSUJI, K., ROSENHECK, K. : The low pH species of bacteriorhodopsin: Structure and proton pump activity. - FEBS Lett. *98* : 368 - 372, 1979.

39979 - TSUJITA, M.J., MURR, D.P., JOHNSON, G. : Leaf senescence of Easter lily as influenced by root/shoot growth, phosphorus nutrition and ancymidol. - Can. J. Plant Sci. *59* : 757 - 761, 1979.

*39980 - TSUNODA, S. : Adaptive differentiation in photosynthetic properties in wheat. - In : Proceedings of the Fifth International Wheat Genetics Symposium. Vol.2. Pp. 916 - 922. Indian Soc. Genet. Plant Breeding, New Delhi 1978.

39981 - TSUNODA, S. : Characteristic of photosynthesis and environmental adaptation of rice. - In : WU, H.P., HSIEH, K.C. (ed.) : Proceedings of the ROC - Japan Symposium on Rice Productivity. Inst. Bot. Acad. Sinica Monograph Ser. 3. Pp. 3 - 8. Taiwan 1979.

39982 - TSUZUKI, M., MIYACHI, S. : Effects of CO_2 concentration during growth and of ethoxyzolamide on CO_2 compensation point in *Chlorella*. - FEBS Lett. *103*: 221 - 223, 1979.

39983 - TU, J.C. : Alterations in chloroplast and cell mebranes associated with cAMP-induced dissociation of starch grains in clover yellow mosaic virus infected clover. - Can. J. Bot. *57* : 360 - 369, 1979.

39984 - TUCKER, C.J. : Red and photographic infrared linear combinations for monitoring vegetation. - Remote Sensing Environm. *8* : 127 - 150, 1979. [Chl.]

39985 - TUCKER, C.J., HOLBEN, B.N., ELGIN, J.H.Jr., McMURTEY, J.E.III : The relationship of red and photographic infrared spectral data to grain yield
variation within a winter wheat field. - NASA tech. Memorandum 80318. Pp.
1 - 21. Goddard Space Flight Center, Greenbelt 1979.

39986 - TUKENDORF, A. : 1. Niektóre funkcje karotenoidów w organizmach fotosynte-
tyzujących. [1.Some carotenoid functions in photosynthesizing organisms.] -
Wiadom. bot. 23 : 169 - 180, 1979. [In Pol.]

39987 - TULBU, G.V., KRENDELEVA, T.E. : Vliyanie razobshchitelei na fotoindutsirovannoe tushenie flyuorestsentsii atebrina khloroplastami gorokha. [Effect
of uncouplers on the photoinduced quenching of atebrine fluorescence by
pea chloroplasts.] - Biol. Nauki 1979 (2) : 110, 1979. [In R.]

*39988 - TULBU, G.V., KRENDELEVA, T.E., KAUROV, B.S., RUBIN, A.B. : Effect of mem-
brane-active substances on the light-induced quenching of atebrin fluorescence in chloroplasts. - Stud. biophys. 62 : 189 - 200, 1977.

39989 - TULLY, R.E., HANSON, A.D. : Amino acids translocated from turgid and water-stressed barley leaves I. Phloem exudation studies. - Plant Physiol.
64 : 460 - 466, 1979. [Photosynthates.]

39990 - TULLY, R.E., HANSON, A.D., NELSEN, C.E. : Proline accumulation in water-
-stressed barley leaves in relation to translocation and the nitrogen budget. - Plant Physiol. 63 : 518 - 523, 1979.

39991 - TURNER, N.C. : Differences in response of adaxial and abaxial stomata to
environmental variables. - In : SEN, D.N., CHAWAN, D.D., BANSAL, R.P. (ed.)
: Structure, Function and Ecology of Stomata. Pp. 229 - 250. Bishen Singh
Mahendra Pal Singh, Dehra Dun 1979.

39992 - TURNER, R.E., WOO, S.W., JITTS, H.R. : Phytoplankton production in a tur-
bid, temperature salt marsh estuary. - Estuar. coast. mar. Sci. 9 : 603 -
- 613, 1979.

39993 - TURVEY, P.M., PATRICK, J.W. : Kinetin-promoted transport of assimilates
in stems of Phaseolus vulgaris L. Localized versus remote site(s) of action. - Planta 147 : 151 - 155, 1979.

39994 - TYAGI, V.V.S., MAYNE, B.C., PETERS, G.A. : Isolation and characterization
of phycobiliproteins from the endophytic cyanobacterium of Azolla. - Plant
Physiol. 63 (Suppl.) : 112, 1979.

39995 - TYANKOVA, L., TSONEV, Ts., DIMITROVA, A., KUDREV, T. : Nutrient deficien-
cy-induced changes in photosynthetic activities and growth of maize plants.
- In : Mineral Nutrition of Plants. Vol. II. Pp. 148 - 152. Publ. House
Central Cooperative Union, Sofia 1979.

39996 - TYLER, J.E. : In situ quantum efficiency of oceanic photosynthesis. - Appl.
Optics 18 : 442 - 445, 1979.

39997 - TYSZKIEWICZ, E., NIKOLIČ, D., POPOVIČ, R., SARIČ, M. : Photophosphorylation
and ultrastructural development in Pinus nigra chloroplasts, grown under
different spectral composition of light. - Physiol. Plant. 46 : 324 - 329,
1979.

39998 - UCHIJIMA, Z., HIRAKI, E., INOUE, K. : [Heat balance characteristics of
single span vinylhouse with cucumber plants.] - Bull. nat. Inst. agr. Sci.,
Ser. A 26 : 89 - 112, 1979. [Foliage development; in Jap., ab : E.]

39999 - UCHIMIYA, H. : Chloroplast adherence to plant protoplasts. Specific interactions of pH, calcium and PEG. - Naturwissenschaften 66 : 314 - 315,
1979.

40000 - UCHIMIYA, H., CHEN, K., WILDMAN, S.G. : Genetic behavior of information
coding for the small subunit polypeptides of Lycopersicon Fraction I protein. - Plant Sci. Lett. 17 : 63 - 66, 1979.

40001 - UCHIMIYA, H., CHEN, K., WILDMAN, S.G. : A micro electrofocusing method for determining the large and small subunit polypeptide composition of Fraction 1 proteins. - Plant Sci. Lett. *14* : 387 - 394, 1979.

40002 - UCHIMIYA, H., WILDMAN, S.G. : Nontranslation of foreing genetic information for fraction 1 protein under circumstances favorable for direct transfer of *Nicotiana gossei* isolated chloroplasts into *N. tabacum* protoplasts. - *In Vitro 15* : 463 - 468, 1979.

*40003 - ULANOVA, E.F. : Izmenenie yadra i khloroplastov v kletkakh tomata, porazhennogo stoiburom. [Changes in nucleus and chloroplasts of tomato plant cells infected by stolbur.] - Izv. Akad. Nauk SSSR, Ser. biol. *1978* (2) : 301 - 305, 1978. [In R, ab : E.]

40004 - UMEHARA, T., TERAO, J., MATSUSHITA, S. : [Photosensitized oxidation of oils with food colors.] - Nippon Nôgeikagaku Kaishi *53* : 51 - 56, 1979. [In Jap., ab : E.]

40005 - URBACH, W., LURZ, G., HARTMEYER, H., URBACH, D. : Induction of reversible tolerance of algal cells to various herbicides. I. Inhibition of photosynthesis by phenol herbicides and dibromothymoquinone, its reversal and development of insensitivity to different herbicides. - Z. Naturforsch. *34 C* : 951 - 956, 1979.

40006 - URMANTSEV, Yu.A. : Sistemnyĭ podkhod k probleme ustoĭchivosti rasteniĭ (na primere issledovaniya zavisimosti soderzhaniya pigmentov v list'yakh fasoli ot odnovremennogo deĭstviya na nee zasukhi i zasoleniya). [Systematic approach to the problem of plant resistance (a study of pigment content in bean leaves as affected by concurrently acting external   drought and salinity).] - Fiziol. Rast. *26* : 762 - 778, 1979. [In R, ab : E.]

40007 - URMANTSEV, Yu.A. : Sistemnyĭ podkhod k probleme ustoĭchivosti rasteniĭ (adekvatnost' i interpretatsiya regressionnykh uravneniĭ zavisimosti soderzhaniya pigmentov v list'yakh fasoli ot odnovremennogo deĭstviya na nee zasukhi i NaCl).[Systematic approach to the problem of plant resistance (adequacy and interpretation of the regression equations of dependence of bean leaf pigment content on external drought and NaCl acted simultaneously).]- Fiziol. Rast. *26* : 1233 - 1244, 1979. [In R, ab : E.]

40008 - URSINO, D.J., HUNTER, D., LAING, R., FOWLER, P., KEIGHLEY, J. : Nitrate modification of carbon assimilation and utilization in young soybean plants.. - Plant Physiol. *63* (Suppl.) : 112, 1979.

40009 - USMANOV, P.D. : Die genetische Kontrolle der Chloroplastenfunktionen. - Arch. Züchtungsforsch. *9* : 3 - 14, 1979.

40010 - USUDA, H. : The regulation of malate-decarboxylation in a CAM plant. - Plant Physiol. *63* (Suppl.) : 37, 1979. [Ps.]

40011 - VACEK, K., LOKAJ, P., URBANOVÁ, M., SLADKÝ, P. : Radiative and nonradiative transitions in subchloroplast particles highly enriched in *P*-700. - Biochim. biophys. Acta *548*: 341 - 347, 1979.

40012 - VADEBONCOEUR, C., MAMET-BRATLEY, M., GINGRAS, G. : Photoreaction center of photosynthetic bacteria. 2. Size and quaternary structure of the photoreaction centers from *Rhodopseudomonas rubrum* strain G9 and from *Rhodopseudomonas sphaeroides* strain 2.4.1. - Biochemistry *18* ; 4308 - 4314, 1979.

40013 - VADEBONCOEUR, C., NOËL, H., POIRIER, L., CLOUTIER, Y., GINGRAS, G. : Photoreaction center of photosynthetic bacteria. 1. Further chemical characterization of the photoreaction center from *Rhodospirillum rubrum*. - Biochemistry *18* : 4301 - 4308, 1979.

40014 - VAKLINOVA, S., MANOLOVA, N. : Issledovanie skorosti vydeleniya kisloroda *in vivo* khronoamperometricheskim metodom u mutanta gorokha, ne soderzhashchego khlorofilla *b* i belka-nositelya FS-II. [A chronoamperometric study

of the rates of oxygen evolution *in vivo* in a chlorophyl *b* deficient and
PS II deficient pea mutant.] - In : VAKLINOVA, S.G., VANKOVA-RADEVA, R.,
VASILEVA, V.S. (ed.) : Fotosinteticheskaya Assimilyatsiya CO_2 i Fotodykha-
nie. Pp. 74 - 79. Izdat. bolg. Akad. Nauk, Sofiya 1979. [In R.]

B40015 - VAKLINOVA, S.G., VANKOVA-RADEVA, R., VASILEVA, V.S. (ed.) : Fotosinteticheskaya Assimilyatsiya CO_2 i Fotodykhanie. [Photosynthetic CO_2 Assimilya-
tion and Photorespiration.] - Izdatel'stvo bolgarskoĭ Akademii Nauk, Sofiya 1979.

40016 - VALANNE, N., PENNANEN, A., VAPAAVUORI, E. : Comparison between the preservation of chloroplast structure in the dark and the turnover rate of chlorophyll-protein complexes in a moss and two varieties of pea. - Plant Cell
Physiol. *20* : 1511 - 1522, 1979.

40017 - VALIKHANOV, M.N., SAGDULLAEV, I.N. : Ob otsutstvii vysokomolekulyarnykh po-
lifosfatov v khloroplastakh khlopchatnika. [Absence of high-molecular po-
lyphosphates in chloroplasts of cotton plants.] - Fiziol. Rast. *26* : 116 -
- 122, 1979. [In R, ab : E.]

40018 - VALLEJOS, C.E. : Genetic diversity of plants for response to low temperature
and its potential use in crop plants. - In : LYONS, J.M., GRAHAM, D., RAISON,
J.K. (ed.) : Low Temperature Stress in Crop Plants: The Role of Membrane. Pp.
473 - 489. Academic Press, New York - San Francisco - London 1979. [Dry-
-matter production.]

40019 - VAN ASSCHE, C.J. : Characterization of a common molecular target for selec-
ted structures of photosynthesis inhibiting herbicide. - In : GEISSBÜHLER,
H. (ed.) : Advances in Pesticide Science. Vol.3. Pp. 494 - 498. Pergamon
Press, Oxford - New York 1979.

40020 - VAN ASSCHE, F., CLIJSTERS, H., MARCELLE, R. : Photosynthesis in *Phaseolus
vulgaris* L., as influenced by supra-optimal zinc nutrition. - In : MARCELLE,
R., CLIJSTERS, H., VAN POUCKE, M. (ęd.) : Photosynthesis and Plant Development. Pp. 175 - 184. Dr.W.Junk bv.-Publ., The Hague - Boston - London 1979.

40021 - VAN BESOUW, A., WINTERMANS, J.F.G.M. : The synthesis of galactosyldiacylgly-
cerols by chloroplast envelopes. - FEBS Lett. *102* : 33 - 37, 1979.

40022 - VAN DEN DRIESSCHE, R., CHEUNG, K.-W. : Relationship of stem electrical impe-
dance and water potential of Douglas-fir seedlings to survival after cold
storage. - Forest Sci. *25* : 507 - 517, 1979. [Growth analysis.]

40023 - VANDEN DRIESSCHE, T. : Phase-shifting effect of IAA on the photosynthetic
circadian rhythm of *Acetabularia*. - In : BONOTTO, S., KEFELI, V., PUISEUX-
-DAO, S. (ed.) : Developmental Biology of *Acetabularia*. Pp. 195 - 204.
Elsevier/North-Holland Biomedical Press, Amsterdam - New York - Oxford
1979.

40024 - VANDEN DRIESSCHE, T., MAGNUSSON, A., GLORY, M., CLAUDE, J.-P. : The circadian rhythms of *Acetabularia* (primarily photosynthesis). - Arch. int.
Physiol. Biochim. *87* : 1053 - 1054, 1979.

40025 - VANDERBILT, V.C., BAUER, M.E., SILVA, L.F. : Prediction of solar irradiance distribution in a wheat canopy using a laser technique. - Agr. Meteorol.
20 : 147 - 160, 1979.

*40026 - VAN DER MEER, J.P. : Hybrid chlorosis in interspecific crosses of *Oenothe-
ra* : polygenic inheritance of the nuclear component. - Can. J. Genet. Cytol.
16 : 193 - 201, 1974.

40027 - VAN DER VELDE, G., GIESEN, T.G., VAN DER HEIJDEN, L. : Structure, biomass
and seasonal changes in biomass of *Nymphoides peltata* (GMEL.) O. KUNTZE
(*Menyanthaceae*), a preliminary study. - Aquat. Bot. *7* : 279 - 300, 1979.

*40028 - VAN GEMERDEN, H., BEEFTINK, H.H. : Specific rates of substrate oxidation
and product formation in autotrophically growing *Chromatium vinosum* cultures. - Arch. Mikrobiol. *119* : 135 - 143, 1978.

40029 - VAN GINKEL, G. : Photoreactivity of isolated Photosystem I particles upon
combination with artificial lipid membranes or Triton X-100 micelles. -
Photochem. Photobiol. *30* : 397 - 404, 1979.

40030 - VAN GINKEL, G., RAISON, J.K. : O_2^- - and $OH^{\cdot}$-formation in systems containing chlorophyll. - Plant Physiol. *63* (Suppl.) : 29, 1979.

40031 - VAN GINKEL, G., RAISON, J.K. : Light-induced formation of oxygen radicals in systems containing chlorophyll. - Carnegie Inst. Year Book *78* : 183 - - 189, 1979.

40032 - VAN HASSELT, P.R., DE KOK, L.J., KUIPER, P.J.C. : Effect of α-tocopherol, β-carotene, monogalactosyldiglyceride and phosphatidylcholine on light-induced degradation of chlorophyll *a* in acetone. - Physiol. Plant. *45* : 475 - - 479, 1979.

40033 - VAN HOLSTEIJN, H.M.C. : A closed system for measurement of photosynthesis, respiration and CO_2 compensation points. - Meded. Landbouwhogesch. Wageningen *79* (10) : 1 - 14, 1979.

40034 - VAN LIERE, L., DE GROOT, G.J., MUR, L.R. : Pigment variation with irradiance in *Oscillatoria agardhii* GOMONT in nitrogen (nitrate)-limited chemostat cultures. - FEMS Microbiol. Lett. *6* : 337 - 340, 1979.

40035 - VAN LIERE, L., MUR, L.R. : Decay of *Oscillatoria agardhii* GOMONT. - Hydrobiol. Bull. *13* : 56 - 60, 1979. [Chl, Bil.]

40036 - VAN LIERE, L., MUR, L.R., GIBSON, C.E., HERDMAN, M. : Growth and physiology of *Oscillatoria agardhii* GOMONT cultivated in continuous culture with a light-dark cycle. - Arch. Microbiol. *123* : 315 - 318, 1979. [Chl, Bil.]

40037 - VAN OORSCHOT, J.L.P. : Types of selective action by herbicides which inhibit photosynthesis. - Z. Naturforsch. *34 C* : 900 - 904, 1979.

40038 - VAN RENSEN, J.J.S. : Action of the herbicide 4,6-dinitro-O-cresol on photosynthetic electron transport and photophosphorylation. - Plant Physiol. *63* (Suppl.) : 42, 1979.

40039 - VAN RENSEN, J.J.S., HOBÉ, J.H. : Mechanism of action of the herbicide 4,6- -dinitro-o-cresol in photosynthesis. - Z. Naturforsch. *34 C* : 1021 - 1023, 1979.

40040 - VAN RENSEN, J.J.S., KRAMER, H.J.M. : Short-circuit electron transport insensitive to diuron-type herbicides induced by treatment of isolated chloroplasts with trypsin. - Plant Sci. Lett. *17* : 21 - 27, 1979.

40041 - VANSTONE, D.E., STOBBE, E.H. : Light requirement of the diphenylether herbicide oxyfluorfen. - Weed Sci. *27* : 88 - 91, 1979. [Chl, Car.]

40042 - VAN VALEN, L. : Switchback evolution and photosynthesis in angiosperms. - Evol. Theory *4* : 143 - 146, 1979.

40043 - VARLET-GRANCHER,C., BONHOMME, R. : Radiation laws in diffusing medium applied to plant canopy. II. Crop trapping solar energy. - Ann. agron. *30* : 1 - 26, 1979.

*40044 - VARTHA, E.W., CLIFFORD, P.T.P. : Growth of new clover cultivars in Canterbury. - New Zeal. J. exp. Agr. *6* : 289 - 292, 1978. [Dry-matter accumulation.]

40045 - VASEV, V.A. : Chistaya produktivnost' fotosinteza v svyazi s èffektivnost'-yu okisleniya u dvukh kukuruznykh linii i ikh gibrida. [Net productivity of photosynthesis in relation to the oxidation efficiency in two maize lines and their hybrids.] - In : VAKLINOVA, S.G., VANKOVA-RADEVA, R., VASILEVA, V.S. (ed.) : Fotosinteticheskaya Assimilyatsiya CO_2 i Fotodykhanie. Pp. 166 - 170. Izdat. bolg. Akad. Nauk, Sofiya 1979. [In R.]

40046 - VASILEVA, V. : Vliyanie khloramfenikola i tsiklogeksimida na aktivnost' RDF- i FEP-karboksilaz v zeleneyushchikh prorostkakh yachmenya v protsesse assimilyatsii nitrata i ammoniya. [Effect of chloramphenicol and cycloheximide on the activity of RuDP and PEP carboxylases in greening barley seedlings during nitrate and ammonium assimilation.] - In : VAKLINOVA, S.G., VANKOVA-RADEVA, R., VASILEVA, V.S. (ed.) : Fotosinteticheskaya Assimilyatsiya CO_2 i Fotodykhanie. Pp. 50 - 56. Izdat. bolg. Akad. Nauk, Sofiya 1979. [In R.]

40047 - VASIN, Yu.A., VERKHOTUROV, V.N. : Issledovanie polyarizatsii fluorestsentsii khlorelly i khloroplastov gorokha, orientirovannykh v magnitnom pole. [Polarized fluorescence of *Chlorella* cells and pea chloroplasts oriented in a magnetic field.] - Biofizika *24* : 260 - 263, 1979. [In R, ab : E.]

40048 - VASIN, Yu.A., VERKHOTUROV, V.N., GULYAEV, B.A. : Osobennosti orientatsii Q_y-perekhodov antennogo khlorofilla fotosistem I i II khlorelly. [The peculiarities of orientation of Q_y-transitions of antenna chlorophyll of the *Chlorella* photosystems I and II.] - Biol. Nauki *1979* (7) : 29 - 33, 1979. [In R.]

40049 - VASSILEV, G.N., SPASSOVSKA, N.H., SPASSOV, A.V., KIMENOV, G.P. : Herbicidal and growth-regulating activity of certain hydrazin-pyrimidines. - Dokl. bolg. Akad. Nauk *32* : 809 - 812, 1979. [Ps.]

40050 - VATER, J., SALNIKOW, J. : Identification of two binding sites of the D-ribulose 1,5-bisphosphate carboxylase/oxagenase from spinach for D-ribulose 1,5-bisphosphate and effectors of the carboxylation reaction. - Arch. Biochem. Biophys. *194* : 190 - 197, 1979.

40051 - VAUGHN, K.C., WILSON, K.G. : Plastome participation in chloroplast development in *Hosta (Liliaceae)*. - Plant Physiol. *63* (Suppl.) : 160, 1979.

40052 - VAUGHN, K.C., WILSON, K.G., REIBACH, P.H. : Analysis of plastome mutants in *Hosta (Liliaceae)* that have apparent C-4 type ultrastructure. - Plant Physiol. *63* (Suppl.) : 63, 1979.

*B40053 - Vazhneǐshie Problemy Fotosinteza v Rastenievodstve. [The Most Important Problems of Photosynthesis in Crop Production.] - Kolos, Moskva 1970. [In R.]

40054 - VECCHI, M., MÜLLER, R.K. : Separation of (3S,3'S)-,(3R,3'R)- and (3S,3'R)--astaxanthin via (-)-camphanic acid esters. - HRC & CC *2* : 195 - 196, 1979.

40055 - VELTHUYS, B.R. : Electron flow through plastoquinone and cytochromes b_6 and f in chloroplasts. - Proc. nat. Acad. Sci. USA *76* : 2765 - 2769, 1979.

40056 - VENEDIKTOV, P.S., RUBIN, A.B., FREǏDLIN, M.I., SHINKAREV, V.P. : Kinetika i termodinamika reaktsiǐ perenosa ĕlektrona v kompleksakh molekul perenoschikov pri fotosinteze. Priblizhennyǐ metod. [Kinetics and thermodynamics of electron transfer reactions in carrier molecules complexes during photosynthesis. Approximation method.] - Biofizika *24* : 1030 - 1034, 1979. [In R, ab : E.]

40057 - VENEDIKTOV, P.S., SHINKAREV, V.P. : O nekotorykh otsenkakh resheniǐ sistemy differentsial'nykh uravneniǐ, opisyvayushchikh protsessy ĕlektronnogo perenosa. [Some evaluations of solutions of the system of differential equations describing electron transfer.] - Biofizika *24* : 382 - 385, 1979. [In R, ab : E.]

40058 - VENKATARAMANA, S., DAS, V.S.R. : Photoactive ATP dependent glutamine synthetase from chloroplasts of *Setaria italica*.BEAUV. - Z. Naturforsch. *34 C* : 210 - 213, 1979.

40059 - VENNESLAND, R., GUERRERO, M.G. : Reduction of nitrate and nitrite. - In : GIBBS, M., LATZKO, E. (ed.) : Photosynthesis II. (Encycl. Plant Physiol. N.S. Vol.6.) Pp. 425 - 444. Springer-Verlag, Berlin - Heidelberg - New York 1979. [Relation to Ps.]

40060 - VENUS, J.C., CAUSTON, D.R. : Plant growth analysis: The use of the Richards function as an alternative to polynomial exponentials. - Ann. Bot. *43* : 623 - 632, 1979.

40061 - VENUS, J.C., CAUSTON, D.R. : Plant growth analysis: A re-examination of the methods of calculation of relative growth and net assimilation rates without using fitted functions. - Ann. Bot. *43* : 633 - 638, 1979.

40062 - VERBELEN, J.P., DE GREEF, J.A. : Leaf development of *Phaseolus vulgaris* L. in light and in darkness. - Amer. J. Bot. *66* : 970 - 976, 1979.

40063 - **VERDIER, G.** : Poly(adenylic acid)-containing RNA of *Euglena gracilis* during chloroplast development 1. Analysis of their complexity by hybridization to complementary DNA. - Europe. J. Biochem. *93* : 573 - 580, 1979.

40064 - **VERDIER, G.** : Poly(adenylic acid)-containing RNA of *Euglena gracilis* during chloroplast development 2. Transcriptional origin of the different RNA. - Europe. J. Biochem. *93* : 581 - 586, 1979.

40065 - **VERHAGEN, J.H.G.** : A model of the plankton dynamics in an eutrophic shallow lake based on field data. - In : JORGENSEN, S.E. (ed.) : State-of-the-Art in Ecological Modelling. Vol.7. Pp. 661 - 673. Pergamon Press, Oxford 1979.

4... 3 - **VERMA, S.B., MOTHA, R.P., ROSENBERG, N.J.** : A comparison of temperature fluctuations measured by a microbead thermistor and a fine wire thermocouple over a crop surface. - Agr. Meteorol. *20* : 281 - 289, 1979.

40067 - **VERMA, S.B., ROSENBERG, N.J.** : Agriculture and the atmospheric carbon dioxide build-up. - Span *22* (2) : 62 - 65, 1979.

40068 - **VERNOTTE, C., ETIENNE, A.L., BRIANTAIS, J.-M.** : Quenching of the system II chlorophyll fluorescence by the plastoquinone pool. - Biochim. biophys. Acta *545* : 519 - 527, 1979.

40069 - **VERSHININ, A.V.** : O prirode genov, formiruyushchikh geterozisnyĭ effekt na osnove khlorofil'noĭ mutatsii u gorokha i veroyatnyĭ mekhanizm etogo effekta. [Nature of genes forming heterosis effect on the basis of chlorophyll mutation in pea and probable mechanism of this effect.] - In : Strukturno-Funktsional'naya Organizatsiya Genoma Eukariot. Pp. 5 - 16. Inst. Tsitologii Genet., Akad. Nauk SSSR, sibir. Otd., Novosibirsk 1979. [In R.]

40070 - **VERSHININ, A.V., SOKOLOV, V.A., SHUMNYĬ, V.K.** : Fiziologo-biokhimicheskie aspekty geteroz isa, poluchennogo na osnove khlorofil'nykh mutantov gorokha. Soobshchenie III. Analiz rosta. [Physiological and biochemical aspects of monohybrid heterosis derived from pea chlorophyll mutants. III. The growth analysis.] - Genetika *15* : 2006 - 2012, 1979. [In R, ab : E.]

40071 - **VIAN, B., ROLAND, J.C.** : The use of ultracryotomy to localize sites of activity of endogenous and exogenous enzymes in plant cells. - In : Ninth International Congress on Electron Microscopy. Vol.II. Pp. 430 - 431. Toronto 1979. [Chloroplast.]

40072 - **VICENTE, C., VALLE, T.** : Variaciones estacionales de algunos parametros fotosinteticos en *Ramalina calicaris*. [Site variations of some photosynthetic parameters in *Ramalina calicaris*.]-Rev. Bryol. Lichénol. *45* : 97 - 102, 1979. [In Span., ab : F.]

40073 - **VIDEAU, C., KHALANSKI, M., PENOT, M.** : Preliminary results concerning effects of chlorine on monospecific marine phytoplankton. - J. exp. mar. Biol. Ecol. *36* : 111 - 123, 1979. [Chl.]

*40074 - **VIDHYASEKARAN, P.** : Possible role of sugars in restriction of lesion development in finger millet leaves infected with *Helminthosporium tetramera*. - Physiol. Plant Pathol. *4* : 457 - 467, 1974. [Ps.]

40075 - **VIDOVIČ, J.** : Vplyv zmeny listového uhla a LAI na radiačný režim porastu a fotosyntézu listov kukurice. [Effect of a change in the leaf angle and LAI on the radiation regime of the maize stand and on the photosynthesis of maize leaves.] - Rostl. Výroba (Praha) : *25* : 1247 - 1256, 1979. [In Slovak, ab : E, G, R.]

40076 - **VIERKE, G.** : Determination of the kinetics of the back reaction of photosystem II in the presence of 3-(3,4-dichlorphenyl)-1,1-dimethylurea from luminescence measurements. - Photochem. Photobiol. *29* : 597 - 604, 1979.

40077 - **VIETOR, D.M., MUSGRAVE, R.B.** : Photosynthetic selection of *Zea mays* L. II. The relationship between CO_2 exchange and dry matter accumulation of canopies of two hybrids. - Crop Sci. *19* : 70 - 75, 1979.

40078 - **VIGH, L., HORVÁTH, I., FARKAS, T., HORVÁTH, L.I., BELEA, A.** : Adaptation of membrane fluidity of rye and wheat seedlings according to temperature. - Phytochemistry *18* : 787 - 789, 1979. [Chloroplast.]

40079 - **VIGNES, D., PLANCHON, C.** : Structure, éclairement et échanges gazeux d'une culture de Soja (*Glycine max* L. MERR.). - Photosynthetica *13* : 136 - 145, 1979.

40080 - **VIIL, J., PÄRNIK, T.** : Parameters of the reductive pentose phosphate cycle and of the glycolate pathway under different concentrations of oxygen. - Z. Pflanzenphysiol. *95* : 213 - 225, 1979.

40081 - **VIIL, J., PÄRNIK, T., VÄRK, E.** : On photorespiration at the expense of glycolic acid and glycine. - In : VAKLINOVA, S.G., VANKOVA-RADEVA, R., VASILEVA, V.S. (ed.) : Fotosinteticheskaya Assimilyatsiya CO_2 i Fotodykhanie. Pp. 121 - 125. Izdat. bolg. Akad. Nauk, Sofiya 1979.

40082 - **VIJAYARAGHAVAN, S.J., SOPORY, S.K., GUHA-MUKHERJEE, S** . : Role of light in the regulation of the nitrate reductase level in wheat (*Triticum aestivum*). - Plant Cell Physiol. *20* : 1251 - 1261, 1979. [Ps.]

40083 - **VINAYA RAI, K.S., MURTHY, K.S.** : Note on the effect of complete submergence on RuBP carboxylase activity in rice seedlings. - Indian J. agr. Res. *13* : 61 - 63, 1979.

40084 - **VINCENT, W.F.** : Mechanisms of rapid photosynthetic adaptation in natural phytoplankton communities. I. Redistribution of excitation energy between photosystems I and II. - J. Phycol. *15* : 429 - 434, 1979.

40085 - **VIRÁGH, K.** : Wachstumsanalyse der Sonnen- und Schattenblätter von *Quercus cerris* und *Quercus petraea* (1973 - 1975). - Acta bot. Acad. Sci. hung. *25* : 143 - 164, 1979.

*40086 - **VITKUS, A.A.** : Dinamika soderzhaniya askorbinovoĭ kisloty i karotina v ĕspartsete posevnom. [Dynamics of contents of ascorbic acid and carotene in sainfoin.] - Liet.TSR Mokslų Akad. Darbai, Ser.C *1969* (1) : 59 - 65, 1969. [In R, ab : E, Lithu.]

*40087 - **VITKUS, A.A.** : Biologicheskie osobennosti i khimicheskiĭ sostav lyutserny zheltoĭ. (2. Askorbinovaya kislota i karotin.) [Biological peculiarities and chemical composition of *Medicago falcata* L. (2. Ascorbic acid and carotene.)] - Liet. TSR Mokslų Akad. Darbai, Ser. C *1975* (4) : 27 - 32, 1975. [In R, ab : E, Lithu.]

*40088 - **VITKUS, A.A., MILYUVENE, S.G.** : Khimicheskiĭ sostav i biologicheskie osobennosti yazvennika obyknovennogo. (1. Askorbinovaya kislota i karotin.) [Chemical composition and biological peculiarities of *Anthyllis vulneraria*. (1. Ascorbic acid and carotene.)] - Liet. TSR Mokslų Akad. Darbai, Ser. C *1977* (1) : 27 - 31, 1977. [In R, ab : E, Lithu.]

40089 - **VLADIMIROVA, M.G., RUDOVA, T.S., SHATILOV, V.R., SALAMATOVA, L.V., NAZAROVA, G.D.** : Sravnitel'naya kharakteristika *Chlorella pyrenoidosa* Chick 82 i *Chlorella pyrenoidosa* Pringsheim 82T v usloviyakh intensivnoĭ kul'tury. [Comparative study of *Chlorella pyrenoidosa* Chick 82 and *Chlorella pyrenoidosa* Pringsheim 82T grown under conditions of intense culture.] - Fiziol. Rast. *26* : 1125 - 1134, 1979. [Ps; in R, ab : E.]

40090 - **VLASYUK, P.A., KLIMOVITSKAYA, Z.M., KHMARA, L.A., PROKOPIVNYUK, L.M.** : Znachenie margantsa v metabolizme rasteniĭ. [The role of manganese in plant metabolism.] - Fiziol. Biokhim. kul't. Rast. *11* : 195 - 208, 1979. [Ps; in R, ab : E.]

40091 - **VLCEK, L.M., GASSMAN, M.L.** : Reversal of α,α'-dipyridyl-induced porphyrin synthesis in etiolated and greening red kidney bean leaves. - Plant Physiol. *64* : 393 - 397, 1979.

40092 - **VOGEL, G., LANCKOW, J.** : Untersuchungen zum Einfluß von directer und diffuser Lichtstrahlung auf die Ertragsleistung von Tomate und Gurke in Gewächshäusern. - Arch. Gartenbau *27* : 3 - 12, 1979. [Dry-matter accumulation.]

40093 - VOIGT, B., LEUPOLD, D., HIEKE, B., HOFFMANN, P. : On the independence of the red shift of chlorophyll absorption *in vivo* of the value of the primary absorption unit of antenna chlorophyll. - Stud. biophys. 25 : 93 - 94, 1979.

40094 - VOĬTSINSKIĬ, V.M., DRACHEV, L.A., KAULEN, A.D., SKULACHEV, V.P. : Fotoèlektricheskie otvety bakteriorodopsina v sistemakh lipid-voda. [The photoelectric responses of bacteriorhodopsin in lipid-water systems.] - Bioorg. Khim. 5 : 1184 - 1195, 1979. [In R, ab : E.]

40095 - VOITURIEZ, B., HERBLAND, A. : The use of the salinity maximum of the Equatorial Undercurrent for estimating nutrient enrichment and primary production in the Gulf of Guinea. - Deep-Sea Res. 26 A : 77 - 83, 1979.

40096 - VOLODARSKIĬ, A.D., OMANN, È., CHAYANOVA, S.S., TIKHONOVSKAYA, N.G. : Immunokhimicheskaya identifikatsiya i otsenka chistoty fermentov na primere ribulozodifosfatkarboksilazy. [Immunochemical identification and enzyme purity evaluation (with ribulose-bisphosphate carboxylase as an example).] - Fiziol. Rast. 26 : 1172 - 1181, 1979. [In R, ab : E.]

40097 - VOLODARSKIĬ, N.I., BYSTRYKH, E.E., NIKOLAEVA, E.K. : Vliyanie azotnogo pitaniya na fotofosforiliruyushchuyu aktivnost' v ontogeneze dvukh sortov pshenitsy raznoĭ produktivnosti. [Effect of nitrogen nutrition on photophosphorylating activity in ontogenesis of two cultivars of wheat of different productivity.] - Biol. Nauki 1979 (7) : 79 - 85, 1979. [In R.]

40098 - VOLODARSKIĬ, N.I., BYSTRYKH, E.E., NIKOLAEVA, E.K. : Aktivnost' fotosinteticheskogo apparata v ontogeneze u razlichnykh po produktivnosti sortov pshenitsy v zavisimosti ot azotnogo pitaniya. [The activity of photosynthetic apparatus in the ontogenesis in wheat cultivars of different performance depending on the nitrogen nutrition.] - Sel'skokhoz. Biol. 14 : 431 - - 435, 1979. [In R.]

40099 - VOLODARSKIĬ, N.I., BYSTRYKH, E.E., NIKOLAEVA, E.K. : O reaktsii fotovosstanovleniya NADF v ontogeneze pshenitsy v syazi s produktivnost'yu. [Photoreduction of NADP during wheat ontogeny in relation to crop productivity.] - Fiziol. Rast. 26 : 35 - 40, 1979. [In R, ab : E.]

40100 - VOLYNETS, A.P., PROKHORCHIK, R.A. : Izmenenie soderzhaniya flavonoidov v khloroplastakh pri obrabotke rasteniĭ zheltogo lyupina 2,4-D i TKhA. [Changes in flavonoid content in chloroplasts induced by treatment of yellow lupine plants with 2,4-D and trichloroacetate.] - Fiziol. Rast. 26 : 259 - - 265, 1979. [In R, ab : E.]

40101 - VOROB'EVA, I.A., GORYUNOVA, S.V., MAKSIMOV, V.N. : Intensivnost' fotosinteza kul'tury mikrovodorosleĭ v norme i pri vozdeĭstvii kadmiya i tsinka (po dannym pH-metrii). [Photosynthetic rate in a culture of microalgae under normal conditions and under the effects of cadmium and zinc (according to the pH-measurement data).] - Gidrobiol. Zh. 15 (5) : 64 - 70, 1979. [In R, ab : E.]

40102 - VOROB'EVA, L.M., SHCHERBAKOVA, I.Yu., KRASNOVSKIĬ, A.A. : Deĭstvie parov organicheskikh rastvoriteleĭ na protokhlorofillovyĭ kompleks ètiolirovannykh list'ev. Usloviya obratimogo i neobratimogo povrezhdeniya. [The effects of organic solvent vapours on the protochlorophyllide complex from etiolated leaves. Conditions for reversible and irreversible destruction.] - Biokhimiya 44 : 880 - 885, 1979. [In R, ab : E.]

40103 - VOS, J. : Effect of temperature and nitrogen on carbon-exchange rates and on growth of wheat during kernel-filling. - In : SPIERTZ, J.H.J., KRAMER, T. (ed.) : Crop Physiology and Cereal Breeding. Pp. 80 - 89. Pudoc, Wageningen 1979.

40104 - VOSKRESENSKAYA, N.P. : Effect of light quality on carbon metabolism. - In : GIBBS, M., LATZKO, E. (ed.) : Photosynthesis II. (Encycl. Plant Physiol. N.S. Vol.6.) Pp. 174 - 180. Springer-Verlag, Berlin - Heidelberg - New York 1979.

B40105 - **VOSKRESENSKAYA, N.P.** : Fotoregulyatornye Aspekty Metabolizma Rasteniĭ.
[Photoregulatory Aspects of Plant Metabolism.] - In : Timiryazevskie Chte-
niya. Vol.38. Pp. 1 - 48. Nauka, Moskva 1979. [Ps; in R.]

*B40106 - **VOZNESENSKIĬ, V.L.** : Konduktometricheskiĭ Pribor dlya Izmereniya Fotosinteza i Dykhaniya Rasteniĭ v Polevykh Usloviyakh. Izd. 2-e. [A Conductome-
tric Apparatus for Measuring Plant Photosynthesis and Respiration in Field
Conditions. 2nd Ed.] - Nauka, Leningrad 1971. [In R.]

40107 - **VOZNYAK, V.M., GANAGO, I.B., MOSKALENKO, A.A., ELFIMOV, E.I.** : Vliyanie
magnitnogo polya na vykhod fluorestsentsii khlorofill-belkovykh kompleksov,
obogashchennykh fotosistemoĭ I. [Magnetic field-induced changes in fluorescence yield of chlorophyll-protein complexes enriched with Photosystem I.]
- Stud. biophys. 77 : 13 - 20, 1979. [In R, ab : E.]

*40108 - **VREMAN, H.J., THOMAS, R., CORSE, J., SWAMINATHAN, S., MURAI, N.** : Cytokinins in tRNA obtained from *Spinacia oleracea* L. leaves and isolated chloroplasts. - Plant Physiol. 61 : 296 - 306, 1978.

40109 - **VRZHESHCH, P.V., ZAĬTSEV, S.V., KUROCHKIN, I.N., VARFOLOMEEV, S.D.** : Vnu-
trennyaya i vneshnaya immobilizatsiya khloroplastov. [Inner and outer immobilization of chloroplasts.] - Biokhimiya 44 : 67 - 73, 1979. [In R, ab :
E.]

40110 - **VSEVOLODOV, N.N., CHEKULAEVA, L.N.** : Spectral transitions in purple membranes from *Halobacterium halobium*. I. Effect of preliminary illumination on
photochemical processes. - J. Bioenerg. Biomembranes 10 : 13 - 22, 1979.

40111 - **VU, C.V., ALLEN, L.H.,Jr., GARRARD, L.A.** : Effects of UV-B radiation on
growth, leaf photosynthetic pigments and proteins, and on activity of RuBP
carboxylase in pea plants (*Pisum sativum* L.). - Plant Physiol. 63 (Suppl.):
126, 1979.

40112 - **VU, C.V., BIGGS, R.H.** : Effects of inhibitors on the biosynthesis of sterols, reducing sugars, and chlorophyll, and the development of isocitrate
lyase in germinating seeds of longleaf pine, *Pinus palustris* MILL. - Plant
Sci. Lett. 16 : 255 - 265, 1979.

40113 - **VUNKOVA-RADEVA, R.V.** : Effect of glycolate on the nitrate reductase activity
induction. - In : VAKLINOVA, S.G., VANKOVA-RADEVA, R., VASILEVA, V.S. (ed.):
Fotosinteticheskaya Assimilyatsiya CO₂ i Fotodykhanie. Pp. 133 - 140. Izdat. bolg. Akad. Nauk, Sofiya 1979.

40114 - **VYARK, É.Ya., KÉÉRBERG, O.F., KÉÉRBERG, Kh.I., PYARNIK, T.R.** : Deĭstvie
sveta raznoĭ intensivnosti i spektral'nogo sostava na metabolizm glikolevoĭ-
-1-¹⁴C kisloty v list'yakh fasoli i kukuruzy. [Influence of light of various intensities and spectral composition upon the metabolism of (1-¹⁴C)
glycolic acid in bean and maize leaves.] - Fiziol. Rast. 26 : 229 - 238,
1979. [In R, ab : E.]

40115 - **VYAS, L.N., GARG, R.K., JINDAL, K.** : Relation between photosynthetic area
and above ground biomass in five tree species of deciduous forests near
Udaipur (Rajasthan), India. - Biológia (Bratislava) 34 : 547 - 553, 1979.

40116 - **VYSHKVARTSEV,D.I., KARAPETYAN, T.Sh.** : Sezonnaya dinamika pervichnoĭ produktsii v mel'kovodnykh bukhtakh zaliva Pos'eta (Yaponskoe more). [Seaso-
nal dynamics of primary production in shallow waters of the Possiet bay
(Sea of Japan).] - Biol. Morya 1979(2) : 28 - 33, 1979. [In R, ab : E.]

40117 - **WACKER, G.** : Einfluß der Halm- und Blatttriebbildung auf Ertrag und Quali-
tät bei Futtergräsern.. - Arch. Züchtungsforsch. 9 : 237 - 244, 1979.

40118 - **WAGGONER, P.E.** : Variability of annual wheat yields since 1909 and among
nations. - Agr. Meteorol. 20 : 41 - 45, 1979.

40119 - **WAGHMODE, A.P., JOSHI, G.V.** : Kranz leaf anatomy & C₄ dicarboxylic acid
pathway of photosynthesis in *Aeluropus lagopoides* L. - Indian J. exp. Biol.
17 : 606 - 607, 1979.

40120 - WAKAMATSU, K., USAMI, K. : Effect of imidazole on the activity of photosystem II in spinach chloroplasts. - Plant Cell Physiol. *20* : 323 - 330,
1979.

40121 - WALBOT, V., COE, E.H. Jr. : Nuclear gene *iojap* conditions a programmed
change to ribosome-less plastids in *Zea mays*. - Proc. nat. Acad. Sci. USA
76 : 2760 - 2764, 1979.

40122 - WALDRON, J.C., ANDERSON, J.M. : Chlorophyll-protein complexes from thylakoids of a mutant barley lacking chlorophyll *b*. - Europe. J. Biochem. *102* :
357 - 362, 1979.

*40123 - WALKER, D.A., ROBINSON, S.P. : Chloroplast and cell. A contemporary view
of photosynthetic carbon assimilation. - Ber. deut. bot. Ges. *91* : 513 -
- 526, 1978.

40124 - WALKER, G.H., IZAWA, S. : Photosynthetic electron transport in isolated
maize bundle sheath cells. - Plant Physiol. *63* : 133 - 138, 1979.

40125 - WALKER, J.R.L., STEVENS, M.M.D. : CO_2-compensation point determination for
C_3 and C_4 plants. - Mauri Ora *7* : 47 - 51, 1979.

40126 - WALKER, R.R., KRIEDEMANN, P.E., MAGGS, D.H. : Growth, leaf physiology and
fruit development in salt-stressed guavas. - Aust. J. agr. Res. *30* : 477 -
- 488, 1979. [Ps.]

40127 - WALLER, S.S., LEWIS, J.K. : Occurrence of C_3 and C_4 photosynthetic pathways in North American grasses. - J. Range Manage. *32* : 12 - 28, 1979.

40128 - WALLSGROVE, R.M., LEA, P.J., MIFLIN, B.J. : Distribution of the enzymes
of nitrogen assimilation within the pea leaf cell. - Plant Physiol. *63* :
232 - 236, 1979. [Chl.]

40129 - WALTER, E., SCHREIBER, S., ZASS, E., ESCHENMOSER, A. : Bakteriochlorophyll
a_{Gg} und Bakteriophäophytin a_p in den photosynthetischen Reaktionszentren
von *Rhodospirillum rubrum G-9*. - Helv. chim. Acta *62* : 899 - 920, 1979.

40130 - WALTER, G. : Zur Dynamik des Phytolgehaltes während des Chlorophyllauf- und
-abbaues in Primärblättern von Weizenkeimpflanzen (*Triticum aestivum* L.). -
Biol. Plant. *21* : 105 - 112, 1979.

40131 - WALTER, G., MEISTER, A. : Zur Photoreduktion des Protochlorophyllid-Holo-
chroms P_{635} *in vivo*. - Photosynthetica *13* : 167 - 174, 1979.

40132 - WALTON, J.D., EARLE, E.D., YODER, O.C., SPANSWICK, R.M. : Reduction of
adenosine triphosphate levels in susceptible maize mesophyll protoplasts
by *Helminthosporium maydis* race T toxin. - Plant Physiol. *63* : 806 - 810,
1979.

40133 - WALTON, P.D., MURCHISON, C. : A plant ideotype for *Bromus inermis* LEYSS.
in Western Canada. - Euphytica *28* : 801 - 806, 1979.

40134 - WALZ, D. : Thermodynamics of oxidation-reduction reactions and its application to bioenergetics. - Biochim. biophys. Acta *505* : 279 - 353, 1979.

40135 - WALZ, D. : Change in aggregation of lecithin due to valinomycin-lipid interaction. - Chimia *33* (2) : 45 - 50, 1979. [Chl.]

40136 - WANG, W. : Photoconversion of photochlorophyllide in the *y-1* mutant of
Chlamydomonas reinhardtii. - Plant Physiol. *63* : 1102 - 1106, 1979.

*40137 - WANGERSKY, P.J. : Mesurement of organic carbon in seawater. - In : GIBB,
T.R.P.Jr.(ed.):Analytical Methods in Oceanography. (Adv.Chem.Ser. Vol.147).
Pp. 148 - 162. Amer. Chem. Society, Washington 1975.

40138 - WANN, M., RAPER, C.D.,Jr. : A dynamic model for plant growth: Adaptation
for vegetative growth of soybeans. - Crop Sci. *19* : 461 - 467, 1979.

40139 - WARD, C.H., KING, J.M. : Effects of simulated hypogravity on respiration
and photosynthesis of higher plants. - In : HOLMQUIST, R. (ed.) : COSPAR.
Life Sciences and Space Research. Vol.17. Pp. 291 - 296. Pergamon Press,
Oxford - New York 1979.

✷40140 - WARDEN, J.T. : Paramagnetic intermediates in photosynthetic systems. - In :
BERLINER, L.J., REUBEN, J. (ed.) : Biological Magnetic Resonance. Vol.1.
Pp. 239 - 275. Plenum Press, New York 1978.

40141 - WARDLE, K., QUINLAN, A., SIMPKINS, I. : Abscisic acid and the regulation
of water loss in plantlets of *Brassica oleracea* L. var. *botrytis* regenerated
through apical meristem culture. - Ann. Bot. *43* : 745 - 752, 1979. [Stomatal
resistance.]

40142 - WAREING, P.F. : Temperature responses and yield in temperate crops. - In :
SCOTT, T.K. (ed.) : Plant Regulation and World Agriculture. Pp. 129 - 139.
Plenum Press, New York - London 1979. [Ps.]

40143 - WAREING, P.F. : Growth regulators and assimilate partition. - In : SCOTT,
T.K. (ed.) : Plant Regulation and World Agriculture. Pp. 309 - 317. Plenum Press, New York - London 1979.

40144 - WAREING, P.F. : Inaugural address: Plant development and crop yield. - In :
MARCELLE, R., CLIJSTERS, H., VAN POUCKE, M. (ed.) : Photosynthesis and Plant
Development. Pp. 1 - 17. Dr.W.Junk bv. Publ., The Hague - Boston - London
1979.

40145 - WAREMBOURG, F.R., PAUL, E.A., RANDELL, R.L., MORE, R.B . : Modèle de répartition du carbone assimilé dans une prairie naturelle. - Oecol. Plant. *14* :
1 - 12, 1979.

40146 - WARMBRODT, R.D., EVERT, R.F. : Comparative leaf structure of several species of homosporous leptosporangiate ferns. - Amer. J. Bot. *66* : 412 - 440,
1979. [Photosynthates.]

40147 - WARMBRODT, R.D., EVERT, R.F. : Comparative leaf structure of six species
of eusporangiate and protoleptosporangiate ferns. - Bot. Gaz. *140* : 153 -
- 167, 1979. [Chloroplast.]

40148 - WARSHEL, A. : Conversion of light energy to electrostatic energy in the
proton pump of *Halobacterium halobium*. - Photochem. Photobiol. *30* : 285 -
- 290, 1979.

40149 - WARSHEL, A. : On the origin of the red shift of the absorption spectra of
aggregated chlorophylls. - J. amer. chem. Soc. *101* : 744 - 746, 1979.

40150 - WARSHEL, A., OTTOLENGHI, M. : Kinetic and spectroscopic effects of protein-
-chromophore electrostatic interactions in bacteriorhodopsin. - Photochem.
Photobiol. *30* : 291 - 293, 1979.

40151 - WATANABE, A. : [Chloroplast genes.] - Kagaku to Seibutsu *17* : 548 - 555,
1979. [In Jap.]

✷40152 - WATANABE, H., KIUCHI, T. : Effect of p-chlorophenyl dimethyl urea on the
degradation of chlorophyll in detached rice leaf in the light. - Soil Sci.
Plant Nutr. *21* : 151 - 159, 1975.

40153 - WATANABE, M.F. : Studies on the metalimnetic blue-green alga *Oscillatoria
mougeotii* in a eutrophic lake with special reference to its population
growth. - Arch. Hydrobiol. *86* : 66 - 86, 1979. [Ps.]

40154 - WATSON, R.L., LANDSBERG, J.J. : The photosynthetic characteristics of apple
leaves (cv. Golden Delicious) during their early growth. - In : MARCELLE,
R., CLIJSTERS, H., VAN POUCKE, M. (ed.) : Photosynthesis and Plant Development. Pp. 39 - 48. Dr.W.Junk bv.-Publ., The Hague - Boston - London 1979.

✷40155 - WEATHERS, P.J., ALLEN, M.M. : Variations in short term products of inorganic carbon fixation in exponential and stationary phase cultures of *Aphanocapsa* 6308. - Arch. Mikrobiol. *116*: 231 - 234, 1978.

40156 - WEBER, J.A., TENHUNEN, J.D., YOCUM, C.S., GATES, D.M. : Variation of photosynthesis in *Elodea densa* with pH and/or high CO_2 concentrations. - Photosynthetica *13* : 454 - 458, 1979.

40157 - WECKESSER, J., DREWS, G., MAYER, H. : Lipopolysaccharides of photosynthetic prokaryotes. - Annu. Rev. Microbiol. *33* : 215 - 239, 1979.

40158 - WEDGE, R., BURRIS, J.E. : Effects of temperature and light intensity on photosynthesis in *Lemna minor* and *Spirodela oligorhiza*. - Plant Physiol. *63* (Suppl.) : 64, 1979.

40159 - WEGMANN, K. : Biochemische Anpassung von *Dunaliella* an wechselnde Salinität und Temperatur. - Ber. deut. bot. Ges. *92* : 43 - 54, 1979. [Ps.]

40160 - WEILAND, R.T., STUTTE, C.A., TALBERT, R.E. : Foliar nitrogen loss and CO_2 equilibrium as influenced by three soybean (*Glycine max*) postemergence herbicides. - Weed Sci. *27* : 545 - 548, 1979. [Ps.]

40161 - WEINBAUM, S.A., GRESSEL, J., REISFELD, A., EDELMAN, M. : Specific depletion of the 32,000 d protein from thylakoids by chloramphenicol. - Plant Physiol. *63* (Suppl.) : 99, 1979.

40162 - WEINBAUM, S.A., GRESSEL, J., REISFELD, A., EDELMAN, M. : Characterization of the 32,000 dalton chloroplast membrane protein. III. Probing its biological function in *Spirodela*. - Plant Physiol. *64* : 828 - 832, 1979.

40163 - WEISSMAN, J.C., BENEMANN, J.R. : Biomass recycling and species competition in continuous cultures. - Biotechnol. Bioeng. *21* : 627 - 648, 1979. [Chl, Bil.]

40164 - WELLBURN, A.R., HAMPP, R. : Appearance of photochemical function in prothylakoids during plastid development. - Biochim. biophys. Acta *547* : 380 - - 397, 1979.

40165 - WELLBURN, F.A.M., WELLBURN, A.R. : Conjoined mitochondria and plastids in the barley mutant "albostrians". - Planta *147* : 178 - 179, 1979.

40166 - WELLS, R., LIEBHARDT, W.C., SVEC, L.V., FRICK, H. : Photosynthetic capacity in potassium stressed soybean: Comparison of CO_2 fixation and O_2 evolution assays. - J. Plant Nutr. *1* : 283 - 293, 1979.

*40167 - WERDAN, K., HELDT, H.W. : Bicarbonate uptake into the chloroplast stroma. - In : AZZONE, G.F., ERNSTER, L., PAPA, S., QUAGLIARIELLO, E., SILIPRANDI, N. (ed.) : Mechanisms in Bioenergetics. Pp. 285 - 292. Academic Press, New York - London 1973.

40168 - WESSELS, C., BIRNBAUM, E. : An improved apparatus for use with the ^{14}C acid-bubbling method of measuring primary production. - Limnol. Oceanogr. *24* : 187 - 188, 1979.

*40169 - WEST, D.W., BLACK, J.D.F. : Irrigation timing - its influence on the effects of salinity and waterlogging stresses in tobacco plants. - Soil Sci. *125* : 367 - 376, 1978. [Resistances.]

40170 - WEST, R.J., LARKUM, A.W.D. : Leaf productivity of the seagrass, *Posidonia australis*, in Eastern Australian waters. - Aquat. Bot. *7* : 57 - 65, 1979.

40171 - WESTERHOFF, H.V., SCHOLTE, B.J., HELLINGWERF, K.J. : Bacteriorhodopsin in liposomes. I. A description using irreversible thermodynamics. - Biochim. biophys. Acta *547* : 544 - 560, 1979.

*40172 - WESTLAKE, D.F. : Productivity in aquatic systems. - In : Solar Energy: Biological Conversion Systems. Pp. 19 - 22. London 1975. [Ps.]

40173 - WETTERN, M. : Lipid metabolism during the regreening of the chaetophoralean green alga *Fritschiella tuberosa* in axening culture. - In : APPELQVIST, L.-Å., LILJENBERG, C. (ed.) : Advances in the Biochemistry and Physiology of Plant Lipids. Pp. 237 - 242. Elsevier/North-Holland Biomedical Press, Amsterdam 1979. [Chl, Car.]

40174 - WETTERN, M., WEBER, A. : Some remarks on algal carotenoids and their interconversion into animal carotenoids. - In : HOPPE, H.A., LEVRING, T., TANAKA, Y. (ed.) : Marine Algae in Pharmaceutical Science. Pp. 551 - 568. W. de Gruyter, Berlin - New York 1979.

40175 - WETZEL, R.G. : The role of the littoral zone detritus in lake metabolism. - Arch. Hydrobiol., Beih. Ergebn. Limnol. *13* : 145 - 161, 1979. [Ps.]

40176 - WETZEL, R.G., PENHALE, P.A. : Transport of carbon and excretion of dissolved organic carbon by leaves and roots/rhizomes in seagrasses and their
epiphytes. - Aquat. Bot. *6* : 149 - 158, 1979.

40177 - WHATLEY, J.M. : Plastid development in the primary leaf of *Phaseolus vulgaris* : variations between different types of cell. - New Phytol. *82* :
1 - 10, 1979.

40178 - WHATLEY, J.M., JOHN, P., WHATLEY, F.R. : From extracellular to intracellular: the establishment of mitochondria and chloroplasts. - Proc. roy.
Soc. London B *204* : 165 - 187, 1979.

*40179 - WHELAN, E.D.P., CHUBEY, B.B. : Chlorophyll content of new cotyledon mutants
of cucumber. - HortScience *8* : 30 - 32, 1973.

40180 - WHITING, B.H., VAN DE VENTER, H.A., SMALL, G.C. : Crassulacean acid metabolism in jointed cactus (*Opuntia aurantiaca* LINDLEY). - Agroplantae *11* :
41 - 43, 1979.

40181 - WHITMAN, W.B., COLLETTI, C., TABITA, F.R. : Activation of spinach ribulose bisphosphate carboxylase by pyridoxal phosphate. - FEBS Lett. *101* :
249 - 252, 1979.

40182 - WHITMARSH, J., CRAMER, W.A. : Cytochrome f function in photosynthetic
electron transport. - Biophys. J. *26* : 223 - 234, 1979.

40183 - WHITMARSH, J., CRAMER, W.A. : Photooxidation of the high-potential iron-
-sulfur center in chloroplasts. - Proc. nat. Acad. Sci. USA *76* : 4417 -
- 4420, 1979.

40184 - WHITNEY, D.E., DARLEY, W.M. : A method for the determination of chlorophyll a in samples containing degradation products. - Limnol. Oceanogr.
24 : 183 - 186, 1979.

40185 - WHITTED, B.E., JOHNSON, J.E., BARR, R., CRANE, F.L. : Lipophilic monofunc-
tional aldehydes inhibit ferricyanide reduction by photosystem II of spinach chloroplasts. - Proc. Indiana Acad. Sci. *88* : 99 - 103, 1979.

40186 - WHITTEN, W.B., PEARLSTEIN, R.M., OLSON, J.M. : New spectral components in
high resolution absorption spectra of green bacterial reaction center complexes at 5 K. - Photochem. Photobiol. *29* : 823 - 828, 1979.

*B40187 - WHITTINGHAM, C.P. : Photosynthesis. Oxford Biology Readers, No.9. - Oxford
University Press, London 1971.

40188 - WHITTINGHAM, C.P. : Photorespiration: Its mechanism and significance. -
Agron. lusit. *39* : 115 - 129, 1979.

40189 - WHITTINGHAM, C.P., KEYS, A.J., BIRD, I.F. : The enzymology of sucrose synthesis in leaves. - In : GIBBS, M., LATZKO, E. (ed.) : Photosynthesis II.
(Encycl. Plant Physiol. N.S. Vol.6.) Pp. 313 - 326. Springer-Verlag, Berlin - Heidelberg - New York 1979.

*40190 - WIDER DE XIFRA, E.A., SANDY, J.D., DAVIES, R.C., NEUBERGER, A. : Control
of 5-aminolaevulinate synthetase activity in *Rhodopseudomonas spheroides*.
- Phil. Trans. roy. Soc. London *273* : 79 - 98, 1976.

*40191 - WIEBE, H.-J. : Zur Übertragung von Ergebnissen aus Klimakammern auf Frei-
landbedingungen mit Hilfe eines Simulationsmodells bei Blumenkohl. - Gartenbauwissenschaft *40* : 70 - 74, 1975.

40192 - WIĘCKOWSKI, S., MACHOWICZ, E., SUBCZYŃSKI, W.K. : Intensity and decay half-
-time of the electron paramagnetic resonance signal-I in bean chloroplasts
at various stages of greening. - J. exp. Bot. *30* : 1179 - 1185, 1979.

40193 - WIEDENROTH, E.M. : Die apparente Photosynthese von Bohnenkeimpflanzen
(*Phaseolus vulgaris* L.) unter dem Einfluss einer kurzzeitigen Hemmung des
Wurzelgaswechsels. - Biol. Plant. *21* : 193 - 200, 1979.

40194 - WIEGAND, C.L., RICHARDSON, A.J., KANEMASU, E.T. : Leaf area index estimates
for wheat from LANDSAT and their implications for evapotranspiration and
crop modeling. - Agron. J. *71* : 336 - 342, 1979.

40195 - WIEGAND, R.C., POND, S. : Fluctuations of chlorophyll and related physical parameters in British Columbia coastal waters. - J. Fish. Res. Board Can. *36* : 113 - 121, 1979.

*40196 - WIENCKE, C., SCHULZ, D. : The development of transfer cells in the haustorium of the *Funaria hygrometrica* sporophyte. - In : Congrès International de Bryologie. Bryophytorum Bibliotheca *13*. Pp. 147 - 167. Bordeaux 1977. [Chloroplast.]

40197 - WIESSNER, W. : Photoassimilation of organic compounds. - In : GIBBS, M., LATZKO, E. (ed.) : Photosynthesis II. (Encycl. Plant Physiol. N.S. Vol.6.) Pp. 181 - 189. Springer-Verlag, Berlin - Heidelberg - New York 1979.

40198 - WIGINTON, J.R., McMILLAN, C. : Chlorophyll composition under controlled light conditions as related to the distribution of seagrasses in Texas and the U.S. Virgin Islands. - Aquat. Bot. *6* : 171 - 184, 1979.

40199 - WILAIPON, B., GIGIR, S.A., HUMPHREYS, L.R. : Apex, lamina and shoot removal effects on seed production and growth of *Stylosanthes hamata* cv. Verano. - Aust. J. agr. Res. *30* : 293 - 306, 1979. [Growth analysis.]

40200 - WILD, A. : Inhibitory effects of the insecticides Allethrin, Lindane and Jacutin-Fogetten sublimate on photosynthetic electron transport. - Z. Naturforsch. *34 C* : 1070 - 1071, 1979.

40201 - WILD, A. : Physiologie der Photosynthese höheren Pflanzen. Die Anpassung an die Lichtbedingungen. - Ber. deut. bot. Ges. *92* : 341 - 364, 1979.

40202 - WILDMAN, S.G. : Aspects of fraction 1 protein evolution. - Arch. Biochem. Biophys. *196* : 598 - 610, 1979.

*40203 - WILDNER, G.F. : The regulation of glucose-6-phosphate dehydrogenase in chloroplasts. - Z. Naturforsch. *30 C* : 756 - 760, 1975.

40204 - WILDNER, G.F., HENKEL, J. : The deactivation and reactivation of ribulose 1,5-bisphosphate carboxylase and oxygenase during air-argon-oxygen transitions. - FEBS Lett. *103* : 246 - 249, 1979.

40205 - WILDNER, G.F., HENKEL, J. : The effect of divalent metal ions on the activity of Mg^{++} depleted ribulose-1,5-bisphosphate oxygenase. - Planta *146* : 223 - 228, 1979.

40206 - WILDNER, G.F., LARSSON, C. : Effects of glycidate on carbon dioxide fixation with isolated spinach chloroplasts. - Plant Physiol. *63* : 887 - 891, 1979.

40207 - WILHELM, E. : Bestandesentwicklung bei Grünkohl *(Brassica oleracea* convar. *acephala* var. *sabellica* L.) II. Modellversuche zur Wirkung von Licht und Temperatur. - Gartenbauwissenschaft *44* : 111 - 118, 1979.

*40208 - WILKINSON, J.F., BEARD, J.B. : Morphological responses of *Poa pratensis* and *Festuca rubra* to reduced light intensity. - In : ROBERTS, E.C. (ed.) : Proceedings of the Second International Turfgrass Research Conference. Pp. 231 - 240. Amer. Soc. Agron., Madison 1974. [Chl.]

40209 - WILLEMOT, C., HOPE, H.J., ST-PIERRE, J.C. : On the inhibition of frost hardening of winter wheat by BASF 13-338, a derivative of pyridazinone. - Can. J. Plant Sci. *59* : 249 - 251, 1979. [Ps.]

40210 - WILLENBRINK, J., KREMER, B.P., SCHMITZ, K., SRIVASTAVA, L.M. : Photosynthetic and light-independent carbon fixation in *Macrocystis, Nereocystis*, and some selected pacific *Laminariales*. - Can. J. Bot. *57* : 890 - 897, 1979.

40211 - WILLENBRINK, J., KREMER, B.P., SCHMITZ, K., WEIDNER, M. : CO_2-Fixierung und Stofftransport in benthischen marinen Algen. - Ber. deut. bot. Ges. *92* : 157 - 167, 1979.

40212 - WILLERT, D.J. von : Vorkommen und Regulation des CAM bei Mittagsblumengewächsen *(Mesembryanthemaceae)*. - Ber. deut. bot. Ges. *92* : 133 - 144, 1979.

40213 - WILLERT, D.J. von, BRINCKMANN, E., SCHEITLER, B., THOMAS, D.A., TREICHEL,S.: The activity and malate inhibition/stimulation of phosphoenolpyruvate-carboxylase in crassulacean-acid-metabolism plants in their natural environment.- Planta *147* : 31 - 36, 1979.

40214 - WILLERT, D.J. von, WILLERT, K. von : Light modulation of the activity of
the PEP-carboxylase in CAM-plants in the *Mesembryanthemaceae*. - Z. Pflanzenphysiol. *95* : 43 - 49, 1979.

☆40215 - WILLIAMS, B.A., AUSTIN, R.B. : Short note: An instrument for measuring the
transmission of short wave radiation by crop canopies. - J. appl. Ecol.
14 : 987 - 991, 1977.

40216 - WILLIAMS, E.J., DALE, J.E., MOORBY, J., SCOBIE, J. : Variation in translocation during the photoperiod: Experiments feeding $^{11}CO_2$ to sunflower. -
J. exp. Bot. *30* : 727 - 738, 1979.

40217 - WILLIAMS, L.E., PHILLIPS, D.A. : Development of nitrogen fixation and photosynthesis in soybeans grown at two levels of irradiance. - Plant Physiol.
63 (Suppl.) : 84, 1979.

40218 - WILLIAMS, M., RANDALL, D.D. : Pyruvate dehydrogenase complex from chloroplasts of *Pisum sativum* L. - Plant Physiol. *64* : 1099 - 1103, 1979.

☆40219 - WILLIAMS, P.A. : Growth, biomass, and net productivity of tall-tussock
(*Chionochloa*) grasslands, Canterbury, New Zealand. - New Zeal. J. Bot.
15 : 399 - 442, 1977.

☆40220 - WILLIAMS, P.A., GRIGG, J.L., NES, P., O'CONNOR, K.F. : Vegetation/soil relationships and distribution of selected macroelements within the shoots
of tall-tussocks on the Murchison Mountains, Fiordland, New Zealand. - New
Zeal. J. Bot. *14* : 29 - 53, 1976. [Dry-matter distribution.]

☆40221 - WILLIAMS, P.A., GRIGG, J.L., NES, P., O'CONNOR, K.F. : Macro-element composition of *Chionochloa pallens* and *C. flavescens* shoots, and soil properties
in the North Island, New Zealand. - New Zeal. J. Bot. *16* : 235 - 246, 1978.
[Dry-matter distribution.]

☆40222 - WILLIAMS, P.A., GRIGG, J.L., NES, P., O'CONNOR, K.F. : Macro-elements within shoots of tall-tussocks (*Chionochloa*), and soil properties on Mt Kaipa-
roro, Wairarapa, New Zealand. - New Zeal. J. Bot. *16* : 255 - 260, 1978.
[Dry-matter distribution.]

☆40223 - WILLIAMS, P.A., NES, P., O'CONNOR, K.F. : Macro-element pools and fluxes
in tall-tussock (*Chionochloa*) grasslands, Canterbury, New Zealand. - New
Zeal. J. Bot. *15* : 443 - 476, 1977. [Dry-matter turnover.]

40224 - WILLIAMS, P.J. leB., RAINE, R.C.T., BRYAN, J.R. : Agreement between the
^{14}C and oxygen methods of measuring phytoplankton production: reassessment
of the photosynthetic quotient. - Oceanol. Acta *2* : 411 - 416, 1979.

40225 - WILLIAMS, R.H., HAYES, J.D. : Relationships between photosynthetic area
and other growth attributes with grain yield in 6- and 2-row barley genotypes. - Ann. appl. Biol. *91* : 391 - 395, 1979.

40226 - WILLIAMS, R.J.P : Some unrealistic assumptions in the theory of chemi-osmo-
sis and their consequences. - FEBS Lett. *102* : 126 - 132, 1979.

40227 - WILLIAMS, R.O., GOSS, J.R. : An assessment of the gasification characteristics of some agricultural and forest industry residues using a laboratory gasifier. - Resour. Recov. Conserv. *3* : 317 - 329, 1979. [Energy contents.]

40228 - WILLIAMS, T.H.L. : An error analysis of the photographic technique for
measuring percent vegetative cover. - Soil Sci. Soc. Amer. J. *43* : 578 -
- 582, 1979.

40229 - WILLNER, I., FORD, W.E., OTVOS, J.W., CALVIN, M. : Photoinduced electron
transfer across a water-oil boundary as a model for redox reaction separation. - Nature *280* : 823 - 824, 1979.

☆40230 - WILSON, A.T. : Pioneer agriculture explosion and CO_2 levels in the atmosphere. - Nature *273* : 40 - 41, 1978. [$\delta^{13}C$.]

40231 - WILSON, K.G., VAUGHN, K.C. : Organelle destruction, a new mechanism to ex-
plain maternal inheritance of plastids and mitochondria in higher plants.
- Plant Physiol. *63* (Suppl.) : 5, 1979.

40232 - WILSON, R.E., OKUBO, A., ESAIAS, W.E. : A note on time-dependent spectra for chlorophyll variance. - J. mar. Res. *37* : 485 - 491, 1979.

*40233 - WILSON, R.T., KRIEG, D.R., DAHL, B.E. : A physiological study of developing pods and leaves of honey mesquite. - J. Range Manage. *27* : 202 - 203, 1974. [Ps.]

40234 - WILSON, W.H., KIEFER, D.A. : Reflectance spectroscopy of marine phytoplankton. Part 2. A simple model of ocean color. - Limnol. Oceanogr. *24* : 673 - 682, 1979.

40235 - WINTER, C. : Die Wirkung verschieden langer Lichtperioden auf die Produktivität einiger Gräser. - Photosynthetica *13* : 401 - 408, 1979.

40236 - WINTER, K. : Effect of different CO_2 regimes on the induction of crassulacean acid metabolism in *Mesembryanthemum crystallinum* L. - Aust. J. Plant Physiol. *6* : 589 - 594, 1979.

40237 - WINTER, K. : $\delta^{13}C$ values of some succulent plants from Madagascar. - Oecologia *40* : 103 - 112, 1979.

40238 - WINTER, K., LÜTTGE, U. : C_3-Photosynthese und Crassulacean-Sauerstoffwechsel bei *Mesembryanthemum crystallinum* L. - Ber. deut. bot. Ges. *92* : 117 - 132, 1979.

40239 - WITHERS, J.R. : Studies on the status of unburnt *Eucalyptus* woodland at Ocean Grove, Victoria. IV. The effect of shading on seedling establishment. - Aust. J. Bot. *27* : 47 - 66, 1979. [Ps.]

40240 - WITHERS, N.J. : Effects of water stress on *Lupinus albus* I. Response of vegetative growth to water stress during a single growth stage at two humidity levels. - New Zeal. J. agr. Res. *22* : 445 - 454, 1979. [Dry-matter accumulation.]

40241 - WITHERS, N.J., EDGE, E.A. : Effects of water stress on *Lupinus albus* IV. Response to high temperature and adequate and restricted water. - New Zeal. J. agr. Res. *22* : 571 - 575, 1979. [Production.]

40242 - WITHERS, N.J., FORDE, B.J. : Effects of water stress on *Lupinus albus* III. Response of seed yield and vegetative growth to water stress imposed during two or three growth stages. - New Zeal. J. agr. Res. *22* : 463 - 474, 1979. [Photosynthates.]

40243 - WITHERS, N.J., FORDE, B.J. : Translocation of ^{14}C in *Lupinus albus*. - New Zeal. J. agr. Res. *22* : 561 - 569, 1979. [Photosynthates.]

40244 - WITHERS, N.W., TUTTLE, R.C. : Effects of visible and ultraviolet light on carotene-deficient mutants of *Crypthecodinium cohnii*. - J. Protozool.*26* : 120 - 122, 1979.

40245 - WITHERS, N.W., TUTTLE, R.C. : Carotenes from mutants of the dinoflagellate, *Crypthecodinium cohnii*. - J, Protozool. *26* : 135 - 138, 1979.

40246 - WITT, H.T. : Charge separation in photosynthesis. - In : GERISCHER, H., KATZ, J.J. (ed.) : Light-Induced Charge Separation in Biology and Chemistry. Pp. 303 - 330. Verlag Chemie, Weinheim - New York 1979.

40247 - WITT, H.T. : Energy conversion in the functional membrane of photosynthesis. Analysis by light pulse and electric pulse methods. The central role of the electric field. - Biochim. biophys. Acta *505* : 355 - 427, 1979.

40248 - WITTENBACH, V.A. : Ribulose bisphosphate carboxylase and proteolytic activity in wheat leaves from anthesis through senescence. - Plant Physiol. *64* : 884 - 887, 1979.

40249 - WITTENBACH, V.A., ACKERSON, R.C. : Ribulose bisphosphate carboxylase and proteolytic activity in soybean leaves from anthesis through senescence. - Plant Physiol. *63* (Suppl.) : 158, 1979.

40250 - WITTWER, S.H. : Agricultural production - research imperatives for the future. - In : SCOTT, T.K. (ed.) : Plant Regulation and World Agriculture. Pp. 11 - 33. Plenum Press, New York - London 1979. [Ps.]

40251 - WITZTUM, A., POSNER, H.B., GOWER, R.A. : Phototactic chloroplast displacement in the photosynthetic mutant, *Lemna paucicostata* strain 1073. - Ann.
Bot. *44* : 1 - 4, 1979.

40252 - WIUM-ANDERSEN, S. : Plankton primary production in a tropical mangrove
bay at the south-west coast of Thailand. - Ophelia *18* : 53 - 60, 1979.

40253 - WOJTASZEK, T., STARZECKI, W., LIBIK, A., MACZEK, W., MYDLARZ, J. : Theoretical and technical aspects of CO_2 enrichment of greenhouse atmosphere in
tomato production. - Phytotron. Newslett. *20* : 56 - 62, 1979. [Ps.]

40254 - WOLEDGE, J. : Effect of flowering on the photosynthetic capacity of ryegrass leaves grown with and without natural shading. - Ann. Bot. *44* : 197 -
- 207, 1979.

40255 - WOLLMAN, F.-A. : Ultrastructural comparison of *Cyanidium caldarium* wild
type and III-C mutant lacking phycobilisomes. - Plant Physiol. *63* : 375 -
- 381, 1979.

40256 - WOLOSIUK, R.A., CRAWFORD, N.A., YEE, B.C., BUCHANAN, B.B. : Isolation of
three thioredoxins from spinach leaves. - J. biol. Chem. *254* : 1627 -
- 1632, 1979.

40257 - WOLOSIUK, R.A., PERELMUTER, M.E., CHEHEBAR, C. : Enhancement of chloroplast
fructose-1,6-bisphosphatase activity by fructose-1,6-bisphosphate and di-
thiothreitol-reduced thioredoxin-*f*. - FEBS Lett. *109* : 289 - 293, 1979.

40258 - WOLVERTON, B.C., McDONALD, R.C. : Water hyacinth (*Eichhornia crassipes*)
productivity and harvesting studies. - Econ. Bot. *33* : 1 - 10, 1979.
[Growth analysis.]

40259 - WONG, D., GOVINDJEE : Antagonistic effects of mono- and divalent cations
on polarization of chlorophyll fluorescence in thylakoids and changes in
excitation energy transfer. - FEBS Lett. *97* : 373 - 377, 1979.

40260 - WONG, D., MERKELO, H., GOVINDJEE : Regulation of excitation transfer by
cations: Wavelength-resolved fluorescence lifetimes and intensities at
77 K in thylakoid membranes of pea chloroplasts. - FEBS Lett. *104* : 223 -
- 226, 1979.

40261 - WONG, J.H.H., BENEDICT, C.R. : $^{14}CO_2$ fixation into sugars in non-photosynthetic endosperm of germinating castor beans. - Plant Physiol. *63* (Suppl.):
17, 1979.

*40262 - WONG, P.Y.O., THROWER, L.B. : Effect of *Colletotrichum lindemuthianum* on
photosynthesis and respiration of *Vigna sesquipedalis*. - Phytopathol. Z.
92 : 88 - 94, 1978.

*40263 - WONG, P.Y.O., THROWER, L.B. : Sugar metabolism and translocation in *Vigna
sesquipedalis* infected by *Colletotrichum lindemuthianum*. - Phytopathol. Z.
92 : 102 - 112, 1978. [Ps.]

40264 - WONG, S.C. : Elevated atmospheric partial pressure of CO_2 and plant growth
I. Interactions of nitrogen nutrition and photosynthetic capacity in C_3
and C_4 plants. - Oecologia *44* : 68 - 74, 1979.

40265 - WONG, S.C., COWAN, I.R., FARQUHAR, G.D. : Stomatal conductance correlates
with photosynthetic capacity. - Nature *282* : 424 - 426, 1979.

40266 - WONG, S.L., CLARK, B. : The determination of desirable and nuisance plant
levels in streams. - Hydrobiologia *63* : 223 - 230, 1979. [Chl.]

40267 - WONG, W.W., BENEDICT, C.R., KOHEL, R.J. : Enzymic fractionation of the
stable carbon isotopes of carbon dioxide by ribulose-1,5-bisphosphate carboxylase. - Plant Physiol. *63* : 852 - 856, 1979.

40268 - WOO, K.C. : Properties and intramitochondrial localization of serine hydro-
xymethyltransferase in leaves of higher plants. - Plant Physiol. *63* : 783 -
- 787, 1979. [Photorespiration.]

40269 - WOOD, A.M. : Chlorophyll *a:b* ratios in marine planktonic algae. - J. Phycol.
15 : 330 - 332, 1979.

40270 - WOODROW, S., O'BRIEN, M.J., EASTERBY, J.S., POWLS, R. : Glyceraldehyde-3-
-phosphate dehydrogenase of *Scenedesmus obliquus*. A kinetic analysis of
the effects of nucleotide and dithiothreitol on the production of NADPH-
-dependent activity. - Europe. J. Biochem. *98* : 425 - 430, 1979.

*40271 - WOODRUFF, D.R., MAWHOOD, R.P. : Yield response of selected Mexican and
Australian wheat cultivars to manipulations of the assimilate supply and
grain number. - Queensland J. agr. anim. Sci. *35* : 95 - 100, 1978.

40272 - WOODRUFF, W.W.,III, WOLFENDEN, R. : Inhibition of ribose-5-phosphate isomerase by 4-phosphoerythronate. - J. biol. Chem. *254* : 5866 - 5867, 1979.

40273 - WOODWARD, F.I. : The differential temperature responses of the growth of
certain plant species from different altitudes. I. Growth analysis of
Phleum alpinum L., *P. bertolonii* D.C., *Sesleria albicans* KIT. and *Dactylis
glomerata* L. - New Phytol. *82* : 385 - 395, 1979.

40274 - WOODWARD, F.I. : The differential temperature responses of the growth of
certain plant species from different altitudes. II. Analyses of the control
and morphology of leaf extension and specific leaf area of *Phleum bertolonii* D.C. and *P. alpinum* L. - New Phytol. *82* : 397 - 405, 1979.

40275 - WOODWARD, F.I., SHEEHY, J.E. : Microclimate, photosynthesis and growth of
lucerne (*Medicago sativa* L.). II. Canopy structure and growth. - Ann. Bot.
44 : 709 - 719, 1979.

40276 - WOODWARD, F.I., YAQUB, M. : Integrator and sensors for measuring photosynthetically active radiation and temperature in the field. - J. appl. Ecol.
16 : 545 - 552, 1979.

*40277 - WOODWELL, G.M., WHITTAKER, R.H., REINERS, W.A., LIKENS, G.E., DELWICHE,
C.C., BOTKIN, D.B. : The biota and the world carbon budget. - Science *199* :
141 - 146, 1978. [Primary productivity.]

40278 - WRAIGHT, C.A. : Electron acceptors of bacterial photosynthetic reaction
centers. II. H+ binding coupled to secondary electron transfer in the quinone acceptor complex. - Biochim. biophys. Acta *548* : 309 - 327, 1979.

40279 - WRAIGHT, C.A. : The role of quinones in bacterial photosynthesis. - Photochem. Photobiol. *30* : 767 - 776, 1979.

40280 - WRIGHT, D., HEBBLETHWAITE, P.D. : Lodging studies in *Lolium perenne* grown
for seed. 3. Chemical control of lodging. - J. agr. Sci. *93* : 669 - 679,
1979. [Dry-matter accumulation.]

40281 - WRIGHT, K., CORBETT, J.R. : Biochemistry of herbicides affecting photosynthesis. - Z. Naturforsch. *34 C* : 966 - 972, 1979.

40282 - WU, Xiang-yu, LI, Xi-jing, WU, Guang-yao, YUAN, Xiao-hua, WANG, Lan-xian,
DENG, Yue-fen : [Studies on fructose-1,6-diphosphatase (FDP-ase) in chloroplasts. I. Isolation of chloroplast FDP-ase and comparison of some kinetic properties of purified chloroplast FDP-ase and that of FDP-ase in
freshly ruptured chloroplasts.] - Acta bot. sin. *21* : 231 - 237, 1979.
[In Chin., ab : E.]

40283 - WUN, C.K., RHO, J., WALKER, R.W., LITSKY, W. : A simplified method for
the simultaneous extraction of phytoplanktonic chlorophyll and fecal ste-
rol from water. - Water Air Soil Pollut. *11* : 173 - 178, 1979.

40284 - WUN, C.K., RHO, J., WALKER, R.W., LITSKY, W. : An XAD-1 column method for
the rapid extraction of phytoplankton chlorophylls. - Water Res. *19* : 645 -
- 649, 1979.

40285 - WYNN, T.E. : The effects of O_2 tension and temperature on the dark release of CO_2 following light fixation. - Plant Physiol. *63* (Suppl.) : 152,
1979. [Ps.]

40286 - WYSS, H.-R., BRUNOLD, C. : Regulation of adenosine 5'-phosphosulfate sulfo-
transferase activity by H_2S and cyst(e)ine in primary leaves of *Phaseolus
vulgaris* L. - Planta *147* : 37 - 42, 1979.

✻40287 - YABUKI, K., AOKI, M. : The effect of wind speed on the photosynthesis of rice field. - In : MONSI, M., SAEKI, T. (ed.) : Ecophysiology of Photosynthetic Productivity. JIBP Synthesis. Vol. 19. Pp. 154 - 159. University of Tokyo Press, Tokyo 1978.

✻40288 - YABUKI, K., AOKI, M., HAMOTANI, K. : Characteristics of the forest microclimate. - In : KIRA, T., ONO, Y., HOSOKAWA, T. (ed.) : Biological Production in a Warm-Temperate Evergreen Oak Forest of Japan. JIBP Synthesis. Vol. 18. Pp. 55 - 64. University of Tokyo Press, Tokyo 1978.

✻40289 - YAGSHIEV, A. : Vliyanie mikroélementov na soderzhanie pigmentov v lyutserne. [Effect of trace elements on pigment contents in alfalfa.] - Izv. Akad. Nauk turkm. SSR, Ser. biol. Nauk 1974 (3) : 24 - 27, 1974. [In R, ab : E, Turkm.]

40290 - YAKOVLEVA, G.A., MOLOTKOVSKIĬ, Yu.G. : Kationnaya pronitsaemost' tilakoidnykh membran. [Cation permeability of thylakoid membranes.] - Fiziol. Rast. 26 : 283 - 293, 1979. [In R, ab : E.]

40291 - YAKSHINA, A.M., MALKINA, I.S. : Uglekislotnyĭ gazoobmen duba chereshchatogo i ego vliyanie na gazovyĭ sostav atmosfery. [Carbon dioxide exchange of Quercus pedunculata EHRH. and its effect on the gas composition of the atmosphere.] - Zh. obshch. Biol. 40 : 926 - 930, 1979. [Ps; in R, ab : E.]

40292 - YAMADA, O., KUROZUMI, A., FUTATSUYA, F., ITO, K., ISHIDA, S., MUNAKATA, K. : Studies on chlorosis-inducing activities and plant growth inhibition of benzophenone derivatives. - Agr. biol. Chem. 43 : 1467 - 1471, 1979.

40293 - YAMAUCHI, T., FRIEND, D.J.C. : Effect of leaf age and irradiance on photosynthesis of Coffea arabica. - Photosynthetica 13 : 271 - 278, 1979.

40294 - YAMAMOTO, Y., KE, B. : A temperature-dependent conformational change in photosystem-II thylakoid membrane. - FEBS Lett. 107 : 137 - 140, 1979.

40295 - YAMAGUCHI, R., MATSUSHITA, S. : Light-induced lipid peroxidation in isolated chloroplasts and role of α-tocopherol. - Agr. biol. Chem. 43 : 2157 - 2161, 1979.

40296 - YANG, Y.-S., HORI, Y. : Studies on retranslocation of accumulated assimilates in 'Delaware' grapevines I. Retranslocation of ^{14}C-assimilates in the following spring after ^{14}C feeding in summer and autumn. - Tohoku J. agr. Res. 30 : 43 - 56, 1979.

40297 - YANKO, R.V., YANKO, V.M. : K voprosu matematicheskogo opisaniya protsessov rosta rasteniĭ. [Mathematic description of plant growth processes.] - Sel'skokhoz. Biol. 14 : 644 - 649, 1979. [In R, ab : E.]

40298 - YANYUSHIN, M.F. : Gidrogenaznaya aktivnost' sinkhronnoĭ kul'tury Chlamydomonas reinhardii. [Hydrogenase activity in Chlamydomonas reinhardii synchronous culture.] - Fiziol. Rast. 26 : 394 - 400, 1979. [In R, ab : E.]

✻40299 - YAO, A.Y.M. : Agricultural potential estimated from the ratio of actual to potential evapotranspiration. - Agr. Meteorol. 13 : 405 - 417, 1974. [Leaf area.]

40300 - YARULLINA, N.A. : Osobennosti formirovanniya pervichnoĭ biologicheskoĭ produktivnosti v pustynnykh soobshchestvakh del'ty Tereka. [Peculiarities of formation of primary biological productivity in desert communities of the Terek river delta.] - Bot. Zh. 64 : 877 - 883, 1979. [In R.]

40301 - YASHINA, S.G., STAKHOV, L.F., POLEVAYA, V.S., MAKAROV, A.D. : Nekotorye osobennosti fotosinteticheskogo élektronnogo transporta v kul'ture tkani Ruta graveolens. [Some features of photosynthetic electron transport in Ruta graveolens tissue culture.] - Fiziol. Rast. 26 : 501 - 505, 1979. [In R, ab : E.]

40302 - YASNIKOV, A.A., VOLKOVA, N.V., KANIVETS, N.P., REĬNGARD, T.A., VASILENOK, L.I., VOLOVIK, O.I., MUSHKETIK, L.S., ZAĬTSEVA, N.A., KHRIPKO, S.S., OSTROVSKAYA, L.K., ALEKSEEVA, T.A. : Ion-radical mechanisms of phosphorylation and light dependent transport of protons in chloroplasts. - Photosynthetica 13 : 439 - 445, 1979.

*40303 - YAVORSKAYA, T.K., KURINNYĬ, F.I., GAVVA, I.A. : Vliyanie usloviĭ kal'tsi-
evogo i magnievogo pitaniya na metabolizm i produktivnost' rasteniĭ sakharnoĭ svekly. [Effect of conditions of calcium and magnesium nutrition on
the metabolism and productivity of sugar beet plants.] - In : Mineral'noe
Pitanie i Produktivnost' Rasteniĭ. Pp. 80 - 85, 319. Naukova Dumka, Kiev
1978. [Chl; in R.]

40304 - YAYOCK, J.Y. : Effects of variety and spacing on growth, development and
dry matter distribution in groundnut (Arachis hypogaea L.) at two locations
in Nigeria. - Exp. Agr. 15 : 339 - 351, 1979. [Leaf area index.]

40305 - YENTSCH, C.S., YENTSCH, C.M. : Fluorescence spectral signatures: The characterization of phytoplankton populations by the use of excitation and
emission spectra. - J. mar. Res. 37 : 471 - 483, 1979.

40306 - YEOH, H.H., STONE, N.E., CREASER, E.H., WATSON, L. : Isolation and characterization of wheat ribulose-1,5-diphosphate carboxylase. - Phytochemistry
18 : 561 - 564, 1979.

40307 - YIN, Hong-zhang, SHEN, Yun-gang, WANG, Tian-duo, SHI, Jiao-nai : [Studying
the photosynthetic apparatus in operation.] - Acta phytophysiol. sin. 5 :
295 - 317, 1979. [In Chin.]

40308 - YODER, J.A. : Effect of temperature on light-limited growth and chemical
composition of Skeletonema costatum (Bacillariophyceae) . - J. Phycol. 15 :
362 - 370, 1979. [Chl.]

40309 - YODER, J.A. : A comparison between the cell division rate of natural populations of the marine diatom Skeletonema costatum (GREVILLE) CLEVE grown
in dialysis culture and that predicted from a mathematical model. - Limnol.
Oceanogr. 24 : 97 - 106, 1979.

40310 - YOKOTA, A., KITAOKA, S. : Occurrence and operation of the glycollate-gly-
oxylate shuttle in mitochondria of Euglena gracilis z. - Biochem. J. 184 :
189 - 192, 1979.

40311 - YOSHIDA, M., SONE, N., HIRATA, H., KAGAWA, Y., UI, N. : Subunit structure
of adenosine triphosphate. Comparison of the structure in thermophilic
bacterium PS3 with those in mitochondria, chloroplasts, and Escherichia
coli. - J. biol. Chem. 254 : 9525 - 9533, 1979.

40312 - YOSHIDA, T. :[Growth analysis of late-summer sown barley.] - Jap. J. Crop
Sci. 48 : 495 - 501, 1979. [In Jap., ab : E.]

40313 - YOSHIDA, T. : Relationship between stomatal frequency and photosynthesis
in barley. - Jap. agr. Res. quart. 13 : 101 - 105, 1979 .

40314 - YOSHIDA, Y. : Nuclear dependency in chloroplast activities. - Sci. Rep. Ni-
igata Univ., Ser. D 16 : 39 - 64, 1979.

40315 - YOSHIURA, M., IRIYAMA, K. : [Methods for the preparation and qualitative
and quantitative analysis of chlorophylls.] - Tanpakushitsu, Kakusan, Koso
[Protein, nucl. Acid, Enzyme] 24 : 612 - 620, 1979. [In Jap.]

40316 - YOSHIURA, M., IRIYAMA, K., SHIRAKI, M., OKADA, A. : Preparation of chloro-
phyll-a and chlorophyll-b by column chromatography with Sephasorb HP ultrafine. - Bull. chem. Soc. Jap. 52 : 2383 - 2385, 1979.

40317 - YOUNG, D.N. : Ontogeny, histochemistry and fine structure of cellular inclusions in vegetative cells of Antithamnion defectum (Ceramiaceae, Rhodophyta). - J. Phycol. 15 : 42 - 48, 1979.

40318 - YOUNG, D.R., SMITH, W.K. : Influence of sunflecks on the temperature and
water relations of two subalpine understory congeners. - Oecologia 43 :
195 - 205, 1979. [Stomatal resistance.]

40319 - YOUNGMAN, R.J., DODGE, A.D. : Mechanism of paraquat action: inhibition of
the herbicidal effect by a copper chelate with superoxide dismutating activity. - Z. Naturforsch. 34 C : 1032 - 1035, 1979. [Chl, Car, chloroplast.]

40320 - YOUNGMAN, R.J., DODGE, A.D., LENGFELDER, E., ELSTNER, E.F. : The role of
superoxide in the herbicidal action of paraquat. - Plant Physiol. 63 (Suppl.)
: 42, 1979.

40321 - **YOUNIS, H.M., BOYER, J.S., GOVINDJEE** : Conformation and activity of chloroplast coupling factor exposed to low chemical potential of water in cells. - Biochim. biophys. Acta *548* : 328 - 340, 1979.

40322 - **YOUNIS, H.M., BOYER, J.S., GOVINDJEE** : Magnesium binding to chloroplast coupling factor correlated with effects of low water potential. - Plant Physiol. *63* (Suppl.) : 40, 1979.

40323 - **YUKIMOTO, M., YAMASHITA, S.** : [Ultrastructural changes in Chinese cabbage chloroplasts induced by phosalone.] - J. Pestic. Sci. *4* : 521 - 524, 1979. [In Jap.]

40324 - **YUN, S., ISHII, R., HYEON, S., SUZUKI, A., MURATA, Y., TAMURA, S.** : Effects of some chemicals on photorespiration and photosynthesis in the excised rice leaves. - Agr. biol. Chem. *43* : 2207 - 2209, 1979.

40325 - **ZAFIROV, I., SALCHEVA, G.** : Vliyanie na CCC v"rkhu tsiklichnoto fotofosforilirane na khloroplasti, izolirani ot fasulevi rasteniya, okhlazhdani pri niski polozhitelni temperaturi. [Effect of CCC on cyclic photophosphorylation of chloroplasts isolated from bean plants cooled at low positive temperatures.] - Fiziol. Rast. (Sofia) *5* (3) : 54 - 61, 1979. [In Bulg., ab : E, R.]

40326 - **ZAGDAŃSKA, B., PACANOWSKA, A.** : Dehydration tolerance of spring wheat and its relation to plant growth and productivity under soil drought conditions. - Biol. Plant. *21* : 452 - 461, 1979.

40327 - **ZÁHUMENSKÝ, L.** : Primárna produkcia fytoplanktónu rybníka č.4 na Železnej Studničke v Bratislave. [Primary production of phytoplankton in the fishpond No.4 at Železná Studnička in Bratislava.] - Biológia (Bratislava) *34* : 583 - 587, 1979. [In Slovak, ab : E.]

*40328 - **ZAIGRAEV, S.A., MEDVEDEVA, N.G.** : Dinamika pigmentov v list'yakh razlichnykh sortov yarovoĭ pshenitsy v usloviyakh Buryatskoĭ ASSR. [Dynamics of pigments in leaves of various varieties of spring wheat in Buryat ASSR.] - In : Vliyanie Mineral'nykh Udobreniĭ na Urozhaĭ i Kachestvo Sel'sko--Khozyaĭstvennykh Kul'tur. Pp. 26 - 28. Irkutsk 1975. [In R.]

40329 - **ZAKARIA, M., SIMPSON, K., BROWN, P.R., KRSTULOVIC, A.** : Use of reversed--phase high-performance liquid chromatographic analysis for the determination of provitamin A carotenes in tomatoes. - J. Chromatogr. *176* : 109 - 117, 1979.

40330 - **ZAKRZHEVSKIĬ, D.A., ANAN'EV, G.M.** : Sootnoshenie mezhdu fazami impul'snoĭ i nepreryvnoĭ kinetik vydeleniya kisloroda u kletok *Chlorella*, vyrashchennykh v razlichnykh fiziologicheskikh usloviyakh. [Relationship between pulsed and continuous oxygen evolution phases in *Chlorella* cells grown under various physiological conditions.] - Fiziol. Rast. *26* : 824 - 831, 1979. [In R, ab : E.]

*B40331 - **ZALENSKIĬ, O.V.** (ed.) : Fotosintez i Ispol'zovanie Solnechnoĭ Radiatsii. [Photosynthesis and Solar Energy Utilization.] - Nauka, Leningrad 1971. [In R, ab : E.]

40332 - **ZALENSKIĬ, O.V., GLAGOLEVA, T.A., ZUBKOVA, E.K., MAMUSHINA, N.S., NIKULINA, G.N., FILIPPOVA, L.A., CHULANOVSKAYA, M.V.** : O modelirovanii énergeticheskikh vzaimootnosheniĭ mezhdu dykhaniem i fotosintezom pri pomoshchi ékzogennoĭ ATF-^{14}C. [Use of exogenous ^{14}C-ATP in modelling energy·interrelationships of respiration·and photosynthesis.] - Fiziol. Rast. *26* : 1076 - 1084, 1979. [In R, ab : E.]

40333 - **ZÁLETOVÁ, E., PAULECH, C.** : Štúdium vhodnosti gravimetrickej metódy na meranie fotosyntézy listov druhu *Prunus laurocerasus* L. [Study on suitability of the gravimetric method for measuring photosynthesis of leaves of *Prunus laurocerasus* L.] - Biológia (Bratislava) *34* : 769 - 773, 1979. [In Slovak, ab : E, R.]

40334 - ZANETTI, G., GOZZER, C., SACCHI, G., CURTI, B. : Modification of arginyl residues in ferredoxin-NADP$^+$ reductase from spinach leaves. - Biochim. biophys. Acta *568* : 127 - 134, 1979.

40335 - ZAVADSKAYA, I.G., ANTROPOVA, T.A. : O "paradoksal'nom" éffekte pri deĭstvii vysokikh temperatur na list'ya nekotorykh vysshikh rasteniĭ. ["Paradoxical" effects after heating the leaves of some higher plants.] - Tsitologiya *21* : 46 - 56, 1979. [Chl; in R, ab : E.]

40336 - ZAVITKOVSKI, J. : Energy production in irrigated, intensively cultured plantations of *Populus* "Tristis #1" and Jack pine. - Forest Sci. *25* : 383 - - 392, 1979. [Energy content.]

40337 - ZEE, D.V., KENNEDY, R.A. : Ultrastructural studies of *Echinocloa oryzicola* shoots germinated under anaerobic conditions. - Plant Physiol. *63* (Suppl.) : 97, 1979. [Chloroplast.]

40338 - ZEHNDER, A.J.B., BROCK, T.D. : Biological energy production in the apparent absence of electron transport and substrate level phosphorylation. - FEBS Lett. *107* : 1 - 3, 1979.

40339 - ZEIGER, E., HEPLER, P.K. : Blue light-induced, intrinsic vacuolar fluorescence in onion guard cells. - J. Cell Sci. *37* : 1 - 10, 1979. [Chl.]

40340 - ZEINALOV, Yu., LITVIN, F.F. : Oxygen evolution after switching off the light and S$_i$-state desactivation in photosynthesizing systems. - Photosynthetica *13* : 119 - 123, 1979.

40341 - ŻELAWSKI, W., LECH, A. : Growth function characterizing dry matter accumulation of plants. - Bull. Acad. pol. Sci.,Sér. Sci. biol. Cl. V *27* : 675 - - 681, 1979.

40342 - ZELENSKIĬ, M.I., MOGILEVA, G.A., SHITOVA, I.P. : Sortovoe raznoobrazie yarovykh pshenits po fotokhimicheskoĭ aktivnosti khloroplastov. [Varietal heterogeneity of spring wheat in photochemical activity of chloroplasts.] - Byull. vsesoyuz. nauch.-issled. Inst. Rastenievod. Im. N.I.Vavilova *87* : 36 - 40, 1979. [In R.]

40343 - ZELENSKIĬ, M.I., MOGILEVA, G.A., SHITOVA, I.P. : Prodolzhitel'nost' perioda vskhody-koloshenie i fotokhimicheskaya aktivnost' khloroplastov pshenitsy. [The duration of the period from shoots to ear formation and the photochemical activity of wheat chloroplasts.] - Sel'skokhoz. Biol. *14* : 202 - - 206, 1979. [In R, ab : E.]

40344 - ZELITCH, I. : Photorespiration: Studies with whole tissues. - In : GIBBS, M., LATZKO, E. (ed.) : Photosynthesis II. (Encycl. Plant Physiol. N.S. Vol. 6.) Pp. 353 - 367. Springer-Verlag, Berlin - Heidelberg - New York 1979.

40345 - ZELITCH, I. : Photosynthesis and plant productivity. - Chem. Eng. News *57* (6) : 28 - 32, 37 - 42, 46 - 48, 1979.

40346 - ZEN'KEVICH, É.I., KOCHUBEEV, G.A., LOSEV, A.P., GURINOVICH, G.P. : Smeshannaya agregatsiya pigmentov s uchastiem bakteriokhlorofilla. [Mixed aggregation of pigments with participation of bacteriochlorophyll.] - Mol. Biol. (Moskva) *13* : 888 - 898, 1979. [In R, ab : E.]

40347 - ZEN'KEVICH, É.I., SARZHEVSKAYA, M.V. : Agregatsiya feofitina *a* i ego proizvodnykh v binarnykh smesyakh rastvoriteleĭ. [Aggregation of pheophytin *a* and its derivatives in the binary solvent mixtures.] - Biofizika *24* : 771, 1979. [In R.]

40348 - ZEYNALOV, Y. : On the spectral dependence of the quantum efficiency of photosynthesis. - Dokl. bolg. Akad. Nauk *32* : 373 - 376, 1979.

40349 - ZEYNALOV, Y. : On the amount of oxygen taken up during the induction period of photosynthesis in green algae. - Dokl. bolg. Akad. Nauk *32* : 679 - 682, 1979.

40350 - ZHANG, Qi-de, LOU, Shi-qing, LI, Tong-zhu, MA, Gui-zhi, ZHANG, Guo-zheng, KUANG, Ting-yun : [Structure and function of chloroplast membrane II. The effects of potassium and magnesium ions on the absorption spectrum and photosystem II function of two types of chloroplast membranes.] - Acta bot. sin. *21* : 250 - 258, 1979. [In Chin., ab : E.]

40351 - **ZHAO, Yu-ju, WANG, Yu-qin, WU, Shao-bo** : [Effect of 6-benzylaminopurine on leaf senescence and protein degradation in detached wheat leaves.] - Acta phytophysiol. sin. *5* : 271 - 277, 1979. [Chloroplast; in Chin., ab : E.]

40352 - **ZHIVKOVA, T.D.** : Deĭstvie peremennykh temperatur na rost kukuruzy i soder-zhanie pigmentov. [Effect of variable temperatures on maize growth and pigment content.] - Fiziol. Biokhim. kul't. Rast. *11* : 123 - 129, 1979. [In R, ab : E.]

40353 - **ZHIVKOVA, T.D.** : Morfometrichno izsledvane na strukturata na khloroplasti ot razlichni khibridi tsarevitsa pod deĭstvieto na niski polozhitelni tem-peraturi. [Morphological investigation on the effect of low positive tem-perature over the structure of chloroplasts from various maize hybrids.] - Fiziol. Rast. (Sofia) *5* (3) : 62 - 71, 1979. [In Bulg., ab : E, R.]

40354 - **ZHIVKOVA, T.D., SILAEVA, A.M., PROTSENKO, D.F., PETROVA, O.V.** : Struktur-nye osobennosti khloroplastov otlichayushchikhsya po kholodoustoĭchivosti gibridov kukuruzy. [Structural peculiarities of chloroplasts differing in cold-resistance of maize hybrids.] - Fiziol. Biokhim. kul't. Rast. *11* : 333 - 338, 1979. [In R, ab : E.]

* 40355 - **ZHUKOV, O.S., PALFITOV, V.F., OSIPOVA, L.V.** : Soderzhanie pigmentov v list'yakh radiomutantov alychi i vishni. [Pigment contents in leaves of radiomutants of *Prunus*.] - Byul. nauch. Inform. tsentr. genet. Labor. Im. I.V.Michurina *23* : 10 - 14, 59, 1976. [In R.]

40356 - **ZHUKOVA, G.Ya.** : Osobennosti plastidnogo apparata zarodysha u khloroêm-briofitov i leĭkoêmbriofitov. [Peculiarities of plastid apparatus of em-bryo in chloroembryophytes and leucoembryophytes.] - In : Aktual'nye Vo-prosy Êmbriologii Pokrytosemyannykh RasteniĬ. Pp. 104 - 119. Nauka, Lenin-grad 1979. [In R.]

*40357 - **ZIEGLER, H.** : Zur Physiologie austrocknungsfähiger Kormophyten. - In : MÜLLER, P. (ed.) : Verhandlungen der Gesellschaft für Ökologie, Saarbrücken 1973. Pp. 65 - 73. Dr. W. Junk, bv.-Publ., The Hague 1974. [Ps.]

40358 - **ZIEGLER, H.** : Diskriminierung von Kohlenstoff- und Wasserstoffisotopen: Zusammenhänge mit dem Photosynthesemechanismus und den Standortbedingungen. - Ber. deut. bot. Ges. *92* : 169 - 184, 1979.

40359 - **ZIELINSKI, R.E., PRICE, C.A.** : Synthesis and assembly of chlorophyll-pro-tein complexes by isolated spinach chloroplasts. - Plant Physiol. *63* (Suppl.) : 98, 1979.

40360 - **ZIMA, J., ŠESTÁK, Z.** : Photosynthetic characteristics during ontogenesis of leaves. 4. Carbon fixation pathways, their enzymes and products. - Photosynthetica *13* : 83 - 106, 1979.

40361 - **ZIMOVÁ, D.** : Vliv morforegulátorů na tvorbu výnosu a kvalitu obilnin. [The effect of morphoregulators on the formation of yields and quality of cereals.] - Stud. Inform. ÚVTIZ, rostl. Výroba (Praha) *1979* (3) : 1 - - 76, 1979. [Chl; in Czech, ab : E, R.]

40362 - **ZINGMARK, R.G.** : Are chlorophyll measurements meaningful indicators of phytoplankton biomass? - J. Phycol. *15* (Suppl.) : 19, 1979.

40363 - **ZSAKÓ, J., CHIFU, E., TOMOAIA-COTISEL, M.** : Rotating rigid-plate model of carotenoid molecules and the behaviour of their monolayers at the air--water interface. - Gaz. chim. ital. *109* : 663 - 668, 1979.

40364 - **ZSOLNAY, A.** : A confirmation of the correlation between hydrocarbons and chlorophyll *a* in the upper euphotic zone. - Mar. Pollut. Bull. *10* : 107 - - 108, 1979.

*40365 - **ZUBRICKÝ, J.** : Možnosti úschovy vzoriek rastlinného materiálu s minimál-nou stratou karoténu. [Possibility of storing samples of plant material with a minimal loss of carotene.] - Folia vet. *20* : 51 - 55, 1976 (1979). [In Slovak, ab : E, R.]

*40366 - ZUNDEL, G., WEIDEMANN, E.G. : Possible mechanism of proton conductivity in biological membranes. - In : BRODA, E., LOCKER, A., SPRINGER-LEDERER, H. (ed.) : Proceedings of the First European Biophysics Congress. Vol.6. Pp. 43 - 47. Verlag Wiener Med. Akad., Wien 1971. [Ps.]

40367 - ZUO, Bao-yu, LI, Shi-yi, WANG, Ren-ru, KUANG, Ting-yun, TUAN, Hsu-chuan : [Structure and function of chloroplast membrane III. The effects of Mg^{++} and K^{+} ions toward the ultrastructure of two kinds of chloroplast thyla-koid membranes.] - Acta bot. sin. *21* : 328 - 333, 1979. [In Chin., ab : E.]

40368 - ZÜRRER, H., BACHOFEN, R. : Hydrogen production by the photosynthetic bac-terium *Rhodospirillum rubrum*. - Appl. environm. Microbiol. *37* : 789 - 793, 1979.

*40369 - ZURZYCKI, J. : Rola błon w procesach fotorecepcji. [Role of membranes in processes of photoreception.] - Postępy Biol. Komórki *2* : 61 - 85, 1975. [In Pol.]

*40370 - ZURZYCKI, J. : Błony purpurowe i nowy rodzaj fotofosforylacji. [Purple membranes and new kind of photophosphorylation.] - Zesz. nauk. Uniw. jagiell. *464* (Prace Biol. mol.4) : 143 - 155, 1977. [In Pol., ab : E.]

40371 - ZURZYCKI, J. : Struktura i własności mechaniczne warstwy kontaktowej chlo-roplast-cytoplazma. [Structure and mechanical properties of the chloro-plast-cytoplasm contact layer.] - Zesz. nauk. Uniw. jagiell. *549* (Prace Biol. mol. 6) : 121 - 132, 1979. [In Pol., ab : E.]

40372 - ZVEREVA, M.G., SHUBIN, L.M., KLIMOVA, L.A., SEMENENKO, V.E. : Obratimoe izmenenie spektrov nizkotemperaturnoĭ fluorestsentsii intaktnykh kletok *Chlorella*, vyzyvaemoe repressieĭ fotosinteticheskogo apparata 2-dezoksi--D-glyukozoĭ. [Reversible change in low-temperature fluorescence spectra of intact *Chlorella* cells caused by repression of the photosynthetic apparatus by 2-deoxy-D-glucose.] - Dokl. Akad. Nauk SSSR *244* : 1244 - - 1247, 1979. [In R.]

40373 - ZWATZ, B. : Antiproduktionsfaktoren. - Pflanzenarzt *32* (7-8) : 78 - 79, 1979. [Ps.]